Rethinking the Man-Land Relations in China

Rethinking the Man-Land Relations in China

Editors

Li Ma
Yingnan Zhang
Muye Gan
Zhengying Shan

MDPI

Basel • Beijing • Wuhan • Barcelona • Belgrade • Novi Sad • Cluj • Manchester

Editors

Li Ma
School of Public Policy and Administration
Chongqing University
Chongqing
China

Yingnan Zhang
School of Public Affairs
Zhejiang University
Hangzhou
China

Muye Gan
The Rural Development Academy
Zhejiang University
Hangzhou
China

Zhengying Shan
Department of Geography
University of Wisconsin–Madison
Madison
United States

Editorial Office

MDPI
St. Alban-Anlage 66
4052 Basel, Switzerland

This is a reprint of articles from the Special Issue published online in the open access journal *Land* (ISSN 2073-445X) (available at: www.mdpi.com/journal/land/special_issues/man_land).

For citation purposes, cite each article independently as indicated on the article page online and as indicated below:

Lastname, A.A.; Lastname, B.B. Article Title. *Journal Name* **Year**, *Volume Number*, Page Range.

ISBN 978-3-0365-8669-4 (Hbk)
ISBN 978-3-0365-8668-7 (PDF)
doi.org/10.3390/books978-3-0365-8668-7

Contents

land

Editorial

Rethinking Man–Land Relations in China: A Multidisciplinary Perspective

Li Ma [1], Yingnan Zhang [2,*], Muye Gan [3] and Zhengying Shan [4]

1 School of Public Policy and Administration, Chongqing University, Chongqing 400044, China; mal.1991@cqu.edu.cn
2 School of Public Affairs, Zhejiang University, Hangzhou 310058, China
3 The Rural Development Academy, Zhejiang University, Hangzhou 310058, China
4 Department of Geography, University of Wisconsin–Madison, Madison, WI 53715, USA; zhengyingshan@pku.edu.cn
* Correspondence: zhangyingnan@zju.edu.cn

Land constitutes a fundamental natural asset, vital for the sustenance, advancement, and ecological balance of human civilization [1]. Through a significant portion of human history, land has remained intricately intertwined with economic expansion, serving as a fundamental element of production. Consequently, the governance and utilization of land frequently become focal points of intense human interactions. The symbiotic connection between humanity and the land mirrors their interdependence and reciprocal influence. China has undergone a profound transformation, characterized by unparalleled urbanization, industrialization, and globalization, ushering in novel multifaceted challenges to the dynamics of man-land relationships [2]. Constructing the built environment to accommodate urban populations and their various endeavors stands as a fundamental pillar of urbanization. This dynamic places additional strain on food systems that could potentially disrupt livelihoods in vulnerable regions. However, this represents just a fraction of the evolving interactions between humans and land in China. The intricate interplay between these two entities spans a broad spectrum of dynamics, encompassing factors like intensified agricultural practices, land degradation, the abandonment of farmlands, the emergence of "hollow villages," land fragmentation, urban renewal, traffic congestion, housing shortages, and numerous other variables [3–5]. The current body of research concerning man-land relationships in China is inadequate. It is imperative to employ diverse perspectives to scrutinize the multifaceted dimensions of human interventions on land utilization systems, as well as the reciprocal impacts of land-use transformations on human welfare. Thus, the reevaluation of man-land relationships within the context of this swiftly evolving era warrants immediate attention and inclusion on the agenda.

Citation: Ma, L.; Zhang, Y.; Gan, M.; Shan, Z. Rethinking Man–Land Relations in China: A Multidisciplinary Perspective. *Land* **2023**, *12*, 1606. https://doi.org/10.3390/land12081606

Received: 7 August 2023
Accepted: 11 August 2023
Published: 15 August 2023

This Special Issue aims to reevaluate the transformations in man-land relationships within transitional China, fostering a fresh perspective on the intricacies of human-environment interactions in both urban and rural contexts. In doing so, it seeks to contribute to the advancement of theories in land-use science, a crucial component of both land management and sustainability science.

The collection of peer-reviewed articles included in this Special Issue comprises twenty research articles in total (Appendix A). The Special Issue is organized in the following format: the papers are presented under four major topics, such as (a) human activities and natural ecosystems, (b) land-use conflicts/trade-offs, (c) man–land coordination and sustainable development, and (d) man–land system coupling and optimal regulation.

Five papers focus on the human activities and natural ecosystems, from the perspectives of carbon emissions, ecological security, soil erosion, desertification, and natural resource accounting.

Yan et al. (2022) employed a carbon emissions model to estimate land-use change-related carbon emissions and utilized the logarithmic mean Divisia index (LMDI) model to investigate the key influencing factors. The findings indicated a significant decline in cropland area, accompanied by substantial growth in built-up areas due to rapid urbanization. Additionally, it was noted that the gross domestic product (GDP) per capita exerted the greatest influence on the rise in carbon emissions within the study area, followed by land-use structure, carbon emission intensity per unit of land, and population size. Conversely, the intensity of land use per unit of GDP exhibited a mitigating effect on carbon emissions.

Hu et al. (2022) developed an evaluation system to assess ecological security (ES) in twenty-five international border counties within Yunnan Province. The researchers utilized an entropy weight TOPSIS model to analyze changes in ES between 2004 and 2019. Furthermore, an obstacle degree model was employed to identify the factors impacting ES. The findings revealed that fixed asset investments, per-capita fiscal revenue, per-capita GDP, food production, and water regulation posed hindrances to achieving a desirable level of ES within the study area.

Zhu et al. (2022) employed the revised universal soil loss equation model to determine the soil erosion modulus and investigated the driving factors and superposition mechanism of farmland soil erosion in the hilly region of Northeast China. To achieve this, they introduced the geographically weighted regression model. The findings underscored the significance of landscape fragmentation as a key driving force behind soil erosion, sediment yield, and sediment transport.

Jia et al. (2023) employed Landsat images from 2010 and 2020 to extract desertification information, subsequently constructing the Albedo–NDVI feature space in the Gonghe Basin. The researchers then utilized Geodetector to analyze the temporal and spatial evolution of desertification and its driving factors within the basin between 2010 and 2020. The findings demonstrated effective control over desertification in the Gonghe Basin, thereby offering a valuable foundation for combatting further desertification in the region.

Tan et al. (2023) conducted a comprehensive analysis of literature pertaining to the evaluation of major functions, natural resource accounting, environmental accounting, ecosystem services, and asset accounting. Their study employed the equivalent factor method and input-output method to establish the correlation between major function accounting and natural resource accounting. The findings highlighted that accounting for major functions and resources can effectively guide regional sustainable management through function positioning, resource comparative advantages, and administrative units closely linked to functional units.

Land-use conflicts, representing the spatial embodiment of human-land contradictions, exert a significant influence on regional sustainability. Six studies focus on examining the consequences of land-use conflicts and trade-offs, highlighting economic, social, spatial, and ecological dimensions of these conflicts.

Wang et al. (2022) employed the propensity value matching technique to assess the impacts of land transfer on poverty alleviation among farm households, focusing on the vulnerability expressed as expected poverty (VEP). The findings revealed that rural land transfers have a notable effect in reducing farm households' VEP, with the magnitude of these effects influenced by factors such as location, household characteristics, and household head. This study's results offer valuable insights for policy formulation concerning land management and poverty reduction in agricultural communities.

Lv et al. (2022) employed an integrated "spatial-functional" framework to study the structure and functionality of cultivated land-use transition (CLUT) in a prominent grain-producing region of southern China. The researchers quantitatively assessed and visually represented the CLUT, revealing a significant increase in the comprehensive CLUT index in the middle and lower reaches of the Yangtze River between 2001 and 2019. The study identified a positive aggregation effect with a 5% significance level during this period, indicating a strengthening of both spatial and functional transitions. The authors proposed that differentiated policies should be formulated by the government to promote sustainable land use through spatial and functional transitions in major grain-producing areas.

Shi and Wang (2022) utilized a PSM-DID approach to examine the association between high-speed rail (HSR) infrastructure and cropland abandonment using Chinese labor force survey data. The findings indicated a significant 20.6% rise in the extent of cropland abandonment due to HSR projects. Moreover, these effects were more pronounced in hilly areas but relatively lower in plain regions. Notably, HSR accessibility exerted a "pull" effect, prompting a shift of rural labor force from agriculture to non-agricultural sectors within the local vicinity.

Liang et al. (2022) assessed the extent of land-use conflicts (LUCs) through landscape ecological risk assessment and investigated the spatiotemporal evolution patterns and potential risks of LUCs in the urban center of Chongqing (UCC) over the past two decades. Employing hot-spot analysis and neighborhood analysis, they found that conversions between the living-production space (LPS) and other areas exhibited the highest frequency. Moreover, the out-of-control zone expanded while the controllable zone diminished. The authors emphasized the need for tailored management strategies and policy recommendations on a regional scale, targeting different LUC zones in the UCC, both at international and national levels.

Han et al. (2023) conducted an analysis on the influence of environmental decentralization on the scale of construction land supply by local governments, utilizing panel data from 30 provinces in China between 2003 and 2015. The findings revealed a positive impact of environmental decentralization on the expansion of urban construction land supply. This effect was attributed to the strengthening of land financial dependence and the distortion of land resource allocation. The study further identified that the impact was more significant in regions facing high financial pressure, economic growth pressure, and low environmental protection pressure. In light of these results, the authors provide policy suggestions to ensure a rational supply of urban construction land within the context of decentralization in China.

Zhao et al. (2023) conducted an evaluation of the Lanzhou-Xining urban agglomeration (LXUA) using a multi-dimensional assessment system that incorporated urbanization quality and ecosystem services. The assessment utilized various methodologies including the efficacy function model, entropy weight method, and Integrated Valuation of Ecosystem Services and Trade-offs (InVEST) model to quantitatively evaluate the developmental state of the subsystems. Additionally, the study employed coupling models (CD) and coordination degree (CCD) models to investigate the coupling coordination relationship and spatiotemporal change characteristics of the composite system.

Four papers in this Special Issue address the coordination of man-land systems and sustainable development:

Zhang et al. (2023) conducted an analysis of the spatial distribution and regional variations of the human appropriation of net primary production (HANPP) in China during 2015. The study also examined how HANPP and its components responded to imbalanced urban-rural development in different regions. The findings shed light on the impact of unbalanced regional development on human-induced biomass occupation, comprehensive urban ecological construction, and rural ecological restoration. Importantly, the study highlights the significance of urban-rural integration development as a means to address increasing ecological pressures in the future.

Wang et al. (2022) conducted an analysis of the elemental composition, structural organization, and functional state of China's northwest arid areas using a human-Earth system approach. The findings revealed a lack of coupling and coordination among humans, the economy, resources, and environmental elements in these regions. However, during the anti-poverty stage, China's northwest arid areas showed innovative efforts in establishing a human-Earth coupling mechanism. Additionally, three pathways were identified to enhance sustainable livelihood, consolidate poverty alleviation achievements, and achieve rural revitalization. Notably, it is essential to establish an endogenous growth mechanism for sustainable poverty alleviation and green development.

Cai et al. (2022) developed an evaluation index system to assess the various functions of cultivated land in oasis areas (OCL), including social, economic, and ecological perspectives. Using this framework, the study quantitatively evaluated the evolution of cultivated land functions (CLFs) and their interactions in Xinjiang from 1990 to 2018. The findings indicated that the evolution of CLFs in Xinjiang initially focused on ecological and social functions but gradually shifted toward economic functions. Additionally, the study revealed a weakening in the synergistic relationship between CLFs and an increase in trade-offs over time. This research expands our understanding of multi-functional studies related to cultivated land and provides valuable insights for decision-making regarding the sustainable utilization and synergistic management of oasis cultivated land in Xinjiang, China.

Zhang et al. (2023) assessed the rural-urban transition in China from 1980 to 2020 by utilizing socio-economic data and a rural-urban transition coordination model. They developed a comprehensive rural-urban development and integration index system to analyze the process. The findings reveal that, since the reform and opening-up, China has witnessed a gradual expansion of the rural-urban development index (URDI) across different regions, while the rural-urban integrated index (URII) initially declined before experiencing subsequent growth. Over the past four decades, the spatial distribution of URDI exhibited a "south high-north low" pattern, whereas the URII demonstrated a more balanced distribution. The study also put forth optimization strategies for each type to further enhance rural-urban integration.

Next, five papers included in this Special Issue discuss man–land system coupling and optimal regulation.

Gong et al. (2022) conducted an assessment of the multi-functions of cultivated land in the grain-producing area of Jilin Province's cultivated black soils over the past three decades. The study employed an improved TOPSIS model to analyze the data. By utilizing the obstacle degree model and Geodetector, the researchers also identified the key limiting and influencing factors of cultivated land's multi-functions. The findings indicated an overall increase in multi-functionality from 1990 to 2020. However, the simultaneous improvement of economic and social functions impeded progress in the ecological function of cultivated land. The analysis also highlighted spatial variations in the functions across different counties. Based on the results, the study put forward several policy recommendations, including reducing regional disparities in cultivated land functions, quantifying the multi-functional value of cultivated land, and providing subsidies for land cultivation. These measures aim to strengthen multi-functional planning and design, enhance ecological utilization, and promote the sustainable use of cultivated land.

Liang et al. (2022) conducted a study on the coupling and coordinated changes of land-use production, living, and ecological functions (PLEFs) in relation to human activity intensity (HAI) in Wanzhou District, China, spanning from 2000 to 2020. The researchers employed the coupling coordination degree (CCD) model to assess the level of coordinated development among PLEFs, while HAI was measured through the equivalent level of construction land. The synchronous development model was utilized to analyze the relationship between these factors. The findings revealed significant spatial distribution variations and evident spatial complementarity among PLEFs in Wanzhou District. Based on the synchronous development state of HAI and CCD of PLEFs, the district was categorized into three development types. This highlights the need to propose regulatory strategies tailored to regions with different development types.

Cheng et al. (2023) employed ecological niche theory, a coupling coordination model, and a trade-off synergy model to construct an evaluation index system. This system was utilized to assess the spatiotemporal evolution characteristics, trade-off synergy, and coupling coordination degree of land-use production, living, and ecological functions (PLEFs) across 38 counties in Chongqing, China. The findings revealed that over the past two decades, Chongqing's "living-production" function transitioned from a trade-off model

to a collaborative development relationship. Additionally, the "living-ecological" function generally exhibited a collaborative development relationship.

Zhou and Jiang (2022) conducted an analysis of the influence of urban development on immigration and labor migration trends in Macau from 1992 to 2019. The study reveals that Macau exhibits a high dependence on short-term migrant workers. Consequently, the paper suggests several measures to address this issue, including reducing the costs associated with city expansion, enhancing economic diversity, and fostering closer collaboration with neighboring mainland cities. Such actions would enable Macau to effectively utilize resources, attract non-local talent, and ensure sustainable urban development.

Guo and Zhong (2023) conducted an analysis to examine the underlying meaning of rural transformation development (RTD). They also explored the spatiotemporal patterns of RTD in the Yanshan-Taihang Mountains and identified the influencing factors through the use of a geographically and temporally weighted regression model. The findings indicated that RTD is a dynamic process characterized by qualitative changes in rural regional systems, which stem from the accumulation of quantitative changes in elements. The measurement of RTD hinges on the analysis of the coupling coordination degree between the quantitative changes of these elements.

The evolution of man-land relationships is closely intertwined with socio-economic development, calling for the application of dialectical thinking and dynamic systems analysis to explore these issues in contemporary China. This Special Issue in the journal Land encompasses a collection of 20 papers that delve into four main themes, namely human activities and natural ecosystems, land-use conflicts/trade-offs, man-land coordination, and sustainable development. Additionally, the studies delve into man-land system coupling and optimal regulation. These research contributions expand the scope and content of man-land relationship research, providing valuable theoretical and practical insights for urban-rural integration, regional sustainable development, rural revitalization, and global poverty reduction in the new era.

Nevertheless, there remains significant potential for advancement in the examination of man-land relationships in China, especially in the context of the papers featured in this Special Issue. While current research offers policy recommendations from various perspectives to enhance regional man-land coordination and sustainable development, a more comprehensive analysis of the intricate impact of interactions between man-land systems at the urban-rural and regional levels is still lacking [6]. Therefore, further research is required to bolster the examination of intricate man-land system coupling, simulation, and prediction.

Driven by rapid economic and social development, as well as advancements in science and technology, the scope and scientific implications of man-land relationship research are expanding. Particularly in the context of economic globalization, there has been a shift in focus from regional systems to spatial network systems of man-land relationships [7,8]. The study of regional systems aims to coordinate man-land relationships, optimizing and regulating global, national, or regional systems through considerations of spatial structure, temporal processes, overall effects, and synergistic complementarity. This provides a theoretical foundation for effective regional development and management. On the other hand, studying spatial network systems emphasizes the interconnectedness and long-distance connections across regions. The increasing interconnectedness brought about by globalization and urbanization, facilitated by the flow of information, capital, goods, and population, reshapes traditional man-land relationships into a complex network.

Against the backdrop of globalization and urbanization, it is crucial to reassess and broaden the theoretical implications of man-land relationship research. Addressing the conflicts and trade-offs between humans and the environment in the modern era necessitates a systemic outlook, strategic thinking, and interdisciplinary, integrated research across multiple scales. Future research directions will involve scrutinizing shifts in perspectives, underlying assumptions, conceptual frameworks, and research methodologies.

Author Contributions: Conceptualization, L.M., Y.Z., M.G. and Z.S.; writing—original draft preparation, L.M.; writing—review and editing, Y.Z., M.G. and Z.S.; supervision, L.M.; project administration, L.M.; funding acquisition, Y.Z., M.G. and Z.S. All authors have read and agreed to the published version of the manuscript.

Funding: This research was supported by the National Natural Science Foundation of China (grant Nos. 42101203, 42201204), Fundamental Research Funds for the Central Universities (grant No. 2022CD-SKXYGG006), and the Natural Science Foundation of Chongqing (grant No. 2018GXNSFDA281032).

Acknowledgments: We would like to thank all of the contributors to this Special Issue for their hard work and commitment to producing this volume, and we would also like to give our sincere thanks to the editors of *Land* for their ever-present support in producing this Special Issue and the reviewers for their constructive engagement with the manuscripts.

Conflicts of Interest: The authors declare no conflict of interest.

Appendix A

1. Yan, H., Guo, X., Zhao, S., Yang, H., 2022. Variation of Net Carbon Emissions from Land Use Change in the Beijing-Tianjin-Hebei Region during 1990–2020. Land. 7, 997.
2. Hu, Z., Qian, M., Teng, X., Zhang, Z., Zhong, F., Cheng, Q., Wen, C., 2022. The Spatiotemporal Evolution of Ecological Security in Border Areas: A Case Study of Southwest China. Land. 6, 892.
3. Zhu, Y., Li, W., Wang, D., Wu, Z., Shang, P., 2022. Spatial Pattern of Soil Erosion in Relation to Land Use Change in a Rolling Hilly Region of Northeast China. Land. 8, 1253.
4. Jia, H., Wang, R., Li, H., Diao, B., Zheng, H., Guo, L., Liu, L., Liu, J., 2023. The Changes of Desertification and Its Driving Factors in the Gonghe Basin of North China over the Past 10 Years. Land. 5, 998.
5. Tan, N., Chang, X., Ma, T., 2023. Study on Production–Living–Ecological Function Accounting and Management in China. Land. 6, 1163.
6. Wang, Z., Yang, M., Zhang, Z., Li, Y., Wen, C., 2022. The Impact of Land Transfer on Vulnerability as Expected Poverty in the Perspective of Farm Household Heterogeneity: An Empirical Study Based on 4608 Farm Households in China. Land. 11, 1995.
7. Lv, T., Fu, S., Zhang, X., Wu, G., Hu, H., Tian, J., 2022. Assessing Cultivated Land–Use Transition in the Major Grain-Producing Areas of China Based on an Integrated Framework. Land. 10, 1622.
8. Shi, J., Wang, F., 2022. The Effect of High-Speed Rail on Cropland Abandonment in China. Land. 7, 1002.
9. Liang, T., Du, P., Yang, F., Su, Y., Luo, Y., Wu, Y., Wen, C., 2022. Potential Land-Use Conflicts in the Urban Center of Chongqing Based on the "Production–Living–Ecological Space" Perspective. Land. 9, 1415.
10. Han, L., Wu, X., Tang, P., 2023. Does Environmental Decentralization Affect the Supply of Urban Construction Land? Evidence from China. Land. 3, 601.
11. Zhao, W., Shi, P., Wan, Y., Yao, Y., 2023. Coupling and Coordination Relationship between Urbanization Quality and Ecosystem Services in the Upper Yellow River: A Case Study of the Lanzhou–Xining Urban Agglomeration, China. Land. 5, 1085.
12. Zhang, S., Zhao, J., Jiang, Y., Cheshmehzangi, A., Zhou, W., 2023. Assessing the Rural–Urban Transition of China during 1980–2020 from a Coordination Perspective. Land. 6, 1175.
13. Wang, Y., Qu, L., Wang, J., Liu, Q., Chen, Z., 2022. Sustainable Revitalization and Green Development Practices in China's Northwest Arid Areas: A Case Study of Yanchi County, Ningxia. Land. 11, 1902.
14. Cai, T., Zhang, X., Xia, F., Lu, D., 2022. Function Evolution of Oasis Cultivated Land and Its Trade-Off and Synergy Relationship in Xinjiang, China. Land. 9, 1399.

15. Zhang, T., Peng, J., Cao, X., 2023. China's Urban and Rural Development Significantly Affects the Pattern of Human Appropriation of Net Primary Production. Land. 5, 1062.
16. Gong, H., Zhao, Z., Chang, L., Li, G., Li, Y., Li, Y., 2022. Spatiotemporal Patterns in and Key Influences on Cultivated-Land Multi-Functionality in Northeast China's Black-Soil Region. Land. 7, 1101.
17. Liang, T., Yang, F., Luo, Y., Fang, M., Huang, X., Zhang, Z., Wen, C., Ren, X., 2022. The Synchronous Development Pattern and Type Division of Functional Coupling Coordination and Human Activity Intensity Based on the "Production–Living–Ecological" Space Perspective: A Case Study of Wanzhou District. Land. 11, 1897.
18. Cheng, L., Cui, H., Liang, T., Huang, D., Su, Y., Zhang, Z., Wen, C., 2023. Study on the Trade-Off Synergy Relationship of "Production-Living-Ecological" Functions in Chinese Counties: A Case Study of Chongqing Municipality. Land. 5, 1010.
19. Zhou, B., Jiang, L., 2022. Unsustainable Urban Development Based on Temporary Workers: A Study on the Changes of Immigration in Macau between 1992 and 2019. Land. 11, 1985.
20. Guo, Y., Zhong, W., 2023. Rural Transformation Development and Its Influencing Factors in China's Poverty-Stricken Areas: A Case Study of Yanshan-Taihang Mountains. Land. 5, 1080.

References

1. Meyfroidt, P.; Abeygunawardane, D.; Ramankutty, N.; Thomson, A.; Zeleke, G. Interactions between land systems and food systems. *Curr. Opin. Environ. Sustain.* **2019**, *38*, 60–67. [CrossRef]
2. Yuan, F.; Wei, Y.D.; Xiao, W. Land marketization, fiscal decentralization, and the dynamics of urban land prices in transitional China. *Land Use Policy* **2019**, *89*, 104208. [CrossRef]
3. Ma, L.; Zhang, Y.; Chen, S.; Yu, L.; Zhu, Y. Environmental effects and their causes of agricultural production: Evidence from the farming regions of China. *Ecol. Indic.* **2022**, *144*, 109549. [CrossRef]
4. Long, H.; Zhang, Y.; Ma, L.; Tu, S. Land Use Transitions: Progress, Challenges and Prospects. *Land* **2021**, *10*, 903. [CrossRef]
5. Xu, M.; Zhang, Z. Spatial differentiation characteristics and driving mechanism of rural-industrial Land transition: A case study of Beijing-Tianjin-Hebei region, China. *Land Use Policy* **2021**, *102*, 105239. [CrossRef]
6. Thomson, A.M., Ellis, E.C.; Grau, H.R.; Kuemmerle, T.; Meyfroidt, P.; Ramankutty, N.; Zeleke, G. Sustainable intensification in land systems: Trade-offs, scales, and contexts. *Curr. Opin. Environ. Sustain.* **2019**, *38*, 37–43. [CrossRef]
7. Deines, J.M.; Schipanski, M.E.; Golden, B.; Zipper, S.C.; Nozari, S.; Rottler, C.; Guerrero, B.; Sharda, V. Transitions from irrigated to dryland agriculture in the Ogallala Aquifer: Land use suitability and regional economic impacts. *Agric. Water Manag.* **2020**, *233*, 106061. [CrossRef]
8. Long, H. Theorizing land use transitions: A human geography perspective. *Habitat Int.* **2022**, *128*, 102669. [CrossRef]

land

Article

Variation of Net Carbon Emissions from Land Use Change in the Beijing-Tianjin-Hebei Region during 1990–2020

Haiming Yan [1,2], Xin Guo [2], Shuqin Zhao [3,*] and Huicai Yang [1,2]

1 Hebei Province Collaborative Innovation Center for Sustainable Utilization of Water Resources and Optimization of Industrial Structure, Hebei GEO University, Shijiazhuang 050031, China; haiming.yan@hgu.edu.cn (H.Y.); huicai.yang@hgu.edu.cn (H.Y.)
2 School of Land Science and Space Planning, Hebei GEO University, Shijiazhuang 050031, China; guoxin@hgu.edu.cn
3 Graduate School, Hebei GEO University, Shijiazhuang 050031, China
* Correspondence: zhaoshuqin1@hgu.edu.cn

Abstract: Global increasing carbon emissions have triggered a series of environmental problems and greatly affected the production and living of human beings. This study estimated carbon emissions from land use change in the Beijing-Tianjin-Hebei region during 1990–2020 with the carbon emission model and explored major influencing factors of carbon emissions with the Logarithmic Mean Divisia Index (LMDI) model. The results suggested that the cropland decreased most significantly, while the built-up area increased significantly due to accelerated urbanization. The total carbon emissions in the study area increased remarkably from 112.86 million tons in 1990 to 525.30 million tons in 2020, and the built-up area was the main carbon source, of which the carbon emissions increased by 370.37%. Forest land accounted for 83.58–89.56% of the total carbon absorption but still failed to offset the carbon emission of the built-up area. Carbon emissions were influenced by various factors, and the results of this study suggested that the gross domestic product (GDP) per capita contributed most to the increase of carbon emissions in the study area, resulting in a cumulative increase of carbon emissions by 9.48 million tons, followed by the land use structure, carbon emission intensity per unit of land, and population size. By contrast, the land use intensity per unit of GDP had a restraining effect on carbon emissions, making the cumulative carbon emissions decrease by 103.26 million tons. This study accurately revealed the variation of net carbon emissions from land use change and the effects of influencing factors of carbon emissions from land use change in the Beijing-Tianjin-Hebei region, which can provide a firm scientific basis for improving the regional land use planning and for promoting the low-carbon economic development of the Beijing-Tianjin-Hebei region.

Keywords: carbon emission; carbon neutrality; land use change; Beijing-Tianjin-Hebei; LMDI

Citation: Yan, H.; Guo, X.; Zhao, S.; Yang, H. Variation of Net Carbon Emissions from Land Use Change in the Beijing-Tianjin-Hebei Region during 1990–2020. *Land* **2022**, *11*, 997. https://doi.org/10.3390/land11070997

Academic Editors: Li Ma, Yingnan Zhang, Muye Gan and Zhengying Shan

Received: 1 June 2022
Accepted: 29 June 2022
Published: 30 June 2022

1. Introduction

The global increasing carbon emission under high carbon emission mechanisms has triggered a series of environmental problems, e.g., global climate anomalies, sea level rise, and frequent extreme weather events, which have greatly affected the production and living of human beings in recent decades [1,2]. Previous studies suggested that the land use system serves as a vital link between the human socio-economic system and the natural ecological environment, and the carbon emissions caused by land use change have been one of the important influencing factors of global warming [1–3]. In fact, various human activities, e.g., social construction, economic development, industrial arrangement, urban expansion, and energy consumption, are all closely related to the carbon emissions, all of which are ultimately implemented in different land use practices [4,5], and relevant land use change has been considered as the second most important influencing factor of the global increasing atmospheric CO_2 content [6,7]. The 14th Five-Year Plan of China proposed to achieve the "peak carbon dioxide emissions" by 2030 and "carbon neutrality" by 2060, that the low-carbon economy should serve as a new economic growth mode in the future, and that new green energy sources should be developed to reduce the dependence of economic growth on the major fossil energy sources such as coal and oil [4]. It is, therefore, of practical significance to explore the regional carbon emissions from land use change to achieve low carbon land use, promote low carbon economic development, and establish a resource-conserving and environment-friendly society in this context [8,9].

Previous studies on carbon emissions from land use change at home and abroad were primarily concentrated on the spatiotemporal variation of different land use types and their relevant effects on carbon emissions, carbon emission accounting, influencing factors of land use change, and carbon emissions [10–13]. The Guidelines for the National Greenhouse Gas Inventories prepared by the Intergovernmental Panel on Climate Change (IPCC) provided a valuable methodological reference for accounting carbon emissions from land use change [13,14]. Besides, some scholars also proposed the emission coefficient method for the carbon emission accounting of cropland, forest land, grassland, and built-up area and explored the effects of land use change on carbon emissions [14–16]. In addition, most scholars have generally explored the influencing factors of carbon emissions with various econometric methods, such as the factor decomposition method [17,18], and some other scholars explored the drivers of carbon emissions with the Laspeyres decomposition method [19–21]. Moreover, more scholars carried out decomposition analyses based on the Logarithmic Mean Divisia Index (LMDI) model to reveal the influencing factors of carbon emissions in various relevant fields, e.g., carbon emissions per capita, carbon emissions due to industrial combustion energy, carbon emissions in the manufacturing industry, and drivers of carbon emissions from energy consumption for different time durations in China [22,23]. The results of these studies provided valuable methodological references for exploring a series of issues related to carbon emissions from land use change [24,25]. However, these existing studies focused more on some major land use types, such as cropland and built-up area, with less consideration of the carbon emissions from some other land use types [25,26]. In particular, there are relatively fewer quantitative studies on the influencing factors of carbon emissions from land use change, especially the studies on the influencing factors of carbon emissions from land use change with the LMDI model [6,27,28].

The Beijing-Tianjin-Hebei region, as one of the major urban agglomerations in China, accounted for 11–12% of the national total carbon emissions, which is higher than the national proportions of both gross domestic product (GDP) and the population of this region [5,12]. In particular, the goal of peak carbon dioxide emissions and carbon neutrality has put forward new and higher requirements for the synergistic development of this region [16,18,29]. This study estimated the net carbon emissions from land use change in the Beijing-Tianjin-Hebei region from the perspective of land use and decomposed the influencing factors of carbon emissions from land use change to quantitatively reveal the effects of these influencing factors on the carbon emissions from land use change, aiming to provide a firm scientific basis for improving the regional land use planning, promoting the low-carbon economic development, and guiding the development of the Beijing-Tianjin-Hebei region into the capital economy circle.

2. Materials and Methods

2.1. Study Area

The Beijing-Tianjin-Hebei region is located in the northern part of China (36°5′–42°40′ N, 113°27′–119°50′ E), which is one of the three major urban agglomerations in China [18,26,30]. There is very complex terrain in this region, where there are mainly higher mountains and plateaus in the northern and western parts and flatter plains in the southern and eastern parts (Figure 1). There is a warm-temperate continental monsoon climate in this region, with higher temperatures and precipitation in the summer. The Beijing-Tianjin-Hebei region is the political, economic, cultural, and scientific center of China and plays a strategically important role in the economic development of China [18]. The Beijing-

Tianjin-Hebei region takes up approximately 2% of the total national land area, but it accounts for approximately 8.1% of the total national population and 9.44% of the total national GDP. However, there is very high energy consumption intensity along with rapid urbanization in the Beijing-Tianjin-Hebei region, leading to very high carbon emission. For example, the carbon emission in the Beijing-Tianjin-Hebei region reached 1.085 billion tons in 2018, accounting for about 1/9 of the total national carbon emissions of China [26,30]. Exploring the long time-series variation of net carbon emissions from land use change can provide valuable information for addressing the pressure of carbon emission reduction in the Beijing-Tianjin-Hebei region, especially in the context of the coordinated development of the Beijing-Tianjin-Hebei region [26,27].

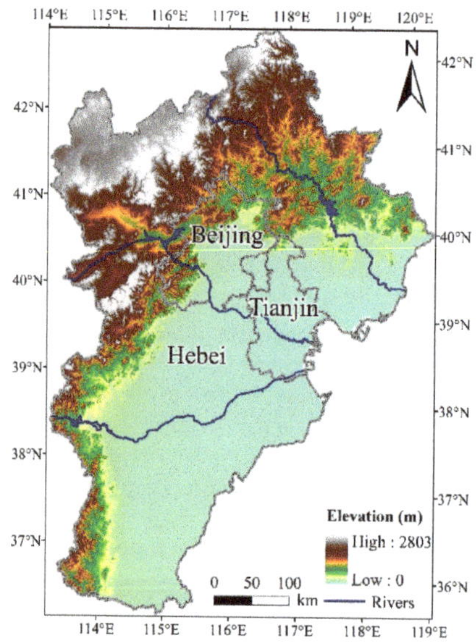

Figure 1. Location of the Beijing-Tianjin-Hebei region.

2.2. Data and Processing

The spatial data used in this study includes the 1-km land use data extracted from the Land Use Remote Sensing Monitoring Data of China in 1990, 1995, 2000, 2005, 2010, 2015, and 2020 (http://www.resdc.cn, accessed on 31 October 2021), which were reverted from Landsat TM/ETM images. The land types were classified into cropland, forest land, grassland, water area, built-up area, and barren land [12,26], based on which this study explored the land use transfer matrix of the Beijing-Tianjin-Hebei region during 1990–2020. Besides, the non-spatial data used in this study mainly included the socioeconomic data (e.g., population and GDP), regional energy consumption data, and carbon emission coefficients of the Beijing-Tianjin-Hebei region, which were obtained from various issues of the China Statistical Yearbook, the China Energy Statistical Yearbook, IPCC reports, and the relevant literature. Finally, these data in different parts of the study area were combined to obtain the regional energy consumption amount and the carbon emission coefficients of different land use types, and these regional data were further summed up to obtain the relevant total data of the whole Beijing-Tianjin-Hebei region.

2.3. Carbon Emission Accounting Model and Carbon Emission Coefficients

The carbon emissions can be categorized into direct and indirect carbon emissions [27]. The former refers to carbon emissions caused by the processes of maintenance and conversion and specific land types, while the latter refers to carbon emissions generated by the land serving as a carrier of production and living processes of human beings [28,29]. A large number of studies have shown that some land use types may be both carbon sources and sinks, and the intensity of carbon sources and sinks generally varies greatly among different land use types [30,31]. This study primarily focused on the carbon emission effects of land use change caused by human activities, i.e., the amounting of carbon emissions and the sequestration of cropland, forest land, grassland, water body, built-up area, and barren land under the influence of human activities, and it finally summarized the carbon emission amount of different land use types. The major crops in the study area include wheat and maize, and these crops on cropland can absorb CO_2 in the air through photosynthesis in general, but most of the crop biomass is then decomposed in the soil and released back into the air in the short term, so there are generally insignificant effects of crop biomass as a carbon sink. Meanwhile, the effects of cropland inputs and soil emissions on the carbon emissions can also be reflected with the carbon emission coefficient of cropland [30]. By contrast, carbon emissions from energy consumption and industrial activities such as housing, mining, and manufacturing and transportation are the main sources of carbon emissions. Thus, the built-up area and cropland generally serve as the carbon sources, with positive carbon emission coefficients, while the forest land, grassland, water body, and barren land generally serve as carbon sinks, which are carbon absorbers with negative carbon emission coefficients. The regional carbon emission can be estimated based on the carbon emission coefficients according to the guidelines of IPCC as follows:

$$E_c = \sum e_i = \sum A_i \times \xi_i \quad (1)$$

where E_c is the total carbon emission (or absorption) amount, e_i is the carbon emission (or absorption) amount from the ith land use type, A_i is the area of the ith land use type, and δ_i is the carbon emission (or absorption) coefficient of the ith land use type.

The carbon emission (or absorption) coefficients of different land use types, which were assumed to keep stable during the study period, were determined as follows. The cropland that provides both carbon emission and carbon absorption serves as both a carbon source and carbon sink [29,30]. It is therefore necessary to take into account the greenhouse gas produced during the crop production and CO_2 absorption of crops during the reproductive period, and the difference between the two can be used to estimate the net carbon emission coefficient of the cropland [22,24]. Previous studies have shown that the carbon emission coefficient and carbon sequestration coefficient of cropland are approximately 0.422 t/hm^2 and 0.007 t/hm^2, respectively [31,32], so this study took the difference between the two as the net carbon emission coefficient of cropland, i.e., 0.415 t/hm^2. In addition, the forest land and grassland are the most important carbon sink and carbon sequestration systems in the terrestrial ecosystem, and previous studies have shown that the carbon emission coefficients of the forest land and grassland were −0.623 t/hm^2 and −0.144 t/hm^2, respectively [33,34], which were also adopted in this study. In addition, previous studies showed that there is very limited carbon absorption of the water body and barren land, generally with very weak impacts on the regional net carbon emissions [27,32]. However, the water body and barren land accounted for 4% of the total area in the Beijing-Tianjin-Hebei region, so this study still considered the carbon emission coefficients of the water body and barren land, which were approximately −0.03 t/hm^2 and −0.05 t/hm^2, respectively, according to the literature survey results [35–37]. Moreover, there are various types of built-up area, e.g., urban land, rural settlements, traffic roads, factories and mines, industrial areas, oil fields, salt fields, and quarries. The built-up area carries a large amount of the energy consumed in the production and living of human beings, and it is unfeasible to calculate the carbon emissions of the built-up area according to only the area share of built-up area [38,39]. It is necessary to estimate the carbon emissions from the built-up area based on the carbon emissions generated by the energy consumption of human beings on the built-up area [40,41], which can be estimated according to the carbon emission coefficient method of the IPCC as follows:

$$EC = \sum m_i \times \beta_i \times \theta_i \quad (2)$$

where EC is the carbon emission from energy consumption on the built-up area, m_i is the consumption amount of various fossil energy sources, β_i is the standard coal conversion coefficient of each energy resource, and θ_i is the carbon emission coefficient of each energy resource. This study used the standard coal conversion coefficient and carbon emission coefficient of each energy resource published by the IPCC guidelines [32,42,43], which are shown in Table 1.

Table 1. Standard coal conversion coefficients and carbon emission coefficients of various energy sources.

Energy Sources	Standard Coal Conversion Coefficient	Carbon Emission Coefficient
Coal	0.7143 (kgce/kg)	0.7559
Coke	0.9714 (kgce/kg)	0.8550
Crude oil	1.4286 (kgce/kg)	0.5857
Gasoline	1.4714 (kgce/kg)	0.5538
Kerosene	1.4714 (kgce/kg)	0.5714
Diesel oil	1.4571 (kgce/kg)	0.5921
Fuel oil	1.4286 (kgce/kg)	0.6185
Natural gas	1.3301 (kgce/m^3)	0.4483
Electric power	0.1229 kg (kgce/kWh)	0.7476

2.4. Decomposition Analysis of Influencing Factors of Carbon Emissions

This study explored the influencing factors of carbon emissions with the LMDI model, which is one of the most widely used methods to explore the influencing factors of energy consumption in the field of low carbon economy due to its advantages such as high operability, full decomposition, no residuals, and unique results [36,37]. Specifically, this study analyzed the effects of different influencing factors on carbon emissions from land use change according to the Kaya identity by introducing the land use factor and establishing the formula of influencing factors of regional carbon emissions from five aspects, i.e., energy consumption structure, land output intensity, land use structure, economic growth, and population scale effect, as follows [41,42]:

$$C = \sum \frac{C_i}{L_i} \times \frac{L_i}{L} \times \frac{L}{G} \times \frac{G}{P} \times P \tag{3}$$

where C is the total carbon emissions from land use change (million tons), C_i is the carbon emission amount of the ith land use type (million tons), L_i is the area of the ith land use type (km^2), L is the total land area of the study area (km^2), G is the GDP (10^8 CNY), and P is the regional population size (10^4 persons).

Then, the regional total carbon emissions can be expressed as follows [40,42]:

$$f_i = \frac{C_i}{L_i};\ S_i = \frac{L_i}{L};\ q = \frac{L}{G};\ g = \frac{G}{P} C = \sum f_i \times S_i \times q \times g \times P \tag{4}$$

where f_i, S_i, q, g, and P refer to the carbon emission intensity per unit of the ith land use type, the effect of the land use structure, land use intensity per unit of GDP, GDP per capita, and population size, respectively. According to this formula, the contribution value and contribution rate of each factor can be further analyzed with the LMDI model. Assuming the carbon emission in the base period and the Tth period are C^0 and C^T, respectively, then the carbon emission change during the study period (0–T) can be expressed as follows [40,43]:

$$\begin{aligned} \Delta C = C^T - C^0 &= \sum_{i=1,2,\ldots\ldots 6} f_i^T \times s_i^T \times q^T \times g^T \times P^T - \sum_{i=1,2,\ldots\ldots 6} f_i^0 \times s_i^0 \times q^0 \times g^0 \times P^0 \\ &= \Delta C_{f_i} + \Delta C_{s_i} + \Delta C_q + \Delta C_g + \Delta C_P + \Delta C_{rsd} \end{aligned} \tag{5}$$

$$D = \frac{C^T}{C^0} = D_f D_s D_q D_g D_P D_{rsd} \tag{6}$$

where ΔC is the carbon emission change during the study period, ΔC_{f_i}, ΔC_{s_i}, ΔC_q, ΔC_g, and ΔC_P are the contribution values of f_i, S_i, q, g, and P, respectively, and ΔC_{rsd} is the decomposition residual. If the obtained contribution value is >0, then the factor has a pulling effect on the carbon emissions during the study period; otherwise, the factor has a suppressing effect on carbon emissions. D is the carbon emission change percentage between the base period and the Tth period; D_f, D_s, D_q, D_g, D_p, and D_{rsd} are the contribution rates of f_i, S_i, q, g, and the residual error, respectively.

The following are the relationships in the additive decomposition mode according to the LMDI model [40,44,45]:

$$\begin{aligned}
\Delta C_{f_i} &= \sum_i \frac{C_i^T - C_i^0}{\ln C_i^T - \ln C_i^0} \times \ln \frac{f_i^T}{f_i^0} \\
\Delta C_{s_i} &= \sum_i \frac{C_i^T - C_i^0}{\ln C_i^T - \ln C_i^0} \times \ln \frac{S_i^T}{S_i^0} \\
\Delta C_q &= \sum_i \frac{C_i^T - C_i^0}{\ln C_i^T - \ln C_i^0} \times \ln \frac{q^T}{q^0} \\
\Delta C_g &= \sum_i \frac{C_i^T - C_i^0}{\ln C_i^T - \ln C_i^0} \times \ln \frac{g^T}{g^0} \\
\Delta C_P &= \sum_i \frac{C_i^T - C_i^0}{\ln C_i^T - \ln C_i^0} \times \ln \frac{P^T}{P^0}
\end{aligned} \quad (7)$$

The following are the relationships in the multiplicative decomposition mode according to the LMDI model [40,45]:

$$\begin{aligned}
&D_f = \exp(W\Delta C_{f_i}); D_s = \exp(W\Delta C_{s_i}) \\
&D_q = \exp(W\Delta C_q); D_g = \exp(W\Delta C_g) \\
&D_P = \exp(W\Delta C_P); D_{rsd} = 1 \\
&W = \frac{\ln D}{\Delta C}
\end{aligned} \quad (8)$$

3. Results

3.1. Land Use Change in the Beijing-Tianjin-Hebei Region

There was remarkable land use change in the Beijing-Tianjin-Hebei region during 1990–2020. There is mainly cropland and forest land in the Beijing-Tianjin-Hebei region, where the cropland is mainly distributed in the central and southeast parts of the study area, and the forest land as well as grassland are mainly distributed in the northeast and western parts (Figure 2). The built-up area is concentrated in the central part and the peripheral zone of the central towns in the study area. The water body is very limited in the study area, which mainly includes rivers near towns and lakes in the northwest. More specifically, cropland as the main land use type accounted for approximately 51.92% of the entire area in 1990. The forest land and grassland ranked second and third, accounting for about 20.65% and 16.49% of the entire area, respectively, while the built-up area, water body and barren land accounted for 10.94% of the entire area in total. There was a decreasing trend of cropland, forest land, grassland, and barren land from 2000–2015, while the built-up area increased significantly due to the accelerated urbanization process. In particular, the cropland decreased significantly from 2015–2020, while the built-up area continued to increase, and other land types only changed slightly.

Table 2 shows the land use transfer in the Beijing-Tianjin-Hebei region from 1990–2020. A total of 73,072 km^2 of land was transferred during 1990–2020, among which the transfer-out area of the cropland ranked first, accounting for about 91.2% of the total transfer-out area. The cropland was mainly transferred into built-up area, grassland, and forest land, with the converted areas of which reaching 6726 km^2 and 4705 km^2 and 17,621 km^2, respectively. Meanwhile the transfer-in area of cropland reached 19,880 km^2, which was mainly converted from grassland and built-up area, accounting for about 69.4% of the total transfer-in area of cropland. The transfer-in area of built-up area increased most significantly during the study area, reaching 20,837 km^2, 84.6% of which was transferred from the cropland. By contrast, watershed and barren land only changed slightly, most of which was converted to cropland, accounting for 49% and 51% of their transfer-out area, respectively.

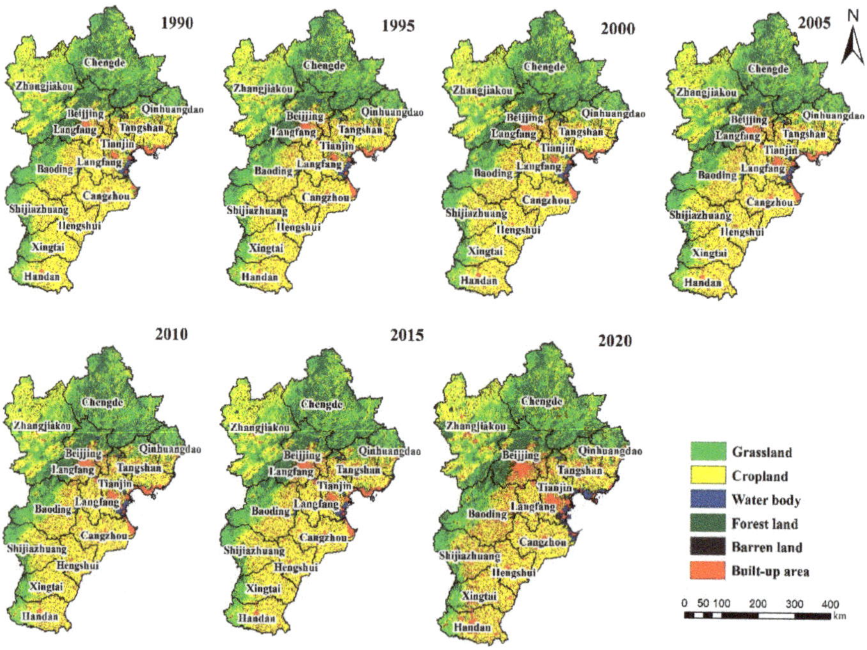

Figure 2. Spatial pattern of land use in the Beijing-Tianjin-Hebei region from 1990–2020.

Table 2. Land use transfer matrix in the Beijing-Tianjin-Hebei region during 1990–2020 (km^2).

1990	2020						
	Grassland	**Cropland**	**Built-up Area**	**Forest Land**	**Water Body**	**Barren Land**	**Total Decrease**
Grassland	19,377	7016	1425	6858	538	223	16,060
Cropland	6726	79,738	17,621	4705	2512	302	31,866
Built-up area	314	6777	6821	190	996	43	8320
Forest land	6713	3454	836	33,103	251	29	11,283
Water body	486	1778	768	361	2288	489	3882
Barren land	320	855	187	75	224	547	1661
Total increase	14,559	19,880	20,837	12,189	4521	1086	73,072
Overall change	−1501	−11,986	12,517	906	639	−575	—

3.2. *Evolution of Net Carbon Emissions in the Beijing-Tianjin-Hebei Region*

The average annual net carbon emission during 1990–2020 was 313.93 million tons, and the net carbon emissions of Hebei Province changed most significantly, increasing by 317.71 million tons during the study period, followed by Tianjin and Beijing. From the perspective of regions within the study area, the changing trend of net carbon emissions of Hebei Province is similar to that of Beijing and Tianjin. Besides, the evolution of carbon emissions in the Beijing-Tianjin-Hebei region over these 30 years was divided into three stages, according to the change of net carbon emissions, the industrial development status, and energy consumption status in the study area, as shown in Figure 3.

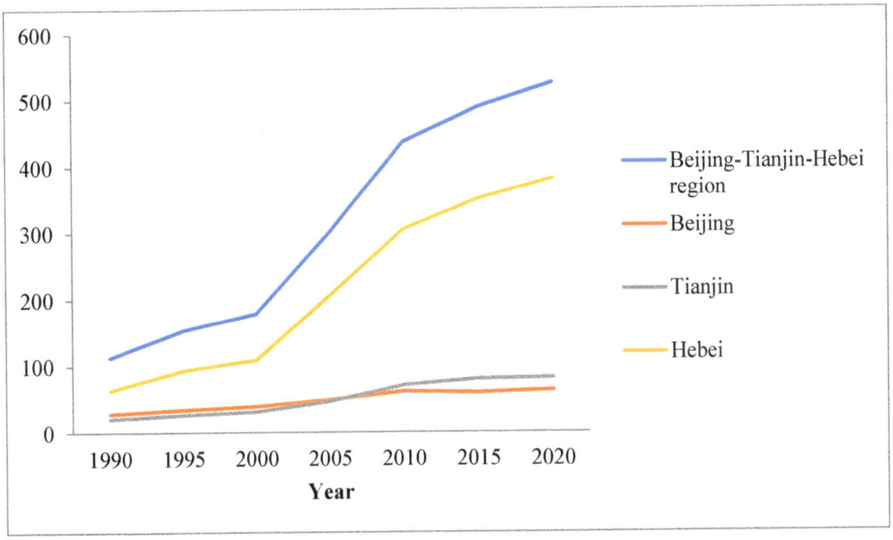

Figure 3. Changing trends of net carbon emissions in the Beijing-Tianjin-Hebei region from 1990–2020 (million tons).

Phase I (1990–2000) is the stage of slow increase in net carbon emissions, which increased from 112.86 million tons to 178.54 million tons but was still significantly lower than the average level of the whole study period. This is mainly due to the fact that the study area was still in the early industrialization stage during this phase, with a low urbanization level and low consumption of various energy sources, which led to relatively lower net carbon emissions.

Phase II (2000–2010) is the stage of rapid growth of net carbon emissions, which rapidly increased from 178.54 million tons to 436.83 million tons, with the average annual growth rate reaching 144.66%. This is mainly due to the national focus on the development of heavy industries and the relatively loose macro production capacity policies during this period, which led to the lower entry requirement of high energy-consuming, high-emission, and low-efficiency enterprises into the study area, thus resulting in the rapid increase of the net carbon emissions.

The third phase (2010–2020) is the stage of steady increase of net carbon emissions, with the total carbon emissions increasing steadily from 436.83 million tons to 525.31 million tons. The average carbon emission level during this phase was 1.58 times that of the second phase, but the average annual growth rate was only 20.25%, which is much lower than that of the second phase. This is mainly due to the improvement of energy efficiency under the influence of the national policies on energy saving and emission reduction and application of advanced technologies. In particular, the implementation of the policies of "peak carbon dioxide emissions" and "carbon neutrality" imposed important limitations on the increase of carbon emissions.

3.3. Variation of Carbon Emissions from Land Use Change

The carbon emissions from land use change are shown in Table 3, which were estimated on the basis of the land use data and energy consumption data of the Beijing-Tianjin-Hebei region during 1990–2020. The carbon emissions from land use change in the Beijing-Tianjin-Hebei region showed an overall increasing trend during 1990–2020, with the total carbon emissions increasing from 112.86 million tons in 1990 to 525.30 million tons in 2020, with an overall growth rate of 365.46% during these 30 years (Table 3). Among the major sources of carbon emissions, the carbon emissions from cropland decreased slowly from 4.66 million tons in 1990 to 4.15 million tons in 2020, with a total decrease of 0.52 million tons and an average annual decrease of 17.12 thousand tons in these 30 years, which is mainly due to the slowly decreasing trend of cropland area in the study area during the study period. However, the carbon emissions from built-up area as another major source of carbon emissions increased rapidly from 111.51 million tons in 1990 to 524.51 million tons in 2020, with an overall growth rate of 370.37% during the study period. In particular, carbon emissions from built-up area accounted for 95.99–99.22% of the total carbon emissions, so built-up area served as the main carbon emission source in the Beijing-Tianjin-Hebei region during the study period. The carbon sinks included the forest land, grassland, water area, and barren land, the effects of which on the carbon absorption shoed a descending order. Specifically, the carbon absorption effect of the forest land was the most significant, accounting for 83.58–89.56% of the total carbon absorption amount of the study area, while the grassland and water body accounted for 9.62–15.49% and 0.47–0.59% of the total carbon absorption amount of the study area, respectively. By contrast, the carbon sequestration effect of barren land was the least significant, accounting for only 0.24–0.35% of the total carbon absorption amount of the study area. It is particularly notable that the ratio of the carbon emissions from the built-up area to the carbon absorption from the forest land ranged between 40.26 and 185 during 1990–2020. In other words, the carbon emissions from the built-up area were so large that the carbon absorption effect of the forest land failed to offset the carbon source effect of built-up area, which is the underlying reason for the continuous increase of the total carbon emissions in the study area during the study period.

Table 3. Carbon emissions from land use change in the Beijing-Tianjin-Hebei region during 1990–2020 (million tons).

Year	1990	1995	2000	2005	2010	2015	2020
Forest land	−2.77	−3.72	−2.79	−2.79	−2.79	−2.78	−2.84
Grassland	−0.51	−0.4	−0.51	−0.5	−0.5	−0.5	−0.49
Water body	−0.02	−0.02	−0.02	−0.02	−0.02	−0.02	−0.02
Barren land	−0.01	−0.01	−0.01	−0.01	−0.01	−0.01	−0.01
Cropland	4.66	4.22	4.55	4.51	4.49	4.46	4.15
Built-up area	111.51	153.74	177.33	299.91	435.66	487.95	524.52
Carbon sink	−3.31	−4.16	−3.32	−3.32	−3.32	−3.31	−3.36
Carbon Source	116.17	157.96	181.87	304.42	440.15	492.41	528.66
Total carbon emissions	112.86	153.8	178.55	301.1	436.83	489.1	525.31

3.4. Effects of Influencing Factors of Carbon Emissions from Land Use Change

The decomposition of the variation of carbon emissions in the Beijing-Tianjin-Hebei region during 1990–2020 with the LMDI model revealed the contribution value and contribution rate of five influencing factors of carbon emissions, i.e., carbon emission intensity per unit of land, land use structure, land use intensity per unit of GDP, GDP per capita, and population size in Figure 4. The results showed that there was obvious annual variation trends of these influencing factors of carbon emissions from land use change in the study area (Figure 4). Besides, Figure 5 shows the cumulative contribution rate of these influencing factors. There were remarkable differences in the contribution value between various influencing factors of the carbon emissions from land use change (Figure 5). The absolute contribution values of each influencing factor to the variation of carbon emissions from land use change during 1990–2020 in a descending order were land use intensity per unit of GDP > GDP per capita > land use structure > carbon emission intensity per unit of land > population size. The land use intensity per unit of GDP was the biggest restraining factor of the increase of carbon emissions, and the remaining four factors all had positive effects on the increase of carbon emissions, among which GDP per capita had the greatest promotion effects on the increase of carbon emissions (Figure 5).

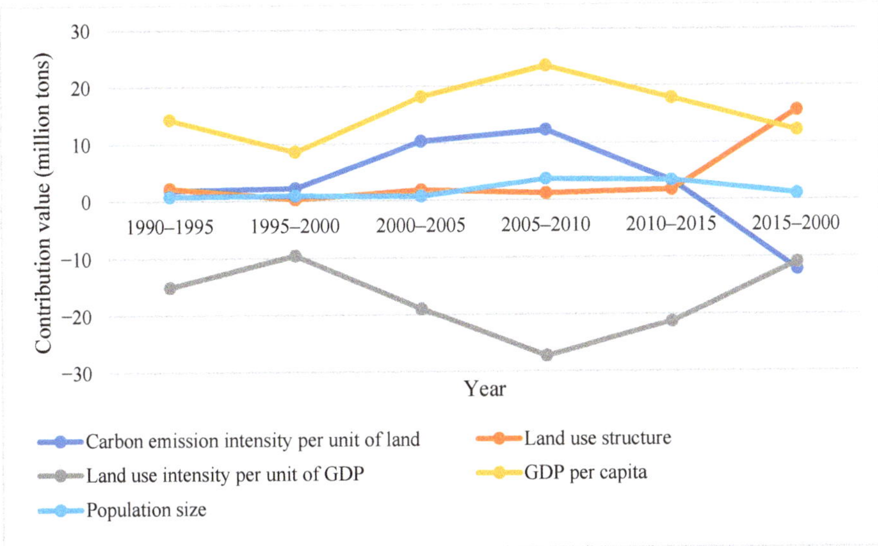

Figure 4. Contribution value of the influencing factors to variation of carbon emissions from land use change in the Beijing-Tianjin-Hebei region.

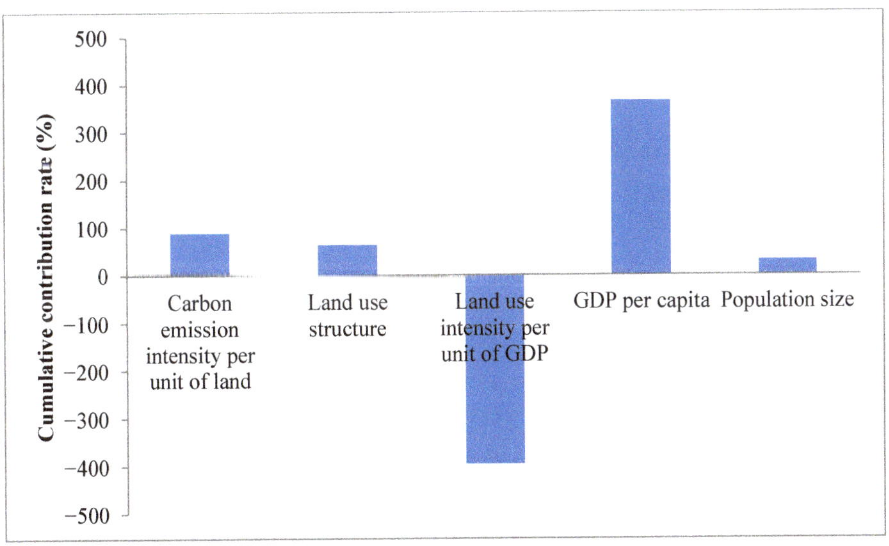

Figure 5. Cumulative contribution rate of influencing factors to the variation of carbon emissions from land use change in the Beijing-Tianjin-Hebei region.

This study suggested that the land use factors played an important role in influencing the carbon emissions. For example, there were extremely unstable effects of the carbon emission intensity per unit of land on the carbon emissions, which showed a promoting effect during 1990–2015 and a restraining effect during 2015–2020, indicating uncertainties of the role of carbon emission intensity per unit of land in influencing the carbon emissions from land use change. Nevertheless, the cumulative contribution rate of the carbon emission intensity per unit of land was 89.06% during the study period. Besides, the land use structure had an overall positive effect on the carbon emissions from land use change during the study period, with a cumulative contribution rate of 64.3%. It is notable that the contribution value of the land use structure showed a significant increase during 2015–2020, which is mainly because the built-up area expanded rapidly during this period, which led to the remarkable increase of carbon emissions from land use change. It is therefore of great significance to the control of carbon emissions to carry out the reasonable layout of land use structure. By contrast, the land use intensity per unit of GDP was the primary restraining factor of the carbon emissions from land use change in the study area, which had a suppressive effect on the net carbon emissions throughout the study period, with a cumulative contribution reaching −385.47%. This indicates that it is feasible to achieve a sustainable development status of the land use by promoting the economic development and adopting some reasonable means.

There was always a positive contribution value of GDP per capita to the increase of carbon emissions, i.e., GDP per capita had played a positive role in promoting the increase of carbon emissions from land use change during the study period. In particular, the cumulative contribution of GDP per capita was the highest among these influencing factors, indicating that GDP per capita played the most important role in promoting the increase of carbon emissions from land use change. In fact, GDP per capita represents the regional economic development level as well as the affluence level, and the rapid economic development not only brings abundant material achievements but also generates a large amount of carbon emissions; it is therefore necessary to pay more attention to the factors of economic development in future research and the practice of carbon emission reduction. Additionally, the population size always had a promoting effect on the increase of carbon emissions during 1990–2015, which is similar to GDP per capita. The cumulative contribution value and cumulative contribution rate of population size reached 1.05 million tons and 30.61%, respectively, indicating that the population size is also one of the most important factors promoting the increase of carbon emissions from land use change. This is primarily because the population growth leads to the increase in energy consumption and further results in more carbon emissions from energy consumption on the land.

4. Discussion

There is an overall high reliability of the results of this study, which were generally consistent with previous studies. For example, some previous studies suggested that the growth of GDP per capita resulting from expansion of coal intensive industries was a major factor driving carbon emissions in China [44,46], and this study also suggested that GDP per capita played a dominant role in promoting the increase of carbon emissions in the Beijing-Tianjin-Hebei region, indicating there was a consistent major driving factor of the Beijing-Tianjin-Hebei region and the whole of China. On the one hand, the Beijing-Tianjin-Hebei region and other parts of China both used to heavily depend on coal, with a slight difference in the technological level of energy utilization among these regions, which led to a similar pattern of carbon emissions of the Beijing-Tianjin-Hebei region and other parts of China. On the other hand, the Beijing-Tianjin-Hebei region, as one of the three major urban agglomerations in China, generally kept pace with the rapid economic development of the whole of China, leading to a similar change of GDP per capita and subsequently carbon emissions. More importantly, this study estimated the carbon emissions with the data extracted from the authoritative statistical yearbooks and carried out a decomposition analysis of influencing factors of carbon emissions with the relatively mature LMDI model, which were both generally consistent with previous studies and therefore guaranteed the reliability of the results of this study.

There are still some uncertainties in the results of this study due to the limitation of data accuracy, and it is especially necessary to further improve the estimation of the carbon emission (or absorption) coefficients based on dynamic observation data in the future. For example, the carbon emission coefficient of the State Grid Corporation of China has been adjusted from 0.6101 tCO_2/MWh to 0.5810 tCO_2/MWh in 2022, according to the latest "Corporate Greenhouse Gas Emissions Accounting Methodology and Reporting

Guidelines for Power Generation Facilities (2022 Revised Edition)". This carbon emission coefficient has declined slightly by only 4.77% after years of technical progress; it is therefore still feasible to assume that the carbon emission coefficient of electricity kept constant in past decades. Nevertheless, it is still necessary to use some dynamic carbon emission coefficients, according to the specific conditions, to more accurately reveal the effects of technical progress and other factors on carbon emissions in the future.

Although there are still some uncertainties, this study still successfully revealed the effects of various influencing factors of carbon emissions from land use change. This study has accordingly proposed the following policy recommendations, which can contribute to promoting the improvement of the lower carbon emission of land use and support the synergetic development of the Beijing-Tianjin-Hebei region.

(1) Optimization of the land use structure and the spatial layout of land use

It is necessary to carry out land use in a more economical and intensive way, making full use of the barren land and a large amount of idle land by giving priority to turning the barren land and idle land into cropland, forest land, and grassland. It is also necessary to carry out a moderate return of cropland to forest land and grass land. It is particularly urgent to control the proportion of built-up area and increase the area of carbon sinks by planting trees and optimizing the spatial layout of public green space, which can contribute to achieving environmental improvement and low-carbon development in the Beijing-Tianjin-Hebei region.

(2) Adjustment of the energy structure and development of new cleaner energy sources

The current energy consumption of the Beijing-Tianjin-Hebei region dominantly depends on fossil energy, especially coal, while the proportion of cleaner energy is still very low, and it is therefore very necessary to adjust the energy structure and reduce the dependence on fossil energy. To achieve this end, it is necessary to promote the development of new cleaner energy sources and to encourage the development of advanced energy-saving technologies, e.g., the Carbon Capture and Storage technologies, which can effectively reduce carbon emissions.

(3) Optimization of the industrial structure and promotion of the development of the tertiary industry

The increase of GDP of the Beijing-Tianjin-Hebei region still heavily depends on the secondary industry, which plays an important role in the increasing the carbon emissions. By contrast, the primary and tertiary industries have relatively small carbon emissions. It is therefore necessary to carry out optimization of the industrial structure by "retreating from secondary industry and developing the tertiary industries", which can make a considerable contribution to the energy saving and carbon emission reduction in the Beijing-Tianjin-Hebei region.

5. Conclusions

This study estimated the carbon emissions in the Beijing-Tianjin-Hebei region during 1990–2020 based on the carbon emission coefficients, and it revealed the quantitative relationship between land use change and carbon emissions with the decomposition analysis of the main influencing factors of carbon emissions from land use change. The major conclusions are as follows: (1) the cropland, forest land, grassland, and barren land in the study area all showed a decreasing trend during 1990–2020, among which the cropland decreased most significantly and the built-up area increased significantly due to the accelerated urbanization. (2) The total carbon emissions in the study area increased from 112.86 million tons in 1990 to 525.31 million tons in 2020, with a growth rate of 365.46%. Built-up area was the main carbon source, the carbon emissions of which increased rapidly from 111.51 million tons in 1990 to 524.52 million tons in 2020, with a growth rate of 370.37%. The forest land accounted for 83.58–89.56% of the total carbon absorption, but it still failed to offset the carbon emissions of the built-up area. (3) GDP per capita contributed most to the increase of the carbon emissions, resulting in a cumulative increase of carbon emissions by 94.78 million tons. While the land use structure, carbon emission intensity per unit of land, and population size led to the increase of carbon emissions by 18.11 million tons, 22.95 million tons, and 8.67 million tons, respectively. By contrast, the land use intensity per unit of GDP had a restraining effect on the carbon emissions, making the carbon emissions decrease by 103.26 million tons in total. This study accurately revealed the variation of net carbon emissions from land use change and the effects of influencing factors of carbon emissions from land use change in the Beijing-Tianjin-Hebei region, and all the conclusions of this study can provide a firm scientific basis for improving the regional land use planning and for promoting the low-carbon economic development of the Beijing-Tianjin-Hebei region.

Author Contributions: H.Y. (Haiming Yan) and S.Z.: investigation, data curation, writing—original draft preparation, writing—review and editing, and funding acquisition; H.Y. (Huicai Yang): conceptualization, methodology, supervision, and project administration; H.Y. (Haiming Yan) and X.G.: software, validation, and visualization. All authors have read and agreed to the published version of the manuscript.

Funding: This study was funded by the Major Project of Humanities and Social Science Research of Hebei Education Department (ZD201907), the Science and Technology Project of Hebei Education Department (BJ2019045), and the National Natural Science Foundation of China (51909052).

Data Availability Statement: Not applicable.

Acknowledgments: We are very grateful for the helpful inputs from the editor and anonymous reviewers.

Conflicts of Interest: The authors declare no conflict of interest.

References

1. Quay, P.; Tilbrook, B.; Wong, C. Oceanic uptake of fossil fuel CO_2: Carbon-13 evidence. *Science* **1992**, *256*, 74–79. [CrossRef] [PubMed]
2. Houghton, R. Magnitude, distribution and causes of terrestrial carbon sinks and some implications for policy. *Clim. Policy* **2002**, *2*, 71–88. [CrossRef]
3. Lau, H.C.; Ramakrishna, S.; Zhang, K.; Radhamani, A.V. The role of carbon capture and storage in the energy transition. *Energy Fuels* **2021**, *35*, 7364–7386. [CrossRef]
4. Chuai, X.; Huang, X.; Wang, W.; Zhao, R.; Zhang, M.; Wu, C. Land use, total carbon emissions change and low carbon land management in Coastal Jiangsu, China. *J. Clean. Prod.* **2015**, *103*, 77–86. [CrossRef]
5. Lai, L.; Huang, X.; Yang, H.; Chuai, X.; Zhang, M.; Zhong, T.; Chen, Z.; Chen, Y.; Wang, X.; Thompson, J.R. Carbon emissions from land-use change and management in China between 1990 and 2010. *Sci. Adv.* **2016**, *2*, e1601063. [CrossRef] [PubMed]
6. Hong, C.; Burney, J.A.; Pongratz, J.; Nabel, J.E.M.S.; Mueller, N.D.; Jackson, R.B.; Davis, S.J. Global and regional drivers of land-use emissions in 1961–2017. *Nature* **2021**, *589*, 554–561. [CrossRef] [PubMed]
7. Zhou, D.; Xiao, J.; Frolking, S.; Zhang, L.; Zhou, G. Urbanization contributes little to global warming but substantially intensifies local and regional land surface warming. *Earth's Future* **2022**, *10*, e2021EF002401. [CrossRef]
8. Yohannes, H.; Soromessa, T.; Argaw, M.; Dewan, A. Spatio-temporal changes in habitat quality and linkage with landscape characteristics in the Beressa watershed, Blue Nile basin of Ethiopian highlands. *J. Environ. Manag.* **2021**, *281*, 111885. [CrossRef]
9. Janizadeh, S.; Chandra Pal, S.; Saha, A.; Chowdhuri, I.; Ahmadi, K.; Mirzaei, S.; Mosavi, A.H.; Tiefenbacher, J.P. Mapping the spatial and temporal variability of flood hazard affected by climate and land-use changes in the future. *J. Environ. Manag.* **2021**, *298*, 113551. [CrossRef]
10. Mukul, S.A.; Halim, M.A.; Herbohn, J. Forest carbon stock and fluxes: Distribution, biogeochemical cycles, and measurement techniques. In *Life on Land, Encyclopedia of the UN Sustainable Development Goals;* Leal Fil-ho, W., Azul, A., Brandli, L., Lange Salvia, A., Wall, T., Eds.; Springer: Cham, Switzerland, 2021; pp. 365–380.
11. Kalnay, E.; Cai, M. Impact of urbanization and land-use change on climate. *Nature* **2003**, *423*, 528–531. [CrossRef]
12. Liang, Y.; Hashimoto, S.; Liu, L. Integrated assessment of land-use/land-cover dynamics on carbon storage services in the Loess Plateau of China from 1995 to 2050. *Ecol. Indic.* **2021**, *120*, 106939. [CrossRef]
13. Lin, Q.; Zhang, L.; Qiu, B.; Zhao, Y.; Wei, C. Spatiotemporal analysis of land use patterns on carbon emissions in China. *Land* **2021**, *10*, 141. [CrossRef]
14. IPCC. *2006 IPCC Guidelines for National Greenhouse Gas Inventories;* Institute for Global Environmental Strategies: Hayama, Japan, 2006.

15. Villamor, G.B.; Le, Q.B.; Djanibekov, U.; van Noordwijk, M.; Vlek, P.L. Biodiversity in rubber agroforests, carbon emissions, and rural livelihoods: An agent-based model of land-use dynamics in lowland Sumatra. *Environ. Model. Softw.* **2014**, *61*, 151–165. [CrossRef]
16. Wu, S.; Hu, S.; Frazier, A.E. Spatiotemporal variation and driving factors of carbon emissions in three industrial land spaces in China from 1997 to 2016. *Technol. Forecast. Soc.* **2021**, *169*, 120837. [CrossRef]
17. Kuang, B.; Lu, X.; Zhou, M.; Chen, D. Provincial cultivated land use efficiency in China: Empirical analysis based on the SBM-DEA model with carbon emissions considered. *Technol. Forecast. Soc.* **2020**, *151*, 119874. [CrossRef]
18. Wang, Q.; Wang, S.; Li, R. Determinants of decoupling economic output from carbon emission in the transport sector: A comparison study of four municipalities in China. *Int. J. Environ. Res. Public Health* **2019**, *16*, 3729. [CrossRef]
19. Bhattacharyya, S.C.; Matsumura, W. Changes in the GHG emission intensity in EU-15: Lessons from a decomposition analysis. *Energy* **2010**, *35*, 3315–3322. [CrossRef]
20. Xie, X.; Shao, S.; Lin, B. Exploring the driving forces and mitigation pathways of CO_2 emissions in China's petroleum refining and coking industry: 1995–2031. *Appl. Energy* **2016**, *184*, 1004–1015. [CrossRef]
21. Lin, B.; Benjamin, N.I. Influencing factors on carbon emissions in China transport industry. A new evidence from quantile regression analysis. *J. Clean. Prod.* **2017**, *150*, 175–187. [CrossRef]
22. Shen, L.; Sun, Y. Review on carbon emissions, energy consumption and low-carbon economy in China from a perspective of global climate change. *J. Geogr. Sci.* **2016**, *26*, 855–870. [CrossRef]
23. Wang, S.; Ma, Y. Influencing factors and regional discrepancies of the efficiency of carbon dioxide emissions in Jiangsu, China. *Ecol. Indic.* **2018**, *90*, 460–468. [CrossRef]
24. Zhao, R.; Huang, X.; Liu, Y.; Zhong, T.; Ding, M.; Chuai, X. Carbon emission of regional land use and its decomposition analysis: Case study of Nanjing City, China. *Chin. Geogr. Sci.* **2015**, *25*, 198–212. [CrossRef]
25. Auffhammer, M.; Carson, R.T. Forecasting the path of China's CO_2 emissions using province-level information. *J. Environ. Econ. Manag.* **2008**, *55*, 229–247. [CrossRef]
26. Yan, H.; Li, W.; Yang, H.; Guo, X.; Liu, X.; Jia, W. Estimation of the rational range of ecological compensation to address land degradation in the Poverty Belt around Beijing and Tianjin, China. *Land* **2021**, *10*, 1383. [CrossRef]
27. Fang, J.; Guo, Z.; Piao, S.; Chen, A. Terrestrial vegetation carbon sinks in China, 1981–2000. *Sci. China Ser. D* **2007**, *50*, 1341–1350. [CrossRef]
28. Chen, Y.; Lu, H.; Li, J.; Xia, J. Effects of land use cover change on carbon emissions and ecosystem services in Chengyu urban agglomeration, China. *Stoch. Environ. Res. Risk Assess.* **2020**, *34*, 1197–1215. [CrossRef]
29. Fang, J.; Zhu, J.; Wang, S. Global warming, carbon emissions and uncertainty. *Sci. China Earth Sci.* **2011**, *10*, 1385–1395.
30. Zhou, Y.; Chen, M.; Tang, Z.; Mei, Z. Urbanization, land use change, and carbon emissions: Quantitative assessments for city-level carbon emissions in Beijing-Tianjin-Hebei region. *Sustain. Cities Soc.* **2021**, *66*, 102701. [CrossRef]
31. Lippke, B.; Perez-Garcia, J. Will either cap and trade or a carbon emissions tax be effective in monetizing carbon as an ecosystem service. *For. Ecol. Manag.* **2008**, *256*, 2160–2165. [CrossRef]
32. Yang, H.; Zheng, H.; Liu, H.; Wu, Q. NonLinear effects of environmental regulation on eco-efficiency under the constraint of land use carbon emissions: Evidence based on a bootstrapping approach and panel threshold model. *Int. J. Environ. Res. Public Health* **2019**, *16*, 1679. [CrossRef]
33. Dunn, J.B.; Mueller, S.; Kwon, H.-y.; Wang, M.Q. Land-use change and greenhouse gas emissions from corn and cellulosic ethanol. *Biotechnol. Biofuels* **2013**, *6*, 51. [CrossRef] [PubMed]
34. Houghton, R.A.; House, J.I.; Pongratz, J.; Van Der Werf, G.R.; Defries, R.S.; Hansen, M.C.; Le Quéré, C.; Ramankutty, N. Carbon emissions from land use and land-cover change. *Biogeosciences* **2012**, *9*, 5125–5142. [CrossRef]
35. Fatichi, S.; Pappas, C.; Zscheischler, J.; Leuzinger, S. Modelling carbon sources and sinks in terrestrial vegetation. *New Phytol.* **2019**, *221*, 652–668. [CrossRef] [PubMed]
36. Zhu, E.; Deng, J.; Zhou, M.; Gan, M.; Jiang, R.; Wang, K.; Shahtahmassebi, A. Carbon emissions induced by land-use and land-cover change from 1970 to 2010 in Zhejiang, China. *Sci. Total Environ.* **2019**, *646*, 930–939. [CrossRef]
37. Guo, R.; Zhao, Y.; Shi, Y.; Li, F.; Hu, J.; Yang, H. Low carbon development and local sustainability from a carbon balance perspective. *Resour. Conserv. Recycl.* **2017**, *122*, 270–279. [CrossRef]
38. Say, N.P.; Yücel, M. Energy consumption and CO_2 emissions in Turkey: Empirical analysis and future projection based on an economic growth. *Energy Policy* **2006**, *34*, 3870–3876. [CrossRef]
39. Nguyen, K.H.; Kakinaka, M. Renewable energy consumption, carbon emissions, and development stages: Some evidence from panel cointegration analysis. *Renew. Energy* **2019**, *132*, 1049–1057. [CrossRef]
40. Fan, R.N. *A Study on the Influencing Factors of Carbon Emissions in Zhejiang Province-Based on Kaya Identity*; Zhejiang University of Finance and Economics: Hangzhou, China, 2016. (In Chinese)
41. Mahony, T.O. Decomposition of Ireland's carbon emissions from 1990 to 2010: An extended Kaya identity. *Energy Policy* **2013**, *59*, 573–581. [CrossRef]
42. Vinuya, F.; DiFurio, F.; Sandoval, E. A decomposition analysis of CO_2 emissions in the United States. *Appl. Econ. Lett.* **2010**, *17*, 925–931. [CrossRef]
43. Han, X.; Yu, J.; Xia, Y.; Wang, J. Spatiotemporal characteristics of carbon emissions in energy-enriched areas and the evolution of regional types. *Energy Rep.* **2021**, *7*, 7224–7237. [CrossRef]

44. Chong, C.; Ma, L.; Li, Z.; Ni, W.; Song, S. Logarithmic mean Divisia index (LMDI) decomposition of coal consumption in China based on the energy allocation diagram of coal flows. *Energy* **2015**, *85*, 366–378. [CrossRef]
45. Wang, W.; Liu, X.; Zhang, M.; Song, X. Using a new generalized LMDI (logarithmic mean Divisia index) method to analyze China's energy consumption. *Energy* **2014**, *67*, 617–622. [CrossRef]
46. Cui, Y.; Li, L.; Chen, L.; Zhang, Y.; Cheng, L.; Zhou, X.; Yang, X. Land-use carbon emissions estimation for the Yangtze River Delta Urban Agglomeration using 1994–2016 Landsat image data. *Remote Sens.* **2018**, *10*, 1334. [CrossRef]

land

MDPI

Article

The Spatiotemporal Evolution of Ecological Security in Border Areas: A Case Study of Southwest China

Zheneng Hu [1], Meijun Qian [1], Xianghe Teng [1], Zhuoya Zhang [2], Fanglei Zhong [3], Qingping Cheng [2] and Chuanhao Wen [1,*]

1 School of Economics, Yunnan University, Kunming 650500, China; zhenenghu@mail.ynu.edu.cn (Z.H.); maggieqian@mail.ynu.edu.cn (M.Q.); tengxh1990@ynu.edu.cn (X.T.)
2 School of Geography and Ecotourism, Southwest Forestry University, Kunming 650224, China; zhuoya@swfu.edu.cn (Z.Z.); cqp@swfu.edu.cn (Q.C.)
3 School of Economics, Minzu University of China, Beijing 100081, China; zfl@muc.edu.cn
* Correspondence: chhwen1972@ynu.edu.cn; Tel.: +86-185-0234-0601

Abstract: Fewer studies on ecological security (ES) in border areas limit the synergistic development of border areas in the context of rapid globalization. The study of ES in border areas of southwest China can enrich the evaluation methods, summarize the knowledge related to ES in border areas, and provide references for similar areas in the world. Therefore, twenty-five international border counties in Yunnan Province were selected to establish a system to evaluate ES; an entropy weight TOPSIS model was used to evaluate the changes in ES from 2004 to 2019. Then, an obstacle degree model was used to diagnose the factors affecting ES. The state of ES was predicted by a gray prediction model (GM) (1,1) in 2025 and 2030. The results show that an improving ES situation presented a spatial distribution pattern of high to low from the southwest to the west and east. Various factors, including fixed assets investment, per-capita fiscal revenue, per-capita GDP, food production, and water regulation, created obstacles to a desirable ES in the study area. Although the ES of border areas will maintain an upward trend under the existing development model, the number of counties that will reach a secure state of ES in 2025 and 2030 is predicted to only be 1 and 2, respectively.

Keywords: border area; ecological security; spatiotemporal evolution; sustainable development; Yunnan Province

Citation: Hu, Z.; Qian, M.; Teng, X.; Zhang, Z.; Zhong, F.; Cheng, Q.; Wen, C. The Spatiotemporal Evolution of Ecological Security in Border Areas: A Case Study of Southwest China. *Land* **2022**, *11*, 892. https://doi.org/10.3390/land11060892

Academic Editors: Li Ma, Yingnan Zhang, Muye Gan and Zhengying Shan

Received: 17 May 2022
Accepted: 9 June 2022
Published: 11 June 2022

Publisher's Note: MDPI stays neutral with regard to jurisdictional claims in published maps and institutional affiliations.

1. Introduction

Rapid globalization has strengthened the political and economic ties between countries [1], which has largely changed the previous disadvantage experienced by border areas of a country [2]. For example, the border areas of the European Union (EU) account for 40% of EU territory, 30% of the EU population, and 30% of the EU gross domestic product (GDP) [3]. The resources and environmental factors of border areas have a more important impact on national security and international economic cooperation than in the past [4–7], and ecologists also pay more attention to the resource management in these regions [8]. Although natural ecosystems are often spatially continuous, the boundary lines indicating the boundaries between the territories of different sovereign states will objectively lead to competition between countries for resources and environmental conditions in border areas [9,10]. As a result, the ecological security (ES) of border areas not only becomes an important part of a country's comprehensive national security system but also becomes the most sensitive part of that system. Because border areas involve political boundaries where environmental changes are complex and uncertain, the ES of border areas not only includes the security of development and the environment but also includes the rational use of resources, border management, ecological maintenance of the area, and geographical cooperation between countries [11]. In this context, studying the ES of border areas is of great significance to promote the cooperation between neighboring countries in the ecological field and maintain the stability of border areas.

The idea of ES first appeared in the land function and land health evaluation in the 1940s. Its concept is based on the theory of environmental security [12]. In 1989, the International Institute for Applied Systems Analysis first proposed the concept of ES when explaining global ecological and environmental problems. With the maturity of the theoretical framework of ES, the concept of ES has better clarity, that is, ES refers to maintaining the health and integrity of an ecosystem, ensuring that the human living environment does not change with the changes in external conditions and states, and keeping the environment in a stable and sustainable state [13]. In recent years, with the continuous and increasing change in the global environment and climate [14,15], regional ES has attracted significant attention and has become an important scientific problem that countries urgently need to solve [16]. Researchers have carried out much research on ES assessment in multiple scale areas and achieved many valuable results. By using a Pressure–State–Response model [17,18], the gray comprehensive evaluation method [19], and an ecological footprint model [20], researchers have evaluated ES at the national [21], provincial [22,23], urban agglomeration [24,25], watershed [18,26], and urban scales [27].

Although these studies provide a theoretical basis for ES analysis, some problems remain. First, existing studies have paid more attention to the ES of inland areas, but little attention has been paid to the field of ES in international border areas. Moreover, scholars often study large-scale administrative units such as countries and provinces, while these studies may not meet the needs of more detailed actual environmental management. Second, these existing studies have mainly focused on the retrospective evaluation of ES, which enriches the current situation and phenomena of ES but lacks the ability to predict the future of ES. Third, most of the existing studies ignore the analysis of the mechanisms that promote ES and lack a discussion of measures designed to improve the level of ES.

Since the 1990s, China has promoted a series of "geo-cooperation initiatives," namely transboundary infrastructure projects such as highways, railways, and oil and gas pipelines [10], which has effectively promoted the sustainable development of border areas. However, these projects can also lead to ecological problems such as land use-cover change [28], landscape fragmentation, and a loss of biodiversity [29] in border areas. To this end, the Chinese government has adopted a series of ecological restoration projects, such as a natural forest protection plan [30], designated ecological red lines [31], and mandated transboundary water resource management [32] and biodiversity conservation [33] to protect the environment of border areas. There are 110 international rivers and lakes in the southwest, northwest, and northeast borders of China. These regions are the birthplace of most Asian rivers, making China the most important upstream country for ES in Asia [34]. In 2015, the United Nations adopted the 2030 Agenda for Sustainable Development and formulated 17 Sustainable Development Goals (SDGs) [15], in which SDG 6 calls for sustainable management of water resources and SDG 15 calls for curbing the loss of biodiversity. Yunnan Province, located in the southwest border of China, is the upstream of Nujiang, Lancang, Honghe, and Daying rivers [35]. It is also one of the hotspots of biodiversity in the world [36]. Studying the ES of border areas in Yunnan Province can promote international cooperation on ES and the sustainable management of water resources, so as to effectively reduce border conflicts and ensure the safety and stability of border areas. However, the existing literature has paid little attention to this field. Considering that counties are the basic administrative unit in China [37] and that they can fully reflect the characteristics of border areas, this study selected the 25 counties in Yunnan Province as the case study area.

The security of international borders is becoming more and more complicated. Therefore, how to reposition the functions of the border areas needs more attention from the academic community. Based on the above literature review and an analysis of existing research gaps, the purpose of this study is as follows: (1) Analyze the temporal and spatial change characteristics of ES in the border areas of Yunnan Province under the background of increasingly close international cooperation. (2) Probe into the main obstacles and factors and factors affecting the ES in border areas. (3) Predict the status of ES in border areas in 2025 and 2030. In addition, policy suggestions on improving the level of ES in border areas are also put forward.

2. Study Area and Data Sources

2.1. Study Area

The border areas analyzed in this paper are located in Yunnan, China (21°08′–29°15′ N, 97°31′–106°11′ E) (Figure 1), with a total area of 9.03 × 104 km^2. This region borders Myanmar, Laos, and Vietnam and includes the upstream regions of the Nujiang, Lancang, Honghe, and Daying rivers [35], as well as plateau wetland and tropical forest ecosystems. This area has a complex terrain as part of the Qinghai–Tibet Plateau, Hengduan Mountains, and the Yunnan–Guizhou Plateau where the elevation gradually decreases from northwest to southeast; the maximum relative elevational difference is approximately 4751 m across the region. The climate is dominated by plateau mountain climate, subtropical monsoon climate, and tropical monsoon climate conditions. The average temperature is 15–23.7 °C, and the annual precipitation falls between 1020 and 3388 mm. The combination of climate and complex terrain makes the border areas one of the global biodiversity hotspots [36,38].

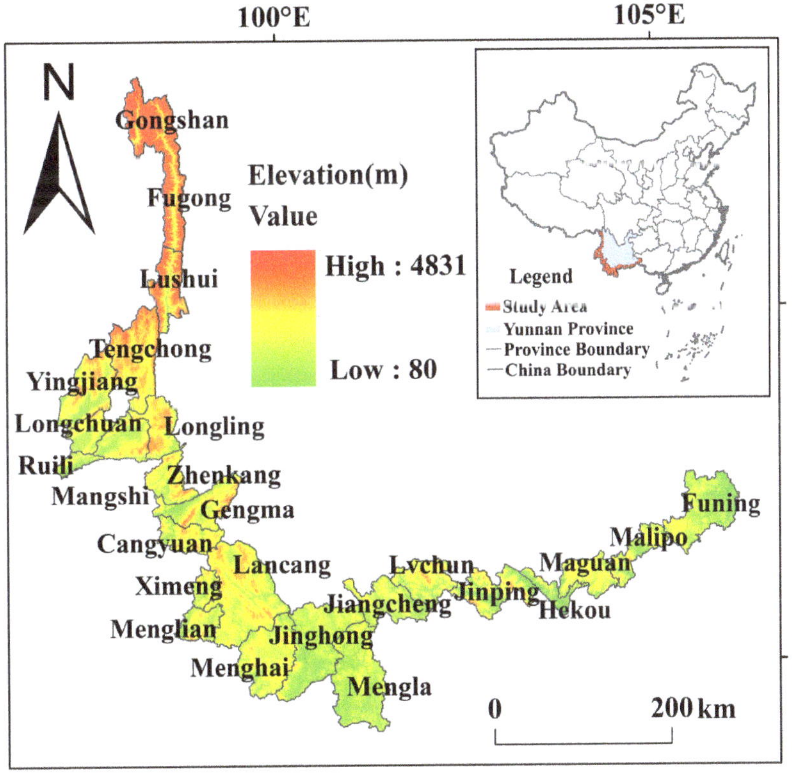

Figure 1. Map of the study area location along the southern and western borders of Yunnan Province. An inset map shows the study area location within the provinces and other administrative areas of China. Note: Map was created by authors using ArcGIS 10.7 based on the digital elevation model (DEM) data from the Resources and Environment Data Center of the Chinese Academy of Sciences (https://www.resdc.cn/, accessed on 4 January 2022).

The study area included the following 25 counties (cities): Cangyuan, Fugong, Funing, Gengma, Gongshan, Hekou, Jiangcheng, Jinping, Jinghong, Lancang, Longling, Longchuan, Lushui, Lvchun, Malipo, Maguan, Mangshi, Menghai, Mengla, Menglian, Ruili, Tengchong, Ximeng, Yingjiang, and Zhenkang. This part of China is relatively less developed than other areas of China and includes areas inhabited by people of various ethnic minorities. The traditional culture of ethnic minorities and rich tourism resources make tourism the main industry in border areas [39]. By 2016, the study area had a permanent population of 6.92 million, and a GDP of 158.48 billion yuan, of which income from tourism comprised 85.99 billion, accounting for 54% of the total GDP. In a word, as the border areas of Yunnan Province are rich in biodiversity and located in the upstream of many international rivers, it is representative to select these regions to study ES.

2.2. Data Sources

We used a digital elevation model (DEM) and multiple datasets, including land use, meteorological, normalized difference vegetation index (NDVI), PM 2.5, and socio-economic data and statistics. The basic data used in this study and the data sources are listed in Table 1.

Table 1. Data information and sources.

Data Type	Time	Data Sources
Land use data	2004–2019	Land use type was classified into cropland, forest, shrub, grassland, water, snow and ice, barren, impervious, and wetland, with a spatial resolution of 30 m based on Yang and Huang [40]. The data were obtained from https://doi.org/10.5281/zenodo.4417810 (accessed on 4 June 2021).
Meteorological data	2004–2019	Derived from the National Science and Technology Infrastructure dataset (http://www.cma.gov.cn/, accessed on 17 July 2021).
Administrative boundary	2015	Obtained from the National Basic Geographic Information Center (http://www.ngcc.cn/, accessed on 6 January 2022); described basic geographic information at a scale of 1:4,000,000.
Digital Elevation Model (DEM)	/	A DEM with a spatial resolution 90 m was provided by the Resources and Environment Data Center of the Chinese Academy of Sciences (https://www.resdc.cn/, accessed on 4 January 2022).
Normalized Difference Vegetation Index (NDVI)	2004–2019	NDVI data were provided by the Resources and Environment Data Center of the Chinese Academy of Sciences (https://www.resdc.cn/, accessed on 14 July 2021).
Atmospheric Particulate Matter content with a diameter of <2.5 μm (PM 2.5)	2004–2019	Data were publicly available as vers. V4.GL.03 via the Atmospheric Composition Analysis Group website at Dalhousie University [41].
Socio-economic data	2004–2019	Data obtained from the Yunnan Statistical Yearbook, county-level socioeconomic statistical yearbooks (China), Chinese agency government work reports, and social development statistical bulletins of each county.

3. Material and Methods

3.1. Research Framework

This study was conducted within a framework of four steps (Figure 2): the first step introduces the realistic needs, theoretical basis, and literature gaps related to ES. Then, the paper describes the study area and data sources, constructs an evaluation index system for ES, and introduces the methods. The third step analyzes the results. The last step expounds the conclusions and policy implications.

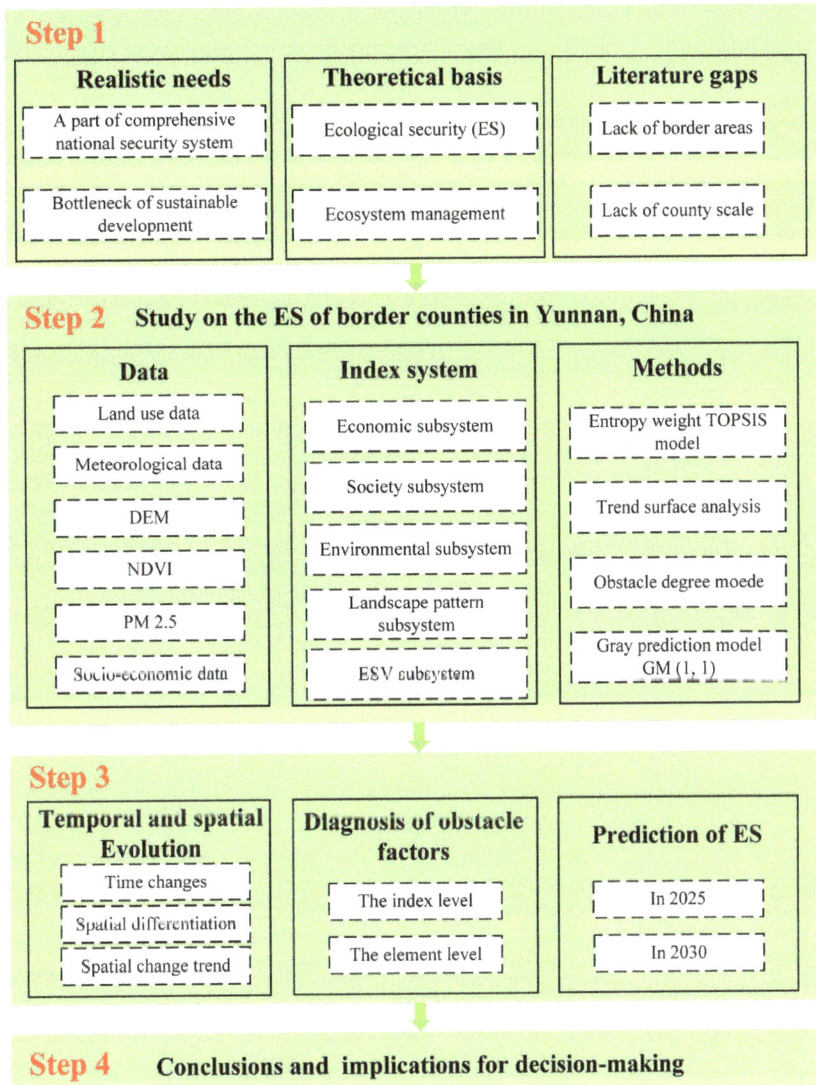

Figure 2. The research framework used in this study.

3.2. Construction of the Index System

Ecological security is a complex system involving nature, the economy, and human society [20,42]; the level of ES in any area is constantly changing [30,43]. In the past, the assessment of ES generally involved the construction of an index system from the perspective of ecosystem structure and function [12], but it did not consider human beings as external factors. In reality, human beings have already become an indispensable part of the Earth's ES system. Supporting the sustainable development of the economy and society is an important reason to study ES [43]. Therefore, the present study evaluates the state of ES in border areas from five subsystems: economy, society, environment, landscape pattern, and ecosystem service value (ESV). The process of calculating the landscape patterns and ESVs is as follows:

(1) Selection and calculation of landscape metrics.

The evolution of landscape pattern leads to spatial changes in the landscape results, which directly reflect the changes in ecosystem structure and composition and finally affect ES [18]. A landscape index can describe the change in landscape pattern. Landscape indices are used to determine the relationship between landscape pattern and the process of landscape change [36]. In this study, the number of patches (NP), patch density (PD), largest patch index (LPI), edge density (ED), landscape shape index (LSI), splitting index (SPLIT), Shannon's diversity index (SHDI), and aggregation index (AI) were selected to reflect the level of landscape pattern subsystem [44,45]. The selected landscape index was calculated using FRAGSTATS 4.2 software.

(2) Calculation of ESV.

Ecological security depends on the level of ecosystem services provided by an ecosystem to human beings [46]. At present, ES has become an important research topic designed to bring ecosystem services into the evaluation of ES systems [46,47]. In 1997, Costanza et al. [48] put forward the method of evaluating global scale ecosystem services with ESV, and Xie, a Chinese scholar [49], summarized an equivalent factor of ESV per unit area according to the actual situation of China. In this study, we referred to their methods and modified the above-mentioned equivalent factor. For the border areas analyzed here, the following standards were used: cropland corresponds to dry land; the equivalent factor for forest is taken as the average value of coniferous, mixed, and broad-leaved forests; grassland corresponds to prairie; snow/ice corresponds to glacier and snow [50]; and impervious areas were assigned "0" [51] (Table S1).

The economic value of one ecological service equivalence factor is 1/7 the grain output value per unit area [49], and the economic value of the equivalent factor in border areas was calculated according to Equation (1). To eliminate the impact of crop price fluctuation on the total value, the area, yield, and average price of the three main crops (rice, corn, and wheat) were selected as the basic data. The calculation process is as follows:

$$VC_0 = \frac{1}{7}\sum_{i=1}^{n} \frac{m_i p_i q_i}{M} (n = 1, 2, 3) \tag{1}$$

According to Equation (1), the economic value of one equivalent factor of ESV in border areas is 1817.76 yuan/ha, and the ESV coefficient per unit area of land use type was obtained (Table S2). Referring to the method of Hu et al. [36], the sensitivity index of the ESV coefficient for all land use types was obtained (Table S3). The sensitivity index of the ESV coefficient was all less than 1, which indicates that the estimated total ESV in the study area is not elastic to the equivalent factor.

The index system of ES is shown in Table 2.

Table 2. The index system of ecological security evaluation in border areas.

Target Layer	Elements Layer	Index Layer	References
ES	Economic subsystem	X1: Annual GDP growth rate (%)	[52]
		X2: Per-capita GDP (yuan)	[22]
		X3: Secondary industry as percentage of GDP (%)	[53]
		X4: Tertiary industry as percentage of GDP (%)	[25]
		X5: Fixed assets investment (10^4 yuan)	[18]
		X6: Per-capita fiscal revenue (yuan)	[25]
	Society subsystem	X7: Population growth rate (%)	[24]
		X8: Population density (person/square kilometer)	[24,52]
		X9: Per-capita food production (tons/person)	[22]
		X10: Number of medical beds per 10,000 persons (pieces/ten thousand people)	[25]
		X11: Urbanization level (%)	[43]

Table 2. *Cont.*

Target Layer	Elements Layer	Index Layer	References
	Environment subsystem	X12: Annual average temperature (°C)	[43]
		X13: Annual precipitation (mm)	[43]
		X14: PM2.5 (μg/m^3)	[24]
		X15: NDVI (/)	[18]
		X16: Forest cover (%)	[25]
		X17: Proportion of construction land (%)	[26]
		X18: Proportion of cultivated land (%)	[25]
	Landscape pattern subsystem	X19: NP (/)	[54]
		X20: PD (/)	[12,43]
		X21: LPI (/)	[43]
		X22: ED (/)	[12]
		X23: LSI (/)	[54,55]
		X24: SPLIT (/)	[12,43]
		X25: SHDI (/)	[12,43]
		X26: AI (/)	[54,55]
	ESV subsystem	X27: Food production (10^4 yuan)	[24,50]
		X28: Raw material (10^4 yuan)	[24,50]
		X29: Water supply (10^4 yuan)	[24,50]
		X30: Air quality regulation (10^4 yuan)	[24,50]
		X31: Climate regulation (10^4 yuan)	[24,50]
		X32: Waste treatment (10^4 yuan)	[24,50]
		X33: Water regulation (10^4 yuan)	[24,50]
		X34: Erosion prevention (10^4 yuan)	[24,50]
		X35: Soil fertility maintenance (10^4 yuan)	[24,50]
		X36: Habitat services (10^4 yuan)	[24,50]
		X37: Cultural services (10^4 yuan)	[24,50]

3.3. Entropy Weight TOPSIS Model

The entropy weight method is an objective weighting method, which uses entropy to indicate the information's size. Generally, the larger the gap between feature values, the larger the size of information it possesses [56]. The technique for order preference by similarity to ideal solution (TOPSIS) is a multi-objective decision-making method. The principle is to rank the evaluation objects according to the closeness of positive and negative ideal solutions. At present, the combination of entropy weight method and TOPSIS method has been widely used in land use planning, sustainable development assessment, and other fields [57,58]. This study selected entropy weight TOPSIS model to evaluate the level of the ES in border areas. The calculation process is as follows:

(1) Data standardization was completed using the following equation:

$$\begin{cases} g_{ij} = \frac{y_{ij}-y_{min}}{y_{max}-y_{min}}, \ where\ g_{ij}\ is\ a\ positive\ indicator \\ g_{ij} = \frac{y_{max}-y_{ij}}{y_{max}-y_{min}}, where\ g_{ij}\ is\ a\ negative\ indicator \end{cases} \quad (2)$$

where g_{ij} is the normalized value; and y_{max} and y_{min} represent the maximum and minimum values of the jth index, respectively.

(2) The information entropy of each index was calculated using the following equation:

$$H_j = -\frac{1}{\ln n}\left(\sum_{i=1}^{n} f_{ij} \ln f_{ij}\right) \quad (3)$$

where $f_{ij} = (1 + g_{ij}) / \left[\sum_{i=1}^{n} (1 + g_{ij})\right]$.

(3) The index weights were determined using the following equation:

$$w_j = \frac{1 - H_j}{m - \sum\limits_{j=1}^{m} H_j} \tag{4}$$

(4) The weight normalization matrix was constructed using the following equation:

$$c_{ij} = g_{ij} * w_j \tag{5}$$

where c_{ij} is the weighted normalized decision matrix; and w_j is the weight value of the jth index.

(5) The positive and negative ideal solutions, C^+ and C^-, respectively, were calculated using the following equation:

$$\begin{cases} C^+ = \left\{ \max\limits_{1 \le i \le m} c_{ij} | i = 1, 2, \cdots, m \right\} = \left\{ c_1{}^+, c_2{}^+, \cdots, c_j{}^+ \right\} \\ C^- = \left\{ \min\limits_{1 \le i \le m} c_{ij} | i = 1, 2, \cdots, m \right\} = \left\{ c_1{}^-, c_2{}^-, \cdots, c_j{}^- \right\} \end{cases} \tag{6}$$

where C^+ and C^- refer to the optimal and the least considered decision schemes, respectively.

(6) The distance from the index value of each evaluation object to C^+ and C^- was calculated using the following equations:

$$\begin{cases} S_j{}^+ = \sqrt{\sum\limits_{j=1}^{n} \left(c_j{}^+ - c_{ij}\right)^2}, i = 1, 2, \cdots, m \\ S_j{}^- = \sqrt{\sum\limits_{j=1}^{n} \left(c_j{}^- - c_{ij}\right)^2}, i = 1, 2, \cdots, m \end{cases} \tag{7}$$

where $S_j{}^+$ and $S_j{}^-$ refer to the distance of the assessment vector to the positive and negative ideal solutions, respectively.

(7) The closeness of each evaluation object to the ideal solution was calculated using the following equation:

$$R_i = \frac{S_j{}^-}{S_i{}^+ + S_i{}^-} \tag{8}$$

where R_i is the closeness of the evaluation object to the optimal solution, and the range of R_i is 0–1. The greater the value of R_i, the higher the ES level. Referring to the research of Cui et al. [59], we divided the level of ES into five levels (Table 3).

3.4. Trend Surface Analysis

Trend surface analysis is a method that can be used to simulate the spatial distribution law and change trend in geographical elements with a smooth mathematical surface [60]. The actual surface is divided into two components: a regional trend and residual values. The regional trend is calculated by using a polynomial surface with continuous power, and the residual values are the arithmetic difference between the original data and the trend surface, indicating local fluctuations. This study used this method to analyze the spatial differentiation trend in ES [61], as follows:

$$Z_i(x_i, y_i) = T_i(x_i, y_i) + \varepsilon_i \tag{9}$$

where $Z_i(x_i, y_i)$, $T_i(x_i, y_i)$, and ε_i represent the observed, trend, and residual values of variable Z at location (x_i, y_i), respectively.

Table 3. Evaluation criteria of ecological security.

Range	Status	Characteristics
[0, 0.25)	Critical	The pressure on the ecosystem is very large, the ecosystem structure is very imperfect, and the ecosystem has a very large risk of collapse.
[0.25, 0.45)	Unstable	The pressure on the ecosystem is large; the ecosystem structure has defects and is in an unstable state.
[0.45, 0.55)	Sensitive	The pressure on the ecosystem is large and close to the threshold; the ecosystem structure is relatively complete and can play the basic function of the ecosystem.
[0.55, 0.75)	Good	The ecosystem has relatively little pressure and perfect functions, and the ecosystem is in a relatively stable state.
[0.75, 1]	Secure	The pressure on the ecosystem is very small, the ecological function and structure are in excellent condition, and the ecosystem is in a very stable state.

3.5. Obstacle Degree Model

Based on the ES evaluation of border areas, identifying the key factors that directly influence the ES can help to present adaptation measures designed to help land managers preferably maintain the ES in border areas and promote sustainable regional development. Therefore, our study used an obstacle degree model to analyze the factors creating obstacles and influencing ES. The calculation process is as follows [62]:

$$P_{ij} = 1 - g_{ij} \tag{10}$$

$$M_{ij} = \frac{R_{ij}P_{ij}}{\sum\limits_{i,j=1}^{m} R_{ij}P_{ij}} \times 100\% \tag{11}$$

where M_{ij} represents the obstacle degree of the ith indicator in year j, and P_{ij} is the degree of deviation for indicator i in year j

3.6. GM (1,1) Gray Prediction Model

The gray system theory was first put forward by Deng [63]. A gray prediction model (GM) (1,1) adopts the basic concept of the gray system theory and has been widely used in the fields of ecology and social economics [64,65]. It has unique advantages in predicting and analyzing objects and process systems with small amounts of data, no obvious change law of data, has an unclear structural relationship, and has an operation mechanism. The prediction calculation process is simple and accurate. Considering that the change in ES has fuzzy and uncertain characteristics [43], and predicting its change is a typical gray evaluation process, GM (1,1) is selected to predict the ES level of border areas in 2025 and 2030. The calculation process is as follows:

(1) Preprocess the data. The original sequence ${Y_i}^{(0)}$ of all data from 2004 to 2019 is set as follows:

$${Y_i}^{(0)} = \left[{Y_i}^{(0)}(1), {Y_i}^{(0)}(2), \cdots, {Y_i}^{(0)}(16)\right] \tag{12}$$

Use Equation (13) to obtain a new sequence ${Y_i}^{(1)}(k)$:

$${Y_i}^{(1)}(k) = \sum_{i=1}^{k} {Y_i}^{(0)} = {Y_i}^{(1)}(k-1) + {Y_i}^{(0)}(k) \tag{13}$$

$${Y_i}^{(1)} = \left[{Y_i}^{(1)}(1), {Y_i}^{(1)}(2), \cdots, {Y_i}^{(1)}(16)\right] \tag{14}$$

(2) Set up the gray differential equation:

$$Y_i^{(1)}(k+1) = \left[Y_i^{(0)} - \frac{b}{a}\right]e^{-ak} + \frac{b}{a} \tag{15}$$

where the values of parameters a and b are calculated by the least squares method [66].

(3) Predict the data. Based on Equation (16), the ES level of border areas in 2025 and 2030 is predicted as follows:

$$\hat{Y}_i^{(0)}(k+1) = \hat{Y}_i^{(1)}(k+1) - \hat{Y}_i^{(1)}(k) \tag{16}$$

(4) Test the accuracy of the prediction data. After the fitting value $\hat{Y}_i^{(0)}$ of ES is obtained using Equation (16), the correctness of the model needs to be tested according to the original $Y_i^{(0)}$ and fitting $\hat{Y}_i^{(0)}$ sequences. The mean absolute percent error (MAPE) is usually used to test the accuracy of the model:

$$MAPE = \frac{1}{n}\sum_{k=1}^{n}\left|\frac{Y_i^{(0)}(k) - \hat{Y}_i^{(0)}(k)}{Y_i^{(0)}(k)}\right| \tag{17}$$

We referred to the studies of Wang et al. [66] and Wang and Li [67]; when the MAPE is less than 10%, the forecasting is highly accurate. After calculation, the MAPE of all the data that were predicted in this paper was less than 10% (Table 4), which meets the requirement needed to verify prediction accuracy.

Table 4. Mean absolute percent error (MAPE) for each county, which was used to test the accuracy of the model; a MAPE > 10% indicates that the predicted results are highly accurate.

County	Mape	County	Mape	County	Mape
Cangyuan	3.47%	Lancang	0.52%	Mengla	2.26%
Fugong	1.51%	Longling	4.47%	Menglian	1.91%
Funing	0.80%	Longchuan	1.41%	Ruili	4.25%
Gengma	3.28%	Lushui	2.03%	Tengchong	2.23%
Gongshan	1.43%	Lvchun	2.81%	Ximeng	1.59%
Hekou	5.93%	Malipo	1.69%	Yingjiang	0.80%
Jiangcheng	0.74%	Maguan	3.14%	Zhenkang	2.98%
Jinping	1.87%	Mangshi	1.45%		
Jinghong	2.71%	Menghai	1.46%		

4. Results

4.1. Temporal Changes in ES

Based on an entropy weight TOPSIS model, we calculated the ES of each border county. From 2004 to 2019, the ES of all border counties showed a positive upward trend (Figure 3). Tengchong, Jinghong, and Ruili ranked among the top three counties in terms of the increase in ES. Among them, the ES of Tengchong increased from 0.4124 in 2004 to 0.6417 in 2019, an increase of 0.2294. The ES of Jinghong increased from 0.4798 in 2004 to 0.6882 in 2019, an increase of 0.2084. The ES of Ruili increased from 0.1994 in 2004 to 0.3537 in 2019, an increase of 0.1544.

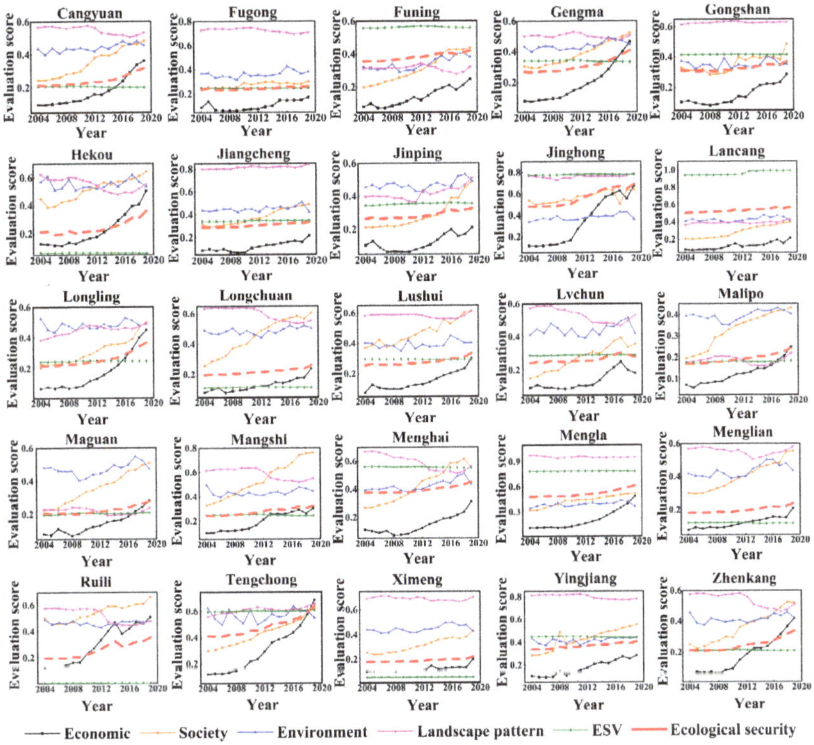

Figure 3. Evaluation scores for temporal changes in ecological security and subsystems (economic, social, environmental, landscape pattern, and ecosystem service value (ESV)) in 25 border counties of Yunnan Province, China.

We also analyzed the time trend in five subsystems, namely, the economic, social, environmental, landscape pattern, and the ecosystem service value subsystems (Figure 3). All border counties generally showed an upward trend for their economic and social subsystems, with the economic subsystem of Tengchong rising the most, from 0.1186 in 2004 to 0.6871 in 2019, an increase of 0.5685. The social subsystem of Mangshi rose the most, from 0.3296 in 2004 to 0.7634 in 2019, an increase of 0.4338. The counties of Cangyuan, Fugong, Funing, Gengma, Jinping, Jinghong, Longchuan, Lvchun, Malipo, Menghai, Mengla, Menglian, and Yingjiang showed a fluctuating upward trend in their environmental subsystems, while the level of the environmental subsystems of other counties generally decreased. Jinping experienced the largest increase in terms of landscape pattern subsystem, from 0.3966 in 2004 to 0.5095 in 2019, an increase of 0.1128. the ESV subsystem of all border counties changed little during the study period.

4.2. Spatial Changes in ES

Our results show the spatial changes in ES in border areas in 2004, 2009, 2014, and 2019 (Figure 4). In 2004, the ES level of all border counties was lower than the good level, among which Lincang, Jinghong, and Mengla had the highest ES level and were classified as being in a sensitive state. In 2009, the level of ES in Mangshi changed from critical to the unstable state. In 2014, the level of ES in Jinghong first reached the good state. Tengchong changed from the unstable to the sensitive state, while Longling, Zhenkang, and Lvchun changed from the critical to the unstable state. In 2019, the number of border counties in the critical state had decreased, with only Ximeng, Menglian, and Malipo in the critical state, showing that the ES level in the southwestern part of the study area was higher than that in the west and east.

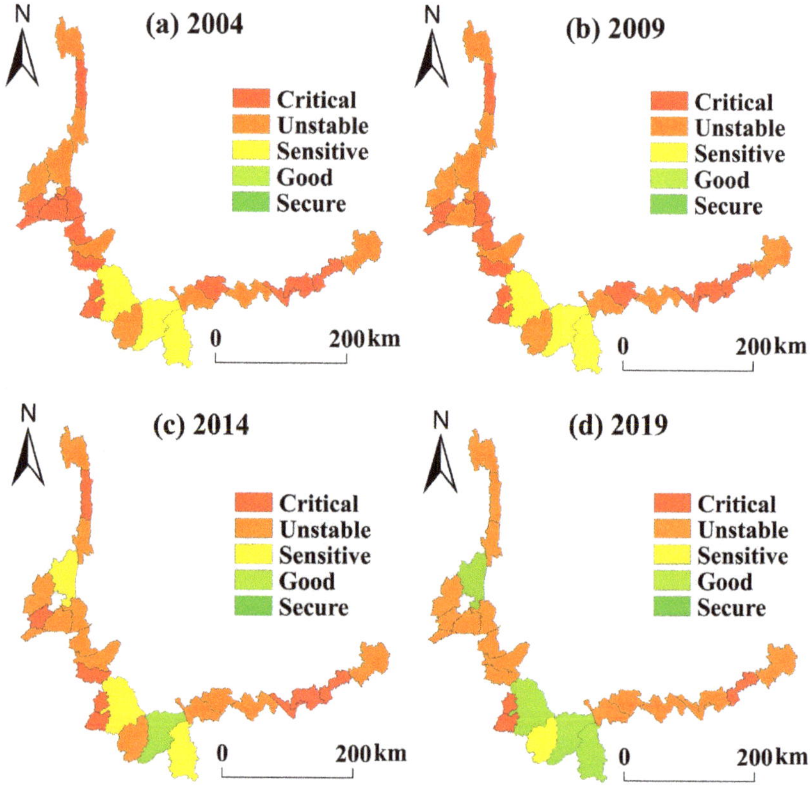

Figure 4. Spatial change in ecological security in 25 border counties in Yunnan Province, China for: (**a**) 2004; (**b**) 2009; (**c**) 2014; and (**d**) 2019. Note: Maps were created by authors using ArcGIS 10.7.

Based on the evaluation of ES in 2004, 2009, 2014, and 2019, the Geostatistical Analyst tool in ArcGIS 10.7 was applied to visualize the spatial representation of ES conditions. A spatial change trend map of the ES was obtained thereby (Figure 5). The X and Y axes indicate the east and north directions, respectively, while the Z-axis indicates the size of the ecological safety assessment value. The green and blue lines in Figure 5 represent the fitting curve of ES in the east–west and north–south directions [68]. In the east–west direction, the trend lines of the four years analyzed here remained stable, and all showed an "inverted U-shaped" distribution (Figure 5), characterized by the central part being higher than the western and eastern parts, while the western part was higher than the eastern part. In the north–south direction, the trend line changed from a "U-shaped" distribution to a "straight line," characterized by higher values in the southern and northern parts than in the central part, while the gap has become significantly narrow; the ES of the northern part has always remained lower than that of the southern part. These results are consistent with the spatial distribution pattern of high in the southwest and low in the east and west.

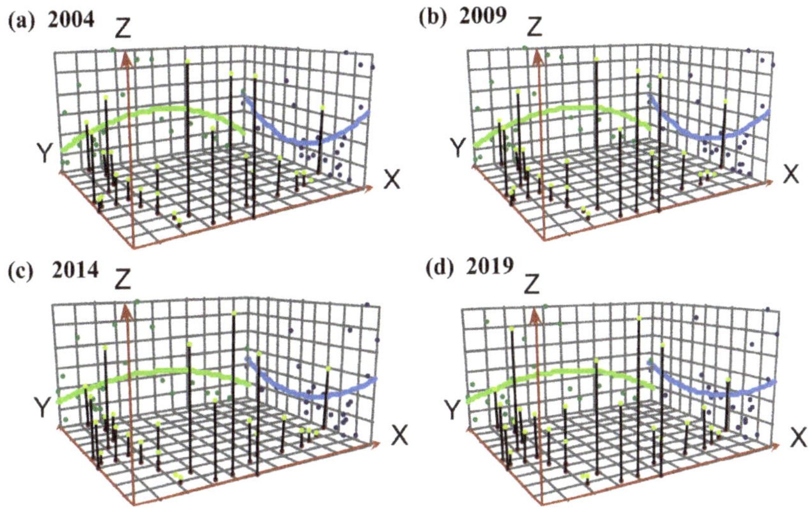

Figure 5. Spatial change trend in ecological security in 25 border counties in Yunnan Province, China for: (**a**) 2004; (**b**) 2009; (**c**) 2014; and (**d**) 2019.

4.3. Diagnosis of Obstacle Factors for ES

4.3.1. Analysis of Obstacle Factors at the Index Level

Based on the degree of obstacle model, the factors creating obstacles that affect ES in the studied border areas were determined (Figure 6; however, only the factors in 2004 and 2019 are listed). By analyzing the degree of obstacles for each factor [69], the five top obstacle factors were found to be the following: fixed asset investment (X5), per capita fiscal revenue (X6), per capita GDP (X2), food production (X27), and water regulation (X33). In 2004, the average obstacle degree of X5, X6, X2, X27, and X33 was 14.11%, 8.86%, 7.90%, 4.74%, and 4.21%, respectively. In 2019, the average obstacle degree of X5, X6, X2, X27, and X33 was 11.10%, 7.07%, 6.31%, 5.42%, and 4.73%, respectively.

4.3.2. Analysis of Obstacle Factors at the Element Level

Based on the calculations of the obstacle degree of each indicator, the obstacle degree of factors at the element level was also obtained (Figure 7). The obstacle degrees at the element level in border areas varied between 2004 and 2019. In 2004, the economic subsystem was the main obstacle to improving the ES in Jinghong, Lancang, and Mengla, while the ESV subsystem was the main obstacle to improving the ES in the other 22 border counties. In 2019, the obstacle degree of the economic subsystem increased in Langcang, while the obstacle degree of the economic subsystem of all the other 24 border counties decreased. The obstacle degrees of social, environmental, and landscape pattern subsystems changed little.

4.4. Prediction of Changes in ES for the Period 2025–2030

According to the respective ES level of each border county from 2004 to 2019, predicted values for ES in 2025 and 2030 were obtained (Figure 8). The ES of border areas is predicted to maintain an upward trend. In 2025, Jinghong would first reach a secure state; Lancang, Mengla, and Tengchong would reach a good state; and Menghai, Ruili, Gengma and Funing would reach a sensitive state. Ximeng would maintain a critical state, and the other 16 counties would be in an unstable state. In 2030, Jinghong and Tengchong would reach a secure state; Lancang, Mengla, and Ruili would reach a good state; and Funing, Gengma, Hekou, Longling, Menghai, Yingjiang, and Zhenkang would reach a sensitive state. Cangyuan, Fugong, and another 10 counties would be in an unstable state. However, Ximeng would remain in a critical state.

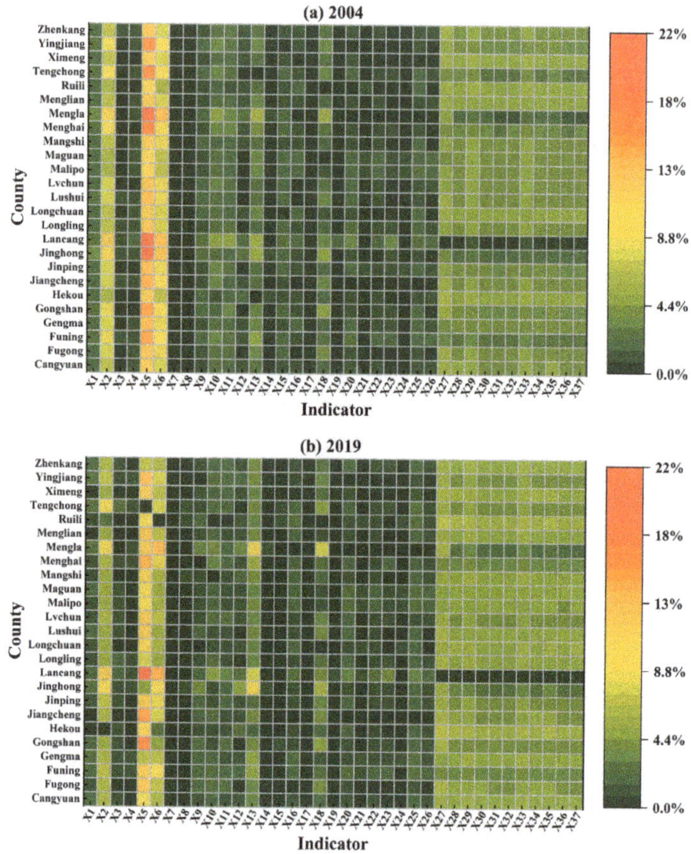

Figure 6. The obstacle degree of each indicator of ecological security in 25 border counties in Yunnan Province, China in: (**a**) 2004 and (**b**) 2019 (%).

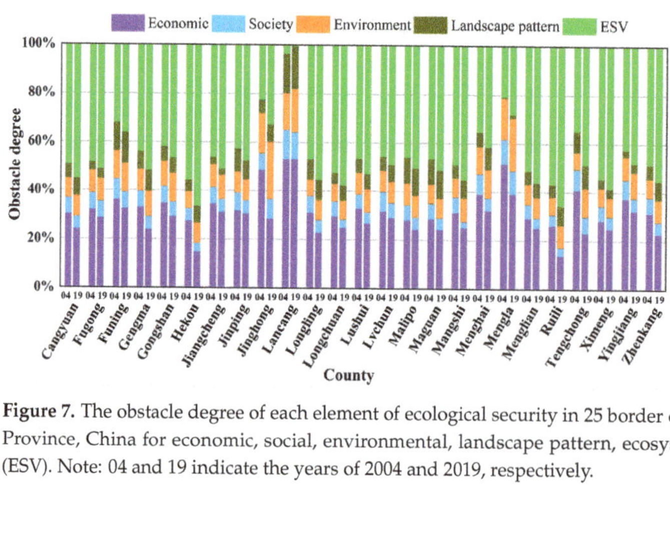

Figure 7. The obstacle degree of each element of ecological security in 25 border counties in Yunnan Province, China for economic, social, environmental, landscape pattern, ecosystem service value (ESV). Note: 04 and 19 indicate the years of 2004 and 2019, respectively.

County	2025	2030	County	2025	2030	County	2025	2030
Cangyuan	0.35	0.41	Lancang	0.59	0.61	Menghai	0.47	0.50
Fugong	0.27	0.28	Longling	0.43	0.52	Mengla	0.65	0.71
Funing	0.45	0.47	Longchuan	0.27	0.30	Menglian	0.25	0.28
Gengma	0.46	0.54	Lushui	0.35	0.38	Ruili	0.47	0.58
Gongshan	0.38	0.41	Lvchun	0.31	0.34	Tengchong	0.73	0.86
Hekou	0.42	0.53	Malipo	0.27	0.30	Ximeng	0.22	0.23
Jiangcheng	0.36	0.38	Maguan	0.30	0.34	Yingjiang	0.43	0.46
Jinping	0.35	0.38	Mangshi	0.37	0.41	Zhenkang	0.40	0.47
Jinghong	0.84	0.98						

Critical Unstable Sensitive Good Secure

Figure 8. Prediction of the ecological security in 25 border counties of Yunnan Province, China in 2025 and 2030.

5. Discussion

5.1. Discussion of Change in ES

The main objective of this study was to analyze the ES of 25 counties with international borders in Yunnan Province, China. First, we selected these 25 counties as the case study area and constructed an index system to evaluate the ES from the five subsystems of economy, society, environment, landscape pattern, and ESV. We found that human and natural systems interact to affect the ES, which is consistent with previous studies [30]. Efforts related to ecological protection need to be coordinated with social and economic development, which can provide more comprehensive guidance for land management by environmental decision makers [70,71]. Second, we found that the ES of border areas showed a spatial distribution pattern of high in the southwest and low in the west and east. Jinghong had the highest level of ES, because Jinghong has adhered to the implementation of a project designed to return rubber plantation land to forest and has delimited an ecological red line, resulting in the ES of Jinghong ranking first among the border counties. Third, we found that from 2004 to 2019, the ES of border areas showed an overall upward trend, indicating that remarkable achievements have been made in ecological protection in Yunnan Province. However, on the premise of maintaining the existing development mode, we predicted the ES status of border areas in 2025 and 2030. We found that only one county will reach the classification of secure status in 2025 and two counties will reach the secure status in 2030, which is still a certain distance from the goal of establishing an ES barrier. Therefore, local governments should continue to pay more attention to the ES in border areas and formulate environmental policies according to the natural endowment and economic development of border areas.

5.2. Discussion on Obstacle Factors

In China, significant differences in socio-economic development and natural conditions among regions have caused the factors influencing ES in different regions to vary [12,17,22,26,72–75] (Figure 9). For the border areas in Yunnan Province, we found that the fixed asset investment, per capita fiscal revenue, per capital GDP, and food production are the main factors creating obstacles to improving the level of ES, which is consistent with previous research results [73]. Fixed asset investment plays a major role in promoting both the economy and the upgrading of industrial infrastructure, so as to reduce the pressure of industrial development on the environment. The higher the per-capita GDP and fiscal revenue, the more capable people are of protecting the environment. In addition, the border areas are located in the upstream regions of the Nujiang, Lancang, and other rivers.

This region serves as an important area for supplying and regulating water resources in Asia [10]. Therefore, the capacity to regulate water availability significantly affects the ES in border areas. In short, economic development and the provision of water-related ecosystem services are the main factors creating obstacles that affect the ES in border areas.

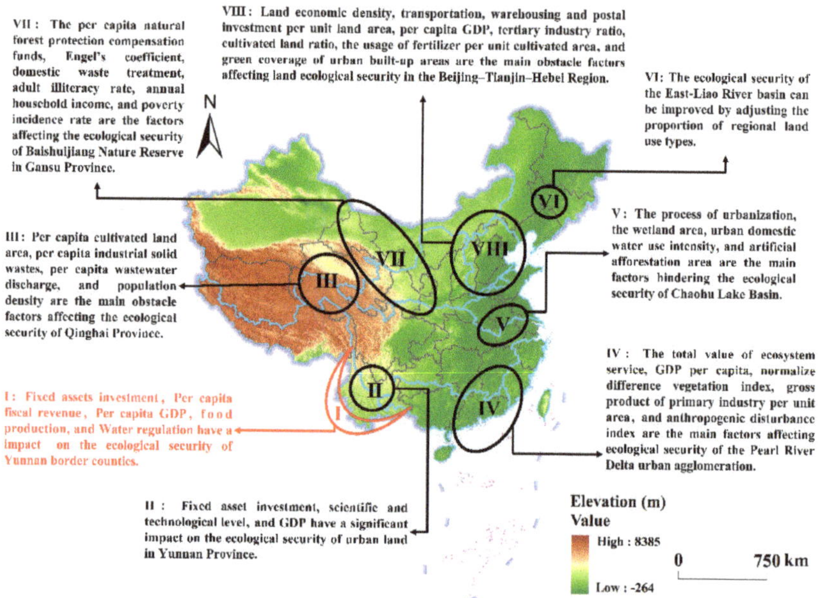

Figure 9. Spatial distribution of influencing factors of ecological security in China. Eight regions are labeled as follows: (I) Border areas in Yunnan Province, (II) Yunnan Province [73], (III) Qinghai Province [22], (IV) the Pearl River Delta Urban agglomeration [12], (V) the Chaohu Lake Basin [17], (VI) the East-Liao River basin [26], (VII) Gansu Province [74], (VIII) the Beijing–Tianjin–Hebei Region [72]. Note: Map was created by authors using ArcGIS 10.7 based on the digital elevation model (DEM) data from the Resources and Environment Data Center of the Chinese Academy of Sciences (https://www.resdc.cn/, accessed on 4 January 2022).

5.3. Limitations and Implications

This study has several limitations and can be improved in future research. First, biodiversity is an important embodiment of ES, especially in border areas of Yunnan Province, a global biodiversity hotspot. However, biodiversity data are difficult to obtain at the county scale. In the future work, the indicators for measuring biodiversity need to be further considered. Second, when predicting the ES of border areas, this study only considered the state that the border areas are predicted to reach in 2025 and 2030 under the situation of maintaining the existing development mode, but did not consider multiple scenarios, such as models of existing development and priorities in economic and environmental development, to simulate the future ES of border areas. Future research should make up for these shortcomings.

In the process of national development, as the border areas were far away from the political and economic center of the country in the past, previous studies often ignored the border areas [76]. With the acceleration in globalization, the border areas have changed from the original inferior states to the area with increasingly close international cooperation. The border areas of Yunnan Province, China, border on Myanmar, Laos, and Vietnam, and belong to the upstream area of many international rivers [35]. The situation in this area is relatively complicated. Selecting this area as the research case can provide reference for researchers to study the ES of border areas. This study finds that the Chinese government has effectively improved the level of ES in the border areas by implementing a series of environmental protection policies [77]. Therefore, other countries can also achieve the purpose of protecting the ecosystem by continuously increasing their attention to the ES in the border areas. However, the national boundaries divided by administrative units are not necessarily the boundaries of natural ecosystems. Carrying out mutual cooperation among countries is also an important way to improve the ES of border areas. In terms of specific implementation, first, it is essential to pay attention to the ecological restoration and protection measures in the most valuable areas [78]. By distinguishing the differences in the level of ES in different regions of the study area, it helps the local government to choose the priority of policy implementation. Second, when implementing environmental protection policies, the local government should identify the obstacle factors affecting the ES in advance, so as to implement environmental policies pertinently. Third, as an important basis for a comprehensive national security system, the amount of research on the ES of border areas needs to be increased, so as to provide reference for local environmental decision making.

6. Conclusions

Improving the ES of border areas is of great significance to local sustainable development [10]. However, few studies have addressed the status of ES in border areas, especially on the county scale [79]. Therefore, 25 border counties in Yunnan Province were selected as the study case. We used an entropy weight TOPSIS model to evaluate the ES conditions of border areas from 2004 to 2019 and used trend surface analysis to evaluate the spatial differentiation trends. Then, we diagnosed the factors creating obstacles that affect ES and used a GM (1,1) model to predict the state of ES in both 2025 and 2030. The results show the following patterns:

(1) From 2004 to 2019, the level of ES in all border counties showed a positive upward trend. Tengchong, Jinghong, and Ruili counties ranked among the top three in terms of the improved ES, with increases of 0.2294, 0.2084, and 0.1544, respectively. In terms of five subsystems (economy, society, environment, landscape pattern, and ecosystem service value), the 25 counties had obvious differences. The levels of the economic and social subsystems showed an overall upward trend; the levels of environmental and landscape pattern subsystems fluctuated continuously; and the overall change in the ESV subsystem was small.

(2) The ES of border areas presented a spatial distribution pattern of high in the southwest and low in the west and east. The level of ES in Lancang was the highest in 2004 and 2009 and that of Jinghong was the highest in 2014 and 2019. The trend of spatial change in ES in border areas generally presented the characteristics of remaining stable in the east–west direction and changing in the north–south direction.

(3) In terms of index level, fixed asset investment, fiscal revenue, per-capita GDP, food production, and water regulation are the top five obstacles to improving the ES in border areas. In terms of element level, the economic subsystem is the main factor creating an obstacle to improving the ES in Jinghong, Lancang, and Mengla, while the ESV subsystem is the main factor affecting improvement in the ES in the other 22 counties.

(4) The gray prediction model GM (1,1) can effectively predict the future situation of border areas in both 2025 and 2030. The level of ES in border areas is predicted to maintain an upward trend. In 2025, the ES in Jinghong will reach 0.84 and will be in a secure state. By 2030, the number of border counties with a secure state of ES will increase to two, namely Jinghong and Tengchong. Their ES level will reach 0.98 and 0.86, respectively.

Our research provides reliable information on the ES of 25 border counties in Yunnan and puts forward targeted policy suggestions based on the research results, which will be necessary if China desires to implement sustainable development planning and management at the smallest administrative scale. In addition, this study can provide a reference for other countries to improve the level of ES in border areas.

Supplementary Materials: The following supporting information can be downloaded at: https://www.mdpi.com/article/10.3390/land11060892/s1, Table S1: Equivalent factor per unit area of ecosystem services in border areas; Table S2: The value coefficient per unit area of ecosystem services in border areas (yuan/ha/yr); Table S3: The sensitivity index of the ESV coefficient.

Author Contributions: Conceptualization, Z.H., Q.C. and C.W.; Writing—original draft preparation, Z.H. and X.T.; Methodology, Z.H., Z.Z. and F.Z.; Software, Z.H.; Data curation, M.Q.; Validation, M.Q., F.Z. and Q.C.; Supervision, C.W. All authors have read and agreed to the published version of the manuscript.

Funding: This research was funded by the Major Project of the National Social Science Fund of China (Grant No. 20&ZD095) and the Scientific Research Foundation of the Department of Education of Yunnan Province (Grant No. 2022Y079).

Institutional Review Board Statement: Not applicable.

Informed Consent Statement: Not applicable.

Data Availability Statement: The data presented in this study are available on request from the corresponding author.

Conflicts of Interest: The authors declare no conflict of interest.

References

1. Lambin, E.F.; Meyfroidt, P. Global land use change, economic globalization, and the looming land scarcity. *Proc. Natl. Acad. Sci. USA* **2011**, *108*, 3465. [CrossRef] [PubMed]
2. Weidenfeld, A. Tourism and cross border regional innovation systems. *Ann. Tour. Res.* **2013**, *42*, 191–213. [CrossRef]
3. Edzes, A.J.E.; van Dijk, J.; Broersma, L. Does cross-border commuting between EU-countries reduce inequality? *Appl. Geogr.* **2022**, *139*, 102639. [CrossRef]
4. Duan, W.L.; Zou, S.; Chen, Y.N.; Nover, D.; Fang, G.H.; Wang, Y. Sustainable water management for cross-border resources: The Balkhash Lake Basin of Central Asia, 1931–2015. *J. Clean. Prod.* **2020**, *263*, 121614. [CrossRef]
5. Su, X.B.; Li, C.S. Bordering dynamics and the geopolitics of cross-border tourism between China and Myanmar. *Political Geogr.* **2021**, *86*, 102372. [CrossRef]
6. Johnson, C.; Jones, R.; Paasi, A.; Amoore, L.; Mountz, A.; Salter, M.; Rumford, C. Interventions on rethinking 'the border' in border studies. *Political Geogr.* **2011**, *30*, 61–69. [CrossRef]
7. Newman, D. The lines that continue to separate us: Borders in our 'borderless' world. *Prog. Hum. Geog.* **2006**, *30*, 143–161. [CrossRef]
8. Laako, H.; Kauffer, E. Between colonising waters and extracting forest fronts: Entangled eco-frontiers in the Usumacinta River Basin. *Political Geogr.* **2022**, *96*, 102566. [CrossRef]
9. Chen, D.N.; Cai, Y.P.; Wang, X.; Li, C.H.; Yin, X.N.; Liu, Q. An inexact modeling approach for supporting water resources allocation under natural and social complexities in a border city of China and Myanmar. *Resour. Conserv. Recycl.* **2021**, *168*, 105245. [CrossRef]
10. He, D.M.; Wu, R.D.; Feng, Y.; Li, Y.G.; Ding, C.Z.; Wang, W.L.; Yu, D.W. REVIEW: China's transboundary waters: New paradigms for water and ecological security through applied ecology. *J. Appl. Ecol.* **2014**, *51*, 1159–1168. [CrossRef]
11. Liu, J.; Wu, R.D.; He, D.M. Transboundary eco-security regulation for geopolitical cooperation in land border areas. *Geogr. Program* **2015**, *34*, 606–616.
12. Li, Z.T.; Yuan, M.J.; Hu, M.M.; Wang, Y.F.; Xia, B.C. Evaluation of ecological security and influencing factors analysis based on robustness analysis and the BP-DEMALTE model: A case study of the Pearl River Delta urban agglomeration. *Ecol. Indicat.* **2019**, *101*, 595–602. [CrossRef]
13. Norton, S.B.; Rodier, D.J.; van der Schalie, W.H.; Wood, W.P.; Slimak, M.W.; Gentile, J.H. A framework for ecological risk assessment at the EPA. *Environ. Toxicol. Chem.* **1992**, *11*, 1663–1672. [CrossRef]
14. Dai, Y.C.; Peng, G.C.; Wen, C.H.; Zahoor, B.; Ma, X.D.; Hacker, C.E.; Xue, Y.D. Climate and land use changes shift the distribution and dispersal of two umbrella species in the Hindu Kush Himalayan region. *Sci. Total Environ.* **2021**, *777*, 146207. [CrossRef]
15. Kong, Y.; He, W.J.; Yuan, L.; Zhang, Z.F.; Gao, X.; Zhao, Y.E.; Mulugeta Degefu, D. Decoupling economic growth from water consumption in the Yangtze River Economic Belt, China. *Ecol. Indicat.* **2021**, *123*, 107344. [CrossRef]
16. Dai, L.; Liu, Y.B.; Luo, X.Y. Integrating the MCR and DOI models to construct an ecological security network for the urban agglomeration around Poyang Lake, China. *Sci. Total Environ.* **2021**, *754*, 141868. [CrossRef]
17. Tang, Y.S.; Zhao, X.Y.; Jiao, J.L. Ecological security assessment of Chaohu Lake Basin of China in the context of River Chief System reform. *Environ. Sci. Pollut. Res.* **2020**, *27*, 2773–2785. [CrossRef]

18. Ma, L.B.; Bo, J.; Li, X.Y.; Fang, F.; Cheng, W.J. Identifying key landscape pattern indices influencing the ecological security of inland river basin: The middle and lower reaches of Shule River Basin as an example. *Sci. Total Environ.* **2019**, *674*, 424–438. [CrossRef]
19. Onkal-Engin, G.; Demir, I.; Hiz, H. Assessment of urban air quality in Istanbul using fuzzy synthetic evaluation. *Atmos. Environ.* **2004**, *38*, 3809–3815. [CrossRef]
20. Wu, Y.Z.; Zhang, T.C.; Zhang, H.; Pan, T.; Ni, X.L.; Grydehøj, A.; Zhang, J.M. Factors influencing the ecological security of island cities: A neighborhood-scale study of Zhoushan Island, China. *Sustain. Cities Soc.* **2020**, *55*, 102029. [CrossRef]
21. Liu, T.; Wang, H.Z.; Wang, H.Z.; Xu, H. The spatiotemporal evolution of ecological security in China based on the ecological footprint model with localization of parameters. *Ecol. Indicat.* **2021**, *126*, 107636. [CrossRef]
22. Fan, Y.P.; Fang, C.L. Evolution process and obstacle factors of ecological security in western China, a case study of Qinghai province. *Ecol. Indicat.* **2020**, *117*, 106659. [CrossRef]
23. Sun, M.Y.; Li, X.H.; Yang, R.J.; Zhang, Y.; Zhang, L.; Song, Z.W.; Liu, Q.; Zhao, D. Comprehensive partitions and different strategies based on ecological security and economic development in Guizhou Province, China. *J. Clean. Prod.* **2020**, *274*, 122794. [CrossRef]
24. Peng, C.C.; Li, B.; Nan, B. An analysis framework for the ecological security of urban agglomeration: A case study of the Beijing-Tianjin-Hebei urban agglomeration. *J. Clean. Prod.* **2021**, *315*, 128111. [CrossRef]
25. Cao, Y.H.; Liu, M.Y.; Zhang, Y.; Chen, C.; Cao, W.D. Spatiotemporal Evolution of Ecological Security in the Wanjiang City Belt, China. *Chin. Geogr. Sci.* **2020**, *30*, 1052–1064. [CrossRef]
26. Zhang, M.X.; Bao, Y.B.; Xu, J.; Han, A.; Liu, X.P.; Zhang, J.Q.; Tong, Z.J. Ecological security evaluation and ecological regulation approach of East-Liao River basin based on ecological function area. *Ecol. Indicat.* **2021**, *132*, 108255. [CrossRef]
27. Ghosh, S.; Das Chatterjee, N.; Dinda, S. Urban ecological security assessment and forecasting using integrated DEMATEL-ANP and CA-Markov models: A case study on Kolkata Metropolitan Area, India. *Sustain. Cities Soc.* **2021**, *68*, 102773. [CrossRef]
28. Ziegler Alan, D.; Fox Jefferson, M.; Xu, J. The Rubber Juggernaut. *Science* **2009**, *324*, 1024–1025. [CrossRef]
29. Foley, S. *Threats to Mekong Forest Ecosystems: Last Chance for the Indo-Burma Biodiversity Hotspot, in Encyclopedia of the World's Biomes*; Goldstein, M.I., DellaSala, D.A., Eds.; Elsevier: Oxford, UK, 2020; pp. 168–184.
30. Wu, X.; Liu, S.L.; Sun, Y.X.; An, Y.; Dong, S.K.; Liu, G.H. Ecological security evaluation based on entropy matter-element model: A case study of Kunming city, southwest China. *Ecol. Indicat.* **2019**, *102*, 469–478. [CrossRef]
31. Bai, Y.; Wong, C.P.; Jiang, B.; Hughes, A.C.; Wang, M.; Wang, Q. Developing China's Ecological Redline Policy using ecosystem services assessments for land use planning. *Nat. Commun.* **2018**, *9*, 3034. [CrossRef]
32. Gao, J.Y.; Castelletti, A.; Burlado, P.; Wang, H.; Zhao, J.S. Soft-cooperation via data sharing eases transboundary conflicts in the Lancang-Mekong River Basin. *J. Hydrol.* **2022**, *606*, 127464. [CrossRef]
33. Zanzanaini, C.; Trần, B.T.; Singh, C.; Hart, A.; Milder, J., DeClerck, F. Integrated landscape initiatives for agriculture, livelihoods and ecosystem conservation: An assessment of experiences from South and Southeast Asia. *Landsc. Urban Plan.* **2017**, *165*, 11–21. [CrossRef]
34. Immerzeel Walter, W.; van Beek Ludovicus, P.H.; Bierkens Marc, F.P. Climate Change Will Affect the Asian Water Towers. *Science* **2010**, *328*, 1382–1385. [CrossRef]
35. Peng, J.; Yang, Y.; Liu, Y.X.; Hu, Y.N.; Du, Y.Y.; Meersmans, J.; Qiu, S.J. Linking ecosystem services and circuit theory to identify ecological security patterns. *Sci. Total Environ.* **2018**, *644*, 781–790. [CrossRef] [PubMed]
36. Hu, Z.N.; Yang, X.; Yang, J.J.; Yuan, J.; Zhang, Z.Y. Linking landscape pattern, ecosystem service value, and human well-being in Xishuangbanna, southwest China: Insights from a coupling coordination model. *Glob. Ecol. Conserv.* **2021**, *27*, e01583. [CrossRef]
37. Zhang, J.; Qu, M.; Wang, C.; Zhao, J.; Cao, Y. Quantifying landscape pattern and ecosystem service value changes: A case study at the county level in the Chinese Loess Plateau. *Glob. Ecol. Conserv.* **2020**, *23*, e01110. [CrossRef]
38. Myers, N.; Mittermeier, R.A.; Mittermeier, C.G.; da Fonseca, G.A.B.; Kent, J. Biodiversity hotspots for conservation priorities. *Nature* **2000**, *403*, 853–858. [CrossRef]
39. Gao, J.; Ryan, C.; Cave, J.; Zhang, C.Z. Tourism border-making: A political economy of China's border tourism. *Ann. Tour. Res.* **2019**, *76*, 1–13. [CrossRef]
40. Yang, J.; Huang, X. The 30 m annual land cover dataset and its dynamics in China from 1990 to 2019. *Earth Syst. Sci. Data* **2021**, *13*, 3907–3925. [CrossRef]
41. Hammer, M.S.; van Donkelaar, A.; Li, C.; Lyapustin, A.; Sayer, A.M.; Hsu, N.C.; Levy, R.C.; Garay, M.J.; Kalashnikova, O.V.; Kahn, R.A.; et al. Global Estimates and Long-Term Trends of Fine Particulate Matter Concentrations (1998–2018). *Environ. Sci. Technol.* **2020**, *54*, 7879–7890. [CrossRef]
42. Yang, Y.J.; Song, G.; Lu, S. Assessment of land ecosystem health with Monte Carlo simulation: A case study in Qiqihaer, China. *J. Clean. Prod.* **2020**, *250*, 119522. [CrossRef]
43. Chen, Y.; Wang, J.L. Ecological security early-warning in central Yunnan Province, China, based on the gray model. *Ecol. Indicat.* **2020**, *111*, 106000. [CrossRef]
44. Yi, D.; Guo, X.; Han, Y.; Guo, J.; Ou, M.; Zhao, X. Coupling Ecological Security Pattern Establishment and Construction Land Expansion Simulation for Urban Growth Boundary Delineation: Framework and Application. *Land* **2022**, *11*, 359. [CrossRef]
45. Liang, F.; Bai, M.; Hu, Q.; Lin, S.-H. Ecological Security and Ecosystem Quality: A Case Study of Xia-Zhang-Quan Metropolitan Area in China. *Land* **2022**, *11*, 707. [CrossRef]

46. Wang, Y.; Pan, J.H. Building ecological security patterns based on ecosystem services value reconstruction in an arid inland basin: A case study in Ganzhou District, NW China. *J. Clean. Prod.* **2019**, *241*, 118337. [CrossRef]
47. Gao, J.B.; Du, F.J.; Zuo, L.Y.; Jiang, Y. Integrating ecosystem services and rocky desertification into identification of karst ecological security pattern. *Landsc. Ecol.* **2021**, *36*, 2113–2133. [CrossRef]
48. Costanza, R.; d'Arge, R.; de Groot, R.; Farber, S.; Grasso, M.; Hannon, B.; Limburg, K.; Naeem, S.; O'Neill, R.V.; Paruelo, J.; et al. The value of the world's ecosystem services and natural capital. *Nature* **1997**, *387*, 253–260. [CrossRef]
49. Xie, G.D.; Lu, C.X.; Leng, Y.F.; Zheng, D.; Li, S.C. Ecological assets valuation of the Tibetan Plateau. *J. Nat. Resour.* **2003**, *18*, 189–196.
50. Xie, G.D.; Zhang, C.X.; Zhen, L.; Zhang, L.M. Dynamic changes in the value of China's ecosystem services. *Ecosyst. Serv.* **2017**, *26*, 146–154. [CrossRef]
51. Jiang, W.; Lü, Y.H.; Liu, Y.X.; Gao, W.W. Ecosystem service value of the Qinghai-Tibet Plateau significantly increased during 25 years. *Ecosyst. Serv.* **2020**, *44*, 101146. [CrossRef]
52. Ke, X.L.; Wang, X.Y.; Guo, H.X.; Yang, C.; Zhou, Q.; Mougharbel, A. Urban ecological security evaluation and spatial correlation research—Based on data analysis of 16 cities in Hubei Province of China. *J. Clean. Prod.* **2021**, *311*, 127613. [CrossRef]
53. Cheng, H.R.; Zhu, L.K.; Meng, J.J. Fuzzy evaluation of the ecological security of land resources in mainland China based on the Pressure-State-Response framework. *Sci. Total Environ.* **2022**, *804*, 150053. [CrossRef] [PubMed]
54. Wang, S.D.; Zhang, X.Y.; Wu, T.X.; Yang, Y.Y. The evolution of landscape ecological security in Beijing under the influence of different policies in recent decades. *Sci. Total Environ.* **2019**, *646*, 49–57. [CrossRef]
55. Yushanjiang, A.; Zhang, F.; Yu, H.Y.; Kung, H.-t. Quantifying the spatial correlations between landscape pattern and ecosystem service value: A case study in Ebinur Lake Basin, Xinjiang, China. *Ecol. Eng.* **2018**, *113*, 94–104. [CrossRef]
56. Shannon, C.E. A mathematical theory of communication. *Bell Syst. Tech. J.* **1948**, *27*, 379–423. [CrossRef]
57. Wang, Y.J.; Chen, H.; Long, R.Y.; Sun, Q.Q.; Jiang, S.Y.; Liu, B. Has the Sustainable Development Planning Policy Promoted the Green Transformation in China's Resource-based Cities? *Resour. Conserv. Recycl.* **2022**, *180*, 106181. [CrossRef]
58. Xiao, Y.; Chai, J.X.; Wang, R.; Huang, H. Assessment and key factors of urban liveability in underdeveloped regions: A case study of the Loess Plateau, China. *Sustain. Cities Soc.* **2022**, *79*, 103674. [CrossRef]
59. Cui, X.Y.; Fang, L.; Wang, X.R.; Kang, J.F. Urban eco-security assessment in the urban agglomerations based on DPSIR model: A case study of Yangtze River Delta, China. *Acta Ecol. Sin.* **2020**, *41*, 302–319.
60. Wang, H.C.; Zuo, R.G. A comparative study of trend surface analysis and spectrum–area multifractal model to identify geochemical anomalies. *J. Geochem. Explor.* **2015**, *155*, 84–90. [CrossRef]
61. Lu, X.H.; Zhang, Y.W.; Lin, C.R.; Wu, F. Analysis and comprehensive evaluation of sustainable land use in China: Based on sustainable development goals framework. *J. Clean. Prod.* **2021**, *310*, 127205. [CrossRef]
62. Sun, Y.; Zhao, T.Y.; Xia, L. Spatial-temporal differentiation of carbon efficiency and coupling coordination degree of Chinese county territory and obstacles analysis. *Sustain. Cities Soc.* **2022**, *76*, 103429. [CrossRef]
63. Deng, J.L. Control problems of grey Systems. *Syst. Control Lett.* **1982**, *1*, 288–294. [CrossRef]
64. Lin, C.S.; Liou, F.M.; Huang, C.P. Grey forecasting model for CO_2 emissions: A Taiwan study. *Appl. Energ.* **2011**, *88*, 3816–3820. [CrossRef]
65. Yang, G.C.; Shang, P.P.; He, L.C.; Zhang, Y.C.; Wang, Y.; Zhang, F.; Zhu, L.; Wang, Y. Interregional carbon compensation cost forecast and priority index calculation based on the theoretical carbon deficit: China as a case. *Sci. Total Environ.* **2019**, *654*, 786–800. [CrossRef] [PubMed]
66. Wang, X.L.; Wei, Y.W.; Shao, Q.L. Decomposing the decoupling of CO_2 emissions and economic growth in China's iron and steel industry. *Resour. Conserv. Recycl.* **2020**, *152*, 104509. [CrossRef]
67. Wang, Z.X.; Li, Q. Modelling the nonlinear relationship between CO_2 emissions and economic growth using a PSO algorithm-based grey Verhulst model. *J. Clean. Prod.* **2019**, *207*, 214–224. [CrossRef]
68. Zheng, H.; Khan, Y.A.; Abbas, S.Z. Exploration on the coordinated development of urbanization and the eco-environmental system in central China. *Environ. Res.* **2022**, *204*, 112097. [CrossRef] [PubMed]
69. Chen, Y.; Zhu, M.K.; Lu, J.L.; Zhou, Q.; Ma, W.B. Evaluation of ecological city and analysis of obstacle factors under the background of high-quality development: Taking cities in the Yellow River Basin as examples. *Ecol. Indicat.* **2020**, *118*, 106771. [CrossRef]
70. Mooney, H.A.; Duraiappah, A.; Larigauderie, A. Evolution of natural and social science interactions in global change research programs. *Proc. Natl. Acad. Sci. USA* **2013**, *110*, 3665–3672. [CrossRef]
71. Wu, X.T.; Wei, Y.P.; Fu, B.J.; Wang, S.; Zhao, Y.; Moran Emilio, F. Evolution and effects of the social-ecological system over a millennium in China's Loess Plateau. *Sci. Adv.* **2020**, *6*, eabc0276. [CrossRef]
72. Guo, D.Y.; Wang, D.Y.; Zhong, X.Y.; Yang, Y.Y.; Jiang, L.X. Spatiotemporal Changes of Land Ecological Security and Its Obstacle Indicators Diagnosis in the Beijing–Tianjin–Hebei Region. *Land* **2021**, *10*, 706. [CrossRef]
73. Yang, R.Y.; Du, W.Y.; Yang, Z.S. Spatiotemporal Evolution and Influencing Factors of Urban Land Ecological Security in Yunnan Province. *Sustainability* **2021**, *13*, 2936. [CrossRef]
74. Liu, Y.Y.; Zhao, C.Y.; Liu, X.M.; Chang, Y.P.; Wang, H.; Yang, J.H.; Yang, X.G.; Wei, Y. The multi-dimensional perspective of ecological security evaluation and drive mechanism for Baishuijiang National Nature Reserve, China. *Ecol. Indicat.* **2021**, *132*, 108295. [CrossRef]

75. Rao, Y.; Dai, J.; Dai, D.; He, Q.; Wang, H. Effect of Compactness of Urban Growth on Regional Landscape Ecological Security. *Land* **2021**, *10*, 848. [CrossRef]
76. Ackleson, J. Constructing security on the U.S.–Mexico border. *Political Geogr.* **2005**, *24*, 165–184. [CrossRef]
77. Sills, J.; Yin, D.; Yuan, Z.J.; Li, J.; Zhu, H. Mitigate human-wildlife conflict in China. *Science* **2021**, *373*, 500–501. [CrossRef]
78. Huang, L.Y.; Wang, J.; Fang, Y.; Zhai, T.L.; Cheng, H. An integrated approach towards spatial identification of restored and conserved priority areas of ecological network for implementation planning in metropolitan region. *Sustain. Cities Soc.* **2021**, *69*, 102865. [CrossRef]
79. Liu, C.; Li, W.; Xu, J.; Zhou, H.; Li, C.; Wang, W. Global trends and characteristics of ecological security research in the early 21st century: A literature review and bibliometric analysis. *Ecol. Indicat.* **2022**, *137*, 108734. [CrossRef]

land

Article

Spatial Pattern of Soil Erosion in Relation to Land Use Change in a Rolling Hilly Region of Northeast China

Yuanli Zhu [1,2,3,*], Wenbo Li [4], Dongyan Wang [4], Zihao Wu [1,2,3] and Peng Shang [5]

1 Research Center for Transformation Development and Rural Revitalization of Resource-Based Cities in China, China University of Mining and Technology, Xuzhou 221116, China
2 School of Public Policy & Management, China University of Mining and Technology, Xuzhou 221116, China
3 Research Center for Land Use & Ecological Security Governance in Mining Area, China University of Mining and Technology, Xuzhou 221116, China
4 College of Earth Sciences, Jilin University, Changchun 130061, China
5 Natural Resources Bureau of Zhanhua District, Binzhou 256899, China
* Correspondence: 6268@cumt.edu.cn

Abstract: The black soil region in Northeast China is one of the major grain-producing areas of China. Soil erosion in the black soil region caused by natural factors and anthropic activities has attracted much attention, especially in a rolling hilly region. Compared with natural factors, the land use factor of cropland encompasses the most easily optimized measures. Jiutai County of Changchun City, located in the hilly areas of Northeast China, was taken as an example to calculate the soil erosion modulus using the Revised Universal Soil Loss Equation model. The overall soil erosion status of cultivated land in the study area was mainly slight and light, the proportion of cultivated land affected by extremely intensive and severe erosion was relatively small, and the average soil erosion modulus was 7.09 $t{\cdot}hm^{-2}{\cdot}a^{-1}$ in 2019. In view of spatial distribution characteristics of soil erosion revealed by the spatial aggregation and hot spot analysis, the most serious soil erosion intensity was concentrated in the southeast and northeast sloping farmland over 8°. With the increase in elevation and topographic slope, the proportion of slight and light soil erosion gradually decreased, which was closely related to the increase in soil erodibility caused by the space–time migration of soil organic carbon caused by the interaction of hydraulic and tillage erosion in complex topographic areas. The Geographically Weighted Regression model was introduced to explore the driving factors and superposition mechanism of farmland soil erosion in the hilly region of Northeast China. Based on the relationship between soil erosion and landscape fragmentation, landscape fragmentation was an important driving force promoting soil erosion, sediment yield, and sediment transport. This paper is committed to providing a basis for accurately deploying regional soil and water conservation measures and formulating macro land management policies.

Keywords: agricultural land; land use; soil erosion; RUSLE; hilly region

Citation: Zhu, Y.; Li, W.; Wang, D.; Wu, Z.; Shang, P. Spatial Pattern of Soil Erosion in Relation to Land Use Change in a Rolling Hilly Region of Northeast China. *Land* **2022**, *11*, 1253. https://doi.org/10.3390/land11081253

Academic Editors: Li Ma, Yingnan Zhang, Muye Gan and Zhengying Shan

Received: 4 July 2022
Accepted: 3 August 2022
Published: 5 August 2022

1. Introduction

Soil erosion has become a major factor driving soil degradation processes [1], affecting soil and nutrient loss at original sites and accumulation at deposition sites [2,3]. Soil erosion results in nutrient loss through the denudation of surface soil and soil organic matter, which not only leads to land degradation and fertility loss but also influences the corresponding biogeochemical cycles (the siltation and eutrophication of water environment, enhancement of flooding, and decrease or increase in CO_2 emissions) [4]. For example, Hancock and Wells using a flume experiment suggested that soil organic carbon (SOC) is enriched in eroded sediment [5]. Soil erosion alters the biological process of SOC mineralization resulting in the loss of C from the soil to the atmosphere [6]. The disturbance of soil erosion in the terrestrial carbon cycle also restricts the buffering effect of the soil ecosystem on climate change and further affects the global ecosystem security [7,8]. China is one of the countries with the worst soil erosion phenomenon in the world [9], according to the Ministry of Water Resources, which issued the second national soil erosion remote sensing survey results showing that the national soil erosion area reached 3.56 million km^2. Moreover, the problem of soil erosion in the black soil area of Northeast China has attracted extensive attention [10,11].

Soil erosion is derived from the joint effect of natural and human influencing factors [12]. Land use is the most important predictor of soil erosion susceptibility, which can slow down or speed up soil erosion, depending on the specific situation [13]. Irrational land use is one of the substantial factors that accelerate regional soil erosion [14]. Understanding the impact of land use changes on soil erosion is crucial for exploring the mechanism of regional erosion and identifying the driving forces of anthropogenic controllable erosion [15–17]. Most studies regarding the response of soil erosion to land use change focus on the estimation of soil erosion of different land use types [18,19], but an exploration of the response of land use intensity and landscape fragmentation to soil erosion is lacking. Therefore, building on the spatial characterization of soil erosion patterns and the identification of erosion hotspots, the impacts of land use intensity and farmland landscape fragmentation on the spatial differentiation characteristics of arable land soil erosion were identified, and scientific land management methods were adopted to control regional soil erosion and improve soil quality, which is conducive to the sustainable use of land resources and achieving regional sustainable development.

The inherently fertile black soils (Mollisols) account for nearly 6.9% of the Earth's land area. Tremendous black soils all over the world are suffering soil erosion and soil degradation due to unreasonable management and prolonged agricultural production [20,21]. However, black soil is also an abundant organic carbon (OC) pool; thus, any loss of black SOC due to global climate change is likely to produce considerable feedback [22]. The black soil region of Northeast China is crucial for grain production in China but has been threatened by intensive and extensive soil erosion due to long-term intense cultivation activity after the transformation from forest to cropland. Therefore, this paper selected the Jiutai District of Changchun City, which is located in the rolling hilly region of Northeast China. In the past few decades, although it is an important commodity grain production base in China, it has experienced severe degradation in physical [23], chemical [24], and biological properties [25], which provides a platform for examining the separated effects and influence mechanism of land use change on soil erosion, and is vital for the subsequent sustainable, optimal management of black soils in Northeast China.

In this paper, Jiutai County of Changchun City, located in the black soil region of Northeast China, was taken as an example area to carry out the soil erosion modulus using the Revised Universal Soil Loss Equation (RUSLE) and then to recognize the erosion hot spots. The Geographical Weighted Regression (GWR) model was introduced to discuss the key driving factors and superposition mechanism of farmland soil erosion in the hilly region of Northeast China and to analyze the control mechanism of land use intensity and landscape fragmentation index on farmland soil erosion. This paper is committed to providing a basis for accurately deploying regional soil and water conservation measures and formulating macro land management policies.

2. Materials and Methods

2.1. Study Area, Soil Sampling and Analysis

Jiutai County of Changchun City is located in typical low mountain and hilly terrain, roughly between 43°50′ N–44° 31′ N latitude and 125°24′ E–126°29′ E longitude. The coverage area of Jiutai County has reached about 3 km^2 and belongs to the monsoon climate zone. The soil in the study site is classified as Mollisols with a texture of clay loam (USDA, Taxonomy).

As the SOC data used in the calculation of the soil erodibility factor in the RUSLE model used in this paper were obtained through remote sensing inversion, to ensure accuracy, the collection range of soil samples was expanded (Figure 1). A total of 240 soil plots were collected by a shovel with depths of 0–30 cm. At each sampling plot within a radius of 2 m, five sites, four sampling points at the corners, and one sampling point at the center of the sampling plot, were selected. The soil samples within the five sampling sites were mixed and placed in a sealed plastic bag to indicate the soil properties of each plot while the spatial coordinates of the plot center were recorded in GPS (Garmin Etrex32X, Bern, Switzerland).

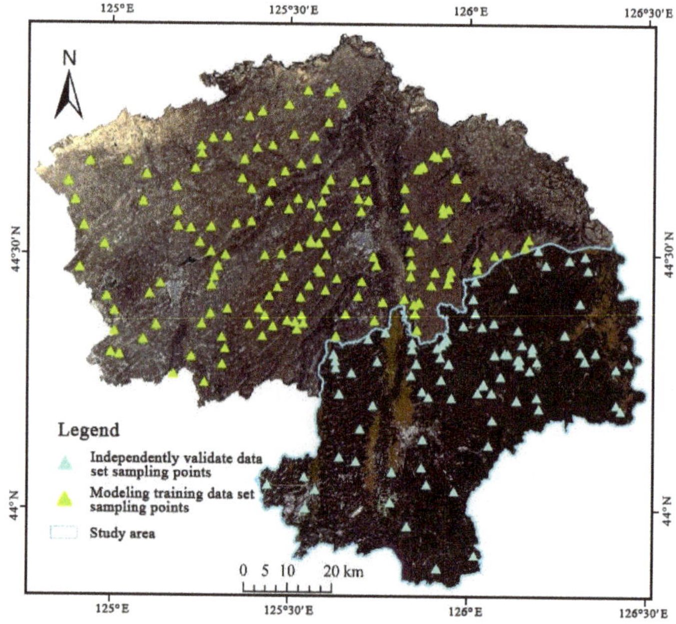

Figure 1. Distribution map of sampling points.

The collected soil samples were dried continuously in an oven at 60 °C for 72 h until constant weight and ground through a 2 mm sieve. The next steps were removing the plant roots and other substances from the sample and further grinding it through a 100 µm screen to determine the SOC content. The determination method of the total carbon content of the soil samples was the dry firing method using a VarioMax CN analyzer (Elementar GmbH, Berlin, Germany). For the soil samples with evident reactions under the treatment of 10% HCl solution, the inorganic carbon content was determined by the pressure calcium meter method. Finally, the SOC content was obtained by subtracting the inorganic carbon content from the total carbon content.

2.2. Data Sources

The Sentinel-2 (S2) remote sensing image data used in this paper were from the European Space Agency and the United States Geological Survey website during 2018–2020. L1C multispectral image data with good quality and less than 10% cloud cover were selected and the Sen2Cor plug-in (ESA released to produce L2A data) was used for atmospheric correction to determine the range of bare cultivated land and SOC inversion. A seamless mosaic tool was employed to merge the multispectral images through Environment for Visualizing Images software. The NDVI is utilized in numerous studies due to its simple estimation, easy availability, and cancellation of noise that is caused due to solar angle, topographic illumination, and clouds [26–28]. In this paper, the normalized difference vegetation index (NDVI) was computed by the S2 remote sensing image dataset established above to represent the vegetation cover condition.

$$NDVI = (Band8 - Band4)/(Band8 + Band4) \quad (1)$$

The time node representing 2019 was selected to calculate the erosion triggering factors and amounts of soil erosion. The rainfall from 1986 to 2015 was provided by WorldClim with a spatial resolution of 5 km. Soil texture data were obtained from the 250 m × 250 m rasterized SoilGrids dataset with soil depths of 0–30 cm. The soil data at depths of 0–30 cm were estimated via the weighted depth average method. To determine the terrain factor in the RUSLE model, ALOS DEM data, which have a spatial resolution of 30 m and are derived from NASA EARTHDATA, were selected. On this basis, topographic raster data such as slope gradient, slope length, and slope aspect in the study area were obtained by QGIS 3.10 software.

Based on the first national detailed land survey in 1996 and the unified land survey database in 2019, the land use maps were visually interpreted by using the images from the United States Geological Survey. The accuracy of land use classification was above 90%, and the Kappa coefficient was approximately 0.85. The land use data of the above two years provided the basis for social and economic data, such as the proportion of residential area and road network density, and land use conditions, such as land use intensity and land use landscape fragmentation in the GWR model. The land use intensity mentioned above refers to the efficiency of land resource utilization. In this paper, land use intensity was quantitatively described as four different grades according to land use types (Table 1). Considering the possibility of multiple land use types distributed in the same grid cell, multiple values of land use intensity are in the same grid cell. Therefore, the land use intensity analysis model should be adopted to calculate the comprehensive land use intensity index of the grid cell as follows:

$$Q_j = 100 \times \left(\sum_{i=1}^{n} q_i \times p_i \right) \quad (2)$$

where Q_j refers to the comprehensive index of land use intensity of the jth grid unit, n is the number of land use intensity grading, q_i refers to the Grade I land use intensity index in the grid unit, and p_i is the ratio of the Grade I land use intensity to the total area of the grid.

Table 1. Land use intensity assignment of different land use types.

Land Use Types	Unused Land	Forest, Grassland and Water	Arable Land	Construction Land
Land use intensity grading index	1	2	3	4

In this paper, "construction land" refers to the land used for building buildings, structures, land as a production base, and living place.

Landscape fragmentation is an index of landscape fragmentation in the landscape pattern index, which can reflect the complexity of landscape spatial structure and the degree of human interference to the landscape to a certain extent. In this paper, patch density, edge density, and aggregation index were calculated by Fragstats 4.2 software. The comprehensive index of cultivated land landscape fragmentation degree was obtained after the weight of three indices was calculated by the analytic hierarchy process.

2.3. RUSLE Model

RUSLE, which has been widely applied in multiple previous studies [29–31], is a statistical relationship model for predicting the amount of soil erosion, which can correlate the amount of soil erosion and its influencing factors (such as climate condition, soil properties, topography conditions, vegetation cover, and human activities), and can be flexibly modified in terms of the local conditions of influencing factors [32]. Five factors are used in the RUSLE model to estimate the annual average soil loss [33] as follows:

$$A = R \times K \times LS \times C \times P \tag{3}$$

where A indicates the annual average soil erosion modulus (t hm^{-2} a^{-1}), R is the rainfall erosivity factor (MJ mm hm^{-2} h^{-1} a^{-1}) linked to the amount of rainfall, and K is the soil erodibility factor (t hm^{2} h hm^{-2} MJ^{-1} mm^{-1}), which is closely related to soil properties. L and S are the slope length and the slope gradient factor, respectively, C is the vegetation cover management factor, and P is the erosion control practice factor. Many researchers in the Northeast Black Soil Region have established local applicable methods for calculating the factors of the RUSLE model [34,35].

2.3.1. Rainfall Erosivity Factor (*R*)

R characterizes the source power of the rain to create erosion. *R* was calculated using the method of Zhang and Fu (2003) [36], which is defined using daily rainfall data as follows:

$$R = \varepsilon X^{\alpha} \tag{4}$$

where *R* indicates the mean rainfall erosivity factor (MJ mm hm^{-2} h^{-1} a^{-1}); X indicates the average annual rainfall (mm); ε and α are model parameters, $\varepsilon = 0.0668$, $\alpha = 1.6266$; and the determination coefficient is 0.828.

2.3.2. Soil Erodibility Factor (*K*)

K represents the soil vulnerability to erosion, which indicates the soil sensitivity to denudation separated under raindrop splash, fluctuation, and transportation by runoff. *K* is sensitive to the texture, structure, OC, hydraulic properties, and wettability of the soil. Many methods for calculating *K* are available, but the most generally used is the EPIC model based on the soil texture and SOC data [37–39]:

$$K = \left\{0.2 + 0.3exp\left[-0.0256SAN\frac{1-SIL}{100}\right]\right\} \times \left(\frac{SIL}{CLA+SIL}\right)^{0.3} \times \left(1 - \frac{0.25C}{C + exp(3.72 - 2.95C)}\right) \times \left(1 - \frac{0.7S_{n1}}{S_{n1} + exp(-5.51 + 22.9S_{n1})}\right) \tag{5}$$

where *SAN*, *SIL*, and *CLA* indicate the sand (0.05–2.0 mm), silt (0.002–0.05 mm), and clay (<0.002 mm) contents (%), respectively, and *C* is the SOC content (%), $S_{n1} = 1 - SAN/100$. By multiplying by the coefficient 0.1317, *K* can be expressed as an SI metric (t hm^{2} h hm^{-2} MJ^{-1} mm^{-1}).

SOC data: S2 remote sensing image data during 2018–2020 were used to construct the bare cropland SOC inversion model. Ten bands covering visible bands (Band2, Band3, and Band4), red-side bands (Band5, Band6, and Band7), near-infrared bands (Band8, Band8A), and shortwave infrared bands (Band11, Band12) were used as an explanatory variable for SOC prediction. The SOC data were the dependent variable in spectral modeling. Coordinate information was used to link soil SOC data from sampling sites with remote sensing spectral information. Model calibration and validation were carried out in R 4.0.3. To evaluate the uncertainty of SOC prediction, different datasets were used for model calibration and cross-validation in 100 repeated simulations. The coefficient of determination (R^2) and root mean square error (RMSE) of the measured and predicted values were used to evaluate the performance of the model.

2.3.3. Slope Length and Steepness Factor (*LS*)

LS is an acceleration factor that reflects the effects of slope length and slope gradient on soil erosion. The estimation method proposed by McCool (1989) [40] and Liu et al.,

1994 [41] was comprehensively adopted and applied in the black soil area of Northeast China. The calculation formula is as follows:

$$S = \begin{cases} 10.8sin\theta + 0.03,\ \theta < 5^\circ \\ 16.8sin\theta - 0.5,\ 5^\circ \leq \theta \leq 10^\circ \\ 21.9sin\theta - 0.96,\ \theta > 10^\circ \end{cases} \tag{6}$$

where θ denotes the slope angle (°).

$$L = \left(\frac{\lambda}{22.13}\right)^m \tag{7}$$

$$m = \begin{cases} 0.2,\ \theta \leq 0.5^\circ \\ 0.3,\ 0.5^\circ < \theta \leq 1.5^\circ \\ 0.4,\ 1.5^\circ < \theta \leq 2.5^\circ \\ 0.5,\ \theta > 2.5^\circ \end{cases} \tag{8}$$

where m and λ are the slope length index and the slope length (m), respectively. Among them, 22.13 is the slope length of the standard plot. The LS factors were calculated by the DEM data, and the specific calculation process was achieved by using ArcGIS 10.3 software.

2.3.4. Cover and Management Factor (C)

C represents the erosion control ability of different surface cover types [42]. In the RUSLE model, C is calculated from various subfactors, namely, prior land use, canopy cover, surface cover, surface roughness, and soil moisture [43]. In this paper, the following calculation method was selected for estimating C factor using vegetation cover value [44]:

$$V = \frac{NDVI - NDVI_{min}}{NDVI_{max} + NDVI_{min}} \tag{9}$$

$$\begin{cases} C = 1, 0 \leq V \leq 0.001 \\ C - 0.6508 - 0.34361 \log V, 0.001 < V \leq 0.783 \\ C = 0, V > 0.783 \end{cases} \tag{10}$$

where V is the vegetation cover, calculated by retrieval NDVIs derived from the S2 images; $NDVI_{max}$ is the maximum value ($NDVI$ value of pure vegetation pixel), and $NDVI_{min}$ is the minimum value ($NDVI$ value of pure bare soil pixel).

2.3.5. Support Practice Factor (P)

P represents an inhibiting factor that reflects the effects of support practices (such as terracing and contour tillage) on soil erosion [45]. Based on previous research results and the actual situation of the research area, the p value of the whole research area was set to 1 in this paper.

2.4. GWR Model

First, before regression weighted analysis, the need for soil erosion risk research and land use factor analysis were considered, and spatial grid units were taken as the basic unit of the study to achieve the purpose of information space statistics. Now, the whole research area was divided into 3 km × 3 km grid cells, and each grid cell was assigned a unique identifier, totaling 445 spatial grid cells.

GWR analysis was first presented by Brunsdon [46] and has been widely used by society and natural scientists [47–49]. The GWR model has been mostly applied in urban planning [47,50], environmental science studies [49], geological and geographical remote sensing [51], agricultural sciences [52], and geosciences in some cases [53].

GWR offers better spatial variation as a local function among dependent and independent variables compared with OLSR as a general trend in the entire study area. The general GWR is defined by the following equation [46]:

$$Z_i = \alpha_0(x_i, y_i) + \sum_{k=1}^{p} \alpha_k(x_i, y_i) w_{ik} + \varepsilon_i \; i = 1, 2, \ldots, n \tag{11}$$

In this paper, Z_i is the explained variable and represents the soil erosion risk index of the ith grid unit. The coordinate of the target region i is (x_i, y_i), which represents the coordinate of the center point of the ith grid cell. $\alpha_0(x_i, y_i)$ is the intercept term, K is the explanatory variables value, and $\alpha_k(x_i, y_i)$ is the first K regression parameter of the ith sampling point and represents the regression parameter of the ith grid cell. w_{ik} is the observed value of explanatory variable w_k at position (x_i, y_i), ε_i is the random error of the ith sampling point.

In the GWR model, the calculation of the spatial weight matrix is the core content, which represents the spatial dependence of data. The calculation of the weight matrix is closely related to the type and bandwidth of the kernel function. In this paper, Bi-Square function, which has more computational advantage when facing regression analysis with a high degree of data dispersion, was selected. The calculation formula is as follows:

$$W_{ij} = e^{\frac{(d_{ij}/b)^2}{2}} \tag{12}$$

$$W_{ij} = \begin{cases} \left(1 - (d_{ij}/b)^2\right)^2, d_{ij} \leq b \\ 0, \; d_{ij} > b \end{cases} \tag{13}$$

where b is the bandwidth, that is, the parameter for calculating the weight value based on the spatial distance; d_{ij} is the actual spatial distance between observation point j and sampling point i.

In terms of optimal bandwidth selection, this paper adopted the AIC criterion with a higher optimization degree but a relatively complex calculation compared with the cross-validation method. The calculation formula is as follows:

$$\mathrm{AIC}_c(b) = 2n \ln \hat{\sigma} + n \ln 2\pi + n\left(\frac{n + tr(S)}{n - 2 - tr(S)}\right) \tag{14}$$

where n is the number of sample points; $\hat{\sigma}$ is the maximum likelihood estimate of the variance of the random error term, $\sigma = \frac{RSS}{n - tr(S)}$. The predicted amount of soil erosion in 2019 was used as the dependent variable with a spatial scale of 3 km × 3 km. The selection of explanatory variables considered the three aspects of nature, social economy, and land use that affect soil erosion. Finally, nine explanatory variables were selected to construct the explanatory index system, namely, elevation (X1), slope (X2), precipitation (X3), SOC content (X4), vegetation coverage (X5), change in the proportion of residential areas (X6), change in road network density (X7), change in land use intensity (X8), and change in landscape fragmentation degree of cropland (X9). X6–X9 indicators indicate the change values, which refer to those change values from 1996 to 2019. The impact of land use intensity and land use landscape fragmentation on the spatial distribution of soil erosion was further studied from the perspective of land use.

In this paper, SPSS 22.0 software was used to conduct preliminary data tests on the nine explanatory variables through the tolerance and variance inflation factor (*VIF*), and the calculation formula is as follows:

$$VIF = \frac{1}{1 - R_i^2} \tag{15}$$

where R_i^2 is the square value of the determination coefficient. The larger the *VIF* is, the smaller the tolerance between explanatory variables.

3. Results and Discussion

3.1. SOC Inversion Based on Multi-Temporal S2 Remote Sensing Image and Composite Soil Pixels

3.1.1. Determination of the Scope of Bare Cultivated Land

NDVImax and NDVImin composite data were established using S2 remote sensing image series data. To determine the critical threshold of bare soil, the NDVI characteristics of farmland, forest, and construction land were statistically calculated. According to the 2019 land use type map, 2000 validation points were randomly selected from each land use category, and the NDVI values of these sampling points were extracted from the NDVI composite data. For the NDVImin combination, the NDVI value representing farmland was between 0.10 and 0.24, which can be clearly distinguished from the forest area. The NDVImax value of construction land was generally lower than that of farmland and forest. Therefore, "NDVImax > 0.75" excluded the construction land area and the range of bare soil was delimited by combining "NDVImax > 0.75" and "0.10 < NDVImin < 0.24" (Figure 2).

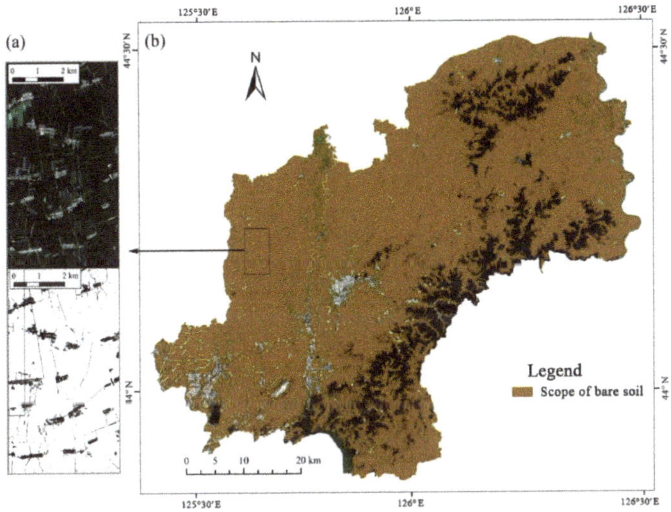

Figure 2. Spatial distribution of bare cultivated land in the study area. (**a**) map is the effect diagram of bare cultivated land identification in part of (**b**) map.

To determine the optimal time window, the first step was to use the planting period of the main crops in the study area determined in the FAO crop calendar. In the second step, according to the investigation of 100 bare croplands in the study area, an NDVI threshold of 0.10–0.24 was set and used to remove green vegetation pixels, which was consistent with the study of Shi et al., 2020 [54]. The third step was that NDVI alone was not sufficient to extract bare land pixels, and the Natural Burn Ratio 2 (NBR2) [55] threshold was used to remove soil pixels contaminated by crop residues. The time evolution map of the NDVI in 2019 was plotted with the 2000 cultivated land sampling points mentioned above. Considering that although a similar NDVI distribution occurred in April and October, corn straw remained in the field after the crop harvest in October, so determining the NBR2 threshold was necessary to remove soil pixels further contaminated by crop residue. By comparing the distribution of NBR2 values in the two periods, the NBR2 threshold of 0.075 was used. Castaldi et al., 2019 showed that an NBR2 value of 0.075 was the most appropriate threshold for building a good SOC prediction model [55]. In addition, a study in northern Germany found that this threshold guaranteed a relatively high bare soil coverage, indicating that the NBR2 = 0.075 threshold was suitable for various environments. The above study proved that NBR2 = 0.075 was practical for removing noise pixels affected by environmental pollution. In this paper, NDVI and NBR2 thresholds were combined to perform the extraction of bare cropland for each single-date data image in a predetermined optimal time window. Then, the processed single-date images were filled into the range of bare farmland, and the multitemporal bare soil composite data were obtained by averaging the reflectance of bare soil repeatedly occurring in each pixel.

3.1.2. SOC Inversion

A scatter plot was made by a one-to-one correspondence of the "GGplot2" function in R4.0.3 software, as shown in Figure 3. Figure 3c reveals the verification results of the PLSR model based on bare soil composite data developed by the multitemporal S2 remote sensing spectrum improved compared with the PLSR model based on single-date S2 remote sensing data, and the SOC prediction model developed by multitemporal bare soil composite data performed better. R^2 = 0.53 (2018), 0.59 (2019), and 0.51 (2020) increased to 0.62 (multitemporal), and the RMSE remained almost unchanged. In this paper, the mathematical relationship between spectral information and soil composition was used, but the effects of agricultural activity factors on the modeling effect were ignored, which is also an aspect where further research needs to be improved. Figure 3c,d reveal that compared with the modeling training dataset, the accuracy test results obtained from the independent validation dataset decreased, R^2 decreased from 0.62 to 0.49, and RMSE increased from 0.17 to 0.23, possibly due to the decreased sample numbers and narrow range of SOC content. Therefore, the prediction results were affected to a certain extent.

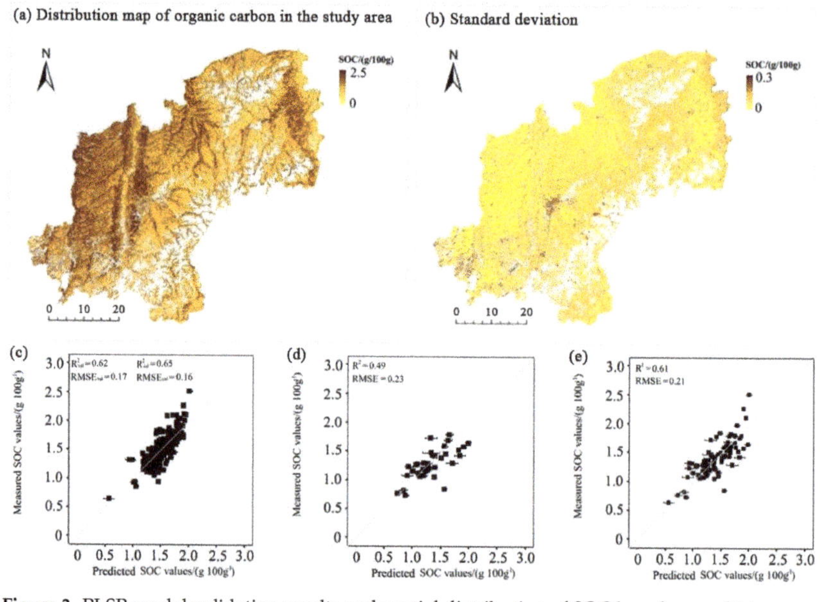

Figure 3. PLSR model validation results and spatial distribution of SOC based on multi-temporal S2 remote sensing spectral data set. (**a**) Spatial distribution map of SOC; (**b**) Spatial distribution of standard deviation of SOC; PLSR model validation results based on (**c**) modeling training data sets; (**d**) independent validation data sets; (**e**) the complete data set.

A total of 80 soil samples (including 35 independent validation sampling points) were collected in the study area. After evaluating the model performance using the complete soil spectral dataset of 80 groups, Figure 3e shows that the prediction accuracy improved with the increase in the number of sampling points (R^2 increased from 0.49 to 0.61, RMSE from 0.23 to 0.21). Compared with the best performance evaluation result (R^2 = 0.32, RMSE = 0.68) in Nascimento et al., 2021 [56] of digital SOC mapping based on remote sensing images, the SOC prediction based on multitemporal S2 remote sensing images in this paper had a more accurate prediction accuracy. Furthermore, the practicability of SOC prediction based on multitemporal remote sensing pixels was illustrated.

3.2. Spatial Mapping of Soil Erodibility Factor and Soil Erosion Risk

The soil texture was fine, and the calculated value of the soil erodibility factor was small (0.0168–0.0203 t hm^2 h hm^{-2} MJ^{-1} mm^{-1}). Clearly, the southeastern region of the study area had a higher SOC content and lower predicted results of soil erodibility factors than the northwestern region.

The SOC spatial distribution data based on S2 high-resolution remote sensing images can improve the spatial resolution of the soil erodibility factor map (Figure 4). Based on the amplification effect of the spatial distribution of the K value (Figure 4), the K value calculated by the SOC data inversion based on S2 can reflect the spatial heterogeneity characteristics of the K value, which is important for refining the spatial difference characteristics of soil erosion. Additionally, proper agricultural practices would maintain sufficient SOC and aggregate structures to help control the development of soil erosion [57].

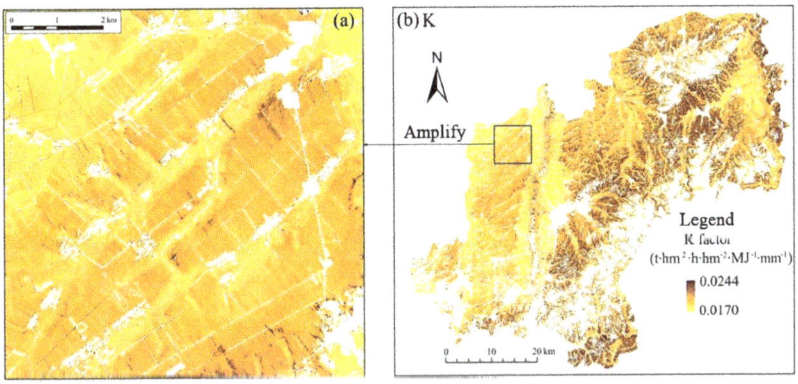

Figure 4. Spatial distribution of soil erodibility factors in the study area. (**a**) map shows the magnified effect of the local area of (**b**) map.

Using the remote sensing inversion techniques, erosion factor calculation equation, and RUSLE model, soil erosion and the interfering factors in the study area in 2019 were assessed and mapped, which provided available information to serve as a primary reference for soil and water conservation management. Based on the soil erosion classification standard in the Technical Standard for Comprehensive Control of Soil and Water Loss in Black Soil Area (SL446-2009) promulgated by the Ministry of Water Resources of China, the soil erosion degree was divided into six grades (Table 2 and Figure 5). From the size of the area eroded, the soil erosion of bare farmland in the study area was mainly slight and light, which accounted for 31.99% and 51.25% of the total area of cultivated land, respectively. Moderate erosion followed, accounting for 11.14% of the total area of bare cultivated land. The proportion of cultivated land with extremely intense and severe erosion was 2.83%. In terms of spatial distribution, the cropland with serious soil erosion was mainly distributed in the sloping farmland with higher elevation and more gullies (southeastern and northeastern hilly areas). In the flat area (central and western regions) where the river flowed, the soil erosion of cultivated land was light, mainly with slight erosion and light erosion. Soil erosion maps can be used as a basic document for rational land planning to avoid potential water and soil losses effectively. The sloping grasslands in the low mountain and hilly area had a high risk of soil and water loss, which required the implementation of effective management and control measures.

Table 2. Statistical results of different soil erosion intensities of cropland.

Soil Erosion Intensity Grade	Soil Erosion Modulus/ t hm^{-2} a^{-1}	Area/km^2	Proportion of Total Arable Land/%
1. Slight	≤2	697.44	31.99
2. Light	2–12	1117.43	51.25
3. Moderate	12–24	242.93	11.14
4. Intense	24–36	60.70	2.78
5. Extremely Intense	36–48	24.18	1.11
6. Severe	>48	37.51	1.72

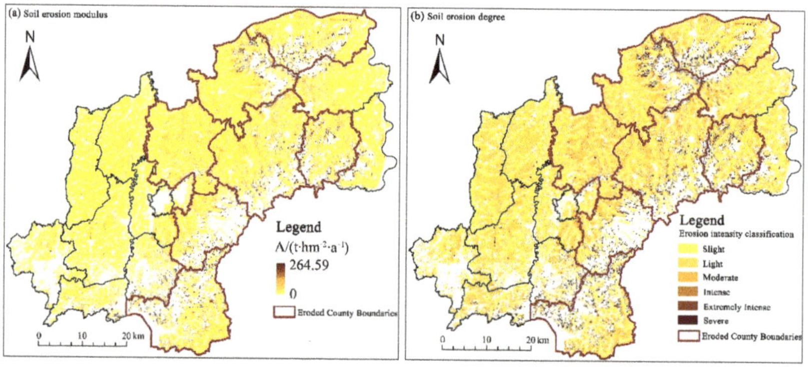

Figure 5. Spatial distribution of soil erosion modulus (**a**) and intensity grade (**b**) in the study area.

Xu and Zhang (2020) [29] showed that the spatiotemporal characteristics in soil erosion could only be uncovered more accurately by spatially quantifying different change methods of triggering factors and their relative importance. Figure 6 shows that topographic factors have the greatest influence on the spatial distribution of soil erosion, and the spatial distribution of topography factors would help explain the change in soil erosion, so the relationship must be determined. According to statistics, with increasing altitude, the proportion of slight and light soil erosion gradually decreased, while the proportion of extreme intensity and severe erosion gradually increases. With the increase in terrain slope, the proportion of slight and light soil erosion gradually decreased, while the proportion of extreme intensity and severe erosion gradually increases. The most serious soil erosion intensity was concentrated in the southeast and northeast sloping farmland over 8°. Finally, overall, soil erosion on the sunny slope was slightly more serious than that on the shady slope possibly because the sunny slope was the windward slope of the southeast monsoon, and the soil water evaporation was larger than that on the shady slope, resulting in a lower soil water content, which reduced the vegetation coverage and made soil erosion more likely. LS was the leading factor of regional soil erosion. Therefore, better spatial optimal allocation of land use is needed to avoid steep areas prone to erosion [29,58]. Arable land was widely distributed in the flat regions with much lower LS factors than in grasslands and forests. The cultivated land distributed around the forestland faced the soil erosion risk due to the terrain and became unstable cultivated land. When the study area was larger, future soil erosion risk from spatial and temporal changes in precipitation should be considered and evaluated by using the future climate change prediction models. The K at the field scale can be calculated by more accurate soil attribute data, which would be needed in water and soil conversation planning.

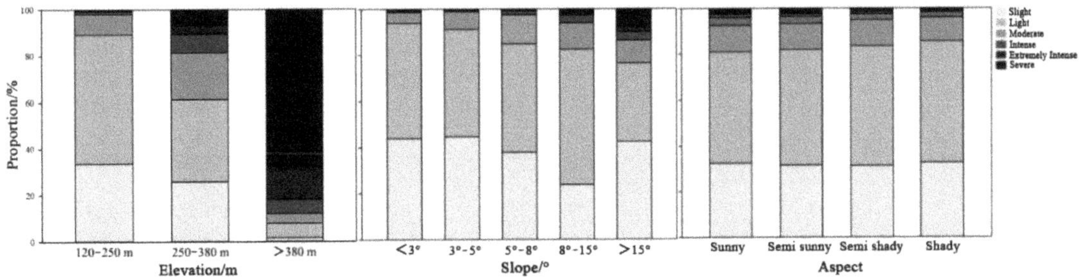

Figure 6. Proportion of soil erosion grades of cultivated land at different altitudes, slopes, and aspects.

3.3. Land Use Factor Analysis of Cropland Soil Erosion Based on the GWR Model

Table 3 shows that the VIF of the independent variables was less than 5 and the condition index was less than 30, which proved that the nine explanatory variables selected passed the multicollinearity test.

Table 3. Multicollinearity test results of independent variables.

Type Layer	Explanatory Variables	Collinearity Test Results	
		Tolerance	VIF
Natural conditions	Elevation (X_1)	0.94	1.07
	Slope (X_2)	0.67	1.49
	Precipitation (X_3)	0.44	2.27
	SOC content (X_4)	0.50	1.99
	Vegetation coverage (X_5)	0.79	1.27
Socioeconomic conditions	Change in the proportion of residential areas (X_6)	0.92	1.09
	Change of road network density (X_7)	0.88	1.14
Land Use conditions	Change of land use intensity (X_8)	0.93	1.08
	Change of landscape fragmentation degree of cropland (X_9)	0.98	1.02

The preliminary statistics of the regression coefficients of each explanatory variable of soil erosion in the GWR model (Table 4) showed that the plus or minus statistical values of the regression coefficients proved the existence of both positive and negative correlations between the explanatory variables and the soil erosion of cultivated land in different grid cells.

Table 4. Statistical result of regression coefficients of explanatory variables.

Explanatory Variables	Minimum	Upper Quartile	Median	Average	Lower Quartile	Maximum	Standard Deviation
Elevation	−1.89	−0.05	0.08	−0.01	0.20	1.01	0.45
Slope	4.03	5.54	5.94	6.18	6.46	9.04	1.09
Precipitation	−1.76	−0.92	−0.55	−0.56	−0.17	0.63	0.51
SOC content	−2.56	−0.72	−0.47	−0.56	−0.30	0.29	0.49
Vegetation coverage	−0.57	0.06	0.16	0.24	0.41	1.54	0.35
Change in the proportion of residential areas	−1.42	0.17	0.83	0.69	1.19	3.01	0.69
Change in road network density	−0.67	0.00	0.11	0.10	0.20	1.46	0.30
Change of land use intensity	−1.98	−0.68	0.49	0.46	0.62	1.39	0.37
Change of landscape fragmentation degree of cropland	−1.92	−0.37	−0.16	−0.17	0.25	1.04	0.59

Chaplot et al., 2005 [59] found that the influence of land use change on soil erosion is greater than that of climate factors, especially in regard to rill erosion. Simonneaux et al., 2015 [60] showed that the increase in soil erosion caused by land use change is much greater than that caused by climate change in Morocco. Compared with forests, farmland has a lower interception rate, resulting in a higher runoff [61].

In this paper, the changes in the proportion of residential area, road network density, and land use intensity, for which the median and mean regression coefficients were positive, explained that the relationship between the above three indices and the cropland soil erosion degree was positively correlated in most grid cells (Figure 7), that is, the increase in the proportion of residential area had a positive influence on the soil erosion severity. The median and mean values of the regression coefficients of farmland landscape fragmentation change were both negative, indicating that the soil erosion degree of farmland decreased with the increase in the explanatory variables of farmland landscape fragmentation change in most grid cells.

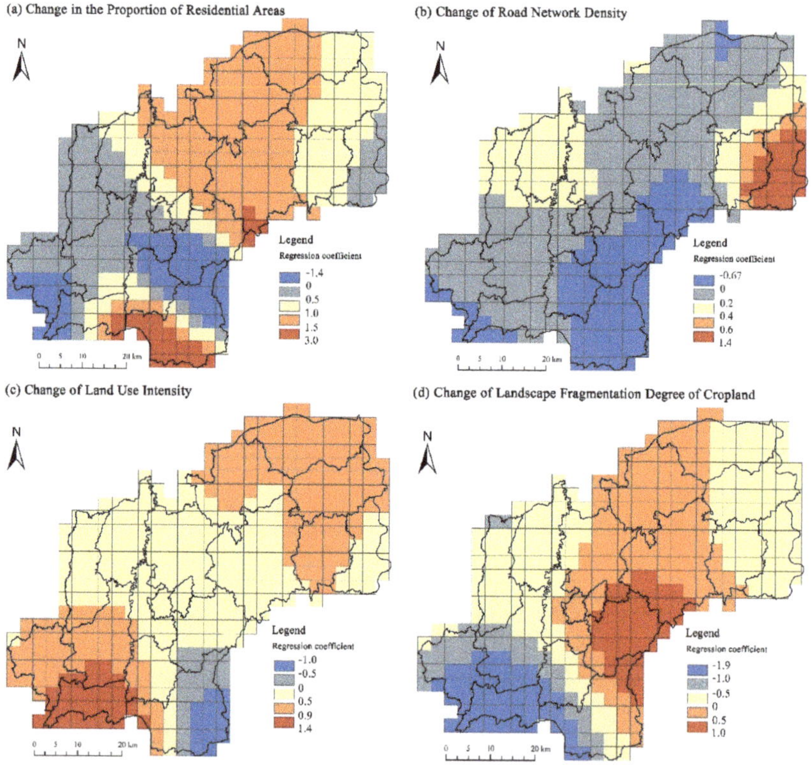

Figure 7. Spatial distribution of regression coefficients of explanatory variables of socioeconomic and land use factors in the GWR model.

The increase in the proportion of residential area, the intensity of land use, and the fragmentation degree of the cultivated land landscape all had a remarkable impact on the soil erosion pattern of surrounding cultivated land to varying degrees (Figure 7). Among them, the proportion of residential area and the fragmentation degree of the cultivated land landscape had the widest impact on the areas with high erosion risk. Therefore, optimizing the development and protection pattern of territorial space with appropriate land regulation means avoiding the further aggravation of soil erosion is necessary.

An increase in residential areas and land use intensity meant an increase in impermeable surface area. In the construction, a large number of materials, such as cement, asphalt, and concrete, were used to cover the soil surface to form an impermeable surface. According to the research, under the conditions of runoff and precipitation, the runoff coefficient of a completely impermeable surface was more than 0.9, that is, almost all precipitation was transformed into surface runoff. The intensification of surface runoff under rainstorm conditions increased the pressure of urban drainage, and the capacity of drainage pipelines usually had difficulty meeting the discharge of a large amount of surface runoff in a short time, resulting in serious surface ponding, especially in low-lying areas. In the area where the land use degree increased, the proportion of impermeable surfaces increase, resulting in an increase in the overall surface runoff in the area. The degree of soil and water loss was more serious in summer with heavy rainfall because the study area is in the monsoon climate area [12]. Wang et al., 2022 found that compared with the western region of the black soil region, the eastern region where the study area is located has a humid climate, and water is the main erosive agent [62].

The area selected in this study belonged to a rolling hilly region with a small area of suitable cultivated land in low mountain and hilly areas, and the overall uphill cultivated land showed the characteristics of high spatial dispersion. To obtain the minimum survival guarantee, only by continuously expanding the cultivated land area, which leads to more barren mountains that are not suitable for farming being reclaimed into cultivated land, accelerates the dispersion and fragmentation of farming space. Compared with natural factors, the land use factor of cropland encompassed the most easily optimized measures. The degree of landscape fragmentation played a positive role in soil erosion. Therefore, to control soil and water loss, the land use pattern can be optimized by enhancing the control effect of the dominant landscape, improving patch uniformity, enriching landscape types, reducing the physical connection between patches, strengthening the aggregation degree of landscape patches, and reducing fragmentation. On a deeper level, as a typical social dominant area, the low hilly agricultural area should always pay attention to the balance/synergy between social services and ecological functions, so as to lay a foundation for sustainable land use in the black soil area [63]

4. Conclusions

In 2019, the overall soil erosion status of cultivated land was mainly slight and light, the proportion of cultivated land affected by the extreme intensity and severe erosion was relatively small, and the average soil erosion modulus was 7.09 $t \cdot hm^{-2} \cdot a^{-1}$. The spatial distribution characteristics of soil erosion based on the results of spatial aggregation and hot spot analysis found that the most serious soil erosion intensity was concentrated in the southeast and northeast sloping farmland. With the increase in elevation and topographic slope, the proportions of slight and light soil erosion gradually decreased while the proportions of extreme intensity and severe erosion gradually increased, which was strongly linked to the increase of soil erodibility caused by the space-time migration and erosion of SOC caused by the interaction of hydraulic and tillage erosion in complex topographic areas.

Based on the relationship between soil erosion and landscape fragmentation using the GWR model in a rolling hilly region, the findings revealed that landscape fragmentation was an important driving force promoting soil erosion, sediment yield, and sediment transport. On this basis, sloping farmland with a high fragmentation degree was effectively integrated to prevent soil erosion in marginal farmland. The prevention and control of soil erosion in hilly areas should be based on the premise of rational utilization of land resources, and the situation of "governing and destroying at the same time" should be completely changed. The finding of this paper can be used as a basis for decision-making in a rolling hilly region, and provide useful information for designing land use planning to regulate the effect of land use change and other soil erosion factors. Land use planning and water and soil conservation planning should be combined to provide a scientific basis for the sustainable land use of land resources and the coordinated development of man-land relationships in hilly areas.

Author Contributions: Y.Z. and W.L. designed the paper; Y.Z. collected the data and contributed to the RUSLE model; Y.Z. wrote the paper; D.W., W.L., P.S. and Z.W. revised the paper. All authors have read and agreed to the published version of the manuscript.

Funding: This paper was supported by the National Natural Science Foundation of China (42071255).

Institutional Review Board Statement: Not applicable.

Informed Consent Statement: Not applicable.

Data Availability Statement: Not applicable.

Conflicts of Interest: The authors declare no conflict of interest.

References

1. Garcia, L.; Celette, F.; Gary, C.; Ripoche, A.; Valdés-Gómez, H.; Metay, A. Management of service crops for the provision of ecosystem services in vineyards: A review. *Agric. Ecosyst. Environ.* **2018**, *251*, 158–170. [CrossRef]
2. Berhe, A.A.; Barnes, R.T.; Six, J.; Marín-Spiotta, E. Role of soil erosion in biogeochemical cycling of essential elements: Carbon, nitrogen, and phosphorus. *Annu. Rev. Earth Planet. Sci.* **2018**, *46*, 521–548. [CrossRef]
3. Zhao, H.L.; Yi, X.Y.; Zhou, R.L.; Zhao, X.Y.; Zhang, T.H.; Drake, S. Wind erosion and sand accumulation effects on soil properties in Horqin Sandy Farmland, Inner Mongolia. *Catena* **2006**, *65*, 71–79. [CrossRef]
4. Quinton, J.N.; Govers, G.; Van Oost, K.; Bardgett, R.D. The impact of agricultural soil erosion on biogeochemical cycling. *Nat. Geosci.* **2010**, *3*, 311. [CrossRef]
5. Hancock, G.R.; Wells, T. Predicting soil organic carbon movement and concentration using a soil erosion and Landscape Evolution Model. *Geoderma* **2021**, *382*, 114759. [CrossRef]
6. Müller-Nedebock, D.; Chivenge, P.; Chaplot, V. Selective organic carbon losses from soils by sheet erosion and main controls Earth Surf Process. *Landforms* **2016**, *41*, 1399–1408. [CrossRef]
7. Pimentel, D.; Harvey, C.; Resosudarmo, P.; Sinclair, K.; Kurz, D.; Mcnair, M.; Crist, S.; Shpritz, L.; Fitton, L.; Saffouri, R. Environmental and Economic Costs of Soil Erosion and Conservation Benefits. *Science* **1995**, *267*, 1117–1123. [CrossRef]
8. Chappell, A.; Baldock, J.; Sanderman, J. The global significance of omitting soil erosion from soil organic carbon cycling schemes. *Nat. Clim. Change* **2016**, *6*, 187–191. [CrossRef]
9. Wang, Z.J.; Jiao, J.Y.; Rayburg, S.; Wang, Q.L.; Su, Y. Soil erosion resistance of "Grain for Green" vegetation types under extreme rainfall conditions on the Loess Plateau, China. *Catena* **2016**, *141*, 109–116. [CrossRef]
10. Yue, Y.; Ni, J.R.; Ciais, P.; Piao, S.L.; Wang, T.; Huang, M.T.; Li, T.H.; Wang, Y.C.; Chappell, A.; Van Oost, K. Lateral transport of soil carbon and land-atmosphere CO_2 flux induced by water erosion in China. *Proc. Natl. Acad. Sci. USA* **2016**, *113*, 6617–6622. [CrossRef]
11. Liu, B.; Xie, Y.; Li, Z.; Liang, Y.; Guo, Q. The assessment of soil loss by water erosion in China. *Int. Soil Water Conserv. Res.* **2020**, *8*, 430–439. [CrossRef]
12. Wan, W.; Liu, Z.; Li, B.G.; Fang, H.Y.; Wu, H.Q.; Yang, H.Y. Evaluating soil erosion by introducing crop residue cover and anthropogenic disturbance intensity into cropland C-factor calculation: Novel estimations from a cropland-dominant region of Northeast China. *Soil Tillage Res.* **2022**, *219*, 105434. [CrossRef]
13. Chen, L.; Wang, J.; Fu, B.; Qiu, Y. Land-use change in a small catchment of northern Loess Plateau, China. *Agric. Ecosyst. Environ.* **2001**, *86*, 163–172. [CrossRef]
14. Alem, B.B. The nexus between land use land cover dynamics and soil erosion hotspot area of Girana Watershed, Awash Basin, Ethiopia. *Heliyon* **2022**, *8*, e08916. [CrossRef]
15. Bagarello, V.; Di Stefano, V.; Ferro, V.; Giordano, G.; Iovino, M.; Pampalone, V. Estimating the USLE Soil Erodibility Factor in Sicily, South Italy. *Trop. Agric.* **2012**, *28*, 199–206. [CrossRef]
16. Klein, J.; Jarva, J.; Frank-Kamenetsky, D.; Bogatyrev, I. Integrated geological risk mapping: A qualitative methodology applied in St. Petersburg, Russia. *Environ. Earth Sci.* **2013**, *70*, 1629–1645. [CrossRef]
17. Adhikary, P.P.; Tiwari, S.P.; Mandal, D.; Lakaria, B.L.; Madhu, M. Geospatial comparison of four models to predict soil erodibility in a semi-arid region of Central India. *Environ. Earth Sci.* **2014**, *72*, 5049–5062. [CrossRef]
18. Garcia-Ruiz, J.M.; Begueria, S.; Nadal-Romero, E.; González-Hidalgo, J.C.; Lana-Renault, N.; Sanjuán, Y. A meta-analysis of soil erosion rates across the world. *Geomorphology* **2015**, *239*, 160–173. [CrossRef]
19. Zare, M.; Panagopoulos, T.; Loures, L. Simulating the impacts of future land use change on soil erosion in the Kasilian watershed. *Land Use Policy* **2017**, *67*, 558–572. [CrossRef]
20. Liu, X.; Burras, C.L.; Kravchenko, Y.S.; Duran, A.; Huffman, T.; Morras, H.; Yuan, X. Overview of Mollisols in the world: Distribution, land use and management. *Can. J. Soil Sci.* **2012**, *92*, 383–402. [CrossRef]

21. Liu, X.; Zhang, S.; Zhang, X.; Ding, G.; Cruse, R.M. Soil erosion control practices in Northeast China: A mini-review. *Soil Tillage Res.* **2012**, *117*, 44–48. [CrossRef]
22. Todd-Brown, K.E.O.; Randerson, J.T.; Hopkins, F.; Arora, V.; Hajima, T.; Jones, C.; Zhang, Q. Changes in soil organic carbon storage predicted by Earth system models during the 21st century. *Biogeosciences* **2014**, *11*, 2341–2356. [CrossRef]
23. Li, H.; Liao, X.; Zhu, H.; Wei, X.; Shao, M. Soil physical and hydraulic properties under different land uses in the black soil region of Northeast China. *Can. J. Soil Sci.* **2019**, *99*, 406–419. [CrossRef]
24. Duan, X.; Xie, Y.; Ou, T.; Lu, H. Effects of soil erosion on long-term soil productivity in the black soil region of northeastern China. *Catena* **2011**, *87*, 268–275. [CrossRef]
25. Yu, Z.; Wang, G.; Jin, J.; Liu, J.; Liu, X. Soil microbial communities are affected more by land use than seasonal variation in restored grassland and cultivated Mollisols in Northeast China. *Eur. J. Soil Biol.* **2011**, *47*, 357–363. [CrossRef]
26. Kumari, N.N.; Srivastava, A.; Dumka, U.C. A Long-Term Spatiotemporal Analysis of Vegetation Greenness over the Himalayan Region Using Google Earth Engine. *Climate* **2021**, *9*, 109. [CrossRef]
27. Matsushita, B.; Yang, W.; Chen, J.; Onda, Y.; Qiu, G. Sensitivity of the enhanced vegetation index (EVI) and normalized difference vegetation index (NDVI) to topographic effects: A case study in high-density cypress forest. *Sensors* **2007**, *7*, 2636–2651. [CrossRef]
28. Pei, F.S.; Zhou, Y.; Xia, Y. Application of Normalized Difference Vegetation Index (NDVI) for the Detection of Extreme Precipitation Change. *Forest* **2021**, *12*, 594. [CrossRef]
29. Xu, E.Q.; Zhang, H.Q. Change pathway and intersection of rainfall, soil, and land use influencing water-related soil erosion. *Ecol. Indic.* **2020**, *113*, 106281. [CrossRef]
30. Hu, T.; Wu, J.S.; Li, W.F. Assessing relationships of ecosystem services on multiscale: A case study of soil erosion control and water yield in the Pearl River Delta. *Ecol. Ind.* **2019**, *99*, 193–202. [CrossRef]
31. Zeng, C.; Wang, S.; Bai, X.; Li, Y.; Tian, Y.; Li, Y.; Wu, L.; Luo, G. Soil erosion evolution and spatial correlation analysis in a typical karst geomorphology using RUSLE with GIS. *Solid Earth* **2017**, *8*, 721–736. [CrossRef]
32. Sun, W.Y.; Shao, Q.Q.; Liu, J.Y.; Zhai, J. Assessing the effects of land use and topography on soil erosion on the Loess Plateau in China. *Catena* **2014**, *121*, 151–163. [CrossRef]
33. Renard, K.G.; Foster, G.R.; Weesies, G.A.; McCool, D.K.; Yoder, D.C. *Predicting Soil Erosion by Water: A Guide to Conservation Planning with the Revised Universal Soil Loss Equation (RUSLE)*; United States Department of Agriculture: Washington, DC, USA, 1997.
34. Tang, Q.; Xu, Y.; Bennett, S.J.; Li, Y. Assessment of soil erosion using RUSLE and GIS: A case study of the Yangou watershed in the Loess Plateau, China. *Environ. Earth Sci.* **2015**, *73*, 1715–1724. [CrossRef]
35. Li, P.F.; Zang, Y.Z.; Ma, D.D.; Yao, W.Q.; Holden, J.; Irvine, B.; Zhao, G.J. Soil erosion rates assessed by RUSLE and PESERA for a Chinese Loess Plateau catchment under land-cover changes. *Earth Surf. Proc. Land* **2020**, *45*, 707–722. [CrossRef]
36. Zhang, W.B.; Fu, J.S. Rainfall erosivity estimation under different rainfall amount. *Resour. Sci.* **2003**, *25*, 35–41.
37. Sharpley, A.N.; Williams, J.R. *EPIC–Erosion/Productivity Impact Calculator: 1. Model Documentation*; USDA Technical Bulletin No. 1768; USDA: Washington, DC, USA, 1990.
38. Zhang, K.L.; Shu, A.P.; Xu, X.L.; Yang, Q.K.; Yu, B. Soil erodibility and its estimation for agricultural soils in China. *J. Arid Environ.* **2008**, *72*, 1002–1011. [CrossRef]
39. Huang, X.F.; Lin, L.R.; Ding, S.W.; Tian, Z.C.; Zhu, X.Y.; Wu, K.R.; Zhao, Y.Z. Characteristics of Soil Erodibility K Value and Its Influencing Factors in the Changyan Watershed, Southwest Hubei, China. *Land* **2022**, *11*, 134. [CrossRef]
40. McCool, D.K.; Foster, G.R.; Mutchler, C.K.; Meyer, L.D. Revised slope length factor for the universal soil loss equation. *Trans. ASAE* **1989**, *32*, 1571–1576. [CrossRef]
41. Liu, B.Y.; Nearing, M.A.; Risse, L.M. Slope gradient effects on soil loss for steep slopes. *Trans. ASAE* **1994**, *37*, 1835–1840. [CrossRef]
42. Guerra, C.A.; Maes, J.; Geijzendorffer, I.; Metzger, M.J. An assessment of soil erosion prevention by vegetation in Mediterranean Europe: Current trends of ecosystem service provision. *Ecol. Ind.* **2016**, *60*, 213–222. [CrossRef]
43. Tanyas, H.; Kolat, C.; Süzen, M.L. A new approach to estimate cover-management factor of RUSLE and validation of RUSLE model in the watershed of Kartalkaya Dam. *J. Hydrol.* **2015**, *528*, 584–598. [CrossRef]
44. Cai, C.; Ding, S.; Shi, Z.; Huang, L.; Zhang, G. Study of applying USLE and geographical information system IDRISI to predict soil erosion in small watershed. *J. Soil Water Conserv.* **2000**, *14*, 19–24. (In Chinese)
45. Phinzi, K.; Ngetar, N.S. The assessment of water-borne erosion at catchment level using GIS-based RUSLE and remote sensing: A review. *Int. Soil Water Conserv. Res.* **2019**, *7*, 27–46. [CrossRef]
46. Brunsdon, C.; Fotheringham, S.; Charlton, M. Geographically weighted regression. *J. R. Stat. Soc. Ser. D* **1998**, *47*, 431–443. [CrossRef]
47. Huang, J.; Huang, Y.; Pontius, R.G.; Zhang, Z. Geographically weighted regression to measure spatial variations in correlations between water pollution versus land use in a coastal watershed. *Ocean Coast. Manag.* **2015**, *103*, 14–24. [CrossRef]
48. See, L.; Schepaschenko, D.; Lesiv, M.; McCallum, I.; Fritz, S.; Comber, A.; Perger, C.; Schill, C.; Zhao, Y.; Maus, V. Building a hybrid land cover map with crowdsourcing and geographically weighted regression. *ISPRS J. Photogramm. Remote Sens.* **2015**, *103*, 48–56. [CrossRef]
49. Wu, S.S.; Yang, H.; Guo, F.; Han, R.M. Spatial patterns and origins of heavy metals in Sheyang River catchment in Jiangsu, China based on geographically weighted regression. *Sci. Total Environ.* **2017**, *580*, 1518–1529. [CrossRef]

50. Gao, J.; Li, S. Detecting spatially non-stationary and scale-dependent relationships between urban landscape fragmentation and related factors using geographically weighted regression. *Appl. Geogr.* **2011**, *31*, 292–302. [CrossRef]
51. Song, W.; Jia, H.; Huang, J.; Zhang, Y. A satellite-based geographically weighted regression model for regional PM 2.5 estimation over the Pearl River Delta region in China. *Remote Sens. Environ.* **2014**, *154*, 1–7. [CrossRef]
52. Zhang, C.; Tang, Y.; Xu, X.; Kiely, G. Towards spatial geochemical modelling: Use of geographically weighted regression for mapping soil organic carbon contents in Ireland. *Appl. Geochem.* **2011**, *26*, 1239–1248. [CrossRef]
53. Zhao, J.; Wang, W.; Cheng, Q. Application of geographically weighted regression to identify spatially non-stationary relationships between Fe mineralization and its controlling factors in eastern Tianshan, China. *Ore Geol. Rev.* **2014**, *57*, 628–638. [CrossRef]
54. Shi, P.; Castaldi, F.; van Wesemael, B.; Van Oosta, K. Vis-NIR spectroscopic assessment of soil aggregate stability and aggregate size distribution in the Belgian Loam Belt. *Remote Sens.* **2020**, *12*, 666. [CrossRef]
55. Castaldi, F.; Hueni, A.; Chabrillat, S.; Ward, K.; Buttafuoco, G.; Bomans, B.; Vreys, K.; Brell, M.; van Wesemael, B. Evaluating the capability of the Sentinel 2 data for soil organic carbon prediction in croplands. *ISPRS J. Photogramm.* **2019**, *147*, 267–282. [CrossRef]
56. Nascimento, C.M.; Mendes, W.D.S.; Silvero, N.E.Q.; Poppiel, R.R.; Sayão, V.M.; Dotto, A.C.; dos Santos, N.V.; Accorsi, M.T.; Demattê, J.A.M. Soil degradation index developed by multitemporal remote sensing images, climate variables, terrain and soil atributes. *J. Environ. Manag.* **2021**, *277*, 111316. [CrossRef]
57. Pimentel, D. Soil Erosion: A Food and Environmental Threat. *Environ. Dev. Sustain.* **2006**, *8*, 119–137. [CrossRef]
58. Sadeghi, S.H.R.; Jalili, K.; Nikkami, D. Land use optimization in watershed scale. *Land Use Policy* **2009**, *26*, 186–193. [CrossRef]
59. Chaplot, V.; Giboire, G.; Marchand, P.; Valentin, C. Dynamic modelling for linear erosion initiation and development under climate and land-use changes in northern Laos. *Catena* **2005**, *63*, 318–328. [CrossRef]
60. Simonneaux, V.; Cheggour, A.; Deschamps, C.; Mouillot, F.; Cerdan, O.; Le Bissonnais, Y. Land use and climate change effects on soil erosion in a semi-arid mountainous watershed (High Atlas, Morocco). *J. Arid Environ.* **2015**, *122*, 64–75. [CrossRef]
61. Fohrer, N.; Haverkamp, S.; Eckhardt, K.; Frede, H.G. Hydrologic Response to land use changes on the catchment scale. *Phys. Chem. Earth B—Hydrol. Oceans Atmos.* **2001**, *26*, 577–582. [CrossRef]
62. Wang, H.; Yang, S.L.; Wang, Y.D.; Gu, Z.Y.; Xiong, S.F.; Huang, X.F.; Sun, M.M.; Zhang, S.H.; Guo, L.C.; Cui, J.Y.; et al. Rates and causes of black soil erosion in Northeast China. *Catena* **2022**, *214*, 106250. [CrossRef]
63. Zhang, Y.N.; Long, H.L.; Tu, S.S.; Ge, D.Z.; Ma, L.; Wang, L.Z. Spatial identification of land use functions and their trade-offs/synergies in China: Implications for sustainable land management. *Ecol. Indic.* **2019**, *107*, 105550. [CrossRef]

land

Article

The Changes of Desertification and Its Driving Factors in the Gonghe Basin of North China over the Past 10 Years

Hong Jia [1,2,3], Rui Wang [1,2,3], Hang Li [1,2,3], Baijian Diao [1,2,3], Hao Zheng [1,2,3], Lanlan Guo [1,2,3], Lianyou Liu [1,2,3] and Jifu Liu [1,2,3,*]

1 Key Laboratory of Environmental Change and Natural Disasters, Ministry of Education, Beijing Normal University, Beijing 100875, China; jiahong@mail.bnu.edu.cn (H.J.); wr21@mail.bnu.edu.cn (R.W.); 201911051106@mail.bnu.edu.cn (H.L.); 202121051161@mail.bnu.edu.cn (B.D.); zhenghao@mail.bnu.edu.cn (H.Z.); guolanlan@bnu.edu.cn (L.G.); lyliu@bnu.edu.cn (L.L.)
2 Engineering Research Center of Desertification and Blown-Sand Control, Ministry of Education, Beijing Normal University, Beijing 100875, China
3 Faculty of Geographical Science, Academy of Disaster Reduction and Emergency Management, Beijing Normal University, Beijing 100875, China
* Correspondence: liujifu@bnu.edu.cn

Abstract: Desertification is one of the most severe environmental and socioeconomic issues facing the world today. Gonghe Basin is located in the monsoon marginal zone of China, is a sensitive area of climate change in the northeastern of the Qinghai-Tibet Plateau in China, desertification issue has become very severe. Remote sensing monitoring provides an effective technical means for desertification control. In this study, we used Landsat images in 2010 and 2020 to extract desertification information to constructed the Albedo-NDVI feature space in the Gonghe Basin. And then analyzed temporal and spatial evolution of desertification and its driving factors using Geodetector in the Gonghe Basin from 2010 to 2020. The main conclusions are as follows: (1) Albedo-NDVI feature space method can accurately classify desertification information with accuracy of more than 90%, which was benefit to quantitative analysis of desertification. (2) The desertification situation in the Gonghe Basin had improved from 2010 to 2020, especially in the west of the basin, desertification land area decreased by 827.46 km^2, and desertification intensity had been obviously reversed. (3) The changes of the desertification in the Gonghe Basin from 2010 to 2020 was affected by both natural and human factors, and the influence of human activities on desertification reversal had increased gradually. The results indicate that the desertification status in the Gonghe Basin had been effectively controlled, and can provide useful basis for the desertification combat in the Gonghe Basin.

Keywords: desertification; Albedo-NDVI; feature space; climate change; human activity; Gonghe Basin

Citation: Jia, H.; Wang, R.; Li, H.; Diao, B.; Zheng, H.; Guo, L.; Liu, L.; Liu, J. The Changes of Desertification and Its Driving Factors in the Gonghe Basin of North China over the Past 10 Years. *Land* **2023**, *12*, 998. https://doi.org/10.3390/land12050998

Academic Editors: Li Ma, Yingnan Zhang, Muye Gan and Zhengying Shan

Received: 11 April 2023
Revised: 25 April 2023
Accepted: 29 April 2023
Published: 1 May 2023

1. Introduction

Desertification is defined as land degradation mainly characterized by aeolian activity in arid and semi-arid areas, due to the inharmonious man-land relationship [1]. Desertification has become one of the most severe ecological environmental issues, which causes economic losses of up to 540 billion RMB annually in the world [2–4]. The Gonghe Basin is one of the centralized distribution regions of desertification in the northeast of the Qinghai-Tibet Plateau [5], in which the ecological environment is fragile. Desertification has become increasingly prominent in the Gonghe Basin due to global climate change and unreasonable human activities, which not only affect the lives of local people but also pose a huge threat to the safety of the Longyangxia Reservoir, has hindered socioeconomic development [6,7]. Thus, it is urgent to strengthen research on desertification in this area.

Gonghe Basin is affected by the Asian monsoon circulation and the mid-latitude westerly circulation, and it is a part of the boundary between the deserts and loess in China [8]. Its unique geographical location provides an ideal research site for exploring the formation and changes of the aeolian activity environment. So far, multiple time scales research had been conducted on the formation, evolution, and driving mechanism of the aeolian activity in the Gonghe Basin [6,9,10]. The studies of desertification of the Gonghe Basin during its geological history mainly focused on geomorphological evolution, sedimentary strata, the history of aeolian activity, and the driving mechanisms [11,12]. Previous studies have shown that the oldest formation age of aeolian sand is 33.5 $\pm$ 2.1 ka BP in the Gonghe Basin, dune fields developed in the early and middle Holocene, and fixed in the late Holocene [10]. The formation of the aeolian sand environment in the Gonghe Basin is influenced by multiple factors, such as the evolution process of the geomorphology, regional climate, and wind strength, the main control factors were different in different periods [13,14].

The research about the modern process of desertification in the Gonghe Basin mainly includes different aspects such as aeolian landforms [15], wind conditions [16], spatial distribution and dynamic monitoring of desertification, driving mechanism of desertification and countermeasures for land desertification prevention and control [6,17–19]. Remote sensing images provide an effective data source for desertification monitoring and information extraction [20,21]. Collado et al. assessed the desertification process in the crop-rangeland boundary of Argentina by using remote sensing data [22]. Qi et al. analyzed the spatiotemporal changes of desertification from1986 to 2003 through supervised classification method in the agropastoral transitional zone of northern Shaanxi Province in China [23]. In the Gonghe basin, Yan et al. used Landsat data from 1975 to 2005 through visual interpretation to explore aeolian desertification trends and driving factors in the Longyangxia Reservoir next to the Yellow River [24]. Ma et al. used TM data from the three periods of 1990, 2000, and 2010 in the Gonghe Basin, it was concluded that the desertification situation was worsening from 1990 to 2010 [25]. However, previous monitoring methods were mainly based on visual interpretation and supervised classification, resulting in low utilization rate and classification accuracy of remote sensing information [24,26,27]. In recent years, many studies have used single indicator (e.g., NDVI, EVI, MSAVI) to assess desertification [28,29]. Nonetheless, due to the complex causes of desertification evolution, using a single index cannot comprehensively reflect desertification information [30]. The Albedo-NDVI feature space basing on the negative correlation between Albedo and NDVI established by Zeng et al. provides an efficient approach for quantitative analysis of desertification [31], which has been used in many desertified land, such as Moulouya basin in Morocco [32], central Mexico [33], Mongolian plateau [34], source region of Yellow River in China [35]. For the Gonghe Basin, there is lack of study on the desertification evolution process over the past 10 years, and short time scale desertification research is of practical significance in assessing the effectiveness of desertification prevention. There is relatively little research on desertification monitoring in the Gonghe Basin compared to other typical desertification regions, and only part of Gonghe Basin was studied, such as around the Longyangxia Reservoir, Gonghe county, and Guinan county [24,36–38]. Additionally, the research on the driving mechanism of desertification evolution in the Gonghe Basin mainly focuses on qualitative research [24,25,38], while quantitative research can better reveal the potential impact factors of desertification process.

In this study, we obtained two desertification monitoring indicators, Albedo and NDVI, to constructed Albedo-NDVI feature space basing on Landsat images. Additionally, we quantitatively explored the spatiotemporal evolution patterns of desertification and the underlying causes of desertification from 2010 to 2020 using Geodetector model, which provided a theoretical basis for desertification combat.

2. Materials and Methods

2.1. Study Area

The Gonghe Basin is surrounded by mountains on three sides, including the Qilian Mountains, Kunlun Mountains and Qinling Mountains (Figure 1), geographic coordinates are 35°27′–36°56′ N, 98°46′–101°22′ E, with an elevation of 2400–3200 m. It is administratively subordinate to Qinghai Province, including Gonghe county, Guinan county, Xinghai county and Wulan county. The dune fields in the Gonghe Basin are mainly spread in the central and eastern parts of the basin, such as Talatan Plain and Mugetan Plain, with moving dunes, sand ridges, and sand belts [9].

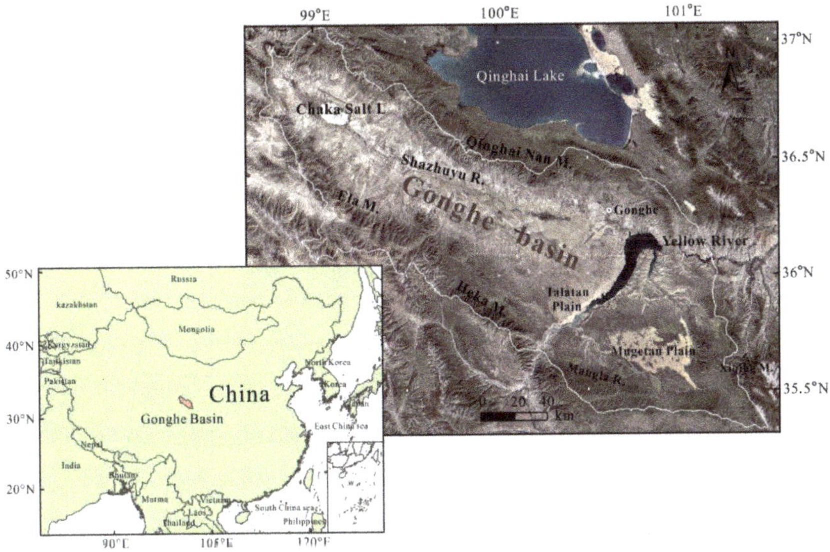

Figure 1. Geographical location of the Gonghe Basin.

The climate type in the Gonghe Basin is a typical alpine and semi-arid climate, the annual average temperature is about 3.7 °C and annual precipitation is about 300 mm. 80% of the precipitation in the Gonghe Basin is mainly concentrated from May to September, accompanying high evaporation. Strong winds prevail in the Gonghe Basin, and the maximum wind velocity reaches 40 m s^{-1} in spring [39]. Aeolian activities are common in this area, resulting in wide dune fields and severe land desertification [10].

2.2. Data Sources

We used the Landsat TM/OLI images to monitor land desertification in the Gonghe Basin in our study, and the images were obtained from the geospatial data cloud platform (http://www.gscloud.cn/ (accessed on 10 February 2023)). A total of 8 images for 2010 (2009–2011) and 2020 were collected during the vegetation growing season (especially in June and August) with cloud coverage of less than 10%. These images were preprocessed mainly using ENVI5.3 software, including radiometric calibration and atmospheric correction. The vector boundary data of the Gonghe Basin was used to clip and mosaic the Landsat images to obtain the entire Landsat image of the Gonghe Basin.

The annual average temperature, annual precipitation and annual average wind velocity data from 2010 to 2019 were calculated basing the ERA5 data set on the Google Earth Engine (GEE) platform. Annual interpolation data for meteorological data and relevant regional socio-economic data included datasets of land use (1:100,000), population density (1 km) and GDP density (1 km) were downloaded from the Resource and Environment Science and Data Center (RESDC, https://www.resdc.cn/ (accessed on 5 March 2023)).

2.3. Methods

2.3.1. Normalized Difference Vegetation Index (NDVI)

NDVI is an important biophysical parameter that reflects the state of surface vegetation, with a range of −1 to 1, and the higher the vegetation coverage, the closer NDVI value is to 1. NDVI can be used to indicate vegetation growth status and reflect vegetation coverage, and we can calculate it using the reflectance of the following two bands in remote sensing images [40].

$$NDVI = (\rho_{nir} - \rho_{red})/(\rho_{nir} + \rho_{red}) \quad (1)$$

where ρ_{nir}, ρ_{red} represent near infrared band and the red band, respectively.

2.3.2. Land Surface Albedo

Land Surface albedo is a physical parameter that reflects the reflection characteristics of the surface to solar shortwave radiation. With the aggravation of desertification, surface vegetation is severely damaged, and surface roughness increases, manifested as an increase in Albedo values in remote sensing images. The value of Albedo is between 0 and 1. In this study, we calculated Albedo using the calculation method proposed by Liang [41].

$$Albedo = 0.356 \times \rho_{blue} + 0.130 \times \rho_{red} + 0.373 \times \rho_{nir} + 0.085 \times \rho_{swir1} + 0.072 \times \rho_{swir2} - 0.0018 \quad (2)$$

where ρ_{blue}, ρ_{red}, ρ_{nir}, ρ_{swir1} and ρ_{swir2} represent blue band, red band, near infrared band and the shortwave infrared bands, respectively.

2.3.3. Data Normalization

The dimensions of NDVI and Albedo are different, so that Albedo-NDVI feature space cannot be directly established, we normalized the values of NDVI and Albedo to between 0 and 1. The NDVI and the Albedo values were normalized using following equations.

$$N = (NDVI - NDVI_{min})/(NDVI_{max} - NDVI_{min}) \quad (3)$$

$$A = (Albedo - Albedo_{min})/(Albedo_{max} - Albedo_{min}) \quad (4)$$

For NDVI, $NDVI_{max}$, $NDVI_{min}$ refer to maximum and minimum values, respectively, N was the normalized value; For Albedo, $Albedo_{max}$ and $Albedo_{min}$ refer to maximum and minimum values, respectively, A was the normalized value.

2.3.4. Albedo–NDVI Feature Space

Zeng et al. [31] conducted research on the feature space composed of NDVI and Albedo, and summarized the desertification situation under different vegetation coverage conditions in an ideal feature space (Figure 2). A, B, C, and D points represent the extreme states in the Albedo-NDVI feature space, respectively. A represents areas with severe drought and no vegetation cover, B represents areas with high water content and no vegetation cover, C represents areas with low water content and high vegetation cover, and D represents areas with high water content and high vegetation cover. The upper boundary AD represents a high albedo line, reflecting drought conditions, the bottom BC is the low albedo line, representing the condition of sufficient surface water. And the distribution of different land cover types presented by NDVI and Albedo has a significant distribution rule in the feature space, which can well distinguish water, high vegetation coverage land, low vegetation coverage land and completely bare land [42]. Overall, there is a significant negative correlation between Albedo and NDVI in the feature space.

Based on the ROI function of ENVI5.3 version, 900 sample points were randomly selected from different degrees of desertification land in the study area [30], extracting the NDVI and Albedo values after normalization in 2010 and 2020, respectively. And then selecting the NDVI values as independent variables, Albedo values as dependent variables, we can construct a linear regression equation between them.

Then the linear regression equation represents negative correlation between Albedo and NDVI was calculated using the following formula:

$$Albedo = k \times NDVI + b \quad (5)$$

where k refers to the slope of the linear expression, and b refers to the parameter.

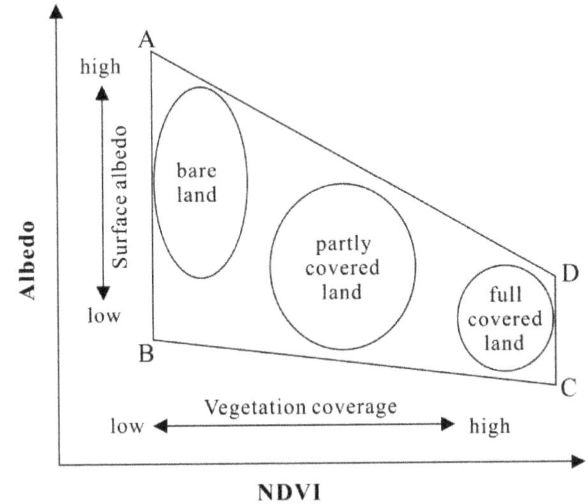

Figure 2. Albedo-NDVI feature space [31].

2.3.5. Desertification Difference Index (DDI)

Based on previous research findings [43], dividing the Albedo-NDVI feature space in the vertical direction representing the trend of desertification change can effectively distinguish different types of desertification land, represented by the Desertification Difference Index (DDI). we can use the following two formulas to calculate the DDI index for 2010 and 2020.

$$\mathrm{k} \times \mathrm{a} = -1 \tag{6}$$

$$\mathrm{DDI} = \mathrm{a} \times \mathrm{NDVI} - \mathrm{Albedo} \tag{7}$$

where a represent the slope of DDI linear expression.

2.3.6. Accuracy Verification

Confusion matrix is also called error matrix, is an effective method for evaluating the accuracy of classification results. In the confusion matrix, each row represents the real category of desertification degree, and each column represents the prediction category [44]. We obtained evaluation indicators, including the overall accuracy (OA), producer's accuracy (PA), user's accuracy (UA), and Kappa coefficient, which can be used to verify the accuracy of the desertification classification results using Albedo-NDVI feature space method. The specific calculation formulas are as follows:

$$OA = \sum_{i=1}^{i=5} X_{ii}/N \tag{8}$$

$$PA = X_{ii}/X_{i+} \tag{9}$$

$$UA = X_{ii}/X_{+i} \tag{10}$$

$$Kappa = \frac{\left[N \times \sum_{i=1}^{i=5} X_{ii} - \left(\sum_{i=1}^{i=5} X_{+i} + X_{+i}\right)\right]}{N^2 - \sum_{i=1}^{i=5} X_{+i} + X_{+i}} \tag{11}$$

where N refers to the total number of samples; X_{ii} refers to the sample quantity in row i and column i, that is, the number of sample points correctly identified for a certain type of desertification degree; X_{i+} refers to the sample quantity in row i, is the total real sample size of a certain type of desertification; and X_{+i} refers to the sample quantity in column i, is the predicted total sample size of a certain type of desertification.

2.3.7. Desertification Land Transfer Matrix

The land transfer matrix has been widely applied in land use change. The land transfer matrix can reflect the transformation area from one degree of desertification land to another degree of desertification land within a certain period of time, and can reflect the transformation relationship between different degrees of desertification land [45]. In this study, we used the land transfer matrix to calculate the conversion areas between different grades of desertification land basing ArcGIS 10.8. The formula used is as follows:

$$S_{ij} = \begin{bmatrix} S_{11} & S_{12} & \cdots & S_{1n} \\ S_{12} & S_{22} & \cdots & S_{2n} \\ \cdots & \cdots & \cdots & \cdots \\ S_{n1} & S_{n2} & \cdots & S_{nn} \end{bmatrix} \tag{12}$$

where i and j represent different grades of land desertification, S_{ij} represents the transition area from the grade i to j (km^2), and n represents the number of desertification grades.

2.3.8. Dynamic Degree of Desertification Land

The dynamic degree indicates the area change of one grade of desertification land within a certain specific time range in a certain research area [46]. The formula is as follows:

$$K = \frac{u_2 - u_1}{u_1} \times \frac{1}{t_2 - t_1} \times 100\% \tag{13}$$

where K represents the dynamic degree from 2010 to 2020, u_1 refers to the initial area (km^2), u_2 represents the final area (km^2), t_1 and t_2 represent the starting and end time, respectively.

2.3.9. Changes in Desertification Degree

The degree of desertification development was divided into five categories [30]: severe deterioration (desertification degree increased by more than one level), deterioration (desertification degree increased by one level), no change (desertification degree remained stable), restoration (desertification degree decreased by one level), obvious restoration (desertification degree decreased by more than one level).

2.3.10. Geodetector Model

The Geodetector is a widely used statistical model that reveal spatial variability and potential driving forces. Geodetector can detect the spatial heterogeneity of a single factor, and can also reveal possible causal relationships between two factors through calculating their consistency of spatial distribution [47]. This study takes the degree of desertification as the dependent variable Y, and selects independent variable indicators including temperature, precipitation, wind velocity, population density, GDP, and land use. The factor interpretation power in Geodetector is represented by the q value. The expressions used are as follows:

$$q = 1 - \frac{\sum_{h=1}^{L} N_h {\sigma_h}^2}{N\sigma^2} = 1 - \frac{SSW}{SST} \tag{14}$$

$$SSW = \sum_{h=1}^{L} N_h {\sigma_h}^2,\ SST = N\sigma^2 \tag{15}$$

where h refers to the stratification of the independent variable; N_h and N represent the number of units within layer h and the entire area, respectively; ${\sigma_h}^2$ and σ^2 represent the discrete variances of layer h and the entire area, respectively; SSW refers to the sum of intralayer variances; SST refers to the regional total discrete variance; q refers to the explanatory power of the independent variable to the degree of desertification, the range of q is between 0 and 1, the larger the q value represents the stronger the explanatory power of the selected factor.

The purpose of interaction detection is to assess whether interaction between two factors can increase the explanatory power of the degree of desertification or whether the impact of these factors on the degree of desertification is independent [48].

3. Results

3.1. Desertification Classification

The scatterplots of Albedo and NDVI in 2010 and 2020 are shown in Figure 3, there was presented a trapezoidal shape in the Albedo-NDVI feature space [46]. The R^2 values of the linear regression equations were 0.7106 and 0.7044, respectively, the results indicated there was a significant negative correlation between Albedo and NDVI. Using Equation (6) to calculate the k value to obtain the final expression of the desertification difference index (DDI) in 2010 and 2020, as shown in Table 1.

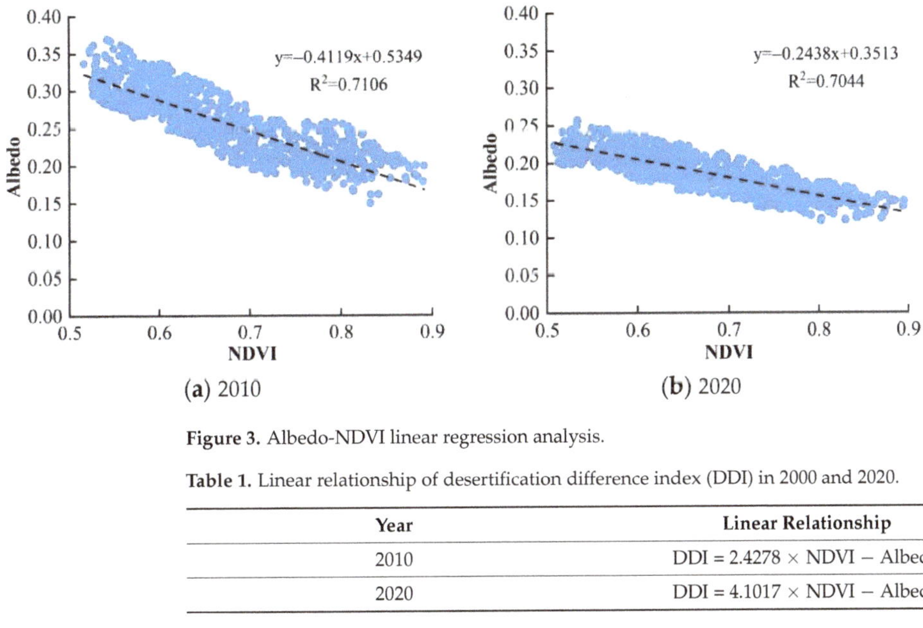

Figure 3. Albedo-NDVI linear regression analysis.

Table 1. Linear relationship of desertification difference index (DDI) in 2000 and 2020.

Year	Linear Relationship
2010	DDI = 2.4278 × NDVI − Albedo
2020	DDI = 4.1017 × NDVI − Albedo

DDI can be used to obtain the desertification classification, we used the natural breaks (Jenks) method combined with field survey data and Google Earth map to classify the desertification intensity into 5 categories, including extremely severe desertification, severe desertification, moderate desertification, slight desertification, and non-desertification. Finally, the spatial distribution maps of desertification degree in 2010 and 2020 were made by using ArcGIS 10.8 (Figure 4).

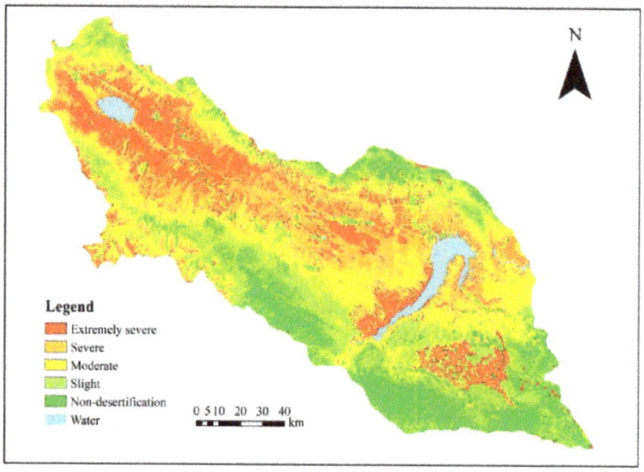

(**a**) 2010

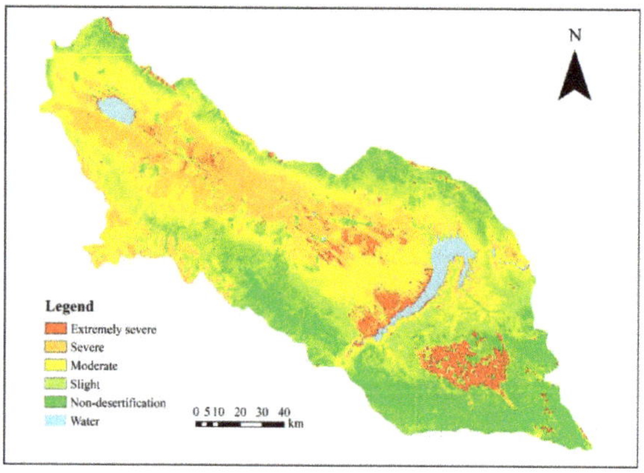

(**b**) 2020

Figure 4. Spatial distribution of desertification in the Gonghe Basin.

To test the accuracy of desertification land classification results, we used Landsat true color image and Google Earth map as reference data, randomly selected 20 sample points in different desertification types (100 points in total) to construct the confusion matrix through visual interpretation. The accuracy evaluation results are shown in Table 2, the overall evaluation accuracy was 94%, and Kappa coefficient was 0.93 in 2010. In 2020, the overall accuracy was 95%, Kappa coefficient was 0.94. The phenomenon of misclassification mainly occurred in slight desertification areas. Overall, the Albedo-NDVI feature space method has certain feasibility and applicability to evaluate desertification level.

Table 2. Classification accuracy of desertification.

Year	Extremely Severe (%)		Severe (%)		Moderate (%)		Slight (%)		Non-Desertification (%)		Kappa Coefficient	OA (%)
	PA	UA	PA	UA	PA	UA	PA	UA	PA	UA		
2010	95	95	94.74	90	94.74	90	86.36	95	100	100	0.93	94
2020	95.24	100	95	95	94.74	90	90.48	95	100	95	0.94	95

3.2. Temporal Distribution Characteristics of Desertification

Table 3 shows the area, proportion, and dynamic changes of different desertification levels in the study area. As shown in Table 3, extremely severe desertification land areas had decreased by 2335.32 km^2 from 2010 to 2020, with the proportion decreasing from 17.4% to 5.6% and the dynamic degree of 6.8%. Severe desertification, moderate desertification, and slight desertification have a slight increasing trend, with the area increasing 592.13 km^2, 687.33 km^2 and 227.40 km^2, respectively, and their proportions had increased 3.0%, 3.5% and 1.1%, respectively. Meanwhile, the non-desertification land areas accounted for 19.9% in 2020, and the area increased by 827.46 km^2. The dynamic degrees of the severe, moderate, slight and non-desertification land area were 1.3%, 1.5%, 0.5% and 2.7%, respectively.

Table 3. Dynamic changes of desertification land area in the Gonghe Basin from 2010 to 2020.

Category	2010		2020		2010–2010
	Area (km^2)	%	Area (km^2)	%	Annual Rate of Change (%)
Extremely severe	3434.27	17.4	1098.95	5.6	−6.8
Severe	4436.59	22.5	5028.72	25.5	1.3
Moderate	4595.28	23.3	5282.61	26.8	1.5
Slight	4153.31	21.1	4381.71	22.2	0.5
Non-desertification	3086.06	15.7	3913.52	19.9	2.7

In the past 10 years, there existed a certain upward trend in the non-desertification area. The moderate desertification had always accounted for a large proportion of the total study area, which was the main type of desertification in the Gonghe Basin. It can be concluded that the desertification status in the Gonghe Basin had generally improved from 2010 to 2020, and the degree of desertification had been mainly reversed from extremely severe to other degrees of desertification.

3.3. Spatial Distribution Characteristics of Desertification

According to the spatial distribution maps of desertification in the Gonghe Basin (Figure 4), we analyzed the dynamic changes in the spatial distribution pattern of desertification in the Gonghe Basin from 2010 to 2020. As shown in Figure 4, desertification was widespread in the Gonghe Basin, the lowlands in the central part of the basin were a concentrated distribution area of desertification, non-desertification areas were mainly spread in the south and southeast areas or on the mountains around the basin.

As shown in Figure 4a, there were large areas of extremely severe desertification around the Shazhuyu River, Mugetan, Talatan and other areas around the Longyangxia Reservoir in 2010. And in the periphery of extremely severe desertification, there were large areas of severe desertification, such as in the center of the basin or around the Longyangxia Reservoir. Moderate desertification was mainly spread in the east of Longyangxia Reservoir, such as Shagou Town, Longyangxia Town, etc. Slight desertification land was spread mainly in the southern and southeastern parts of the study area, such as the northern part of Heka Town and the northwest part of the mobile dunes in Mugetan. Non-desertification was distributed mainly in the Heka Town, Taxiu town, and Senduo town. Which were in the southern edge and southeast of the basin.

Compared with 2010, the overall desertification area had obviously reduced in 2020 (Figure 4b). The extremely severe desertification land spread in the western of the basin had been reduced significantly, mainly reversed to severe or moderate desertification. Meanwhile, the slight desertification and non-desertification land in the Gonghe Basin expanded to the south and southeast, and the non-desertification areas distributed around the northern marginal region increased during the study period.

3.4. Changes in Desertification Intensity

We obtained the transition matrix of desertification in this study from 2010 to 2020, as shown in Table 4. During the research period, the transformation of desertification intensity occurred between different levels of desertification. The conversion area accounted for 83.69% of the total land area, which was 8959.19 km^2. The main desertification transformations were mainly from extremely severe to severe, from severe to moderate, from moderate to slight, and from slight to non-desertification, covering land areas of 2213.94 km^2, 1736.12 km^2, 1418.13 km^2, and 1256.56 km^2, respectively, accounting for 24.71%, 19.38%, 15.83% and 14.03% of the land area. Extremely severe desertification significantly decreased by 2335.32 km^2, and severe, moderate, slight and non-desertification increased by 592.13 km^2, 687.33 km^2, 228.4 km^2, and 827.46 km^2, respectively. This showed that the overall desertification condition in the Gonghe Basin had a great improvement from 2010 to 2020, and sand prevention and control achieved effective results.

Table 4. The transition matrix of desertification in 2010–2020.

2020 / 2010	Extremely Severe	Severe	Moderate	Slight	Non-Desertification	Total (Reduced)
Extremely severe	827.92	2213.94	335.75	27.56	29.11	3434.27
Severe	185.30	2402.03	1736.12	85.14	28.00	4436.59
Moderate	35.58	355.75	2675.32	1418.13	110.51	4595.28
Slight	25.53	35.19	484.33	2351.70	1256.56	4153.31
Non-desertification	24.62	21.82	51.09	499.18	2489.35	3086.06
Total (increased)	1098.95	5028.72	5282.61	4381.71	3913.52	19,705.51

As shown in the changes in desertification intensity from 2010 to 2020 (Figure 5), the degree of desertification development was divided into five categories. The desertification grade unchanged land was sporadically scattered, mainly in the mobile sand dunes of Mugetan and Talatan, accounting for 42.6% of the land in basin (Table 5). The deterioration areas were primarily spread in the southwest, southeast, and northeast of the Gonghe Basin. The restoration and obvious restoration areas were mainly distributed around Chaka Salt Lake and the east of Longyangxia Reservoir. Additionally, the proportion of desertification deterioration and restoration areas were 6.8% and 50.6%, respectively. Desertification restoration areas were 11,067 km^2 larger than desertification deterioration areas.

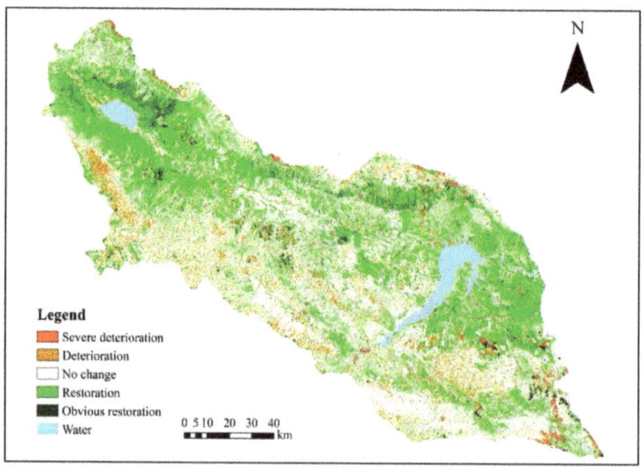

Figure 5. Changes in desertification intensity from 2010 to 2020.

Table 5. Area and proportion of changes in desertification intensity.

2010–2020	Severe Deterioration	Deterioration	No Change	Restoration	Obvious Restoration
Area (km^2)	193.83	1524.56	10,746.31	6624.75	6160.63
Percentage (%)	0.8	6.0	42.6	26.2	24.4

3.5. The Influencing Factors of Desertification

In this research, we selected temperature, precipitation and wind velocity as natural factor indicators, population density, GDP, and land use as human factor indicators, used Geodetector for single factor and interactive factors analysis to explore the explanatory power of different factors on desertification in the Gonghe Basin.

3.5.1. Singer Factor

As shown in Figure 6, for single factor analysis, the order of explanatory power of different factors on desertification in 2010 was precipitation > land use > GDP > population density > wind velocity > temperature. The precipitation was the main interfering factor of desertification, followed by human activities such as land use, GDP, and the impacts of the population density, wind velocity and temperature were relatively weak. The explanatory power of the q value on precipitation reached 0.29, but the explanatory power of temperature was only 0.03. In 2020, the explanatory power of q values on different factors was precipitation > land use > temperature > wind velocity > population density > GDP. The precipitation factor still had the greatest explanatory power on desertification, with a value of 0.22. The explanatory power of temperature and land use increased relatively, with values of 0.18 and 0.21, respectively. The q value of GDP had decreased to 0.02.

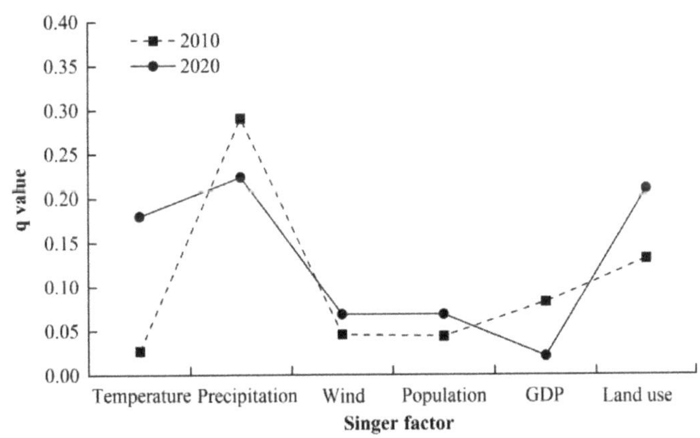

Figure 6. Comparisons of q value for different factors (2010 and 2020).

3.5.2. Interactive Factors

In this study, the influence between two factors was non-linear enhanced after interaction (Figures 7 and 8). In 2010, the order of explanatory power of interaction factors on desertification was precipitation ∩ land use > precipitation ∩ wind velocity > precipitation ∩ population intensity > temperature ∩ precipitation > precipitation ∩ GDP intensity > GDP intensity ∩ land use > population density ∩ GDP intensity. The dominant interactive factor was precipitation ∩ land use (0.392), followed by precipitation ∩ wind velocity (0.365) and precipitation ∩ population intensity (0.345). The q value of temperature ∩ population density was the smallest, which is 0.079. In 2020, the order of explanatory power of interaction factors on desertification was precipitation ∩ land use > temperature

∩ land use > temperature ∩ precipitation > precipitation ∩ wind velocity > temperature ∩ population intensity > precipitation ∩ population intensity > precipitation ∩ GDP intensity. Among them, the precipitation ∩ land use also had the greatest explanatory power on desertification evolution, the q value increased to 0.447, followed by temperature ∩ land use and temperature ∩ precipitation, their q values were 0.351 and 0.340, respectively. All in all, the precipitation ∩ land use was the essential factor influencing the spatiotemporal distribution of desertification in the Gonghe Basin during the study period.

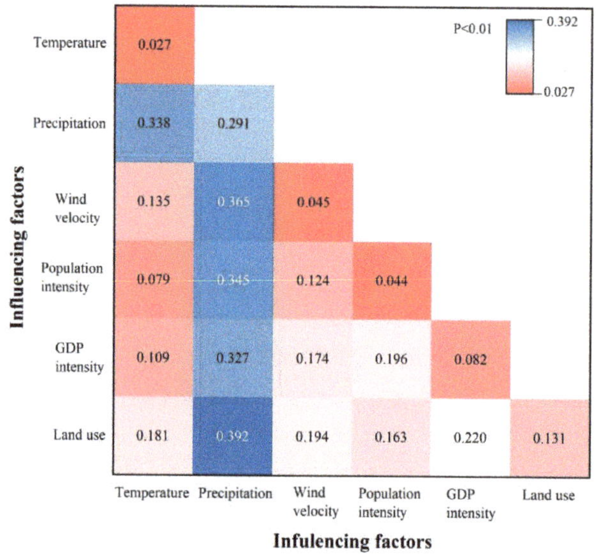

Figure 7. The q values of interactive factors in 2010 (P < 0.01).

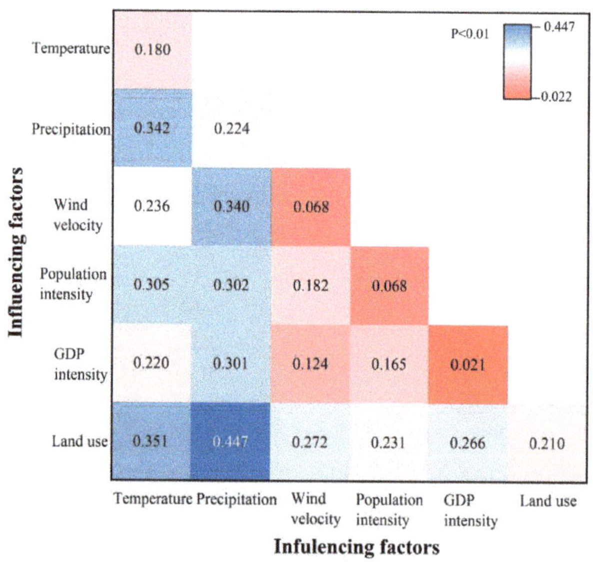

Figure 8. The q values of interactive factors in 2020 (P < 0.01).

4. Discussion

Previous studies have shown that desertification evolution is influenced by natural factors and human activities [26,27,49]. For natural factors, the terrain of the Gonghe Basin is flat and open, providing a good deposition site for the aeolian activity. The basin contained a large amount of Quaternary loose sediments, such as fluvial-lacustrine sediments and ancient aeolian sand [9], which are easily eroded by wind [26], which provided material sources for aeolian activity, causing widespread distribution of desertification land [12]. Among all natural factors, climate change has an essential impact on the development of desertification, temperature, precipitation and wind velocity are the main influencing factors [50,51]. The explanatory power of precipitation factors on desertification was highest among all factors in 2010 and 2020, the explanatory power of temperature and wind velocity on desertification evolution increased between 2010 and 2020, with q values increasing by 0.15 and 0.02, respectively. In northwestern China, the climate is arid all year with scarce precipitation [15], so the precipitation factor played a vital role in the desertification evolution. As shown in Figures 9 and 10, in the Gonghe Basin, we can see a fluctuating downward trend in the annual average temperature from 2010 to 2019, but there was an upward trend in the annual precipitation, which revealing that the climate had become colder and more humid over the past 10 years. The decrease in temperature could effectively reduce evaporation, while accompanied by the increase of precipitation, the improvement of hydrothermal conditions was conducive to the vegetation recovery, affecting the efficiency of sand material acquisition, which reduced aeolian activity [12,52–55]. The annual mean wind velocity also presented a relative downward trend (Figure 11), and reduced wind strength led to the weakening of aeolian activity [56]. In general, these favorable natural factors changes were beneficial to the desertification reversal. Among the changes in human factors, the q values of land use and population density increased by 0.08 and 0.03 respectively, while the value of GDP density decreased by 0.06. From this, it can be seen that the impact of human activities gradually increased during the desertification evolution of the Gonghe Basin from 2010 to 2020.

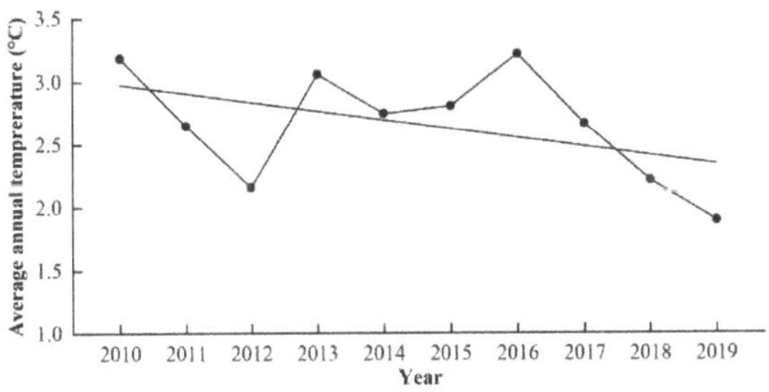

Figure 9. Average annual temperature in from 2010 to 2019.

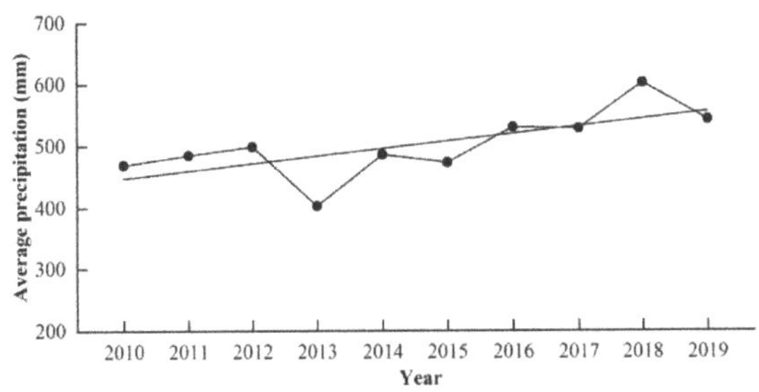

Figure 10. Annual precipitation from 2010 to 2019.

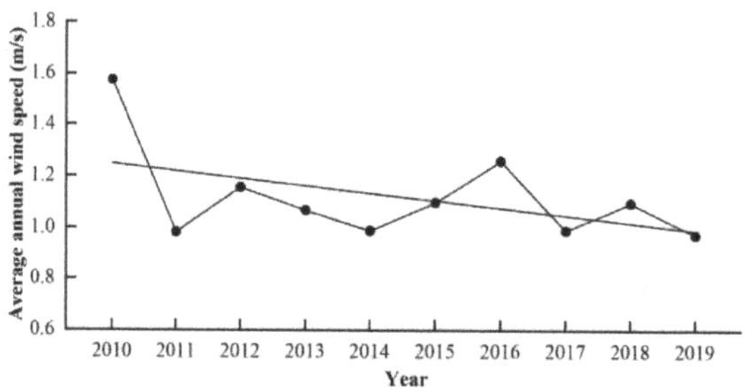

Figure 11. Average annual wind velocity from 2010 to 2019.

However, the interaction between two different impact factors will increase the explanatory power on desertification compared to single factors [57]. The dominant interactive factor in 2010 and 2020 was precipitation ∩ land use, with an increase of 0.055. In 2020, The explanatory power of temperature ∩ land use on desertification had significantly increased, with an increase of 0.17 compared to 2010. In 2010 and 2020, the q values of temperature ∩ precipitation were relatively high, both greater than 0.3. The results showed that the natural factors such as precipitation ∩ temperature played a fundamental role in the desertification change. Furthermore, the improvement of desertification conditions was the combination consequences of natural and human factors, the impact of human activity intensity had been increasing over the past 10 years. This is consistent with previous research on the Qinghai Tibet Plateau [26,27], the source of the Yellow River [35], and the surrounding areas of Qinghai Lake [30]. However, the highest explanatory power of precipitation and land use on desertification in this study is only 0.447, which may be related to the specific geological environment in the Gonghe Basin and the limited factors selection [35,58]. Since 1991, numerous measures have been applied to combat desertification. Tree planting and return the grain plots to forestry reforestation projects can help increase vegetation coverage and improve ecological diversity [59,60]. The photovoltaic power generation base and closed protection zone were established in Talatan [61,62], it is conducive to prevent wind and fix sand, thereby reducing local aeolian activities.

5. Conclusions

In this paper, we used the Albedo-NDVI feature space method based on Landsat images to explore the spatiotemporal evolution of desertification and its driving mechanism in the Gonghe Basin over the past 10 years, and then provide some scientific references for desertification prevention. The main conclusions are as follows:

(1) Desertification in the Gonghe Basin was divided into 5 categories by constructing the Albedo-NDVI feature space. There was high accuracy in the desertification classification by using the feature space method, reaching 94% in 2010 and 95% in 2020.

(2) From 2010 to 2020, the desertification situation in the Gonghe Basin generally improved, especially in the western part of the basin. The proportion of desertification area decreased from 84.3% in 2010 to 80.1% in 2020. The transformation from extremely severe desertification to severe desertification is the main form of desertification reversal.

(3) The improvement of desertification in the Gonghe Basin from 2010 to 2020 is a result of the combined effects of natural and human factors. In natural factors, precipitation played an important role in desertification evolution, and the impact of human factors was gradually increasing.

However, our study still has some limitations. Due to limited data in this study, there are some errors in the classification results of desertification. It is crucial to explore the dominant driving mechanism of desertification on different time scales, and provide targeted suggestions for desertification control in the Gonghe Basin.

Author Contributions: Conceptualization, H.J. and J.L.; data curation and investigation, H.J., R.W., H.L., B.D. and H.Z.; methodology, H.J., L.G., L.L. and J.L.; writing—original draft, H.J.; writing—review and editing, H.J. and J.L.; All authors have read and agreed to the published version of the manuscript.

Funding: This research was funded by the the National Natural Science Foundation of China (No. 41730639), and the Second Tibetan Plateau Scientific Expedition and Research Program (STEP, Grant No. 2019QZKK0906).

Institutional Review Board Statement: Not applicable.

Informed Consent Statement: Not applicable.

Data Availability Statement: The source of relevant data acquisition has been described in the text.

Conflicts of Interest: The authors declare no conflict of interest.

References

1. Wang, T.; Zhu, Z.D. Study on Sandy Desertification in China–1. Definition of Sandy Desertification and its Connotation. *J. Desert Res.* **2003**, *23*, 209–214.
2. UNCED. Managing Fragile Ecosystems: Combating Desertification and Drought. In Proceedings of the United Nations Conference on Environment and Development, Rio de Janeiro, Brazil, 3–14 June 1992.
3. Zhu, Z.D.; Chen, G.T. *Sandy Desertification in China*; Science Press: Beijing, China, 1994; pp. 20–33.
4. Adger, W.N.; Benjaminsen, T.A.; Brown, K.; Svarstad, H. Advancing a Political Ecology of Global Environmental Discourses. *Dev. Change.* **2001**, *32*, 681–715. [CrossRef]
5. Dong, Y.X. Progresses and Problems in Research on Sandy Desertification in Qinghai-Xizang Plateau. *J. Desert Res.* **1999**, *19*, 54–58.
6. Zhang, D.S. Quantitative Analysis of Influential Factors on Land Desertification in Qinghai Gonghe Basin. *J. Desert Res.* **2000**, *20*, 60–63.
7. Li, S.; Dong, Y.X.; Dong, G.R.; Yang, P.; Zhang, C.L. Regionalization of Land Desertification on Qinghai-Tebet Plateau. *J. Desert Res.* **2001**, *21*, 418–427.
8. Lu, H.Y.; Miao, X.D.; Zhou, Y.L.; Mason, J.; Swinehart, J.; Zhang, J.F.; Zhou, L.P.; Yi, S.W. Late Quaternary aeolian activity in the Mu Us and Otindag dune fields (north China) and lagged response to insolation forcing. *Geophys. Res. Lett.* **2005**, *32*, L21716. [CrossRef]
9. Xu, S.Y.; Xu, D.F.; Shi, S.R. Aeolian Sand Deposits in The Gonghe Basin, Qinghai Province. *J. Desert Res.* **1982**, *2*, 5–12.
10. Qiang, M.R.; Chen, F.H.; Song, L.; Liu, X.X.; Li, M.Z.; Wang, Q. Late Quaternary Aeolian Activity in Gonghe Basin, Northeastern Qinghai-Tibetan Plateau, China. *Quat. Res.* **2013**, *79*, 403–412. [CrossRef]
11. Xu, S.Y. Depositional Period and Sedimentary Environment of Gonghe Series in the Qinghai Province, China. *J. Lanzhou Univ. (Nat. Sci.)* **1987**, *23*, 109–119.
12. Qiang, M.R.; Jin, Y.X.; Liu, X.X.; Song, L.; Li, H.; Li, F.S.; Chen, F.H. Late Pleistocene and Holocene Aeolian Sedimentation in Gonghe Basin, Northeastern Qinghai-Tibetan Plateau: Variability, Processes, and Climatic Implications. *Quat. Sci. Rev.* **2016**, *132*, 57–73. [CrossRef]
13. Liu, L. The Formation and Evolution of Aeolian Environment in the Gonghe Basin. Master's Thesis, Lanzhou University, Lanzhou, China, 2018.
14. Liu, B.; Jin, H.L.; Sun, Z.; Su, Z.Z.; Zhang, C.X. Spatiotemporal Evolution of Sandy Land in the Gonghe Basin during the Holocene Period and Its Possible Driving Mechanism. *J. Desert Res.* **2014**, *34*, 982–991.
15. Dong, G.R.; Gao, S.Y.; Jin, J.; Li, B.S.; Zhou, G.C. Land Desertification and Its Control in Gonghe Basin, Qinghai Province. *J. Desert Res.* **1989**, *9*, 61–71.
16. Chen, Z.Y.; Dong, Z.B.; E, C.Y.; Wang, Q.C. Characteristics of wind regime and its variation trend in the Gonghe Basin from 1971 to 2015. *J. Lanzhou Univ. (Nat. Sci.)* **2020**, *56*, 224–230.
17. Yang, S.Q.; Gao, W.S.; Sui, P.; Chen, Y.Q. Quantitative research on factors of soil desertification in Gonghe basin. *Acta Ecol. Sin.* **2005**, *25*, 3182–3187.
18. Guo, L.Y.; Ding, S.X.; Wang, Q.C. Climate change and human activities of Gonghe Basin and its impact on eco-environment. *Agric. Res. Arid Areas.* **2009**, *27*, 219–226.
19. Li, X.Y.; Yao, Z.Y.; Dong, Z.B. Driving Mechanism of Aeolian Desertification in Gonghe Basin of Qinghai Province. *Bull. Soil Water Conserv.* **2018**, *38*, 337–344.
20. Hostert, P.; Roder, A.; Jarmer, T.; Udelhoven, T.; Hill, J. The potential of remote sensing and GIS for desertification monitoring and assessment. *Ann. Arid Zone* **2001**, *40*, 103–140.
21. Runnstrom, M.C. Rangeland development of the Mu Us sandy land in semiarid China: An analysis using Landsat and NOAA remote sensing data. *Land Degrad. Develop.* **2003**, *14*, 189–202. [CrossRef]

22. Collado, A.D.; Chuvieco, E.; Camarasa, A. Satellite remote sensing analysis to monitor desertification processes in the crop-rangeland boundary of Argentina. *J. Arid Environ.* **2002**, *52*, 121–133. [CrossRef]
23. Qi, Y.B.; Chang, Q.R.; Jia, K.L.; Liu, M.Y.; Liu, J.; Chen, T. Temporal-Spatial Variability of Desertification in an Agro-Pastoral Transitional Zone of Northern Shaanxi Province, China. *Catena* **2012**, *88*, 37–45. [CrossRef]
24. Yan, C.Z.; Song, X.; Zhou, Y.M.; Duan, H.C.; Li, S. Assessment of Aeolian Desertification Trends from 1975′s to 2005′s in the Watershed of the Longyangxia Reservoir in the Upper Reaches of China's Yellow River. *Geomorphology* **2009**, *112*, 205–211. [CrossRef]
25. Ma, Y.J.; Sha, Z.J.; Chen, X.J.; Fu, G.Y.; Ye, M.; Ma, J.L.; Zhai, Y.L. Desertificated land changes in Gonghe Basin from 1990 to 2010. *J. Arid Land* **2016**, *30*, 176–181.
26. Zhang, C.L.; Li, Q.; Shen, Y.P.; Zhou, N.; Wang, X.S.; Li, J.; Jia, W.-R. Monitoring of Aeolian Desertification on the Qinghai-Tibet Plateau from the 1970s to 2015 Using Landsat Images. *Sci. Total. Environ.* **2018**, *619–620*, 1648–1659. [CrossRef] [PubMed]
27. Li, Q.; Zhang, C.L.; Shen, Y.P.; Jia, W.R.; Li, J. Quantitative Assessment of the Relative Roles of Climate Change and Human Activities in Desertification Processes on the Qinghai-Tibet Plateau Based on Net Primary Productivity. *Catena* **2016**, *147*, 789–796. [CrossRef]
28. Javier, T.; Silva, P.; Barbosa, A.A.; Rodriguez, D.A.; De, O.; Sestini, M.F. Desertification trends in the northeast of Brazil over the period 2000–2016. *Int. J. Appl. Earth Obs.* **2018**, *73*, 197–206. [CrossRef]
29. Ding, C.; Huang, W.; Li, Y.; Zhao, S.; Huang, F. Nonlinear changes in dryland vegetation greenness over east Inner Mongolia, China, in recent years from satellite time series. *Sensors* **2020**, *20*, 3839. [CrossRef]
30. Zhou, Y.Q.; Hu, Z.W.; Geng, Q.Q.; Ma, J.R.; Liu, J.Y.; Wang, M.; Wang, Y.C. Monitoring and analysis of desertification surrounding Qinghai Lake (China) using remote sensing big data. *Environ. Sci. Pollut. Res.* **2023**, *30*, 17420–17436. [CrossRef]
31. Zeng, Y.N.; Xiang, N.P.; Feng, Z.D.; Hu, H. Albedo-NDVI space and remote sensing synthesis index models for desertification monitoring. *J. Sci. Geogr. Sin.* **2006**, *26*, 75–81.
32. Lamaamri, M.; Lghabi, N.; Ghazi, A.; El Harchaoui, N.; Adnan, M.S.G.; Shakiul Islam, M. Evaluation of Desertification in the Middle Moulouya Basin (North-East Morocco) Using Sentinel-2 Images and Spectral Index Techniques. *Earth Syst. Environ.* **2022**, 1–20. [CrossRef]
33. Becerril-Piña, R.; Díaz-Delgado, C.; Mastachi-Loza, C.A.; González-Sosa, E. Integration of remote sensing techniques for monitoring desertification in Mexico. *Hum. Ecol. Risk Assess* **2016**, *22*, 1323–1340. [CrossRef]
34. Wei, H.S.; Wang, J.L.; Cheng, K.; Li, G.; Ochir, A.; Davaasuren, D.; Chonokhuu, S. Desertification Information Extraction Based on Feature Space Combinations on the Mongolian Plateau. *Remote. Sens.* **2018**, *10*, 1614. [CrossRef]
35. Guo, B.; Wei, C.X.; Yu, Y.; Liu, Y.F.; Li, J.L.; Meng, C.; Cai, Y.M. The Dominant Influencing Factors of Desertification Changes in the Source Region of Yellow River: Climate Change or Human Activity? *Sci. Total. Environ.* **2022**, *813*, 152512. [CrossRef] [PubMed]
36. Li, S.; Yan, C.Z.; Song, X.; Xie, J.L.; Duan, H.C. Remote Sensing Investigation of Aeolian Desertification in Longyangxia Reservoir in Rencent 30 Years. *J. Desert Res.* **2011**, *31*, 836–841.
37. Qi, Y.H.; Pan, M.H.; Hao, Z.W.; Yang, A.N.; Xue, W.X. Variations in Aeolian Landform Patterns in the Gonghe Basin over the Last 30 Years. *J. Mt. Sci.* **2021**, *18*, 2034–2047. [CrossRef]
38. Ma, H.R.; Xiao, F.J.; Dong, Z.B.; Nan, W.G.; Chen, H. Change of sandy desertification land and its driving factors in Gonghe county, Qinghai province from 2000 to 2020. *J. Arid. Land Resour. Environ.* **2022**, *36*, 139–148.
39. Dong, G.R.; Gao, S.Y.; Jin, J. *Desertification and Controls in the Gonghe Basin, Qinghai Province*; Science Press: Beijing, China, 1993; pp. 1–166.
40. Carlson, T.N.; Ripley, D.A. On the relation between NDVI, fractional vegetation cover, and leaf area index. *Remote Sens. Environ.* **1997**, *62*, 241–252. [CrossRef]
41. Liang, S.L. Narrowband to broadband conversions of land surface albedo I: Algorithms. *J. Remote Sens. Environ.* **2001**, *76*, 213–238. [CrossRef]
42. Pan, J.H.; Qin, X.J. Extracting desertification from landsat imagery using a feature space composed of vegetation index and albedo-a case study of Zhangye oasis and its adjacent areas. *Sci. Surv. Mapp.* **2010**, *35*, 193–195.
43. Verstraete, M.M.; Pinty, B. Designing Optimal Spectral Indexes for Remote Sensing Applications. *IEEE Trans. Geosci. Remote Sens.* **1996**, *34*, 1254–1265. [CrossRef]
44. Feng, Y.; Wang, S.; Zhao, M.; Zhou, L. Monitoring of Land Desertification Changes in Urat Front Banner from 2010 to 2020 Based on Remote Sensing Data. *Water* **2022**, *14*, 1777. [CrossRef]
45. Huang, H.; Zhou, Y.; Qian, M.; Zeng, Z. Land Use Transition and Driving Forces in Chinese Loess Plateau: A Case Study from Pu County, Shanxi Province. *Land* **2021**, *10*, 67. [CrossRef]
46. Wei, W.; Yu, X.; Zhang, M.Z.; Zhang, J.; Yuan, T.; Liu, C.F. Dynamics of desertification in the lower reaches of Shiyang River Basin, Northwest China during 1995-2018. Chin. *J. Appl. Ecol.* **2021**, *32*, 2098–2106.
47. Wang, J.F.; Xu, C.D. Geodetector: Principle and prospective. *Acta Geogr. Sin.* **2017**, *72*, 117–134.
48. Wang, X.J.; Liu, G.X.; Xiang, A.C.; Xiao, S.M.; Lin, D.R.; Lin, Y.B.; Lu, Y. Terrain gradient response of landscape ecological environment to land use and land cover change in the hilly watershed in South China. *Ecol. Indic.* **2023**, *146*, 109797. [CrossRef]
49. Xu, D.Y.; Li, C.L.; Song, X.; Ren, H.Y. The Dynamics of Desertification in the Farming-Pastoral Region of North China over the Past 10years and Their Relationship to Climate Change and Human Activity. *Catena* **2014**, *123*, 11–22. [CrossRef]

50. Lancaster, N.; Helm, A.P. Test of a Climatic Index of Dune Mobility Using Measurements from the Southwestern United States. Earth Surf. Process. *Landforms* **2000**, *25*, 197–207. [CrossRef]
51. Wang, X.M.; Dong, Z.B.; Yan, P.; Zhang, J.W.; Qian, G.Q. Wind Energy Environments and Dunefield Activity in the Chinese Deserts. *Geomorphology* **2005**, *65*, 33–48. [CrossRef]
52. Gao, Y.H.; Zhou, X.; Wang, Q.; Wang, C.Z.; Zhan, Z.M.; Chen, L.F.; Yan, J.X.; Qu, R. Vegetation Net Primary Productivity and Its Response to Climate Change during 2001–2008 in the Tibetan Plateau. *Sci. Total. Environ.* **2013**, *444*, 356–362. [CrossRef]
53. Wang, X.M.; Hua, T.; Lang, L.L.; Ma, W.Y. Spatial Differences of Aeolian Desertification Responses to Climate in Arid Asia. *Glob. Planet. Chang.* **2017**, *148*, 22–28. [CrossRef]
54. Duan, H.C.; Wang, T.; Xue, X.; Yan, C.Z. Dynamic Monitoring of Aeolian Desertification Based on Multiple Indicators in Horqin Sandy Land, China. *Sci. Total. Environ.* **2019**, *650*, 2374–2388. [CrossRef] [PubMed]
55. Li, J.; Yao, Q.; Zhou, N.; Li, F. Modern Aeolian Desertification on the Tibetan Plateau under Climate Change. *Land Degrad. Dev.* **2020**, *32*, 1908–1916. [CrossRef]
56. Wu, Z. *Geomorphology of Wind-Drift Sands and Their Controlled Engineering*; Science Press: Beijing, China, 2003.
57. Chen, A.; Yang, X.C.; Guo, J.; Xing, X.Y.; Yang, D.; Xu, B. Synthesized remote sensing-based desertification index reveals ecological restoration and its driving forces in the northern sand-prevention belt of China. *Ecol. Indic.* **2021**, *131*, 108230. [CrossRef]
58. Zhang, J.; Guan, Q.; Du, Q.; Ni, F.; Mi, J.; Luo, H.; Shao, W. Spatial and temporal dynamics of desertification and its driving mechanism in Hexi region. *Land Degrad. Dev.* **2022**, *33*, 3539–3556. [CrossRef]
59. Xu, Z.W.; Hu, R.; Wang, K.X.; Mason, J.A.; Wu, S.Y.; Lu, H.Y. Recent Greening (1981-2013) in the Mu Us Dune Field, North-Central China, and Its Potential Causes. *Land Degrad. Dev.* **2018**, *29*, 1509–1520. [CrossRef]
60. Wang, X.M.; Ge, Q.S.; Geng, X.; Wang, Z.S.; Gao, L.; Bryan, B.A.; Chen, S.Q.; Su, Y.N.; Cai, D.W.; Ye, J.S.; et al. Unintended Consequences of Combating Desertification in China. *Nat. Commun.* **2023**, *14*, 1139. [CrossRef] [PubMed]
61. Yin, D.Y.; Ma, L.; Qu, J.J.; Zhao, S.P.; Yu, H.; Tan, L.H.; Xiao, J.H. Effect of Large Photovoltaic Power Station on Microclimate of Desert Region in Gonghe Basin. Bull. *Soil Water Conserv.* **2017**, *37*, 15–21.
62. Xiao, J.H.; Si, J.H.; Liu, C.; Li, X.J.; Xi, H.Y.; Yu, T.F.; Zhang, C.Q.; Zhao, C.Y.; Zhu, M.; Jia, B. Concept, connotation and development model of Desert Energy Ecosphere. *J. Desert Res.* **2021**, *41*, 11–20.

land

Article

Study on Production–Living–Ecological Function Accounting and Management in China

Nairong Tan [1,*], Xiaoying Chang [2] and Tao Ma [1,*]

[1] School of Management, Harbin Institute of Technology, Harbin 150001, China
[2] School of Humanities, Social Sciences & Law, Harbin Institute of Technology, Harbin 150001, China; changxiaoying@hit.edu.cn
* Correspondence: tannairong@hit.edu.cn (N.T.); matao@hit.edu.cn (T.M.)

Abstract: The current insufficient quantification and evaluation of major functions fundamentally affected regional sustainable management and policy implementation. This study focused on the problem that no effective quantitative accounting relationship has been established between development activities and resource utilization. In order to establish the relationship between major function accounting and natural resource accounting, we analyzed the relevant studies on the evaluation of major functions, natural resource accounting, environmental accounting, ecosystem services, and assets accounting. The efficiency comparison of different functions was completed using the equivalent factor method for ecosystem service value measurement and the input–output method for water footprint measurement. We found that the accounting of major functions and resources can guide regional sustainable management by using function positioning and resource comparative advantages. In addition, administrative units were linked to functional units, providing the possibility of cross-regional comparison of total functional resources, efficiency, and structure of major functions.

Keywords: production-living-ecological function; land use; water use; ecosystem service value

Citation: Tan, N.; Chang, X.; Ma, T. Study on Production–Living–Ecological Function Accounting and Management in China. *Land* **2023**, *12*, 1163. https://doi.org/10.3390/land12061163

Academic Editors: Li Ma, Yingnan Zhang, Muye Gan and Zhengying Shan

Received: 16 April 2023
Revised: 28 May 2023
Accepted: 29 May 2023
Published: 31 May 2023

1. Introduction

In order to cope with the tightening of resource constraints, serious environmental pollution, and ecosystem degradation, China proposed the Major function zoning, which is considered a major theoretical innovation in China's regional development theory and spatial planning. China's resource and environmental carrying capacity can hardly support high-speed industrialization and urbanization [1]. Major function zoning is a sustainable planning and management system in China designed to guide the spatial division of production, living, and ecological functions, and to coordinate sustainable goals through national resources and red line control [2,3]. Regional coordination and policy evaluation are more complex due to the increase in development activities across management units. Various regional strategies, plans, and policies have multifaceted and multilayered impacts, while indicators and methodological differences add to the complexity of monitoring and evaluation [4]. The differentiation of financial, regional, and industrial development policies inevitably leads to unequal interests between four types of functional zones. Therefore, there is an urgent need to realize unified quantification and comparison across regional scales for major functions [5].

Quantifying the imbalance between production, living demand and ecological supply is a prerequisite for spatial governance, and the quantification of major functions must take into account spatial heterogeneity, functional diversity, and complexity. The basic evaluation of major functions and spatial zoning is "double evaluation" (i.e., resource and environmental carrying capacity evaluation and territorial development suitability evaluation) [6,7]. The indicator system and method of "double evaluation" are closely related to the major function types, but there is no unified evaluation of production, living, and ecological functions of all types of areas [8]. The "double evaluation" provides a set of important indicators to support the delineation and optimization of function zoning, and provides a more accurate evaluation of the local environmental carrying capacity and spatial suitability through the grid cells. With the increase in cross-regional, cross-level and cross-period development activities, links between regions and feedback in policy management need to be strengthened in "double evaluation".

The current insufficient quantification and evaluation of major functions fundamentally affect regional sustainable management and policy implementation. Existing studies on the major function zoning have focused on the impact on land use changes [9], optimization of production–living–ecological space [10,11], the impact of carbon emissions [12], relationship between planning and economic growth [13]. The quantified objects of these studies were not the production, living, and ecological functions themselves, but some indirect measurements of impacts. Most quantitative studies of spatial function have focused on the local scale of cities, provinces, or basins, and few have accounted for different functions at the national scale. There is relatively little discussion of the linkage between functional units and administrative units. The quantitative relationships have not linked natural resource demand and function in the region closely enough, leading to some inefficient resource use and regional imbalances.

Therefore, the establishment of a basic quantitative accounting relationship between regional natural resource use and major function objectives is a key issue in the downscaling and regional management of the major function strategy to achieve the sustainable use of limited resources in the national land space [14]. The implementation of the major functional areas urgently needs to incorporate new trends and requirements such as uncertainty, complexity, and non-linearity in the use of natural resources, and to establish basic quantitative accounting relationships. The transmission mechanism between regional sustainable management system, economic activities, and resource utilization activities is not clear enough. The major function zoning is based on the evaluation of resource and environmental carrying capacity, which supports the delineation of three control lines of cropland redline, ecological redline, and urban development boundary [15]. Among them, the ecological red line combines the ecosystem service function and ecological vulnerability to delineate different levels of ecological functional areas. The role of guidance and constraint of the sustainability system is much greater than the direct allocation role, and a more effective quantitative accounting relationship between major functions and resources is needed. According to the different objectives of regional major functions, regional resource allocation will change accordingly, and accounting relationships are needed to connect ecological-economic systems and resources-functions in order to balance the contradictory issues between cross-regional ecological protection and economic development.

The goal of ecological–economic accounting is to effectively construct and assess the interaction between economic activities, resource use, and environmental-carrying capacity. Research and application of natural resource, environmental, and ecological accounting has developed and enriched with the concept of sustainable development. The development of resource economics, environmental economics, and ecological economics reflects the expansion of the boundaries of human understanding of the relationship between nature: from the acquisition of resources by nature to the impact and feedback of human behavior on the environment, to the integration of human beings and ecosystems. This led to the expansion of accounting objects from products and raw materials, to resources, wastes, pollutants to ecological service functions, according to the development of knowledge and management needs. In 2021, the United Nations Statistical Commission published the Environmental-Economic Accounting-Ecosystem Accounting (SEEA-EA) [16]. SEEA expands the non-monetization and physical boundaries, then SEEA-EA expands the boundaries of production and consumption [17]. Scholars have conducted a lot of research for ecosystem accounting application and improvements to advance the valuation of ecosystem services, and in recent years it has started to become a quantitative approach to sustainable management in China. The need for resource and environmental sustainability has also driven the development of carrying capacity research. The concept of carrying capacity was first applied to ecological carrying capacity [18]. With the resource and environmental problems caused by population growth and economic development, resource carrying capacity focuses more on the relationship between population, food and resources (land, water, etc.). To further measure human impact on the ecological environment, the concept of ecological footprint was proposed as a complement and refinement to resource carrying capacity. Subsequently, a series of footprint indicators [19] such as water footprint [20], carbon footprint [21], phosphorus footprint [22], and nitrogen footprint [23] has been developed.

In order to discuss the quantification and comparison of major function zoning, we applied the ecosystem service and footprint accounting to major function accounting, and discussed the regional sustainable evaluation and management in China. By establishing the accounting relationship between resources and major functions (Figure 1), we analyzed the distribution and efficiency of production–living–ecological functions and functional resources in different functional and administrative units.

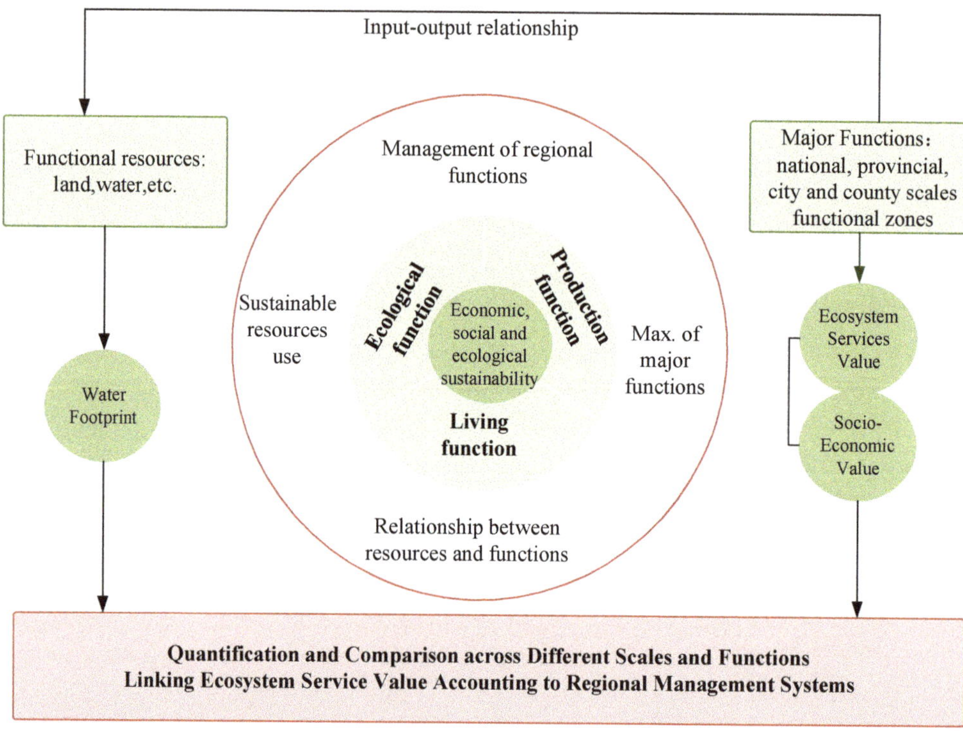

Figure 1. The accounting relationship of major function and resources.

2. Materials and Methods

2.1. Study Area: Major Function Zoning in China

The major function zoning divided China's territorial space into different major functional zones according to the development methods and development contents (Table 1). Functional units and administrative units overlapped but did not correspond exactly. In order to establish the linkage between functional units and administrative units, we collected the major functions of 2852 county-level units according to the national and provincial major function planning, and used them as the basic units to account for the products realized by the functions. There were also some prohibited development zones scattered in other functional areas, such as national parks and historical sites, which are not considered separately. In the paper, production, living, and ecological functions were directly used as the accounting objects, and the results based on the raster data realized the functional accounting and integration of different administrative and functional units. In order to quantify the relationship of resource use between regions, the city scale was chosen as the accounting unit to examine the input–output relationship of water resource function. The city scale defined in this paper included densely populated and industrialized urbanized areas as well as subordinate administrative units-districts and counties, which was a multifunctional spatial unit of production, living, and ecological functions. The accounting results covered the functions of 309 cities nationwide, and analyzed the differences in the amount of production, living, and ecological functions and function structures of the country and different types of development zones. The water accounting data source uses the city-scale multi-regional input–output table of the China Carbon Accounting Database (CEADs) for water footprint measurement. The table covers 313 administrative units in China, including 309 cities (prefecture-level administrative units and municipalities directly under the central government) and 4 provinces (Hainan, Yunan, Xizhang, Qinghai), covering more than 95% of the country's population and more than 97% of its GDP.

Table 1. Priority order of goals for different function zones.

	Classification	First-Order Major Function	Second-Order Major Function	Development Intensity	Protection Intensity
According to development method	optimized development zone	production/living	ecological	very high	very low
	prioritized development zone	Production/living	ecological	high	low
	restricted development zone	agricultural production/ecological	living	low	high
	prohibited development zone	ecological	production/living	very low	very high
According to development content	urbanized zone	industrial products and services	agricultural products/ecological products	high	high
	main agricultural production zone	agricultural products	ecological/industrial products and services	low	high
	key ecological function zone	ecological products	agricultural/industrial products, services	low	high

2.2. Definition of Key Concepts

2.2.1. Production, Living, and Ecological Functions of Territorial Space

Territorial space has basic natural geographical characteristics as well as economic and political characteristics. Spatial function referred to the division of responsibilities undertaken by spatial units in the ecological, economic, and social system at a certain stage of development, relying on its own development foundation and potential. Spatial function is manifested as land use at the microscopic scale and more as dominant function at the macroscopic scale. Typically, each space is multifunctional and different factors combine in different ways to reflect different functional characteristics, and function and space are complex many-to-many relationships [24].

The functions provided by the national space are mainly the three major functions of production, living, and ecological, as well as other national strategic support functions. They correspond to production space, living space, ecological space, and other functional spaces, respectively. Other spaces include land space for transportation facilities, water conservancy facilities, national defense, and religion, etc. The major functional products provided by different spaces vary, and the same space has multiple functions [25]. For example, urbanized space is a high degree of superposition and overlap between living space and production space, and its major function is to provide industrial goods and service products, as well as to provide the living function carried by the urban population and the ecological function of the urban green space system.

The spatial overlap makes the constraints and support relationships between functions more complex. Ecological functions provide material conditions for living and production functions, but at the same time resource and environmental thresholds limit development. The same function may provide several products and services simultaneously, and there is no one-to-one correspondence. For example, the agricultural products of farmland ecosystems are part of the product provisioning services of ecological functions and are also part of agricultural production functions. The relationship between functions is not a simple either/or relationship, so it is necessary to sort out the relationship between each function and product service.

2.2.2. Functional Resources: Land and Water

Based on the resource input and function output relationship, the required resources increase with the increase in regional functions. Natural resources, which play a fundamental supporting role for the major functions, are also the key limiting factors for regional development. Water and land resources are the most important resource indicators and constraints in the initial division and subsequent evaluation of major functional zones. Territorial spatial development at the micro level can be seen as land use and structure, and land use types and functions are closely related. As a basic natural resource and irreplaceable factor, water resources are characterized by limited total amount, difficulty of spatial transfer, and multifunctionality, which are crucial to the division and realization of major functions. The constraint target of resource overconsumption assessment refers to the excessive consumption of various natural resources in the accounting period due to the development needs of the region, including two types of natural overconsumption, where the resource utilization exceeds its renewal capacity, and policy overconsumption, where the resource utilization exceeds the red line of various policies.

2.3. *Accounting Methods and Data Sources*

2.3.1. Major Function Accounting

Major function accounting is explained as the total accounting of products and services (Y) produced by the use of natural resources and non-natural resources in the economic–social–ecological system in a specific spatial unit within a certain period of time [26,27].

$$Y = Y_P + Y_L + Y_E \tag{1}$$

where Y_P is the industrial and service products provided by the production function; Y_L is the amount of products and services required by residents' living provided by the living function; and Y_E is the value of ecological services provided by the ecological function. The three functions are defined as output results and quantified objects of territorial spatial development activities, which can cover almost all types of products and services produced by human production, living activities, and ecological impacts, thus establishing the input–output relationship with resource utilization.

The three major functions are accounted for based on the definition of accounting. The production and living functions are accounted for using the input–output method. The living function is first reflected by the population carrying quantity. Second, the final consumption products can also characterize the quantitative and structural differences of regional living functions.

Table 2. Indicators and data sources for land, water resources, and major functions.

Accounting Indicators		Total Amount Indicators	Efficiency Indicators	Data Sources
resource input	Land	Ecological functional land (LD_E) Production-ecological functional land ($LD_{P\text{-}E}$) Production-living functional land ($LD_{P\text{-}L}$)	Ecological function per unit (y_E)	The 30 m annual China Land Cover Dataset [28]
			Production function per unit (y_P)	The data set of 1 km Grid GDP in China 2015 & 2019 *
	Water	Ecological functional water (W_E) Production functional water (W_P) Living functional water (W_L)	Living function land per unit (y_L)	The data set of 1 km grid population in China 2015 & 2019 *
function output	Ecological function	Ecological function value (Y_E)	Water per ecological function (w_E)	National, Provincial and Municipal Water Resources Bulletins 2015
	Production function	Production function value (Y_P)		
	Living function	Living function value (Y_L)	Water per production functional (w_P)	China Statistical Yearbook 2015 & 2019
		Amount of population (POP)	Water per living function (w_L)	China City-level MRIO Table 2015 [29]

* Data citation: Data Registration and Publication System, Resource and Environmental Sciences and Data Center, Chinese Academy of Sciences. (http://www.resdc.cn/DOI), 2017, accessed on 7 August 2022.

Ecological function accounting used the ecosystem service value calculation method. In order to make comparisons at different spatial scales, the accounting method refers to Xie's value equivalent factor method and constructs a table of ecosystem service values per unit area [30]. The regional ecological function values at the city and county levels were calculated by adjusting [31] according to the biomass factor of each province and using the 30 m annual China Land Cover Dataset. The specific calculation equation is as follows:

$$E_a = \frac{1}{7}\left(\sum_{i=1}^{n} m_i p_i q_i\right)\frac{1}{S} \quad (2)$$

where E_a is the value of food production services per unit area of cropland ecosystem in area a (yuan/hm^2); n is the main food crop species (rice, wheat, corn) in the study area; m_i is the sown area of crop i (hm^2); p_i is the total yield of crop i (kg); q_i is the average price of crop i (yuan/kg); S is the total sown area of food crops (hm^2); 1/7 means that the economic value provided by natural ecosystems without human inputs is 1/7 of the economic value of food production services of existing farmland ecosystems per unit area. Statistics on production, prices, and area of crop are taken from the National Compilation of Cost and Benefit Information on Agricultural Products and the China Statistical Yearbook. The equivalent factors for each province in the country are shown in Appendix A Table A1.

$$ESV = \sum_{i=1}^{m}\sum_{j=1}^{n} A_j E_{ij} \quad (3)$$

where ESV is the ecosystem service value (yuan); A_j is the area of the land type j (hm^2); E_{ij} is the value of the ecological service i of the land type j (yuan/hm^2). Calculate the value of ecosystem services based on the Ecosystem service equivalent value per unit area.

2.3.2. Functional Resource Accounting

Natural resources are the material basis for the realization of ecological, production, and living functions. The sustainability of regional development means that it depends on the regeneration and replacement of natural resources and the support and improvement of ecosystem services [32]. When the relationship between human and nature shifted from passive adaptation to active utilization, the concept of resource carrying capacity based on the relationship between population and resources was proposed, and its prominent representatives are the carrying capacity of land resources and water resources. We integrated the carrying capacity and footprint to discuss the relationship between key natural resources and functions.

The amount of functional resource input can quantify the difference in the structure of resource utilization in the region, which is manifested by the different amount of resources inputted in different functions and resources. By decomposing and accounting the input resources of the region by functional types, the amount of input resources corresponding to the realization of the function can be obtained, which can characterize the regional resource input situation and the functional resource utilization structure. Usually, the resource utilization structure is closely related to the industrial structure and functional positioning of the spatial scale.

The amount of resource input per unit of function can quantify the amount of resources required to realize one unit of function in a regional unit, i.e., the ratio of the amount of resource input to the amount of function, which can be used to characterize the efficiency of resource utilization in the region and make cross-regional comparison. For example, the corresponding accounting relationships for the amount of water (W) and land (LD) resources required as inputs for the realization of production functions are established as follows.

$$Y_P = F\,(W_P, LD_P, \ldots) \tag{4}$$

The amount of water resources to be input to achieve a unit of production function in region i can be calculated as

$$w_P^i = W_P^i / Y_P^i \tag{5}$$

$$y_{wP}^i = Y_P^i / W_P^i \tag{6}$$

The amount of output per unit of resource can be used to characterize the efficiency of resource use in the region. If the functional efficiency of water resources in region i is higher than in region j, region i has more output per unit of water resources than region j, i.e., $w_P^j > w_P^i$, $y_{wP}^i > y_{wP}^j$.

The total amount of natural resources that can be used in region i is set, and usually sustainable management goals set thresholds for resource use. For example, the resource boundaries for land and water use are set in the planetary boundaries [33], and the upper line of water use is set in the spatial planning and regional management of China. In which, the total water resource use constraint of region i can be expressed as:

$$W^i = \sum\left(W_P^i, W_L^i, W_E^i\right) \le W_{boundary\ or\ upperline}^i \le W_{total}^i \tag{7}$$

That is, the input of water resources in the process of realizing the production, living, and ecological functions of region i should not exceed the upper line of water resource use and the total amount of resources in the region. Region i is usually not a closed space, and resources are imported or exported in the region to support the realization of functions within the region through direct water transfer or indirect virtual water trade exchange. Thus, when discussing water constraints, the flow of imported water (W_{IM}^i) and exported water (W_{EX}^i) through the region i should be considered.

$$W_P^i + W_L^i + W_E^i \le W^i + W_{IM}^i - W_{EX}^i \tag{8}$$

The data used in this paper are all publicly available information and datasets, as detailed in Table 2. Considering the availability of data, 2015 data were used to compare the positioning of major functions and the amount and efficiency of the major functions at the city level. The division of functional land follows the priorities of the major functions and the disturbance level of the ecosystem and is divided into three categories (Table 3). The key intermediate results obtained during the collation and calculation of the data are in the Appendix A (Tables A1 and A2). The raster data were calculated using ArcGIS and the input–output analysis of the water footprint was implemented using Matlab. The result maps were based on the standard map provided by the China's Ministry of Natural Resources (mnr.gov.cn).

Table 3. Classification of functional land.

Functional Land	Corresponding Land Classification
Ecological functional land (LD_E)	forest shrub grassland wetland water snow and ice barren
Production–ecological functional land ($LD_{P\text{-}E}$)	cropland
Production–living functional land ($LD_{P\text{-}L}$)	impervious

Functional water was water footprint based on input–output analysis, and the types were divided into three categories: production functional water, living functional water, and ecological water. The quantification of inter-regional water resources and the inputs and outputs realized by different functions is measured on the basis of a multi-regional input–output table [34], which is applicable to water resources accounting at the provincial and district to local and municipal levels, and can consider the regional allocation of water resources from a comprehensive perspective, and verify that the regional functional positioning matches the actual water use structure including virtual water, and that the regional resource endowment constraints match the direct water use structure. The 42 sectors of the table are combined into 8 sectors, as shown in Table A3. In addition, considering the difficulty of deploying physical water between different regions than the difficulty of exchanging product virtual water through trade, and the difference of regional scarcity of water resources, the regional value of water resources can also be reflected by the amount of functional water resources input.

3. Results

3.1. Ecological, Production, and Living Functions

In order to describe the situation of the major functions at different scales of the national space, quantitative accounting was conducted using the total amount and efficiency of production, living and ecological functions, and the provincial accounting results are shown in Table 4.

The distribution of ecological function was significantly different from the distribution of production and living function nationwide. In 2019, Guangdong Province still ranked first, with a total function value of 10,626.90 billion yuan of Y_P and 1029.19 billion yuan of Y_E.

The top ten provinces in total production function were Guangdong, Jiangsu, Shandong, Zhejiang, Henan, Sichuan, Hubei, Fujian, Hunan, and Shanghai, accounting for 62.12% of the national total function. The top ten provinces in total ecological functions were Tibet, Sichuan, Hunan, Inner Mongolia, Jiangxi, Guangdong, Xinjiang, Heilongjiang, Yunnan, and Guangxi, accounting for 62.12% of the national total ecological functions.

Table 4. Results of ecological–production–living function accounting by province.

Province	Ecological Function (Y_E, Billion Yuan)	Compared to 2015	Production Function (Y_P, Billion Yuan)	Compared to 2015	Living Function ($Y_{L\ (pop)}$, Million)	Compared to 2015
Beijing	50.74	0	3554.95	0.64	21.45	−0.01
Tianjin	28.35	−0.1	1406.30	−0.25	15.62	0.01
Hebei	429.32	−0.04	3173.36	0.18	75.30	0
Shanxi	181.47	−0.03	1672.33	0.4	37.27	0.02
Inner Mongolia	1154.89	−0.03	1686.06	−0.18	25.08	−0.02
Liaoning	379.21	0.08	2349.24	−0.08	42.51	0
Jilin	536.77	0.02	905.45	−0.26	26.91	−0.02
Heilongjiang	956.07	−0.02	1259.10	−0.08	36.58	0
Shanghai	54.54	−0.17	3821.74	0.48	24.31	0.04
Jiangsu	684.80	−0.08	9391.76	0.32	81.67	0.01
Zhejiang	786.03	−0.06	5950.02	0.44	57.11	0.02
Anhui	519.49	−0.07	3482.24	0.87	63.65	0.06
Fujian	813.59	−0.04	4244.68	0.66	39.78	0.04
Jiangxi	1058.74	−0.04	2167.36	0.38	46.09	−0.13
Shandong	364.62	−0.05	6857.98	0.21	100.69	0.04
Henan	415.06	−0.01	5432.44	0.48	96.31	0.03
Hubei	902.80	−0.04	4253.30	0.5	59.26	0
Hunan	1577.44	−0.04	4039.30	0.3	68.75	0.01
Guangdong	1029.19	−0.07	10,626.90	0.67	114.46	0.09
Guangxi	913.33	−0.04	2104.77	0.27	49.54	0.03
Hainan	92.83	−0.08	498.71	0.5	9.32	0.02
Chongqing	327.91	−0.01	2351.57	0.6	31.21	0.01
Sichuan	2111.79	−0.02	4658.34	0.54	83.70	0.03
Guizhou	378.69	−0.01	1677.85	0.45	36.26	0.04
Yunnan	946.11	−0.03	2272.52	0.69	48.53	0.03
Tibet	2959.11	−0.01	160.33	3.56	3.46	0.08
Shaanxi	332.93	−0.02	2486.29	0.48	37.59	−0.02
Gansu	295.20	−0.02	849.71	0.28	26.29	0.02
Qinghai	708.92	−0.02	289.92	0.42	6.44	0.09
Ningxia	72.45	−0.04	373.80	0.32	6.93	0.09
Xinjiang	995.52	−0.02	1429.51	0.39	27.27	0.15
total amount	22,057.90	−0.03	95,427.83	0.37	1399.34	0.02

3.1.1. Results of Ecological Function Accounting

Accounting for ecological functions with the value of ecosystem services in 2019, the total value of ecosystem services in 2019 nationwide was 22,042.95 billion yuan. There are differences in the ecological functions provided by different ecosystem types (Table 5). From the perspective of the major functions, croplands mainly provided agricultural products, accounting for 7.60% of the total Y_E, while forests and grasslands accounting for 51.46% and 21.04% of total Y_E, respectively. From the perspective of the types of ecosystem services, the value of ecological regulating services was the largest, at 15,561.08 billion yuan, accounting for 70.55% of the total Y_E, while product supply services were 1309.52 billion yuan, accounting for only 5.94% of the total Y_E, indicating that the supply of physical products in ecological functions was not the major function, and ecological regulating services such as hydrological regulation and climate regulation were important value-accounting content.

In terms of the total amount of ecological functions, there are clear differences at the regional level (Figure 2). The regions with higher total ecological values were distributed in the southwestern, northeastern, and southeastern provinces. The top five provinces were Tibet, Sichuan, Hunan, Inner Mongolia, and Jiangxi. In terms of the efficiency of ecological functions, the south of China was higher than the north, and the Yangtze River Delta region as well as Hunan and Fujian were among the top regions in the country in terms of ecological value per unit. The addition of ecological function values changed the total amount and structure of functions nationwide, especially in provinces where the scale of economic development was relatively small. The total ecological function of Tibet is 2959.11 billion yuan, accounting for 94.86% of the total functions. And among them, the ecological value mainly came from grassland ecosystem. Grassland covered 73.38% of Tibet's area. It provided an ecological value of 1776.33 billion yuan, accounting for 60.04% of Y_E. The hydrological regulation and climate regulation functions accounted for 35.40% and 20.25% of Tibet, respectively. The areas with a high ecological value per unit were concentrated in the forest ecological zone of the southeastern Tibetan Plateau and the desert ecological zone of the northwestern Tibetan Plateau, which was consistent with its positioning as a national key ecological function zone.

Table 5. The ecosystem function value provided by different types of ecosystems.

Ecosystem Service		Cropland	Forest	Grassland	Wetland	Barren	Water	Snow/Ice	Total Value/Billion Yuan	Function Share/%
Provisioning Services	Food production	4687.85	1462.76	897.86	1.47	11.43	262.55	0	7323.92	3.32
	Raw materials production	1039.39	3371.07	1321.13	1.44	34.3	75.48	0	5842.81	2.65
	Water supply	−5536.33	1747.8	731.11	7.44	22.87	2720.63	234.97	−71.5	−0.03
Regulating Services	Gas regulation	3775.73	11,093.94	4643.2	5.46	148.64	252.7	19.58	19,939.25	9.04
	Climate regulation	1972.71	33,175.53	12,274.98	10.35	114.34	751.54	58.74	48,358.18	21.92
	Environmental Purification	572.72	9652.11	4053.18	10.35	468.79	1821.41	17.41	16,595.96	7.52
	Water regulation	6342.38	20,710.6	8991.39	69.65	274.41	33,553.35	775.62	70,717.4	32.06
Support Services	Soil formation	2206.05	13,502	5656.49	6.64	171.51	305.21	0	21,847.9	9.9
	Nutrient cycling	657.57	1034.4	436.1	0.52	11.43	22.97	0	2163	0.98
	Biodiversity	721.21	12,289.37	5143.43	22.62	160.07	836.86	1.09	19,174.67	8.69
Cultural Services	Aesthetic Landscapes	318.18	5386.74	2270.29	13.6	68.6	620.26	9.79	8687.46	3.94
Total Value		16,757.47	113,426.33	46,419.16	149.52	1486.39	41,222.97	1117.2	220,579.04	100
Value share/%		7.6	51.46	21.06	0.07	0.67	18.7	0.51	-	-
Area/million hm^2		188.23	241.30	279.59	0.18	189.51	14.92	7.49	921.23	-

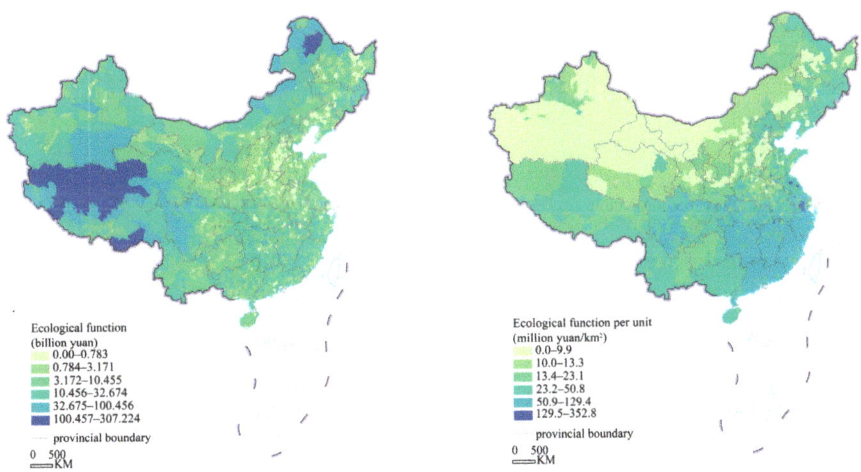

(a) (b)

Figure 2. Ecological function distribution in 2019. (**a**) The distribution of total ecological functions; (**b**) distribution of ecological function per unit.

3.1.2. Results of Production and Living Function Accounting

The spatial distribution of production and living function shows an obvious southeastern regional concentration (Figures 3 and 4). In 2019, the total national production function was 95,427.83 billion yuan, and the living function was 1399.34 million people. The top five provinces with the highest total production function were Guangdong, Jiangsu, Shandong, Henan, and Sichuan, and the top five provinces with the highest total living function are Guangdong, Shandong, Henan, Sichuan, and Jiangsu. In terms of efficiency, the distribution of production function per unit area and living function per unit area also showed a regional concentration in Southeast China. Developed provinces with a high degree of urbanization have the major function of providing industrial goods and services, and the production and living functions are much larger than the ecological functions. Among them, Shanghai, Beijing and Tianjin, as mega cities and national optimized development zones, were at the top in terms of scale and efficiency of production and living functions.

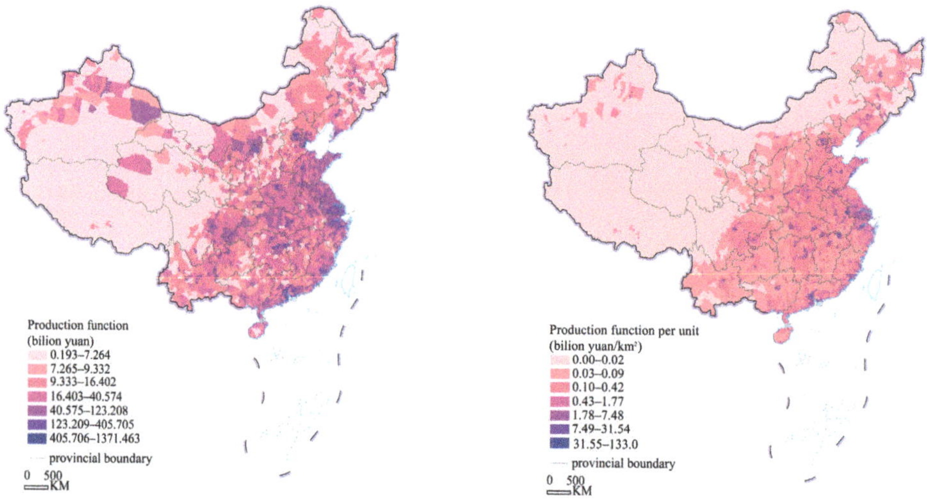

(a) (b)

Figure 3. Production function distribution in 2019. (**a**) The distribution of total production functions; (**b**) distribution of production function per unit.

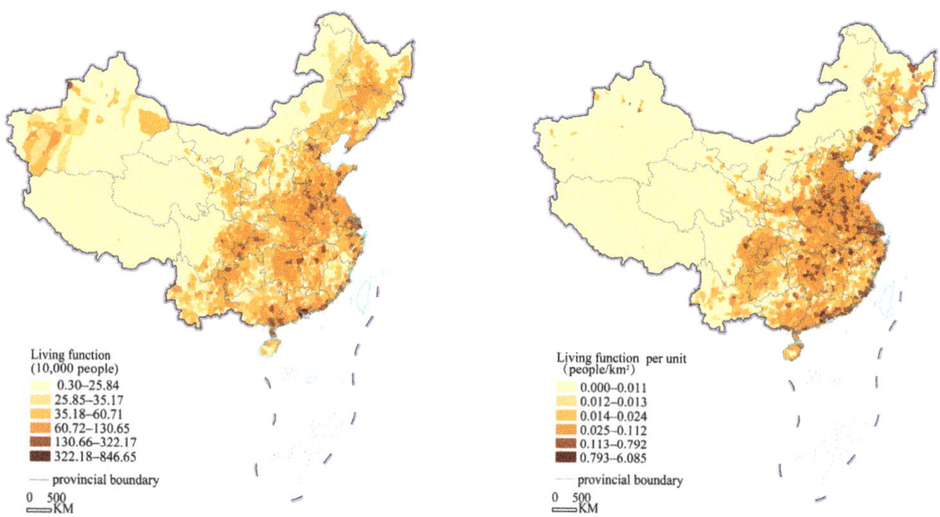

(a) (b)

Figure 4. Living function distribution in 2019. (**a**) The distribution of total living functions; (**b**) distribution of living function per unit.

3.1.3. Results for Different Functional Zones

1. Optimized development zones and prioritized development zones

According to major function zoning, optimized development areas refer to urbanized areas with more developed economy, more dense population, higher development intensity, more prominent resources, and environmental problems, and thus should be optimized for industrialized urbanization and development. Optimized development areas at the national level mainly include the Bohai Sea region, Yangtze River Delta, and Pearl River Delta regions. They are related to regional development strategies for the Beijing-Tianjin-Hebei region, the Yangtze River Delta region, and the Guangdong-Hong Kong-Macao Greater Bay Area, which form the strategic pattern of China's regional development.

According to the accounting results in Table 6, the optimized development zones accounted for only 2.41% of the national land area, but carried 30.47% of the production function and 17.66% of the living function. Prioritized development zones are urbanized areas that have a certain economic foundation, strong resource and environmental carrying capacity, higher development potential, and better conditions for population and economic, and thus should focus on industrialized urbanization and development. The prioritized development zones accounted for 15.83% of the national land area, and carry 40.30% of the production function and 38.41% of the living function.

Table 6. Accounting results of different function zones.

Function Zone	Y_E (Billion Yuan)	y_E (Million Yuan/km^2)	Y_P (Billion Yuan)	y_P (Million Yuan/km^2)	Y_L (Million People)	y_L (Thousand People/km^2)	Area (km^2)
Optimized development zones	1068.29	4.68	29,075.04	127.47	247.17	1.08	228,089.79
Prioritized development zones	4561.02	3.05	38,453.92	25.69	537.50	0.36	1,497,019.89
Main agricultural production zones	5416.98	2.39	17,792.52	7.84	407.20	0.18	2,268,399.03
Key ecological function zones	11,011.83	2.02	10,106.54	1.85	207.46	0.04	5,462,384.94

Prioritized development zones are functionally positioned as important growth poles for national economic growth and dense areas for population and economy. The optimized development zones and prioritized development zones are both urbanized areas with the same development content in general, but different development intensity and development methods. In terms of functional efficiency, the production function per unit area in the optimized development zone was 127.5 million yuan/km^2, much higher than that in the key development area, which was 25.7 million yuan/km^2.

2. Main agricultural production zones and key ecological function zones

The main agricultural production zones with better agricultural production conditions at the national level, with the main function of providing agricultural products and other functions of providing ecological products, service products and industrial products, need to be restricted from large-scale and high-intensity industrialized urbanized development in order to maintain the agricultural production capacity.

The key ecological function zones with restricted development at the national level are areas where the ecosystem is very important and related to the ecological security of the whole country or a larger area, and where the ecosystem is currently degraded, and need to be restricted from large-scale and high-intensity industrialized urbanized development in order to maintain and improve the ecological supply capacity.

From the perspective of functional land, the main agricultural production zones accounted for 23.99% of the national land area, of which the proportion of cropland accounted for 43.38% of the country, in line with its main functional positioning of providing mainly agricultural products. Key ecological function zones accounted for 68.94% of the country's ecological function land, carrying 111.833 billion yuan of ecological functions, accounting for 49.92% of the country's ecological function ratio. The highest proportion of ecological land was occupied by grassland, forest land, and barren land, accounting for 36.62%, 25.07%, and 26.89% of the functional areas, respectively. The main ecosystem services provided by different types of land use differ. Cropland has a stronger capacity for agricultural product supply services and a weaker capacity for regulating, cultural, and support services; forest has a stronger capacity for regulating and support services and a weaker capacity for product supply services. Therefore, the efficiency represented by the major function shows a differential distribution.

3.2. Input–Output Analysis of Water Resources and Major Functions

As a key factor of sustainable development, water resources must be fully considered to support and constrain the realization of the major functions of the region. In this section, the relationship between the use of water resources and the realization of the major function has been portrayed through the quantitative accounting of the major function and water resources, identifying and sorting out the use patterns of water resources in the realization of different major functions at different scales in regions, provinces, and cities, and accounting for the amount of the major function and the amount of water resources required for the realization of the function in each region.

According to the water footprint accounting results, the 313 selected regions used 568.65 billion m^3 of direct water in 2015, including 393.23 billion m^3 of direct water for agriculture, 128.09 billion m^3 of direct water for industry, 63.55 billion m^3 of water for residential use, and 12.24 billion m^3 of water for ecological functions (Figure 5). According to the water footprint accounting results (Results for only 309 cities in Table A4), the total water footprint of the production function was 565.44 billion m^3, including 140.65 billion m^3 for the agricultural sector, 265.90 billion m^3 for the industrial sector, and 89.81 billion m^3 for the service sector.

Functional water use, by flowing through the economic and social ecosystem, simultaneously provides support for production, living and ecological functions, reflecting the multi-functionality of water resources. According to provincial administrative regions, three provinces with water consumption more than 40 billion m^3 are Xinjiang, Jiangsu, and Guangdong, while five provinces with water consumption less than 5 billion m^3 are Tianjin, Qinghai, Tibet, Beijing, and Hainan. Agricultural water consumption accounted for more than 75% of the total water consumption in Xinjiang, Tibet, Ningxia, Heilongjiang, Gansu, Qinghai, and Inner Mongolia; industrial water consumption accounted for more than 35% of the total water consumption in Shanghai, Jiangsu, Chongqing, and Fujian; domestic water consumption accounted for more than 20% of the total water consumption in Beijing, Chongqing, Zhejiang, Shanghai, and Guangdong.

In terms of direct water use, the largest share of water was used for industrial and agricultural production in 2015. Water for production supports different types of industries by flowing through production activities, while combined with intermediate products containing virtual water from other sectors, supporting the production function of the whole region. Part of the final product corresponding to the production function was used for local consumption, supporting the living function; agricultural production supported by produced water also provided the ecological function. There was a gap between direct water inputs and the accounting results of water inputs for functional use. More water flows through the input–output system through virtual water than through direct water use. Cities in developed regions consumed more functional water through products with large water footprints, such as food and other intermediate products imported from other regions.

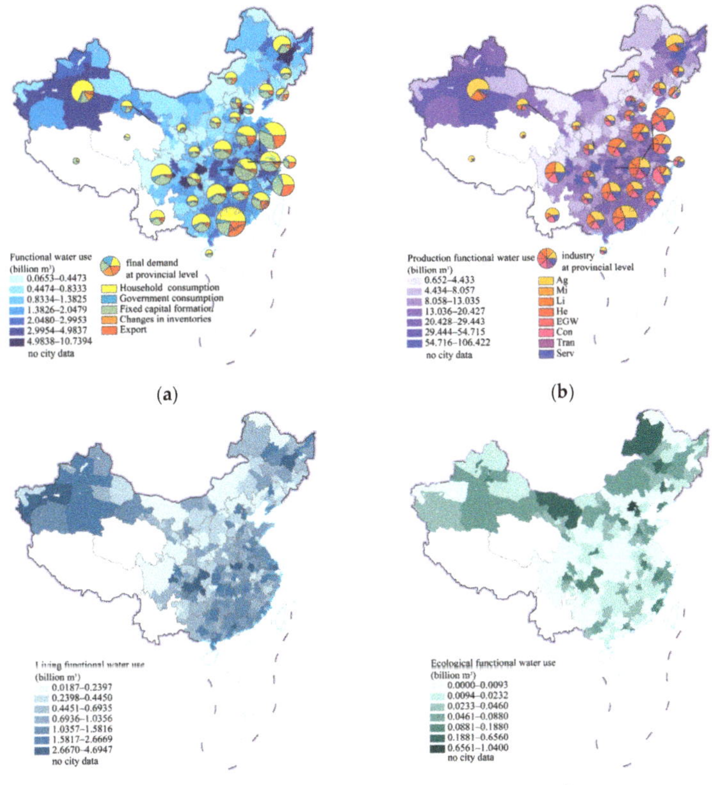

Figure 5. Accounting results of water use for production–living–ecological functions at city level. (**a**) Functional water use and final demands; (**b**) production functional water use and industrial water use; (**c**) living functional water use; (**d**) ecological functional water use.

The top provinces in terms of total water footprint of the production function were mainly concentrated in the southeast (Figure 5b), with the largest total water footprint being Guangdong with 58,558 billion m^3, followed by Jiangsu, Xinjiang, Zhejiang and Hubei. The sectoral structure of Xinjiang's water footprint was much higher than that of other provinces in terms of agricultural water use, accounting for 62.61%, while the other top provinces mainly used water in the industrial sector. In the megacities of Beijing and Shanghai, the service sector accounted for 49.83% and 33.76% of water use, respectively. The bottom five provinces in terms of total amount were Qinghai, Tibet, Hainan, Ningxia, and Tianjin, with production water footprints of 25.68, 29.893, 35.81, 5.658, and 6.261 billion m^3, respectively, of which, except for Tianjin, where the water use sector is concentrated in light industry and construction, all other provinces are small-scale and have the highest proportion of water use in agriculture.

The top provinces in terms of total water footprint for living functions are Xinjiang, Guangdong, Jiangsu, Sichuan, and Heilongjiang. Water for domestic consumption was the main source of water demand (Figure 5c), and in Xinjiang and Heilongjiang, water for domestic consumption accounted for 71.75% and 66.58% of the total water footprint, respectively. Export water footprint was also a major source in Guangdong, Zhejiang, Shanghai, Jiangsu, and Shandong, accounting for 38.06%, 38.96%, 38.61%, 25.22%, and 24.55% of the total, respectively.

The structure of major functional resources (land and water) was consistent with its main functional structure, and there were regional distribution differences in efficiency. According to the water per unit function of the accounting results (Table A4), the optimized development zone has the lowest water consumption per unit of production and living function, 4.00 m^3/thousand yuan and 6.30 m^3/ thousand yuan, respectively. The proportion of industrial and service water consumption was much higher than that of agricultural water consumption. This was related to the dense economic population in the optimized development zone and the high efficiency of water use for production and living functions. At the same time, the water input for ecological functions in the optimized development zones was 3.137 billion m^3, accounting for 30.52% of the national ecological water consumption, while the water consumption per unit of ecological function (1.86 m^3/thousand yuan) was much higher than that of other regions. In contrast, the water input of key ecological functions for ecological functions was relatively small, but provided a large amount of ecological functions, and the water consumption per unit of ecological function was only 0.205 m^3/thousand yuan, indicating that the demand for human allocation of water resources in the process of ecosystem function output in ecological function areas was weak, and from the perspective of providing ecological functions, key ecological function zones have resource advantages among all kinds of function zones. Such areas should ensure and maintain the good operation of ecosystem functions, rather than excessive allocation of water resources.

4. Discussion

In the process of achieving regional sustainable development strategic goals, different regions have different dominant or advantageous functions, and the dominant or advantageous functions may change dynamically with the adjustment of development strategic goals, the depletion of advantageous resources, or the emergence of certain types of emerging resources. The major function zoning is inseparable from the control of some key resource related to the development of land space. This study focused on the problem that no effective accounting relationship has been established between development activities and resource utilization. The value of ecosystem services has been linked with resource accounting and economic accounting to establish a relational framework that meets the needs of economic and social development and management.

Therefore, this paper attempts to solve the quantification and comparison of heterogeneous functions at different spatial scales, taking the key resources of water and land as examples. The cross-regional ecological function evaluation and conversion have become the practical and theoretical problems in the promotion of major function zoning. Land use change and human activities can directly or indirectly affect the trade-offs and synergistic issues of ecosystem services [35]. In order to establish the relationship between major function accounting and natural resource accounting, we synthesized the relevant studies on the evaluation of major functions, natural resource accounting, environmental accounting, ecosystem services, and assets accounting [36]. Chinese scholars systematically studied ecosystem service functions [37], introduced the "ecosystem service valuation" proposed by Costanza et al. [38] and established a unit area value table for terrestrial ecosystem services in China [39]. This method has been widely used to assess ecosystem service at regional scales in China by determining standard equivalence factors and establishing equivalence factor tables for different ecosystem services [40,41]. Based on this method, we also conducted a quantitative and comparative analysis of functions in the national-provincial-municipal-county administrative units and four types of functional zones.

The results showed that the accounting of production, living, and ecological functions were basically consistent with the major function positioning. The production and living functions were generally consistent in terms of spatial distribution, and the production functions were more concentrated than the living functions. The cities in the optimized development zones accounted for 4.03% of the national land area, and their production and living functions accounted for 35.14% and 36.63% of the value of production and living functions, respectively, while the population carrying accounted for 21.02% of the country, while the ecological function was 7.41%, which is in line with the main function positioning of the optimized development zones as the provision of industrial goods and service products. Key ecological function zones provide 10.63% of production function, 9.68% of population carrying and 54.00% of ecological service value with 63.85% of the national land area.

There were differences between the country as a whole and the regions in terms of total size and efficiency across functions. At the national level, the Yangtze River Delta, Pearl River Delta, Beijing-Tianjin-Hebei, and other city agglomerations are national optimization development zones with high overall economic level and dense population [42,43]. However, within the optimization development zones, each region performs different major functions, and there are differences in both scale and structure. Among them, within Beijing–Tianjin–Hebei, according to the accounting results, Beijing and Tianjin were highly dense in production and living functions, accounting for 50% of the production and living functions in Beijing–Tianjin–Hebei with 13.1% of region area. Meanwhile, Hebei Province provided the most important ecological functions in the Beijing–Tianjin–Hebei region, and its functional value reached 446.468 billion yuan, 84.44% of the total ecological functions in the Beijing-Tianjin-Hebei region, which was related to the forests, grasslands, and farmlands in northern Hebei. Therefore, there are differences in the comparative advantages and relative function efficiency within and outside the major functions of the region.

Based on the accounting relationship of major function and resources in Figure 1, this paper considered land and water as natural resources that play a fundamental supporting role for major functions, and are also key constraints for regional development. Due to the limited total water resources and redlines, the accounting of major functions and the efficiency of functional resource input are carried out, and the comparative advantages of different functions in terms of structure and resource efficiency are coordinated to support the maximum allocation of resources with regional functions. We found that the accounting of major functions and resources can guide regional sustainable management by using function positioning and resource comparative advantages. The results showed that the targets set for major functions were basically consistent with the current structure of resource use. There were differences in comparative advantages and relative function realization efficiency of regional major functions. By analyzing the water consumption of 42 sectors in 309 cities using the multi-regional input–output table, we compared the total amount and water efficiency of major functions among regions. In addition, based on grid cells and county-level accounting, administrative units were linked to functional units, providing the possibility of cross-regional comparison of total functional resources, efficiency, and structure of major functions.

In the analysis of water and major functions, it was found that there were regional and functional heterogeneity differences in the supporting and constraining effects of resources on functions [44,45]. Water consumption for ecological functions requires further consideration of the comparative advantages of local ecosystems and resources, as the efficiency of water supporting the realization of ecological functions largely depends on the efficiency of local ecosystems, rather than the input–output efficiency of traditional economic water use. Therefore, regional resource allocation must fully take into account the efficiency of relative function realization of resources [46]. In the quantitative relationship between resources and functions, the resource allocation goal of maximizing functions is achieved [47].

5. Conclusions

This paper established an accounting relationship between two kinds of key functional resources—land and water, and the three major functions of production, living, and ecology. Based on this, it further discussed the way to apply to the sustainable management of major functions, which is a regional and divisional sustainable management system. The ecosystem service value approach provides a basic quantitative basis for major function accounting. If it is to be applied to regional management, it must be linked to administrative and functional units.

Aiming to achieve the sustainable use of natural resources, this paper has established an accounting relationship framework and indicators for the value of major functions and the input of functional resources, aiming at the quantitative relationship between resources and major functions. This approach basically realizes the computability, decomposability, and comparability of the functional structure and efficiency among the three major functions in different spatial units.

In this paper, the major function accounting process combined the economic and ecosystem service accounting methods to reflect the functional output and the footprint analysis of resource use. It established an analysis of resource input and functional value, which is complementary for the unclear two-way relationship between resources and regional development goals, and is helpful for further discussion of the relationship between regional functional management and resource use optimization in future research.

Based on the major function accounting to guide the allocation and optimization of resources, this paper analyzed the use of water resources in 42 sectors of 309 cities in the multi-regional input–output table, the relationship between the major function and the efficiency of functional resources was calculated from the perspective of limited constraints on water resources. Water and land as natural resources that have a fundamental supporting role for the major functions is also a key constraint factor for regional development. Finally, it provides a quantitative basis for cross-regional comparison of the total amount, efficiency, and structure of resources and major functions between different regions.

The limitations of this research are as follows: (1) In order to reflect the structure and flows of water use among regions, the study selected the city-level multi-regional input–output method to measure the water footprint. The advantage of this method was that it can compare the resource consumption of different sectors and the final demand at the city level, but this method relied more on the availability of data and is very restrictive for continuous year analysis. In addition, there was a partial lack of statistical data on water resources at both the city and industry levels. (2) Many cross-system accounting methods did not fully consider the reality of resource needs in the process of achieving goals such as economic development and strategic security. The application of the major function classification in the management of national sustainable development goals requires the coordination of various existing accounting methods. At the same time, the trade-off between development and protection in different administrative units and functional units should also be reflected in the evaluation benchmarks when downscaling the results of the major function zoning from the national level to the provincial, municipal, and county levels. (3) Only the land area has been considered in this study, and the functions of the sea area should be included in the future. China's economically developed regions are also adjacent to the sea, the high-quality growth of the developed regions must be adjusted to the protection of the marine ecological environment.

Author Contributions: Conceptualization, N.T. and T.M.; methodology, N.T.; software, N.T.; formal analysis, N.T.; writing—original draft preparation, N.T.; writing—review and editing, N.T., X.C. and T.M.; visualization, N.T.; supervision, X.C. and T.M.; project administration, N.T., X.C. and T.M.; funding acquisition, X.C. and T.M. All authors have read and agreed to the published version of the manuscript.

Funding: This research was funded by [the National Natural Science Foundation of China] grant number [71974046], [the National Natural Science Foundation of China] grant number [71950001].

Data Availability Statement: Publicly available datasets were analyzed in this study. These data citations can be found here:

Xu, X. 2017. Spatial distribution of China's GDP on 1 km grid dataset; Data Registration and Publication System, Resource and Environmental Sciences and Data Center, Chinese Academy of Sciences (http://www.resdc.cn/DOI), accessed on 7 August 2022; https://doi.org/10.12078/2017121102.

Xu, X. 2017. Spatial distribution of the Chinese population on a 1 km grid dataset; Data Registration and Publication System, Resource and Environmental Sciences and Data Center, Chinese Academy of Sciences (http://www.resdc.cn/DOI), accessed on 7 August 2022; https://doi.org/10.12078/2017121101.

Reference 28: Yang, J.; Huang, X. The 30 M Annual Land Cover Dataset and Its Dynamics in China from 1990 to 2019. Earth Syst. Sci. e Data 2021, 13, 3907–3925. https://doi.org/10.5194/essd-13-3907-2021.

Reference 29: Heng, H.; Többen, J.; Dietzenbacher, E.; Moran, D.; Meng, J.; Wang, D.; Guan, D. Entropy-Based Chinese City-Level MRIO Table Framework. Econ. Syst. Res. 2021, 34, 519–544. https://doi.org/10.1080/09535314.2021.1932764.

Acknowledgments: This work was supported in part by the Independent Topic Selection Project for Young Talents in Ideological and Cultural Publicity (Publicity Department of the CPC Central Committee), the Computational Social Science Research and Cultivation Project (Harbin Institute of Technology) and the Young Scholars Development Grant Program (Harbin Institute of Technology). We would like to thank the reviewers and editors for their useful comments and feedback, which helped us to improve the paper.

Conflicts of Interest: The authors declare no conflict of interest.

Appendix A

Table A1. Standard equivalent factor values across the country.

Provinces	Biomass Factor	Standard Equivalent Factor Values (Yuan/hm^2)			
		Year 2010	Year 2015	Year 2019	Year 2020
Beijing	1.04	1975.79	2384.98	2316.34	2578.95
Tianjin	0.85	1614.83	1949.26	1893.16	2107.79
Hebei	1.02	1937.80	2339.11	2271.80	2529.35
Shanxi	0.46	873.91	1054.89	1024.54	1140.69
Inner Mongolia	0.44	835.91	1009.03	979.99	1091.09
Liaoning	0.9	1709.82	2063.92	2004.53	2231.78
Jilin	0.96	1823.81	2201.52	2138.16	2380.57
Heilongjiang	0.66	1253.87	1513.54	1469.99	1636.64
Shanghai	1.44	2735.71	3302.28	3207.24	3570.85
Jiangsu	1.74	3305.65	3990.25	3875.42	4314.77
Zhejiang	1.76	3343.65	4036.12	3919.96	4364.37
Anhui	1.71	3248.66	3921.46	3808.60	4240.38
Fujian	1.56	2963.69	3577.47	3474.51	3868.42
Jiangxi	1.51	2868.70	3462.81	3363.15	3744.43
Shandong	1.38	2621.73	3164.68	3073.61	3422.06
Henan	1.39	2640.72	3187.62	3095.88	3446.86
Hubei	1.27	2412.75	2912.43	2828.61	3149.29
Hunan	1.95	3704.61	4471.84	4343.14	4835.52
Guangdong	1.4	2659.72	3210.55	3118.15	3471.66
Guangxi	0.98	1861.81	2247.38	2182.71	2430.16
Hainan	0.72	1367.86	1651.14	1603.62	1785.42
Chongqing	1.21	2298.76	2774.83	2694.97	3000.50
Sichuan	1.35	2564.73	3095.89	3006.79	3347.67
Guizhou	0.63	1196.87	1444.75	1403.17	1562.25
Yunnan	0.64	1215.87	1467.68	1425.44	1587.04
Tibet	0.75	1424.85	1719.94	1670.44	1859.82
Shaanxi	0.51	968.90	1169.56	1135.90	1264.68
Gansu	1.42	2697.72	3256.41	3162.70	3521.25
Qinghai	0.4	759.92	917.30	890.90	991.90
Ningxia	0.61	1158.88	1398.88	1358.62	1512.65
Xinjiang	0.58	1101.88	1330.08	1291.81	1438.26
National value	1	1899.80	2293.25	2227.25	2479.76

Table A2. Different ecosystem service functions at the provincial level Correspondence of multi-regional input–output tables.

Ecosystem Service	Provisioning Services				Regulating Services				Support Services		Cultural Services
	Food Production	Raw Materials Production	Water Supply	Gas Regulation	Climate Regulation	Environmental Purification	Water Regulation	Soil Formation	Nutrient Cycling	Biodiversity	Aesthetic Landscapes
Beijing	16	14	−2	49	124	39	141	54	5	47	21
Tianjin	15	4	−1	14	14	13	201	10	2	8	5
Hebei	260	137	−171	474	965	309	1250	469	58	375	168
Shanxi	88	59	−47	201	453	141	429	213	23	177	78
Inner Mongolia	327	333	29	1154	2893	983	2685	1328	117	1177	522
Liaoning	213	117	−129	399	839	268	1176	396	49	320	144
Jilin	268	163	−142	554	1229	389	1599	566	65	467	209
Heilongjiang	429	284	−191	961	2215	699	2828	1002	111	843	379
Shanghai	18	4	15	15	15	24	420	11	2	12	8
Jiangsu	339	81	18	290	269	297	5019	200	47	175	112
Zhejiang	189	193	87	642	1711	572	2710	728	67	660	301
Anhui	281	118	−100	406	737	304	2573	372	53	304	147
Fujian	161	234	75	775	2192	659	1870	911	76	819	363
Jiangxi	293	270	59	902	2337	772	3548	1005	96	897	408
Shandong	382	101	−304	362	358	173	2023	256	57	157	80
Henan	403	144	−355	502	760	253	1524	422	71	293	133
Hubei	342	223	−40	753	1745	612	3521	787	87	683	316
Hunan	513	440	−54	1473	3724	1179	4676	1618	160	1412	634
Guangdong	272	265	69	883	2322	759	3345	993	93	889	403
Guangxi	245	268	4	893	2397	724	2198	1017	92	897	397
Hainan	29	27	−3	90	232	71	247	100	10	87	39
Chongqing	140	99	−60	335	792	245	905	354	38	298	133
Sichuan	657	627	−62	2130	5422	1691	4987	2402	223	2106	935
Guizhou	136	120	−49	401	1015	301	836	441	43	378	166
Yunnan	246	283	7	944	2551	764	2108	1083	97	956	422
Tibet	443	637	935	2224	5994	2257	10,477	2694	207	2550	1173
Shaanxi	108	103	−22	348	887	270	720	391	37	339	150
Gansu	102	88	−16	309	729	265	661	347	32	302	134
Qinghai	115	151	207	533	1387	562	2547	645	50	612	282
Ningxia	32	21	−11	76	166	57	193	82	8	69	31
Xinjiang	262	234	182	848	1883	947	3299	955	85	865	395

Table A3. Correspondence of multi-regional input–output tables.

3 Sectors	8 Sectors		42 Sectors
Agriculture	Agriculture (Ag)	1	"Agriculture, Forestry, Animal Husbandry, and Fishery"
Industry	Mining industry (Mi)	2	"Mining and washing of coal"
		3	"Extraction of petroleum and natural gas"
		4	"Mining and processing of metal ores"
		5	"Mining and processing of nonmetal and other ores"
	Light industry (Li)	6	"Food and tobacco processing"
		7	"Textile industry"
		8	"Manufacture of leather, fur, feather, and related products"
		9	"Processing of timber and furniture"
		10	"Manufacture of paper, printing and articles for culture, education and sport activity"
	Heavy industry (He)	11	"Processing of petroleum, coking, processing of nuclear fuel"
		12	"Manufacture of chemical products"
		13	"Manuf. of non-metallic mineral products"
		14	"Smelting and processing of metals"
		15	"Manufacture of metal products"
		16	"Manufacture of general purpose machinery"
		17	"Manufacture of special purpose machinery"
		18	"Manufacture of transport equipment"
		19	"Manufacture of electrical machinery and equipment"
		20	"Manufacture of communication equipment, computers and other electronic equipment"
		21	"Manufacture of measuring instruments"
		22	"Other manufacturing"
		23	"Comprehensive use of waste resources"
		24	"Repair of metal products, machinery and equipment"
	Electricity, gas and water production and supply (EGW)	25	"Production and distribution of electric power and heat power"
		26	"Production and distribution of gas"
		27	"Production and distribution of tap water"
	Construction (Con)	28	"Construction"
Service	Transportation (Tran)	30	"Transport, storage, and postal services"
	Service industry (Serv)	29	"Wholesale and retail trades"
		31	"Accommodation and catering"
		32	"Information transfer, software, and information technology services"
		33	"Finance"
		34	"Real estate"
		35	"Leasing and commercial services"
		36	"Scientific research and polytechnic services"
		37	"Administration of water, environment, and public facilities"
		38	"Resident, repair, and other services"
		39	"Education"
		40	"Health care and social work"
		41	"Culture, sports, and entertainment"
		42	"Public administration, social insurance, and social organizations"

Table A4. Four types of functional zones production–life–ecological functions and water resources accounting results.

Four Types of Functional Zones		Optimized Development Zones	Prioritized Development Zones	Main Agricultural Production Zones	Key Ecological Function Zones
production function (Y_P, billion yuan)		24,435.45	27,056.12	7391.06	10,658.62
living function	Y_L (billion yuan)	7392.28	7704.73	1952.87	3133.82
	Population (million)	288.08	533.01	234.77	314.88
ecological function (YE, billion yuan)		1686.89	5342.71	12,291.65	3439.26
total functions (Y, billion yuan)		33,514.62	40,103.56	21,635.57	17,231.70
	Y_P share	72.91%	67.47%	34.16%	61.85%
	Y_L share	22.06%	19.21%	9.03%	18.19%
	Y_E share	5.03%	13.32%	56.81%	19.96%
production functional water (billion m^3)		97.66	202.38	85.90	87.41
	agriculture	12.92	46.71	42.45	26.97
	industry	57.43	120.68	32.07	48.07
	service	27.31	35.00	11.39	12.37
living functional water (billion m^3)		46.56	104.25	54.63	51.74
ecological functional water (billion m^3)		3.14	3.27	2.52	1.35
total functional water (billion m^3)		147.36	309.90	143.05	140.51
production functional water share		66.27%	65.31%	60.05%	62.21%
	agriculture	8.77%	15.07%	29.67%	19.20%
	industry	38.97%	38.94%	22.42%	34.21%
	service	18.54%	11.29%	7.96%	8.80%
living functional water share		31.60%	33.64%	38.19%	36.83%
ecological functional water share		2.13%	1.06%	1.76%	0.96%
water per production function (billion m^3)		4.00	7.48	11.62	8.20
water per living function (billion m^3)		6.30	13.53	27.97	16.51
water per ecological function (m^3/thousand yuan)		1.86	0.61	0.21	0.39

References

1. Lu, N.; Tian, H.; Fu, B.; Yu, H.; Piao, S.; Chen, S.; Li, Y.; Li, X.; Wang, M.; Li, Z.; et al. Biophysical and Economic Constraints on China's Natural Climate Solutions. *Nat. Clim. Chang.* **2022**, *12*, 847–853. [CrossRef]
2. Fan, J.; Wang, Y.; Wang, C.; Chen, T.; Jin, F.; Zhang, W.; Li, L.; Xu, Y.; Dai, E.; Tao, A.; et al. Reshaping the Sustainable Geographical Pattern: A Major Function Zoning Model and Its Applications in China. *Earth's Future* **2019**, *7*, 25–42. [CrossRef]
3. Wang, Y.; Fan, J. Multi-Scale Analysis of the Spatial Structure of China's Major Function Zoning. *J. Geogr. Sci.* **2020**, *30*, 197–211. [CrossRef]
4. Xu, K.; Wang, J.; Wang, J.; Wang, X.; Chi, Y.; Zhang, X. Environmental Function Zoning for Spatially Differentiated Environmental Policies in China. *J. Environ. Manag.* **2020**, *255*, 109485. [CrossRef] [PubMed]
5. Zhong, Y.; Su, X. Spatial Selectivity and Intercity Cooperation between Guangdong and Hong Kong. *Urban Stud.* **2019**, *56*, 3011–3029. [CrossRef]
6. Li, K.; Jin, X.; Ma, D.; Jiang, P. Evaluation of Resource and Environmental Carrying Capacity of China's Rapid-Urbanization Areas—A Case Study of Xinbei District, Changzhou. *Land* **2019**, *8*, 69. [CrossRef]
7. Wang, G.; Yang, D.; Xia, F.; Zhong, R.; Xiong, C. Three types of spatial function zoning in key ecological function areas based on ecological and economic coordinated development: A case study of tacheng basin, China. *Chin. Geogr. Sci.* **2019**, *29*, 689–699. [CrossRef]
8. Lin, G.; Jiang, D.; Fu, J.; Zhao, Y. A Review on the Overall Optimization of Production–Living–Ecological Space: Theoretical Basis and Conceptual Framework. *Land* **2022**, *11*, 345. [CrossRef]
9. Nie, X.; Lu, B.; Chen, Z.; Yang, Y.; Chen, S.; Chen, Z.; Wang, H. Increase or Decrease? Integrating the CLUMondo and InVEST Models to Assess the Impact of the Implementation of the Major Function Oriented Zone Planning on Carbon Storage. *Ecol. Indic.* **2020**, *118*, 106708. [CrossRef]
10. Zou, L.; Liu, Y.; Yang, J.; Yang, S.; Wang, Y.; Cao, Z.; Hu, X. Quantitative Identification and Spatial Analysis of Land Use Ecological-Production-Living Functions in Rural Areas on China's Southeast Coast. *Habitat Int.* **2020**, *100*, 102182. [CrossRef]
11. Li, S.; Zhao, X.; Pu, J.; Miao, P.; Wang, Q.; Tan, K. Optimize and control territorial spatial functional areas to improve the ecological stability and total environment in karst areas of Southwest China. *Land Use Policy* **2021**, *100*, 104940. [CrossRef]
12. Zhu, J.; Li, X.; Huang, H.; Yin, X.; Yao, J.; Liu, T.; Wu, J.; Chen, Z. Spatiotemporal Evolution of Carbon Emissions according to Major Function-Oriented Zones: A Case Study of Guangdong Province, China. *Int. J. Environ. Res. Public Health* **2023**, *20*, 2075. [CrossRef]
13. Cao, Y.; Kong, L.; Zhang, L.; Ouyang, Z. The Balance between Economic Development and Ecosystem Service Value in the Process of Land Urbanization: A Case Study of China's Land Urbanization from 2000 to 2015. *Land Use Policy* **2021**, *108*, 105536. [CrossRef]

14. Bateman, I.J.; Harwood, A.R.; Mace, G.M.; Watson, R.T.; Abson, D.J.; Andrews, B.; Binner, A.; Crowe, A.; Day, B.H.; Dugdale, S.; et al. Bringing Ecosystem Services into Economic Decision-Making: Land Use in the United Kingdom. *Science* **2013**, *341*, 45–50. [CrossRef]
15. Wang, L.; Li, F.; Gong, Y.; Jiang, P.; Huang, Q.; Hong, W.; Chen, D. A Quality Assessment of National Territory Use at the City Level: A Planning Review Perspective. *Sustainability* **2016**, *8*, 145. [CrossRef]
16. United Nations. System of Environmental-Economic Accounting—Ecosystem Accounting (SEEA EA) [EB/OL]. UN, 2021. Available online: https://seea.un.org/ecosystem-accounting (accessed on 24 February 2023).
17. Hein, L.; Bagstad, K.J.; Obst, C.; Edens, B.; Schenau, S.; Castillo, G.; Soulard, F.; Brown, C.; Driver, A.; Bordt, M.; et al. Progress in Natural Capital Accounting for Ecosystems. *Science* **2020**, *367*, 514–515. [CrossRef]
18. Wackernagel, M.; Onisto, L.; Bello, P.; Callejas Linares, A.; Susana López Falfán, I.; Méndez García, J.; Isabel Suárez Guerrero, A.; Guadalupe Suárez Guerrero, M. National Natural Capital Accounting with the Ecological Footprint Concept. *Ecol. Econ.* **1999**, *29*, 375–390. [CrossRef]
19. Wiedmann, T.; Allen, C. City Footprints and SDGs Provide Untapped Potential for Assessing City Sustainability. *Nat. Commun.* **2021**, *12*, 3758. [CrossRef]
20. Hoekstra, A.Y.; Mekonnen, M.M.; Chapagain, A.K.; Mathews, R.E.; Richter, B.D. Global Monthly Water Scarcity: Blue Water Footprints versus Blue Water Availability. *PLoS ONE* **2012**, *7*, e32688. [CrossRef]
21. Hoekstra, A.Y.; Wiedmann, T.O. Humanity's Unsustainable Environmental Footprint. *Science* **2014**, *344*, 1114–1117. [CrossRef]
22. Wang, F.; Sims, J.T.; Ma, L.; Ma, W.; Dou, Z.; Zhang, F. The Phosphorus Footprint of China's Food Chain: Implications for Food Security, Natural Resource Management, and Environmental Quality. *J. Environ. Qual.* **2011**, *40*, 1081–1089. [CrossRef] [PubMed]
23. Leach, A.M.; Galloway, J.N.; Bleeker, A.; Erisman, J.W.; Kohn, R.; Kitzes, J. A Nitrogen Footprint Model to Help Consumers Understand Their Role in Nitrogen Losses to the Environment. *Environ. Dev.* **2012**, *1*, 40–66. [CrossRef]
24. Duan, Y.; Wang, H.; Huang, A.; Xu, Y.; Lu, L.; Ji, Z. Identification and spatial-temporal evolution of rural "production-living-ecological" space from the perspective of villagers' behavior—A case study of Ertai town, Zhangjiakou City. *Land Use Policy* **2021**, *106*, 105457. [CrossRef]
25. Feng, C.; Zhang, H.; Xiao, L.; Guo, Y. Land Use Change and Its Driving Factors in the Rural–Urban Fringe of Beijing: A Production–Living–Ecological Perspective. *Land* **2022**, *11*, 314. [CrossRef]
26. Ma, T.; Tan, N. Study on the quantitative relationship between major functions and natural resources. *China Popul. Resour. Env.* **2020**, *30*, 30–40.
27. Tan, N.; Wang, X.; Wang, H.; Gao, Z.; Chang, X.; Ma, T. Downscaling of planetary boundaries and Sustainability Management: A Nexus Analysis of water, land and major functions at the National-Provincial Level. *Sustain. Horiz.* **2022**, *3*, 100028. [CrossRef]
28. Yang, J.; Huang, X. The 30 M Annual Land Cover Dataset and Its Dynamics in China from 1990 to 2019. *Earth Syst. Sci. Data* **2021**, *13*, 3907–3925. [CrossRef]
29. Zheng, H.; Többen, J.; Dietzenbacher, E.; Moran, D.; Meng, J.; Wang, D.; Guan, D. Entropy Based Chinese City Level MRIO Table Framework. *Econ. Syst. Res.* **2021**, *34*, 519–544. [CrossRef]
30. Xie, G.D.; Zhang, C.X.; Zhang, L.M.; Chen, W.H.; Li, S.M. Improvement of the Evaluation Method for Ecosystem Service Value Based on Per Unit Area. *J. Nat. Resour.* **2015**, *30*, 1243–1254.
31. Xie, G.D.; Xiao, Y.; Zhen, L.; Lu, C.X. Study on ecosystem services value of food production in China. *Chin. J. Eco-Agric.* **2005**, *13*, 10–13.
32. Luo, Y.; Li, X.; Cai, G.; Zhao, D. A Study on Atmospheric Environmental Resource Accounting: A Case of SO_2 Capacity Resources in Chinese Provinces. *J. Environ. Manag.* **2019**, *249*, 109432. [CrossRef]
33. Steffen, W.; Richardson, K.; Rockstrom, J.; Cornell, S.E.; Fetzer, I.; Bennett, E.M.; Biggs, R.; Carpenter, S.R.; de Vries, W.; de Wit, C.A.; et al. Planetary Boundaries: Guiding Human Development on a Changing Planet. *Science* **2015**, *347*, 1259855. [CrossRef]
34. Hoekstra, A.Y.; Chapagain, A.K. Water Footprints of Nations: Water Use by People as a Function of Their Consumption Pattern. *Water Resour. Manag.* **2006**, *21*, 35–48. [CrossRef]
35. Jiang, B.; Bai, Y.; Chen, J.; Alatalo, J.M.; Xu, X.; Liu, G.; Wang, Q. Land Management to reconcile ecosystem services supply and demand mismatches—A case study in Shanghai Municipality, China. *Land Degrad. Dev.* **2020**, *31*, 2684–2699. [CrossRef]
36. Ma, G.; Wang, J.; Yu, F.; Yang, W.; Ning, J.; Peng, F.; Zhou, X.; Zhou, Y.; Cao, D. Framework Construction and Application of China's Gross Economic-Ecological Product Accounting. *J. Environ. Manag.* **2020**, *264*, 109852. [CrossRef]
37. Ouyang, Z.Y.; Wang, X.K.; Miao, H. A primary study on Chinese terrestrial ecosystem services and their ecological-economic values. *Acta Ecol. Sin.* **1999**, *19*, 607–613.
38. Costanza, R.; d'Arge, R.; de Groot, R.; Farber, S.; Grasso, M.; Hannon, B.; Limburg, K.; Naeem, S.; O'Neill, R.V.; Paruelo, J.; et al. The Value of the World's Ecosystem Services and Natural Capital. *Nature* **1997**, *387*, 253–260. [CrossRef]
39. Xie, G.D.; Lu, C.X.; Leng, Y.F.; Zheng, D.U.; Li, S.C. Ecological assets valuation of the Tibetan Plateau. *J. Nat. Resour.* **2003**, *18*, 189–196.
40. Xu, J.; Xie, G.; Xiao, Y.; Li, N.; Yu, F.; Pei, S.; Jiang, Y. Dynamic Analysis of Ecological Environment Quality Combined with Water Conservation Changes in National Key Ecological Function Areas in China. *Sustainability* **2018**, *10*, 1202. [CrossRef]
41. Xie, G.; Zhang, C.; Zhen, L.; Zhang, L. Dynamic Changes in the Value of China's Ecosystem Services. *Ecosyst. Serv.* **2017**, *26*, 146–154. [CrossRef]

42. Gao, L.; Ma, C.; Wang, Q.; Zhou, A. Sustainable use zoning of land resources considering ecological and geological problems in Pearl River Delta Economic Zone, China. *Sci. Rep.* **2019**, *9*, 16052. [CrossRef] [PubMed]
43. Gao, Z.; Tan, N.; Geddes, R.R.; Ma, T. Population distribution characteristics and Spatial Planning Response Analysis in metropolises: A Case Study of Beijing. *Int. Rev. Spat. Plan. Sustain. Dev.* **2019**, *7*, 134–154. [CrossRef] [PubMed]
44. Zhang, Q.; Shen, J. Spatiotemporal Heterogeneity and Driving Factors of Water Resource and Environment Carrying Capacity under High-Quality Economic Development in China. *Int. J. Environ. Res. Public Health* **2022**, *19*, 10929. [CrossRef] [PubMed]
45. Hong, S.; Wang, H.; Cheng, T. Analysis of water use characteristics in industrial sectors in Beijing based on an input-output method. *Water Supply* **2019**, *20*, 219–230. [CrossRef]
46. Wang, S.; Chen, B. Unraveling Energy–Water Nexus Paths in urban agglomeration: A case study of Beijing–Tianjin–Hebei. *Appl. Energy* **2021**, *304*, 117924. [CrossRef]
47. Wang, A.; Liao, X.; Tong, Z.; Du, W.; Zhang, J.; Liu, X.; Liu, M. Spatial-temporal dynamic evaluation of the ecosystem service value from the perspective of "production-living-ecological" spaces: A case study in Dongliao River Basin, China. *J. Clean. Prod.* **2022**, *333*, 130218. [CrossRef]

land

Article

The Impact of Land Transfer on Vulnerability as Expected Poverty in the Perspective of Farm Household Heterogeneity: An Empirical Study Based on 4608 Farm Households in China

Zheng Wang [1,2], Mingwei Yang [1,3,*], Zhiyong Zhang [1,4], Yingjuan Li [1] and Chuanhao Wen [5]

1 Research Center for Economy of Upper Reaches of the Yangtse River, Chongqing Technology and Business University, Chongqing 400067, China
2 School of Economics and Management, Taiyuan Normal University, Taiyuan 030619, China
3 Qinghai University Library, Qinghai University, Xining 810016, China
4 Institute of Nationalities, Guizhou Academy of Social Sciences, Guiyang 550002, China
5 School of Economics, Yunnan University, Kunming 650091, China
* Correspondence: 2014990005@qhu.edu.cn

Abstract: Poverty eradication is one of the global challenges, and land transfer provides an effective path to address farmers' poverty; however, the effect of poverty reduction can show heterogeneity depending on the location, household, and head of household. This study employs the propensity value matching technique to compare the effects of the land transfer on the future alleviation of poverty among farm households, based on the vulnerability as expected poverty, using data from 4608 household tracking surveys. The findings point to the following: In general, rural land transfers can significantly lessen farm households' VEP. In terms of regional variations, the positive effects of land transfers on farm households' VEP are mainly in the west. In terms of the differences among households, it was found that land transfers contribute to lower VEP for non-poor, non-financing-constrained, and government-subsidized farm households. With regard to differences in household headship, land transfers have abating effects on the VEP of self-employed heads of farm households. The results of the study can provide a useful reference for policy-making on land management and poverty reduction among farmers

Keywords: land transfer; vulnerability as expected poverty; farm households; heterogeneity

Citation: Wang, Z.; Yang, M.; Zhang, Z.; Li, Y.; Wen, C. The Impact of Land Transfer on Vulnerability as Expected Poverty in the Perspective of Farm Household Heterogeneity: An Empirical Study Based on 4608 Farm Households in China. *Land* **2022**, *11*, 1995. https://doi.org/10.3390/land11111995

Academic Editors: Li Ma, Yingnan Zhang, Muye Gan and Zhengying Shan

Received: 5 October 2022
Accepted: 4 November 2022
Published: 7 November 2022

1. Introduction

Land is essential to the survival of farmers, as it serves multiple purposes, including production, livelihood, and social security [1,2]. However, unlike other commodities, rural land in China is collectively owned and therefore cannot be traded freely [3,4]. Historically, the dated land system has stifled agricultural productivity and created obstacles to the movement of rural labor to industries and regions that provide better economic opportunities [5]. In order to accommodate the rapid growth of China's economy and technology, the government has continuously enhanced the land property rights system [6]. In 2014, the Central Committee's No. 1 document proposed the "three rights division" for contracted land. Farmers are able to protect their land rights and interests and obtain the right to dispose of and benefit from land management rights, and land management rights can be freely traded on the market, thereby activating the asset function of land. In 2015, the "Decision of the Central Committee of the Communist Party of China and the State Council on Winning the Battle of Poverty Alleviation" proposed that farmers' cooperatives and other business entities be supported to increase the income of poor households through land transfer methods such as land trusteeship and the absorption of farmers' land management rights as shares. Through land transfer, the disadvantage of land fragmentation is eliminated, creating conditions for the realization of agricultural scale and modernization [7,8].

Land transfer is a crucial component of China's efforts to deepen the reform of its rural land system [9–11]. By granting farmers more property rights, land transfer has activated the efficiency of market resource allocation, enabled the scale and modernization of agriculture, and become a crucial factor in farmers' ability to escape poverty and become affluent [12].

China's household contracted cultivated land circulation area reached 530 million mu by the end of 2018, representing 35% of the country's cultivated land area. Consequently, a large number of surplus laborers were transferred to cities, which promoted urbanization and industrialization, boosted the overall social welfare level, and improved rural conditions by coordinating the integrated development of urban and rural areas, thereby becoming an effective practice for alleviating poverty in China's rural areas [13–15]. In 2018, the per capita disposable income of rural residents in impoverished areas was CNY 10,371, which was 1.99 times higher than in 2012 and grew by 12.1% annually on average. The alleviation of poverty in China's land transfer can serve as a valuable theoretical reference for farmers in other developing countries seeking to eradicate poverty.

Land transfer in this article refers to the transfer of land use rights from farmers with contracted land management rights to other farmers or economic organizations [16–18]. The essence of rural land transfer is that, in the context of multiple factors in the agricultural economy, the land rental market transfers land use rights from those with lower land valuations to those who are more eager to increase their production value through a mechanism of price equilibrium [19,20], and it assists farmers with varying land labor endowments in re-adjusting their marginal products [21,22]. In addition, land transfer can facilitate the transfer of rural surplus labor from agriculture to other industries [23]. This is the internal mechanism for improving the income of farmers through land transfer [24] (Figure 1).

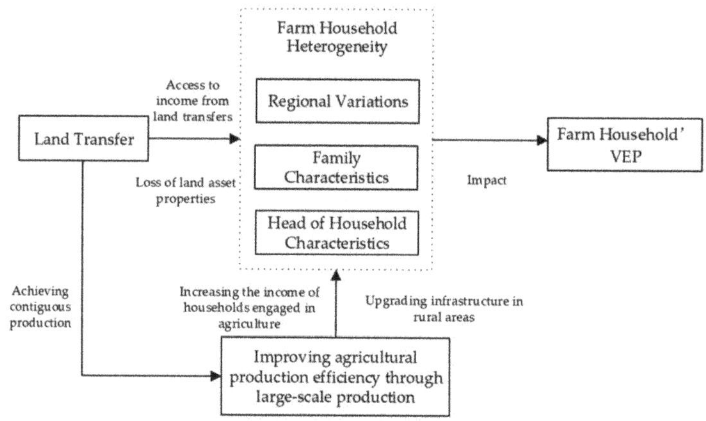

Figure 1. Mechanisms of the effect of land transfer on farm households' VEP.

The effect of the land transfer system on reducing poverty has attracted the attention of numerous scholars [25–27]. Land transfer has a leveling effect on marginal output, transaction income, and the Pareto effect, and it has become a crucial method for efficiently allocating rural land elements [28]. It has a significant effect on the multidimensional poverty of rural households in poor villages through a mechanism known as the "preventive saving motive" [29]. Kassie et al. believe that land transfer not only is conducive to reducing agricultural costs, but also encourages non-agricultural employment of farmers, improves the income structure of farmers, and improves the overall welfare of farmers [30]. It is well acknowledged that "preventing" poverty is far more important than "governing" poverty [31], and "preventing" poverty requires measuring farmers' susceptibility to poverty [32]. The poverty index measures only the welfare level at a static point in time; examines whether farmers are in the ex-post state of poverty, which cannot reflect the poverty risk that has not yet occurred [33]; and disregards the long-term impact of rural land transfer [34]. Land is the most important self-owned resource of poor households [35], has the material function of providing a means of subsistence for poor households, and is deeply embedded in these families' production and living processes [36]. In comparison to the income obtained from the land transfer, the loss of land management rights will have a negative impact on the future employment, social security, and mental health of farmers, thereby increasing the likelihood of future poverty [37]. The 2002 World Development Report of the World Bank used vulnerability as expected poverty (VEP) to measure the likelihood of an individual or family falling into poverty in the future [38]. This research makes use of VEP to calculate the likelihood of a farmer falling into poverty in the future.

In recent years, a number of scholars [39,40] have examined the relationship between land transfer and rural household poverty from the perspective of poverty vulnerability. Data from a field survey of 1682 farmers in Hubei Province by Peng et al. revealed that land transfer can significantly lessen a farmer's vulnerability to poverty and that this vulnerability declines as the area of land transferred increases [41]. Sun et al. found that the poverty vulnerability of land transfer households was 5.13% lower than that of non-transfer households [42]. Nonetheless, some scholars are concerned that the loss of land management rights will increase the likelihood of poverty among farmers [43,44]. Zhang et al. conducted an empirical study based on the survey data of 1386 rural households in southern Xinjiang and discovered that land transfer can significantly increase the income level of farmers, but cannot effectively reduce their susceptibility to poverty [45].

Presently, the academic community has not reached a relatively consistent conclusion regarding the effect of land transfer on farmers' vulnerability to poverty. In light of this, the formulation and implementation of rural land policies should vary from person to person, based on how well they take into account the differences between farmers. This paper will add to the existing body of knowledge in the following ways: We explore the general findings of the impact of land transfer on the poverty vulnerability of farm households using data from a national sample survey to supplement the existing studies that are restricted to a particular province or region. We classify farm households according to the characteristics of land transfer, and we verify the impact of land transfer on poverty vulnerability under different characteristics by regression, so as to identify what groups of characteristic farm households can reduce poverty vulnerability through land transfer, carve out the groups of farm households suitable for land transfer, and provide theoretical support for the implementation policy of land transfer policy classification.

Based on previous studies, this paper empirically analyzes the impact of land transfer on the VEP of farm households using a logit model based on the 2018 Chinese Family Panel Studies (CFPS) data; uses the stepwise regression method to select variables with significant effects; and regresses each of the five dimensions of regional distribution, poverty level, financing constraints, government subsidies, and nature of work. On this basis, propensity score matching (PSM) is used to test the robustness of the study results.

2. Materials and Methods

2.1. Data

The information in this article is drawn from the most recent (2018) Chinese Family Panel Studies (CFPS) (URL: http://www.isss.pku.edu.cn/cfps/; accessed on 15 October 2021). The database survey aims to track and investigate data at three levels: individual, family, and community, reflecting societal, economic, and population changes in China [46]. The survey covers a variety of topics, including family finances, education, health, and child-rearing. The 2018 CFPS database survey targets a sample size of 14,218 households across 31 provinces (excluding Hong Kong, Macao, and Taiwan). In order to acquire high-quality research data, the data were efficiently screened. First, all urban household data were eliminated and only rural household registration data were retained; second, the characteristic data of the household head corresponding to the "financial respondent" were matched and the individual data of non-heads of households were eliminated; finally, the missing values, outliers, and samples missing important variables were deleted; and finally, a valid sample of 4608 households was obtained, including 780 households with land transfer and 3828 households without land transfer.

2.2. Method

2.2.1. Vulnerability as Expected Poverty

Chaudhrui et al. proposed the concept and measurement method of vulnerability as expected poverty [47]. VEP allows for the precise identification of households that may fall into poverty in the future. In particular, this approach quantifies the likelihood that a family will either enter or remain in poverty as a result of the risk of a sudden economic shock [48]. If the likelihood exceeds the predetermined threshold for vulnerability, the family is considered to be vulnerable to poverty. The concept of poverty vulnerability proposed by the VEP measurement method is simple to comprehend, reflects the dynamic characteristics of poverty, and can be effectively applied to cross-sectional data; consequently, it is widely utilized in academia. The formula for calculating poverty vulnerability is as follows:

$$\hat{V}_i = Prob(lnc_i < lnz|X_i) = \Phi\left[(lnz - X_i\hat{\beta}_{FGLS})/\sqrt{X_i\hat{\theta}_{FGLS}}\right] \quad (1)$$

where $\hat{V}_i$ is the estimated value of the probability of poverty of farmer i in the future, c_i is the per capita consumption of the household, z is the poverty line, Φ is the cumulative distribution function of the normal distribution, and $\hat{\beta}_{FGLS}$ and $\hat{\theta}_{FGLS}$ denote the expected value and variance of the family's future consumption estimated by the feasible generalized least squares (FGLS) method; compared with the ordinary least squares, FGLS can effectively eliminate the heteroscedasticity of the model and improve the accuracy of the estimation results. X_i is an observable variable, mainly including family characteristic variables (including income, population, assets, liabilities, employment, and education) and household head characteristic variables (including age, gender, marriage, health, and occupation).

In order to assess a family's VEP, this study uses per capita household consumption. With respect to poverty based on consumption, two observations can be made: First, income is easily underestimated in micro-surveys, while consumption may better reflect the family's level of well-being, and second, using income as an explanatory variable can easily lead to strong endogenous problems in the measurement model. Concerning the choice of the poverty line, there are primarily two standards of per capita daily consumption of USD 1.9 and USD 3.1 proposed by the World Bank in 2015; based on China's average purchasing power and CPI index, we convert them into CNY 2800 and CNY 4570 per capita annual consumption at the end of 2018 [49]. This paper primarily measures the farmers' VEP based on their USD 1.9 per day per capita consumption, in line with existing research. Regarding the vulnerability threshold, this paper refers to the work of Gunther and Ward and sets the vulnerability line at 0.29; i.e., if the probability of a rural household falling into poverty in the future is greater than 0.29, set to 1; less than 0.29, set to 0 [50,51].

2.2.2. Econometric Model

In order to comprehensively examine the impact of land transfer on the poverty vulnerability of farmers, the logit model is constructed as follows:

$$\begin{cases} log\left(\frac{p}{1-p}\right) = \beta_0 + \beta_1 transfer_i + \beta_2 X_i + \varepsilon_i \\ p = prob(v_i = 1) \end{cases} \quad (2)$$

where v_i is the household poverty vulnerability status of farmer i calculated based on the per capita consumption level of the household, $v_i = 1$ represents a poor and vulnerable household, and $v_i = 0$ represents a non-poor and vulnerable household. $transfer_i$ indicates whether farmer i participates in land transfer, and X_i indicates a series of control variables, mainly including family characteristics and household head characteristics.

2.2.3. Stepwise Regression

Since there are many control variables affecting the dependent variable in the model, and the influence of individual variables on the dependent variable is small, the sum of squared errors does not decrease when these variables are included, but on the contrary, the error becomes larger due to the increase in the number of variables, which affects the overall stability. To address this problem, the stepwise regression method is used to select the variables with significant effects from many control variables to establish a regression model. The specific model is as follows:

Given the set of candidate control variables $T = \{x_1 \cdots x_m\}$, from which a subset $T_1 \in T$, the sum of squared errors of the regression model constructed from T_1 and the dependent variable is Q, and then the square of the remaining standard deviation of the model is Equation (3).

$$T^2 = \frac{Q}{n - l - 1} \tag{3}$$

X in the formula is the data sample size. The selected subset T_1 should minimize T as a quantitative criterion for variable selection.

Then determine an initial subset, each time from the subset outside the impact of significant variables to introduce a maximum impact on the dependent variable, and then the original subset of variables to test, from the variables that become insignificant to eliminate a minimum impact, until it cannot be introduced and eliminated. Meanwhile, there are highlights worth noting in this model. First, the significance level a_m for the introduced variables and a_{out} for the excluded variables should be selected appropriately; obviously, the larger the a_m is, the more variables are introduced, and the larger the a_{out} is, the fewer variables are excluded. The second highlight is that due to the correlation between individual variables, the introduction of a new variable will make a variable originally considered significant insignificant and thus be dropped, so we selected variables that are as independent of each other as possible.

2.2.4. PSM Method

Whether land transfer reduces the poverty vulnerability of farm households is a non-randomized experimental self-selection problem that is highly susceptible to selective error, which can be effectively addressed by propensity score matching (PSM). PSM is a non-parametric analysis method of counterfactual inference, which can effectively reduce selectivity bias and endogeneity by analytically processing non-experimental and observational data, and it is commonly applied in the evaluation of policy effects. PSM is used to test the robustness of the logit regression results. The logit model is used to calculate the conditional probability fitting value of the sample farmers' land transfer, which is the propensity score (PS).

$$PS_i = \Pr[D_i = 1|\mathrm{X_i}] = \mathrm{E}[D_i = 0|\mathrm{X_i}] \tag{4}$$

$$ATT = \frac{1}{N^t} \sum\nolimits_{i \in I^t \cap S} \left\{ Y_i - \sum\nolimits_{j \in I^c \cap S} W_{ij} Y_j \right\} \tag{5}$$

where N^t is the number of samples of land transfer households, I^t is the sample set of the disposal group (participating in land transfer), I^c is the sample set of the control group (not involved in land transfer), Y_i is the observed value of the sample of the disposal group, and Y_j is the sample of the control group. S is the common support domain set, W_{ij} is the matching weight, and ATT is the average disposition effect. The main method is to match the samples of the control group and the disposal group according to the propensity value to ensure that there is no significant difference in their main characteristics. Then, the control group is used to estimate the counterfactual state of the treatment group (i.e., no participation in the transfer) and calculate the poverty caused by the land transfer and the net treatment effect of vulnerability ATT.

2.3. Variables

Since the focus of this paper is on whether the loss of land management rights increases the risk of poverty for the farmers, the key variable defined is whether farmers transferred their farmland. Farmers who transferred their farmland are controlled for using question FS2 "whether they lease their land to others", with a value of 1 if the household transferred its farmland and 0 otherwise. A significance of $p < 0.1$ is set, and 10 control variables are identified by excluding insignificant variables through stepwise regression. Descriptive statistics are provided in Table 1.

Table 1. Variable definition and descriptive statistics.

Variables	Variable Meaning	Calculation Method	Mean	Std
explanatory variable				
transfer	land transfer	land transfer = 1, other = 0	0.169	0.375
family characteristic variables				
lnfincome	per capita household income	logarithm of per capita household income	10.483	1.072
lnasset	family assets	logarithm of family assets	12.028	1.188
lnagri	agricultural machinery assets	logarithm of family farming machinery assets	3.231	4.062
lnralat	social capital	logarithm of the family relationship expenditure	7.116	2.388
lnsize	family size	logarithm of household size	3.908	1.979
subside	government subsidy	government subsidy = 1, other = 0	0.657	0.474
house	housing property	more than 1 property = 1, other = 0	0.140	0.347
household head characteristic variables				
lnage	age of head of household	logarithm of 2018 respondent age	52.696	13.316
marry	head of household marriage	married = 1, other = 0	0.858	0.348
edu	head of household education	education level	6.310	4.176

3. Results

3.1. Baseline Regression and Sub-Regional Regression

The results of the baseline regression of the impact of the land transfer on the VEP of farm households are presented in Table 2. The results indicated that across the sample, all other factors being equal, land transfer effectively decreases the VEP of farm households, in that the transfer of land management rights does not increase the likelihood of farm households falling into poverty in the future. This conclusion is generally consistent with Deng and Wang et al.'s findings [52,53].

Table 2. Baseline regression and sub-regional regression results.

Variables	All Area	East	Middle	West
transfer	−0.00605 ***	−0.00227	−0.00523 **	−0.0102 ***
	(0.00123)	(0.00196)	(0.00206)	(0.00229)
lnfincome	−0.00508 ***	−0.00642 ***	−0.00523 ***	−0.00443 ***
	(0.000679)	(0.00111)	(0.00124)	(0.00116)
lnasset	−0.00130 **	−0.00280 ***	−0.000120	−0.000356
	(0.000571)	(0.000887)	(0.00109)	(0.000998)
lnagri	−0.000576 ***	−0.000191	−0.000308	−0.00114 ***
	(0.000140)	(0.000236)	(0.000251)	(0.000242)
lnralat	−0.00102 ***	−0.000659 *	−0.00121 ***	−0.00143 ***
	(0.000251)	(0.000358)	(0.000453)	(0.000493)
lnsize	0.0391 ***	0.0314 ***	0.0396 ***	0.0471 ***
	(0.00121)	(0.00190)	(0.00220)	(0.00222)
subside	−0.00236 **	−0.00158	−0.00219	−0.00296
	(0.00118)	(0.00178)	(0.00218)	(0.00219)
house	−0.00318 *	−0.00032	−0.00459	−0.00628 **
	(0.00167)	(0.00253)	(0.00290)	(0.00317)
lnage	0.0434 ***	0.0343 ***	0.0356 ***	0.0544 ***
	(0.00218)	(0.00384)	(0.00394)	(0.00368)
marry	−0.0230 ***	−0.0188 ***	−0.0215 ***	−0.0282 ***
	(0.00174)	(0.00279)	(0.00333)	(0.00294)
edu	−0.00125 ***	−0.00113 ***	−0.00149 ***	−0.00127 ***
	(0.000145)	(0.000254)	(0.000260)	(0.000244)
Constant	−0.0936 ***	−0.0251	−0.0770 ***	−0.154 ***
	(0.0121)	(0.0207)	(0.0228)	(0.0204)
R-squared	0.275	0.256	0.283	0.306
N	4608	1511	1350	1747

Note: Standard errors are in parentheses; * $p < 0.10$, ** $p < 0.05$, *** $p < 0.01$. Same below.

It is widely accepted in the academic literature that regional economic development is a crucial factor in determining whether or not rural households are able to reduce their poverty levels [54]. There are obvious gaps in the economic development of eastern, central, and western China, and land prices are relatively low in economically underdeveloped regions [55]. Rapidly advancing industrialization and urbanization have substantially increased urban labor compensation, causing a large number of rural laborers to migrate to cities [56]. The level of economic development, industrial structure, employment capacity, and wage level of cities have an immediate impact on the income of urban farmers [57]. Generally, the non-agricultural supply of labor pattern of Chinese agricultural families exhibits specific geographical characteristics [58], and this pattern is driven by the development of non-agricultural enterprises outside the village [59]. In other words, the growth of these industries will have an obvious effect on rural labor migration but will have no discernible effect on the mobility of agricultural property rights [60]. However, within the sub-regional sample, the effect of land transfer on household VEP is statistically insignificant in the eastern region, but statistically significant at the 5% and 1% levels in the central and western regions. There is significant heterogeneity in the estimated coefficients for land transfer between the central and western regions. This may be due to the fact that the development of tertiary industry in the western region lags behind and cannot provide enough alternative employment opportunities outside agriculture, and that the livelihood conversion costs incurred by farm households after the transfer of farmland are more expensive; therefore, the impact of the land transfer on reducing VEP of farm households is not as significant in the western region as it is in the central region.

The results for the control variables show the following: (1) For household characteristics, the estimated coefficients for the variables household income, household assets, agricultural assets, social capital, and housing property were negatively significant, while the estimated coefficients for household size and government subsidies were positively significant. This suggests that higher household income, increased assets, the accumulation of social capital, participation in non-farm businesses, and improved housing conditions will all contribute to reducing household poverty vulnerability. The greater the household size, the greater the risk of future poverty shocks to the farm household. Land transfer would reduce the risk of poverty for households receiving government subsidies; a possible explanation is that the funds obtained from land transfers are an important source of income for such households. (2) Regarding the characteristics of the household head, the marital status and education of the household head had a significant negative effect on the VEP of the household, while the age of the household head had a significant positive effect. VEP is influenced by many factors, such as natural resources, ability, and economic cycles, and only 10 control variables were selected for our study, resulting in a regression with an R^2 between 0.2 and 0.4, but it is sufficient to explain the impact of land transfer.

3.2. Heterogeneity in the Impact of Land Transfer on Household Poverty Vulnerability across Different Types of Farm Households

The fundamental situation, economic source, and capital composition of the family have a substantial bearing on the impact of land transfer. Farmers experience the most economic pressure and the least social pressure; farmers have the highest adaptability to social capital and the lowest adaptability to financial capital [61]. Cultivated land serves as a social security function for farmers, and land assets not only represent agricultural production means in rural China, but also provide employment security for farmers and can be used as collateral for loans [62], allowing farmers to apply for bank loans [63]. With the improvement of the farmland transfer market, the availability of non-agricultural labor has increased, the average non-agricultural labor time of rural households has risen, and the average household income has increased [64]. The family characteristics of farmers, such as income, sources of income, and social capital, have a direct effect on the likelihood of future poverty [65,66]. Different types of farm households were grouped and regressed in clusters to assess the impact of the land transfer on their VEP.

3.2.1. Clustering Farmer Households According to Their Poverty Level

According to World Bank poverty thresholds of USD 1.9 and USD 3.1 per capita per day, households with per capita daily consumption of USD 1.9 or less are classified as extremely poor, those with a per capita daily consumption of USD 1.9 to USD 3.1 are classified as relatively poor, and those with a per capita daily consumption of USD 3.1 or more are classified as non-poor. Consequently, the impact of the land transfer on VEP of three types of farm households, namely absolutely poor, relatively poor, and non-poor, was investigated. In Table 3, columns 1 to 3 display the regression results of land transfer on VEP of absolute poverty, relative poverty, and non-poverty households. The number of absolute poverty in the sample is 0, so column 1 is empty. The results indicated the following: (1) Land transfer can substantially reduce the household VEP of non-poor and relatively poor farmers. This suggests that non-poor and relatively poor farmers can use land transfer to revitalize their household assets to increase their income, reduce poverty, and improve their ability to resist poverty in the future. (2) The further comparison revealed that the greater absolute value of the estimated coefficient of land transfer on VEP of households with relative poverty suggests that land transfer is more effective at preventing the occurrence of poverty in this population. Relatively poor households are even more lacking in the original accumulation of capital to escape poverty. The poverty trap theory proposes that the vicious cycle of low income–low savings–low capital formation–low output–low income is the root cause of poor households' inability to escape poverty [67]. The short-term rental income effect brought by land transfer can provide relatively poor households with the necessary capital accumulation to break out of the poverty trap.

Table 3. Regression results grouped by household poverty level.

Variables	Absolute Poverty	Relative Poverty	Non-Poverty
transfer		−0.00797 * (0.00463)	−0.00535 *** (0.000820)
control variable	control	control	control
Constant		−0.295 *** (0.0470)	−0.0625 *** (0.00804)
R-squared		0.392	0.276
Observations	0	965	3643

3.2.2. Clustering by Criteria of Farm Household Financing Constraints and Government Subsidies

It is common for farmers in rural areas to be constrained by financing, which has become a significant factor impeding their ability to increase their income and reduce poverty. Farmers are divided into financing-constrained households and non-financing-constrained households based on the results of the questionnaire item "Have you ever been denied a loan or credit?" On this basis, the effect of land transfer on the poverty and vulnerability of farmers with varying financial constraints is investigated. Government subsidies refer primarily to whether or not farmers have received various cash or in-kind subsidies from the government, such as subsistence allowances, subsidies for returning farmland to forests, and agricultural subsidies. Government subsidies are an essential policy instrument for alleviating poverty. Farmers were classified as government-subsidized and non-government-subsidized farmers, and the disparate effects of land transfer on the poverty and vulnerability of these two groups were investigated.

The regression results in Table 4 indicate the following: (1) Land transfer has no significant effect on the VEP of rural households with financing constraints, but it has a significant inhibitory effect on the VEP of households without financing constraints. Possible causes include a rise in the demand for credit due to the diversification of farmers' sources of income away from agriculture after the transfer of their land, particularly in the form of non-farm self-employment activities. Non-financing-constrained households can more easily meet credit needs and quickly convert household income to non-agricultural operations, whereas this is difficult for financing-constrained households. (2) Land transfer can effectively reduce the VEP of government-subsidized households while having no significant effect on households that have not received government subsidies. The government's planned provision of targeted vocational education, property development and alleviation programs, and other preferential policies may be the cause of this smooth resolution of the employment issue facing rural land transfer households.

Table 4. Regression results grouped by financing constraints and government subsidies.

Variables	Financing Constraints		Government Subsidies	
	(1) Yes	(2) No	(3) Yes	(4) No
transfer	−0.00155 (0.00342)	−0.00510 *** (0.00168)	−0.00441 *** (0.00170)	−0.00419 (0.00291)
control variable	control	control	control	control
Constant	−0.100 *** (0.0269)	−0.0912 *** (0.0137)	−0.131 *** (0.0144)	−0.0464 ** (0.0219)
R-squared	0.255	0.279	0.324	0.217
Observations	1001	3607	3029	1579

3.3. *Heterogeneity in Head of Household Characteristics*

The individual circumstances of the household's primary breadwinner are a significant factor in that household's level of financial well-being and, by extension, its degree of independence from subsistence vulnerability and external risks, and they vary widely [68]. The job category of the head of household will lead to a widening of the poverty and income gap, and the impact of human capital on household economic growth is significantly greater than that of physical capital [69]. When farmers retire or become too old to work in agriculture, they will sublease their cultivated land in order to maintain their daily consumption [70]. According to the results of a survey conducted by Glauben et al. [71], educational attainment plays a significant role in reducing poverty. The vulnerability of structural poverty is the primary source of vulnerability in rural China, and compulsory education has significantly decreased the vulnerability to poverty [72]. Universal primary education reduces structural poverty vulnerability more effectively than temporary poverty vulnerability [73]. Further analysis proves that compulsory education primarily improves various abilities to obtain a permanent income, such as cognitive ability and participation in non-agricultural work, which are crucial means of reducing structural poverty [74]. Consequently, it is essential to identify these characteristics that influence household head income and incorporate them into the assessment of land transfer effects, as we will do in this paper.

The relationship between land transfer and a family's VEP will vary depending on the nature of work of the head of household, who is typically the primary decision-maker and backbone of the family. Farmers are classified into three categories based on the employment status of the household's head: self-employed farmers, farmers with unstable employment, and farmers with stable employment. Farmers whose primary source of income comes from their own agricultural operations are considered self-employed, farmers whose primary source of income comes from both agriculture and non-agricultural jobs are considered unstably employed, and farmers whose primary source of income comes from non-agricultural jobs with a stable employer are considered employed farmers. Accordingly, the variability of the impact on farm household VEP with different employment choices of the household head is examined. The regression results are shown in Table 5.

Table 5. Regression results grouped by the nature of work of the household head.

Variables	The Nature of the Householder's Work		
	(1) Self-Employed	**(2) Not Stably Employed**	**(3) Stably Employed**
transfer	−0.00527 **	−0.0108 **	−0.000913
	(0.00222)	(0.00465)	(0.00144)
control variable	control	control	control
Constant	−0.150 ***	−0.107 **	−0.0282 **
	(0.0173)	(0.0528)	(0.0139)
R-squared	0.301	0.348	0.132
Observations	3125	121	610

Table 5 presents regression results for the effect of land transfer on household VEP for each of the three household head employment choices. Land transfer can significantly reduce the VEP of self-employed farmers but has no significant effect on the VEP of employed farmers. Possible explanations include the fact that the family income structure of self-employed farmers is single, the transfer of farm land creates contiguous production that increases the efficiency of agricultural output, and the income increase effect is obvious, thereby effectively reducing VEP.

3.4. PSM Robustness Analysis

PSM was developed to test the validity of the preceding conclusions. The farmer households whose farmland was transferred out were assigned to the disposal group, while those who did not participate in the farmland transfer were assigned to the control group. The disposal group and the control group were matched based on their propensity scores, and a balance test was conducted to ensure that there were no significant differences in their main characteristics. The causal relationship between land transfer and VEP of rural households was then investigated. At the same time, based on the family poverty level, financing constraints, government subsidies, and type of work of the household head as criteria, the subsamples are divided for group testing and the average elimination effect is calculated using two matching methods: nuclear matching and nearest neighbor matching. The results of the PSM robustness test are shown in Table 6.

Table 6. PSM method robustness test results.

Sample Classification	Matching Method	ATT	Std. Err.	Sample Classification	Matching Method	ATT	Std. Err.
full sample	kernel matching	−0.027 **	0.0118	financing constraints	kernel matching	0.0093	0.0240
	neighbor matching	−0.0385 *	0.0208		neighbor matching	0.0408	0.0420
east	kernel matching	−0.0110	0.0190	non-financial constraints	kernel matching	−0.036 ***	0.0137
	neighbor matching	−0.0338	0.0362		neighbor matching	−0.0538 **	0.0248
middle	kernel matching	−0.0230	0.0177	government subsidies	kernel matching	−0.040 ***	0.0126
	neighbor matching	−0.0179	0.0353		neighbor matching	−0.0235	0.0266
west	kernel matching	−0.037*	0.0199	non-government subsidies	kernel matching	0.0018	0.0182
	neighbor matching	−0.0216	0.0367		neighbor matching	0.0001	0.0256
absolute poverty	kernel matching	null	null	self-employed	kernel matching	−0.0271 *	0.0157
	neighbor matching	null	null		neighbor matching	−0.0483	0.0324
relative poverty	kernel matching	−0.0076	0.0433	not stably employed	kernel matching	−0.0786	0.0843
	neighbor matching	−0.0444	0.0774		neighbor matching	−0.0277	0.1170
non-poverty	kernel matching	−0.035 ***	0.0102	stably employed	kernel matching	−0.0156	0.0116
	neighbor matching	−0.0372 *	0.0206		neighbor matching	−0.0167	0.0211

Note: The average treatment effect on the treated(ATT) is a participant average treatment effect.

Table 6 demonstrates that the ATT of land transfer on VEP of rural households under the two matching methods is negative and statistically significant. In general, land transfer reduces the VEP of farm households by 2.7% to 3.8%. A possible explanation is that China's per capita arable land is small, and the corresponding land transfer area for farm households is small, so the measured land transfer does not have a significant impact on VEP. The ATT in the western region is significantly negative, and land transfer can reduce the VEP of rural households in the western region by about 3.7%. The average disposal effect in the central region is negative but not statistically significant, which is significantly inconsistent with the negative regression coefficients in Table 2, indicating that the effect of land transfer on the VEP of rural households in the central region requires additional investigation. The average disposition effect in the eastern region is also not significant, which partially corroborates the robustness of the regression results in Table 2.

The ATT of absolute and relatively poor households is not statistically significant, which is consistent with the conclusion of the regression analysis presented in Table 3. The average disposal effect of non-poor households is notably negative, and land transfer can reduce the VEP of non-poor households by 3.5% to 3.7%. This verifies the validity of the regression results presented in Table 3.

From the standpoint of financing constraints and government subsidies, the average disposal effect of households with financing constraints is insignificant, whereas the ATT of households without financing constraints is significantly negative. The transfer of land will reduce the VEP of households without financial constraints by 3.6% to 5.4%. The ATT of farmers who receive government subsidies is significantly negative, and land transfer can reduce the VEP of government-subsidized families by 4%, whereas the average disposal effect of farmers who do not receive government subsidies is positive. The effect is inconsequential. This verifies the validity of the regression results in Table 4.

Based on the household head's job nature grouping, the ATT is significantly negative for self-employed farmers, and land transfer can significantly reduce the poverty risk of self-employed farmers by 2.7%. However, the ATT does not have a significant impact on reducing the VEP of the two types of farm households. The robustness of the regression conclusions presented in Table 5 is evident.

In conclusion, most of the research conclusions presented in Tables 2–5 have passed the robustness test.

4. Discussion, Conclusions, and Implications

4.1. Discussion

Transferring rural land is conducive to enhancing the utilization efficiency of rural land resources, boosting the competitiveness and comprehensive economic benefits of agriculture, and ensuring the continuous increase in farmers' income. However, land transfer compensates agricultural households primarily for the production function of the land, but insufficiently for its security function and asset function [75]. Farmers who transfer land may become "uncultivated land, insecure, and unemployable" if they engage in new labor, participate in labor market competition [76,77], and adapt to urban life [78]. We believe that the issue of the impact of the land transfer on VEP cannot be generalized and needs to be discussed according to the heterogeneity of farm households.

According to the poverty trap theory, regional differences, material resources, educational level, social capital, and financial constraints all have an effect on poverty [79]. In terms of regional heterogeneity, the higher the level of regional economic development and the more comprehensive the infrastructure, the more conducive they are to reducing the likelihood of future poverty in the region; otherwise, the region will fall into persistent poverty [80,81]. This also applies to the effect of land transfer on reducing poverty. Heterogeneity in family characteristics, family income, financial constraints, government subsidies, and other characteristics will result in greater income and labor dividend heterogeneity resulting from land transfer [82–84]. There are disparities in family wealth, and the primary factor is the household head [85]. For the majority of peasant families, the household head is the determining factor in the family's income [86]. The education and occupation of the household head have a strong explanatory power for the family's wealth [87–89]. According to the findings of this study, land transfer can encourage the diversification of farmers' livelihood strategies, thereby increasing farmers' income. Land transfer facilitates large-scale agricultural production and is a crucial means for farmers to increase their income. The study portrayed the group of farmer households able to reduce VPE through land transfer.

Compared with existing studies, this paper uses the VEP indicator to measure the future poverty risk faced by farming households, overcoming the fact that current studies only measure the welfare level of individuals or households at a certain point in time; it takes into account the heterogeneity of farming households and explores the poverty reduction effect of land transfer on households with different characteristics separately; the use of the PSM method effectively solves the endogeneity problem of VEP.

In addition, there are some shortcomings in this research, which future research would need to address: (1) The modest reduction in poverty vulnerability in the PSM test results may be due to the small size of the transferred land area, which needs to be focused on in the next surveys and studies. (2) The effect of land transfer on the VEP of rural households in the east and central region has not yet been conclusively demonstrated, and further research is necessary to analyze the effects of land transfer on poverty reduction in different regions, disaggregated by level of economic development and conditions of intensive land use. (3) Due to a lack of data, the selection of factors influencing VEP is not comprehensive enough, for example, the distance between the village and the main town can be used as a variable to measure how land values affect the VEP. In the future, we will conduct in-depth research on the aforementioned topics in an effort to arrive at more meaningful conclusions and offer more instructive recommendations for practice.

4.2. Conclusions and Implications

This paper conducted a categorical regression of farm households through a multidimensional perspective and applied the PSM method to test the effect of land transfer on farm household VEP and its heterogeneity. The findings indicate the following: (1) In rural China, land transfer has been shown to have abatement effects on VEP of between 2.9% and 4.2% for farm households. (2) In the western region, land transfer significantly reduces the VEP of farm households, while in the eastern and central regions, no such reduction is seen, and the effect on the VEP of farm households in the central region remains to be demonstrated. (3) In terms of household characteristics subgroups, land transfer has no significant effect on the VEP of farming households already in poverty (absolute and relative poverty), while it can significantly reduce the VEP of non-poor farming households. (4) Land transfer was effective in promoting VEP reduction among farm households without financial constraints and with government subsidies, but had no effect on VEP among farm households with financing constraints and without government subsidies. (5) When considering the subgroup of characteristics associated with household heads, land transfer primarily has a significant abating effect on the VEP of self-employed farmers, while having no significant impact on the VEP of employed farmers. These results provide empirical evidence for government land management, agricultural development, and the decisions of farmers.

The preceding findings have significant policy ramifications: The first step is to expedite the process of reforming the system that governs rural land and then to standardize and methodically guide the flow of rural land. Currently, the small amount of arable land per person in China's rural areas, the difficulty of structural adjustment, and the high cost of agricultural production are significant factors limiting the efficiency of agriculture, the increase in farmer income, and the revitalization of rural areas. Transferring rural land is conducive to enhancing the efficacy of rural land resource utilization, boosting the competitiveness and comprehensive economic benefits of agriculture, and ensuring farmers' income growth is sustainable. In addition, the loss of land management rights does not result in severe poverty shocks for farming households, as land transfer enables farmland-transferring households to harvest land transfer rents, boosts the capital accumulation of farming households, encourages the diversification of farmers' occupations, and enhances the structure of farming households' household income. Secondly, rural land transfer must adhere to the local, village-based, and household-based principles. Due to the varying natural conditions in different regions and the unbalanced rural economic development that has resulted in significant differences among villages and farmers, the promotion of land transfer should be tailored to local conditions, meaning that it should be different from village to village and different from household to household. Furthermore, it should not be carried out in a manner that is too hasty, as this would infringe on the lawful rights and interests of farmers. The third phase is to put in place a system to facilitate the transfer of land in rural areas. The government should propose a series of targeted financial and human capital service protection policies, such as financial assistance for eligible new agricultural businesses to take precedence in agriculture-related projects. Furthermore, issues such as a lack of funds for relocating families and achieving diversification of livelihood strategies and sustainable livelihoods should be addressed. In the meantime, the government should offer targeted vocational and special skills training in non-farming fields to increase farmers' employability outside of agriculture.

Author Contributions: Conceptualization, Z.W.; methodology, Z.W., Y.L. and M.Y.; software, M.Y. and Z.W.; validation, Z.W. and M.Y.; formal analysis, Z.W. and Z.Z.; investigation, Z.W.; resources, M.Y.; data curation, M.Y. and Y.L.; writing—original draft preparation, Z.W. and M.Y.; writing—review and editing, M.Y., C.W., Y.L. and Z.Z.; visualization, M.Y. and Z.W.; supervision, M.Y. and C.W. All authors have read and agreed to the published version of the manuscript.

Funding: This research was financially supported by the National Social Science Foundation of China (Grant No. 20&ZD095), Pan Jiahua Expert Workstation (Grant No. 2021GZZH01), Chongqing Social Science Planning Project (Grant No. 2020ZDSC07, 2020ZDGL07), Chongqing Municipal Education Commission Science and Technology Plan Project (Grant No. KJZD-K20210080), Shanxi Provincial Philosophy and Social Science Planning Project (Grant No. 2021YY04), Chongqing Postgraduate Research Innovation Project (Grant No. CYB22271), and Humanities and Social Sciences Research project of Chongqing Municipal Commission of Education (Grant No. 21SKGH346).

Institutional Review Board Statement: Not applicable.

Informed Consent Statement: Not applicable.

Data Availability Statement: Not applicable.

Acknowledgments: The authors thank the Institute of Social Science Survey for providing data.

Conflicts of Interest: The authors declare no conflict of interest.

References

1. Augusseau, X.; Nikiéma, P.; Torquebiau, E. Tree biodiversity, land dynamics and farmers' strategies on the agricultural frontier of southwestern Burkina Faso. *Biodivers. Conserv.* **2006**, *15*, 613–630. [CrossRef]
2. Meert, H.; Van Huylenbroeck, G.; Vernimmen, T.; Bourgeois, M.; Van Hecke, E. Farm household survival strategies and diversification on marginal farms. *J. Rural Stud.* **2005**, *21*, 81–97. [CrossRef]
3. Brandt, L.; Huang, J.; Li, G.; Rozelle, S. Land rights in rural China: Facts, fictions and issues. *China J.* **2002**, 67–97. [CrossRef]
4. Tian, L. The chengzhongcun land market in China: Boon or bane?—A perspective on property rights. *Int. J. Urban Reg. Res.* **2008**, *32*, 282–304. [CrossRef]
5. Wu, Z.; Liu, M.; Davis, J. Land consolidation and productivity in Chinese household crop production. *China Econ. Rev.* **2005**, *16*, 28–49. [CrossRef]
6. Luo, B. 40-year reform of farmland institution in China: Target, effort and the future. *China Agric. Econ. Rev.* **2018**, *10*, 16–35. [CrossRef]
7. Hartvigsen, M. Land reform and land fragmentation in Central and Eastern Europe. *Land Use Policy* **2014**, *36*, 330–341. [CrossRef]
8. Long, H.; Tu, S.; Ge, D.; Li, T.; Liu, Y. The allocation and management of critical resources in rural China under restructuring: Problems and prospects. *J. Rural Stud.* **2016**, *47*, 392–412. [CrossRef]
9. Liu, Y.S.; Zang, Y.Z.; Yang, Y.Y. China's rural revitalization and development: Theory, technology and management. *J. Geogr. Sci.* **2020**, *30*, 1923–1942. [CrossRef]
10. Rithmire, M.E. Land institutions and Chinese political economy: Institutional complementarities and macroeconomic management. *Politics Soc.* **2017**, *45*, 123–153. [CrossRef]
11. Wang, Q.; Zhang, X. Three rights separation: China's proposed rural land rights reform and four types of local trials. *Land Use Policy* **2017**, *63*, 111–121. [CrossRef]
12. Zhou, Y.; Li, X.; Liu, Y. Rural land system reforms in China: History, issues, measures and prospects. *Land Use Policy* **2020**, *91*, 104330. [CrossRef]
13. Cai, F. The great exodus: How agricultural surplus laborers have been transferred and reallocated in China's reform period? *China Agric. Econ. Rev.* **2018**, *10*, 3–15. [CrossRef]
14. Li, X.; Hui, E.C.-m.; Lang, W.; Zheng, S.; Qin, X. Transition from factor-driven to innovation-driven urbanization in China: A study of manufacturing industry automation in Dongguan City. *China Econ. Rev.* **2020**, *59*, 101382. [CrossRef]
15. Ye, J. Stayers in China's "hollowed-out" villages: A counter narrative on massive rural–urban migration. *Popul. Space Place* **2018**, *24*, e2128. [CrossRef]
16. Wang, J.; Lin, Y.; Glendinning, A.; Xu, Y. Land-use changes and land policies evolution in China's urbanization processes. *Land Use Policy* **2018**, *75*, 375–387. [CrossRef]
17. Wang, Y.; Li, X.; Xin, L.; Tan, M.; Jiang, M. Spatiotemporal changes in Chinese land circulation between 2003 and 2013. *J. Geogr. Sci.* **2018**, *28*, 707–724. [CrossRef]
18. Zhou, Y.; Guo, L.; Liu, Y. Land consolidation boosting poverty alleviation in China: Theory and practice. *Land Use Policy* **2019**, *82*, 339–348. [CrossRef]
19. Zhang, Y. Rural Land System and Rights. In *Insights into Chinese Agriculture*; Springer: Beijing, China, 2018; ISBN 9789811310508.
20. Guanghua, Q.; Bao, Z.; Jing, Z.; Brown, C.J.T.R.J. Land rental, prices and the management of China's grasslands: The case of Inner Mongolia Autonomous Region. *Rangel. J.* **2018**, *40*, 231–240. [CrossRef]
21. Feng, S.; Heerink, N.; Ruben, R.; Qu, F. Land rental market, off-farm employment and agricultural production in Southeast China: A plot-level case study. *China Econ. Rev.* **2010**, *21*, 598–606. [CrossRef]
22. Ragkos, A.; Koutsou, S.; Theodoridis, A.; Manousidis, T.; Lagka, V. Labor management strategies in facing the economic crisis. Evidence from Greek livestock farms. *New Medit* **2018**, *17*, 59–71. [CrossRef]
23. Feng, S.; Heerink, N. Are farm households' land renting and migration decisions inter-related in rural China? *NJAS-Wagening. J. Life Sci.* **2008**, *55*, 345–362. [CrossRef]
24. Peng, K.; Yang, C.; Chen, Y. Land transfer in rural China: Incentives, influencing factors and income effects. *Appl. Econ.* **2020**, *52*, 5477–5490. [CrossRef]
25. He, Q.; Deng, X.; Li, C.; Kong, F.; Qi, Y.J.L. Does land transfer improve farmers' quality of life? Evidence from rural China. *Land* **2021**, *11*, 15. [CrossRef]
26. Deininger, K.W. *Land Policies for Growth and Poverty Reduction*; Oxford University Press: Oxford, UK, 2003; ISBN 0821350714.
27. Varga, M.J.W.D. Poverty reduction through land transfers? The World Bank's titling reforms and the making of "subsistence" agriculture. *World Dev.* **2020**, *135*, 105058. [CrossRef]
28. Harvey, J.; Jowsey, E. *Urban Land Economics*; Bloomsbury Publishing: London, UK, 2019.
29. Pradhan, K.C.; Mukherjee, S. Covariate and idiosyncratic shocks and coping strategies for poor and Non-poor rural households in India. *J. Quant. Econ.* **2018**, *16*, 101–127. [CrossRef]

30. Kassie, G.W.; Kim, S.; Fellizar, F.P., Jr. Determinant factors of livelihood diversification: Evidence from Ethiopia. *Cogent Soc. Sci.* **2017**, *3*, 1369490. [CrossRef]
31. Collier, P.; Dollar, D. *Globalization, Growth, and Poverty: Building an Inclusive World Economy*; World Bank Publications: Washington, DC, USA, 2002.
32. Bebbington, A. Capitals and capabilities: A framework for analyzing peasant viability, rural livelihoods and poverty. *World Dev.* **1999**, *27*, 2021–2044. [CrossRef]
33. Fafchamps, M. *Rural Poverty, Risk and Development*; Edward Elgar Publishing: London, UK, 2003.
34. Yesuf, M.; Bluffstone, R.A. Poverty, risk aversion, and path dependence in low-income countries: Experimental evidence from Ethiopia. *Am. J. Agric. Econ.* **2009**, *91*, 1022–1037. [CrossRef]
35. Muraoka, R.; Jin, S.; Jayne, T.S. Land access, land rental and food security: Evidence from Kenya. *Land Use Policy* **2018**, *70*, 611–622. [CrossRef]
36. Ado, A.M.; Savadogo, P.; Abdoul-Azize, H.T. Livelihood strategies and household resilience to food insecurity: Insight from a farming community in Aguie district of Niger. *Agric. Hum. Values* **2019**, *36*, 747–761. [CrossRef]
37. Liang, F.; Lin, C.; Lin, S.-H. Farmers' livelihood, risk expectations, and homestead withdrawal policy: Evidence on Jinjiang Pilot of China. *Int. J. Strateg. Prop. Manag.* **2022**, *26*, 56–71. [CrossRef]
38. Gillis, M.; Shoup, C.; Sicat, G.P. *World Development Report 2000/2001-Attacking Poverty*; The World Bank: Washington, DC, USA, 2001.
39. Zhang, Y.; Wang, W.; Feng, Y. Impact of different models of rural land consolidation on rural household poverty vulnerability. *Land Use Policy* **2022**, *114*, 105963. [CrossRef]
40. Cao, M.; Xu, D.; Xie, F.; Liu, E.; Liu, S. The influence factors analysis of households' poverty vulnerability in southwest ethnic areas of China based on the hierarchical linear model: A case study of Liangshan Yi autonomous prefecture. *Appl. Geogr.* **2016**, *66*, 144–152. [CrossRef]
41. Peng, J.; Wu, H.; Qin, X. Study on the Impact of Land Transfer on the Poverty Vulnerability of Farmers. *China Land Sci.* **2019**, *33*, 67–75. [CrossRef]
42. Qing, S.X.L. study on the poverty reduction effect and heterogeneity of land transfer from multi-dimensional perspective—Based on the analysis of CFPS2018 micro data. *Chin. J. Agric. Resour. Reg. Plan.* **2022**, *43*, 259–268. [CrossRef]
43. Fei, R.; Lin, Z.; Chunga, J. How land transfer affects agricultural land use efficiency: Evidence from China's agricultural sector. *Land Use Policy* **2021**, *103*, 105300. [CrossRef]
44. He, S.; Liu, Y.; Webster, C.; Wu, F. Property rights redistribution, entitlement failure and the impoverishment of landless farmers in China. *Urban Stud.* **2009**, *46*, 1925–1949. [CrossRef]
45. Zhang, Y.; Yang, J. Poverty reduction effect of land transfer in deep poverty areas of southern Xinjiang. *J. Arid Land REsources Environ.* **2020**, *34*, 16–22. [CrossRef]
46. Xie, Y.; Hu, J.W. An Introduction to the China Family Panel Studies (CFPS). *Chin. Sociol. Rev.* **2014**, *47*, 3–29. [CrossRef]
47. Chaudhuri, S.; Jalan, J.; Suryahadi, A. *Assessing Household Vulnerability to Poverty from Cross-Sectional Data: A Methodology and estimates from Indonesia*; Department of Economics Discussion Papers 0102-52; Columbia University: New York, NY, USA, 2002. [CrossRef]
48. Ligon, E.; Schechter, L. Measuring vulnerability. *Econ. J.* **2003**, *113*, C95–C102. [CrossRef]
49. Wu, D.; Liu, J.; Zhang, X. Statistical Measurement of the Overall Scale of Urban Poverty in China. *Sci. Soc. Res.* **2022**, *4*, 104–110. [CrossRef]
50. Günther, I.; Harttgen, K. Estimating Households Vulnerability to Idiosyncratic and Covariate Shocks: A Novel Method Applied in Madagascar. *World Dev.* **2009**, *37*, 1222–1234. [CrossRef]
51. Ward, P.S. Transient Poverty, Poverty Dynamics, and Vulnerability to Poverty: An Empirical Analysis Using a Balanced Panel from Rural China. *World Dev.* **2016**, *78*, 541–553. [CrossRef]
52. Wang, W.; Luo, X.; Zhang, C.M.; Song, J.H.; Xu, D.D. Can Land Transfer Alleviate the Poverty of the Elderly? Evidence from Rural China. *Int. J. Environ. Res. Public Health* **2021**, *18*, 11288. [CrossRef]
53. Deng, X.; Xu, D.D.; Zeng, M.; Qi, Y.B. Does outsourcing affect agricultural productivity of farmer households? Evidence from China. *China Agric. Econ. Rev.* **2020**, *12*, 673–688. [CrossRef]
54. Long, H.; Liu, Y.; Li, X.; Chen, Y. Building new countryside in China: A geographical perspective. *Land Use Policy* **2010**, *27*, 457–470. [CrossRef]
55. Xie, K.; Ding, M.; Zhang, J.; Chen, L. Trends towards Coordination between Grain Production and Economic Development in China. *Agriculture* **2021**, *11*, 975. [CrossRef]
56. Kuang, W.; Liu, J.; Dong, J.; Chi, W.; Zhang, C. The rapid and massive urban and industrial land expansions in China between 1990 and 2010: A CLUD-based analysis of their trajectories, patterns, and drivers. *Landsc. Urban Plan.* **2016**, *145*, 21–33. [CrossRef]
57. Wu, Z.; Yao, S. Intermigration and intramigration in China: A theoretical and empirical analysis. *China Econ. Rev.* **2003**, *14*, 371–385. [CrossRef]
58. Deng, X.; Zeng, M.; Xu, D.D.; Qi, Y.B. Why do landslides impact farmland abandonment? Evidence from hilly and mountainous areas of rural China. *Nat. Hazards* **2022**, *113*, 699–718. [CrossRef]
59. Lu, H.; Xie, H.; Yao, G. Impact of land fragmentation on marginal productivity of agricultural labor and non-agricultural labor supply: A case study of Jiangsu, China. *Habitat Int.* **2019**, *83*, 65–72. [CrossRef]

60. Xu, D.; Yong, Z.; Deng, X.; Zhuang, L.; Qing, C.J.L. Rural-urban migration and its effect on land transfer in rural China. *Land* **2020**, *9*, 81. [CrossRef]
61. Yang, X.; Guo, S.; Deng, X.; Xu, D. Livelihood Adaptation of Rural Households under Livelihood Stress: Evidence from Sichuan Province, China. *Agriculture* **2021**, *11*, 506. [CrossRef]
62. Wang, H.H. Economic value of tradable farmland use rights and mortgage loans in China. *Agric. Financ. Rev.* **2008**, *68*, 289–300. [CrossRef]
63. Galak, J.; Small, D.; Stephen, A.T. Microfinance decision making: A field study of prosocial lending. *J. Mark. Res.* **2011**, *48*, S130–S137. [CrossRef]
64. Xie, H.; Lu, H. Impact of land fragmentation and non-agricultural labor supply on circulation of agricultural land management rights. *Land Use Policy* **2017**, *68*, 355–364. [CrossRef]
65. Gautam, Y.; Andersen, P. Rural livelihood diversification and household well-being: Insights from Humla, Nepal. *J. Rural Stud.* **2016**, *44*, 239–249. [CrossRef]
66. Mbiba, M.; Collinson, M.; Hunter, L.; Twine, W. Social capital is subordinate to natural capital in buffering rural livelihoods from negative shocks: Insights from rural South Africa. *J. Rural Stud.* **2019**, *65*, 12–21. [CrossRef]
67. Kraay, A.; McKenzie, D. Do Poverty Traps Exist? Assessing the Evidence. *J. Econ. Perspect.* **2014**, *28*, 127–148. [CrossRef]
68. Anderloni, L.; Bacchiocchi, E.; Vandone, D. Household financial vulnerability: An empirical analysis. *Res. Econ.* **2012**, *66*, 284–296. [CrossRef]
69. Van den Broeck, A.; De Cuyper, N.; De Witte, H.; Vansteenkiste, M. Not all job demands are equal: Differentiating job hindrances and job challenges in the Job Demands–Resources model. *Eur. J. Work Organ. Psychol.* **2010**, *19*, 735–759. [CrossRef]
70. Wang, D. China's Urban and Rural Old Age Security System: Challenges and Options. *China World Econ.* **2006**, *14*, 102–116. [CrossRef]
71. Glauben, T.; Herzfeld, T.; Rozelle, S.; Wang, X. Persistent Poverty in Rural China: Where, Why, and How to Escape? *World Dev.* **2012**, *40*, 784–795. [CrossRef]
72. Yang, Y.; Guo, X. Universal basic education and the vulnerability to poverty: Evidence from compulsory education in rural China. *J. Asia Pac. Econ.* **2020**, *25*, 611–633. [CrossRef]
73. Chambers, R. Poverty and livelihoods: Whose reality counts? *Environ. Urban.* **1995**, *7*, 173–204. [CrossRef]
74. Torun, H. Compulsory schooling and early labor market outcomes in a middle-income country. *J. Labor Res.* **2018**, *39*, 277–305. [CrossRef]
75. Guo, Y.; Liu, Y. Poverty alleviation through land assetization and its implications for rural revitalization in China. *Land Use Policy* **2021**, *105*, 105418. [CrossRef]
76. Li, Y.; Jia, L.; Wu, W.; Yan, J.; Liu, Y. Urbanization for rural sustainability–Rethinking China's urbanization strategy. *J. Clean. Prod.* **2018**, *178*, 580–586. [CrossRef]
77. Qian, Z. Land acquisition compensation in post-reform China: Evolution, structure and challenges in Hangzhou. *Land Use Policy* **2015**, *46*, 250–257. [CrossRef]
78. Liu, Y.; Fang, F.; Li, Y. Key issues of land use in China and implications for policy making. *Land Use Policy* **2014**, *40*, 6–12. [CrossRef]
79. Carter, M.R.; Barrett, C.B. The economics of poverty traps and persistent poverty: An asset-based approach. *J. Dev. Stud.* **2006**, *42*, 178–199. [CrossRef]
80. Kim, R.; Mohanty, S.K.; Subramanian, S.V. Multilevel Geographies of Poverty in India. *World Dev.* **2016**, *87*, 349–359. [CrossRef]
81. Michalek, A.; Madajova, M.S. Identifying regional poverty types in Slovakia. *Geojournal* **2019**, *84*, 85–99. [CrossRef]
82. Meng, X. Labor Market Outcomes and Reforms in China. *J. Econ. Perspect.* **2012**, *26*, 75–102. [CrossRef]
83. Ma, W.; Grafton, R.Q.; Renwick, A. Smartphone use and income growth in rural China: Empirical results and policy implications. *Electron. Commer. Res.* **2020**, *20*, 713–736. [CrossRef]
84. Chen, H.; Wang, X. Exploring the relationship between rural village characteristics and Chinese return migrants' participation in farming: Path dependence in rural employment. *Cities* **2019**, *88*, 136–143. [CrossRef]
85. Riphahn, R.T.; Serfling, O. Item non-response on income and wealth questions. *Empir. Econ.* **2005**, *30*, 521–538. [CrossRef]
86. Janssens, A. The Rise and Decline of the Male Breadwinner Family? An Overview of the Debate. *Int. Rev. Soc. Hist.* **1997**, *42*, 1–23. [CrossRef]
87. Gounder, R.; Xing, Z. Impact of education and health on poverty reduction: Monetary and non-monetary evidence from Fiji. *Econ. Model.* **2012**, *29*, 787–794. [CrossRef]
88. Joassart-Marcelli, P. Working poverty in southern California: Towards an operational measure. *Soc. Sci. Res.* **2005**, *34*, 20–43. [CrossRef]
89. Abbas, K.; Li, S.X.; Xu, D.Y.; Baz, K.; Rakhmetova, A. Do socioeconomic factors determine household multidimensional energy poverty? Empirical evidence from South Asia. *Energy Policy* **2020**, *146*, 111754. [CrossRef]

land

MDPI

Article

Assessing Cultivated Land–Use Transition in the Major Grain-Producing Areas of China Based on an Integrated Framework

Tiangui Lv [1,2], Shufei Fu [1], Xinmin Zhang [2], Guangdong Wu [3,*], Han Hu [1] and Junfeng Tian [3]

1 School of Tourism and Urban Management, Jiangxi University of Finance and Economics, Nanchang 330013, China
2 Institute of Ecological Civilization, Jiangxi University of Finance and Economics, Nanchang 330013, China
3 School of Public Policy and Administration, Chongqing University, Chongqing 400044, China
* Correspondence: gd198410@cqu.edu.cn

Abstract: The cultivated land-use transition (CLUT) is the morphological result of changes in the cultivated land-use mode over time, and the result of the interaction and mutual restriction of the human land system. This paper applies a "spatial–functional" integrated framework to understand the structure and functioning of CLUTs, and quantitatively evaluates and visualizes CLUTs in the major grain-producing area in southern China. The results show that (1) the comprehensive CLUT index in the middle and lower reaches of the Yangtze River changed from 0.0480 to 0.0711 from 2001 to 2019 and indicated significant differences in the transition index between different regions. (2) The CLUT identified a positive aggregation effect under a 5% significance during the period, and the agglomeration degree of the spatial and functional transitions strengthened, which increased from 0.3776 to 0.4673 and from 0.2127 to 0.2952, respectively. (3) The gravity center of the CLUT demonstrated a pattern of migration from the southwest to the northeast, and the migration speed of the gravity center decreased from 2.9401 km/year to 1.2370 km/year. The migration direction of the gravity center for the spatial transition is opposite to the functional transition, and the migration speed of the gravity center for the spatial and functional transitions decreased from 8.3573 km/year to 1.0814 km/year, and from 3.2398 km/year to 1.0254 km/year, respectively. To address this transition, policymakers should formulate differentiated policies to promote the sustainable use of cultivated land through the spatial and functional transition of major grain-producing areas.

Keywords: cultivated land-use transition (CLUT); spatial morphology; functional morphology; drive mechanism; main grain-producing areas

Citation: Lv, T.; Fu, S.; Zhang, X.; Wu, G.; Hu, H.; Tian, J. Assessing Cultivated Land–Use Transition in the Major Grain-Producing Areas of China Based on an Integrated Framework. *Land* **2022**, *11*, 1622. https://doi.org/10.3390/land11101622

Academic Editor: Francisco Manzano Agugliaro

Received: 19 August 2022
Accepted: 20 September 2022
Published: 22 September 2022

1. Introduction

Cultivated land is a key component to achieve sustainable development and ensure food security [1]. China has the third largest cultivated area in the world, but the per capita level is less than 1/2 of the global average. Since 1978, when socioeconomic reforms were implemented, China has experienced rapid development [2,3]. In 2000, in the opening year of the Tenth Five-Year Plan, the Chinese government stated that China's conditions for promoting urbanization were gradually maturing, and it was necessary to accelerate the implementation of an urbanization strategy [4,5]. Therefore, Chinese urbanization has entered an accelerated development stage. The urbanization levels increased from 36.20% to 60.60%, and the average annual urbanization growth rate remained at 1.28% from 2000 to 2019 [6]. However, this achievement also led to corresponding social, economic, and ecological problems. These problems include disorderly construction land expansion, inefficient land use [7], abandoned villages [8,9], and the overexploitation of ecological resources, which lead to a significant depletion of cultivated land resources and serious non-agriculturalization and pollution [10]. With this context of the times, China's cultivated land conditions are undergoing drastic changes. Accordingly, the cultivated land-use transition (CLUT) has become one of the most urgent issues to be addressed by the academic and governmental sectors. Land-use transition (LUT) originated in the 1990s at the University of Aberdeen in Scotland as the geographer Mather's exploration of the forest transition hypothesis, which involves corresponding changes in land-use morphology during social development [11,12]. The research on LUT has expanded from forests to grasslands, waters, and cultivated land. Meanwhile, from the research perspective of land change from land-use/cover change (LUCC) to land-change science (LCS), LUT has become a frontier hotspot of land-system science [13].

In fact, CLUT is the main component of LUT, which continues the research results of LUT in terms of concept and connotation, and is a trend turning point of cultivated land-use morphology under long-term changes. In other words, it is a process of transition from one stable state to another stable state [14]. Existing studies usually decompose cultivated land-use morphology into dominant and recessive [15]. The former includes the amount of cultivated land, planting structure, and landscape pattern attributes. The latter includes attributes that are not easy to be directly observed, such as cultivated land quality, property rights, and management methods, and can only be obtained through investigation, testing, and analysis [16]. This applies, for example, to the soil chemical content of cultivated land. The dominant morphology of cultivated land is the direct expression of its space utilization status; therefore, the dominant morphology is also called the spatial morphology, and the two are essentially the same [17]. In order to present the recessive morphology of cultivated land more intuitively, this study starts from the functional morphology of cultivated land for transition diagnosis. The functional morphology perspective of cultivated land corresponds to the human demand for cultivated land and is a comprehensive manifestation of recessive morphology [18].

Based on the definition of cultivated land pattern, the current research in this field generally covers the CLUT path, index construction, spatiotemporal differentiation characteristics, and driving mechanism [19] and its relationship to social and economic activities [20]. Regarding the path of CLUT, most studies mainly focus on the expression of a single path of cultivated land morphology. From the perspective of spatial pattern, it focuses on the study of land-use/cover change [21]. Research on the recessive pattern has a wide range of perspectives, such as changes in cultivated land productivity, monitoring of soil quality, and conversion of property rights [22–25]. There is still a gap in the research on combining spatial morphology and functional morphology to comprehensively evaluate CLUT. In terms of index construction, the evaluation of CLUT is mainly to quantify the morphological changes of cultivated land [26]. For different scholars, the understanding of CLUT performance is different, and the selection of indicators is also prone to differences, which reduces the comparability between different studies. For instance, when choosing the evaluation index of the social function of arable land, some researchers consider it to be the ability of cultivated land to solve farmers' work problems, and they select indicators such as agricultural income and the labor-carrying capacity of cultivated land; to the contrary, some researchers think that it is the faculty of cultivated land to produce food to feed the population, so the per capita food availability is selected for evaluation [27]. At the analysis level of the driving mechanism, the spatial econometric model is mainly used for quantitative research, and variables are selected from social and natural factors. However, qualitative research is the foundation of quantitative research, and the current logical analysis of the driving mechanism from a qualitative perspective is still relatively weak. In terms of the relationship with society and the economy, there are studies on the coupling and interaction mechanism of food production, agricultural–economic development, and rural labor transfer [28]. At the same time, CLUT is also usefully explored from the perspective of its ecological environment effect. At the regional level, the research mainly focuses on the national, provincial, and municipal regions. The cultivated land is mainly used for food production, and CLUT has a more profound impact on the main grain-producing areas than other areas. However, scholars have neglected the issue of

CLUT in the main grain-producing areas, which limits the guidance and reference value for formulating policies for sustainable use of regional cultivated land. To sum up, it can be seen that the relevant research on CLUT still has an incomplete evaluation framework, inconsistent index selection criteria, weak qualitative analysis of the driving mechanism, and gaps in regional research [29].

The main grain-producing areas have faced the various tasks of stable grain production, economic growth, industrial upgrading, and ecological protection [30]. The population and economy here are highly dense, which puts enormous pressure on the local land supply and ecological environment. Undoubtedly, urban development has changed the morphology of cultivated land. Therefore, this paper takes the middle and lower reaches of the Yangtze River as the research area. In addition, we attempt to construct a "spatial–functional" integrated framework to evaluate the CLUT. In particular, the study aims to: (1) identify and construct the connotation of CLUT from multiple dimensions; (2) build a comprehensive evaluation index system to achieve a scientific and effective quantitative evaluation of a regional CLUT; (3) reveal the spatiotemporal variation laws and characteristics of the CLUT in the main grain-producing areas; and (4) put forward targeted strategies for the optimal utilization and reasonable transition of cultivated land.

The marginal contributions of this study are mainly in the following four parts. (1) Based on the perspective of the comprehensive "spatial–functional" pattern to explore the spatiotemporal characteristics of CLUT, which expands the definition of cultivated land-use pattern and enriches the related literature of CLUT; (2) the selection of CLUT indicators under the two morphologies is subdivided and explained, which provides an evaluation system that can be used for reference; (3) the driving mechanism of CLUT is described from the theoretical level. The driving principles of natural and socioeconomic policy, technology, and other factors are analyzed. It lays a theoretical foundation for carrying out qualitative research; (4) considering the particularity of the main grain-producing areas, it provides a research basis for the formulation of cultivated land-use policies in the study area and similar areas.

2. CLUT Interpretation Framework

As the social stage develops from point A to point B (Figure 1), which is mainly affected by the natural environment, socioeconomic policy, and technology [31], to meet the needs of the social development stage, the cultivated land users will adjust their cultivated land utilization behavior through the actual conditions of the cultivated land, and the cultivated land-use morphology will change accordingly. When the social stage develops toward a higher point C, the original cultivated land-use morphology at point B should continue to be adjusted accordingly [32]. Therefore, for different scales, the CLUT is an iterative process of dynamic balancing of the human–earth system interactions and joint constraints [33,34].

Thus, this study divides cultivated land-use morphology into two aspects: namely, spatial and functional. Spatial morphology consists of the amount of cultivated land, landscape, and crop planting structure [35]. Functional morphology reflects the types of cultivated land functions, which is an inherent attribute [36]. According to the classification of land-use functions (LUFs), the functions of cultivated land are divided into the three basic functions of crop production, life support, and ecological maintenance [37]. Therefore, this study constructs a "spatial–functional" integration framework to identify and construct the mechanism of CLUT from multiple dimensions, which is helpful to comprehensively identify and evaluate CLUT.

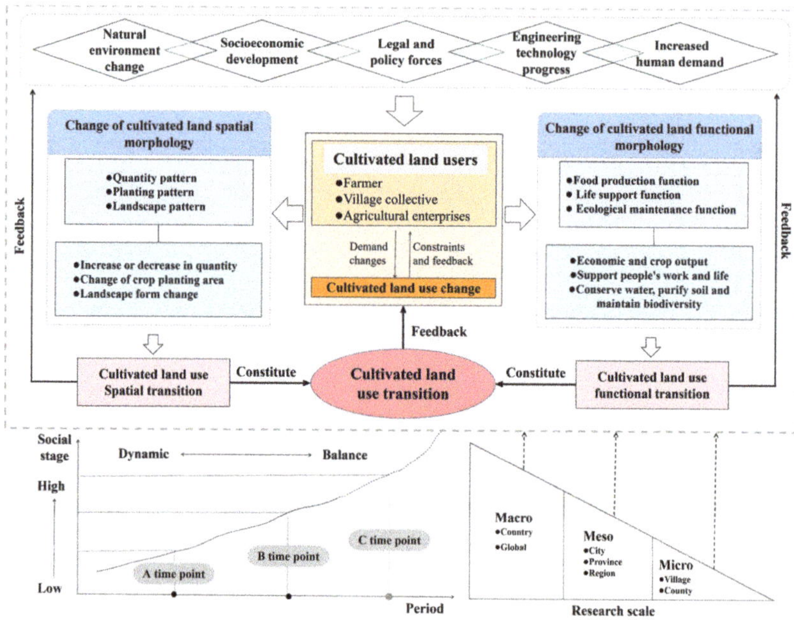

Figure 1. Interpretation framework of CLUT.

(1) Cultivated land-use spatial transition: When the social development level is low, due to the unclear property rights of cultivated land and underdeveloped agricultural technology, the social awareness of cultivated land protection is relatively weak and there is a disorderly expansion in cities. To ensure people's rations, food crops have always been the main crops grown on cultivated land. With the progress of society and the gradual improvement of the agricultural system, people's skills in using cultivated land will continue to increase, and more attention will be given to promoting the sustainability of cultivated land use. The national government will take relevant measures to curb the reduction of the cultivated land area and ensure quality degradation so that the amount of cultivated land may show a trend of "decrease–stable–increase" [38]. For example, China currently implements strict cultivated farmland requisition–compensation-balance policies to ensure the stability of the amount of cultivated land. To facilitate agricultural production, people constantly adjust the distribution pattern of cultivated land plots through engineering means, and the landscape pattern of cultivated land will develop in a concentrated, continuous, and regular direction [39]. Moreover, against the background of enriched consumer demand, and driven by the economic interests of cultivated land operators, the arable land planting structure will show the characteristics of diversified development. However, if the proportion of grain planting is too low, then the government will introduce and implement relevant systems to control the trend of non-grain cultivated land. In general, the quantity, planting, and landscape patterns of cultivated land change in the course of social development.

(2) Cultivated land-use functional transition: For most farmers, agricultural labor can provide employment and income, and the harvested crops can satisfy rations, which maintains the farmers' livelihood, prevents the occurrence of social risks such as hunger and unemployment, and ensures the stable development of society [40]. At this time, an increased use of chemicals is often relied on to increase food production. In this way, it is easy to cause the decline of cultivated land quality and biodiversity, which restricts the ecological function of cultivated land.

The average patch area of cultivated land will increase, the variety of crops will be diversified, and the quality will be increasingly higher so that the production function will be enhanced. In the urbanization process, in the pursuit of higher income, the rural surplus labor force will be transferred to nonagricultural industries, which results in the reduction of farmland employment opportunities for farmers and ultimately affects the life-support function of cultivated land [41]. Moreover, as the extensive use of agriculture leads to ecological destruction, human beings have begun to understand and adjust their relationship with nature [42], reduce the use of pesticides and fertilizers, and promote green fertilizers. Thus, the ecological maintenance function of cultivated land has received much attention and has improved greatly.

Comprehensively, CLUT is a continuous and cyclic process [43] that not only depends on the influence of the social economy, policy, technology, and other factors, as well as the promotion of cultivated land-user behavior, but also demonstrates the influencing factors and the behavior of cultivated land users. The cultivated land user decides whether to adjust the decision-making and behavior mode of cultivated land use according to their satisfaction with the result of a CLUT. At the same time, the feedback of the CLUT can promote the change and generation of the driving factors. For example, to satisfy the endless demands of mankind, the cultivated land users try to continuously add chemical fertilizers in exchange for the output of cultivated land. However, the result may be the destruction of the cultivated land ecosystem [44]. To ensure sustainable development, policymakers need to issue corresponding policies to restrict the input of chemical fertilizers. Therefore, in the overall operation of the CLUT mechanism, the driving factors and transition results promote and restrict one another through the subject of cultivated land use to form a "multifactor drive → subject behavior change → CLUT → transition result feedback → new wheel drive" cyclical interaction process, which is also the internal model of the spatiotemporal dynamic evolution of cultivated land-use morphology.

3. Methodology and Data

3.1. Study Area

The main grain-producing area is situated in Southern China at 24°29′–35°20′ N and 108°21′–121°57′ E. The study included 71 prefecture-level cities under the jurisdiction of the five provinces of Jiangxi, Hunan, Hubei, Anhui, and Jiangsu (Figure 2). The total area measures 81.19×10^4 km^2, which is approximately 8.46% of China's total land area. The terrain here is flat, and it has abundant cultivated land resources. Meanwhile, the mild climate and abundant rainfall have laid a good foundation for China's agricultural production. In 2019, the urban area was 37,672.74 km^2, and the cultivated land of this region was 21.77×10^4 km^2, which account for approximately 4.64% and 26.82% of the total area, respectively. The grain output reached $15{,}617.47 \times 10^4$ tons, which is nearly one-quarter of the country's total grain output (National Bureau of Statistics of China (NBSC), 2020). With the improvement in the degree of economics and urbanization, and the change in social demand, the structure and methods of cultivated land use have undergone tremendous changes.

3.2. Explanation of the CLUT Assessment Indicators

Based on the "spatial–functional" integration of the CLUT, the evaluation indicator system of the CLUT was established from the perspective of morphology. The indicators were selected based on a scientific analysis of cultivated land morphology and collaboration with scholars. The evaluation system consists of six factor layers and 18 indicator layers in spatial and functional morphology. The following is a brief summary of the reasons for choosing these indicators (Table 1).

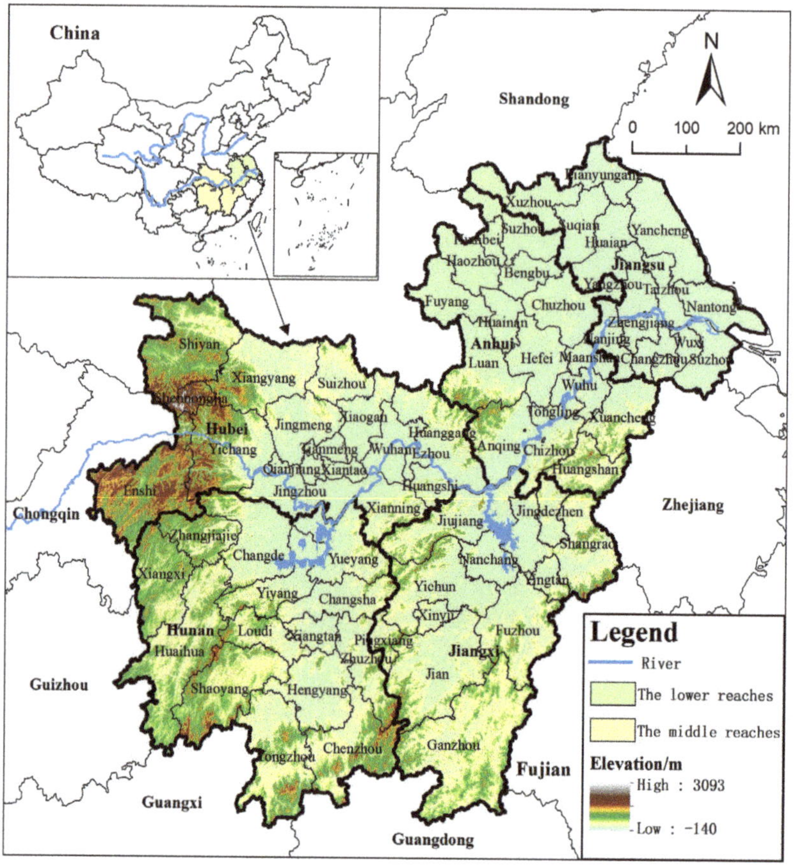

Figure 2. The study area includes Anhui, Hubei, Hunan, Jiangsu, and Jiangxi Provinces.

3.2.1. Selection of the Evaluation Index for the Spatial Transition

(1) The index of the pattern of cultivated land quantity considers its total resources, area-change rate, and utilization degree [45]. The total area of cultivated land can reflect the resource endowment conditions of a region, and the area-change rate can embody the change in the amount of cultivated land. The reclamation rate can embody the change in the cultivated land-use area. Therefore, the quantity pattern of cultivated land is measured by three indicators: namely, the total area of cultivated land (x_1), the rate of area change (x_2), and the reclamation rate (x_3).

(2) Cultivated land planting pattern evaluation indicators consider the spatial change in the cultivated land per capita planting area, the planting intensity, and the planting structure. The per capita cultivated land area owned by planting industry employees can reflect the area of cultivated land that can be planted per capita. The proportion of the grain and cash crop planting area can reflect the planting structure of arable land. The multiple cropping index can measure the planting intensity of cultivated land. The expansion of the sown area of arable land will lead to the exponential growth of multiple cropping. Therefore, we choose the per capita cultivated land area (x_4), the proportion of the planting area for food crops and cash crops (x_5), and the multiple cropping index (x_6) to characterize the changes in the cultivated land planting pattern.

Table 1. Evaluation index system of CLUT.

Target	Factor	Indicators	Unit	Weights	Direction①	Description	Min	Max	Mean
Spatial transition	Quantity pattern	Total area of cultivated land (x_1)	khm^2	0.0603	+	Total area of arable land.	6.3210	843.2165	277.3511
		Rate of area change (x_2)	%	0.0322	+	Newly increased (decreased) cultivated land area/total cultivated land area of the previous year.	−8.2665	5.4520	−0.7233
		Reclamation rate (x_3)	%	0.1430	+	Cultivated land area/total land area.	0.0194	0.6405	0.2914
	Planting pattern	Per capita cultivated land area (x_4)	hm^2/person	0.0547	+	Cultivated land area/employees in planting industry.	0.0334	0.3332	0.1034
		Proportion of the planting area for food crops and cash crops (x_5)	%	0.0292	+	The planting area of food crops/the planting area of cash crops.	0.5423	9.8029	2.2209
		Multiple cropping index (x_6)	%	0.0965	+	Total sown area of crops/cultivated land area.	1.0352	2.9817	1.9745
	Landscape pattern	Fragmentation degree (x_7)	/	0.0195	−	Total number of farmland patches/total farmland landscape area.	0.0010	0.1167	0.0287
		Aggregation degree (x_8)	/	0.0136	+	The number of adjacent patches/the total number of cultivated land patches.	56.2128	99.3243	91.6753
		Landscape patterns index (x_8)	/	0.0169	−	E/min E.	2.8757	83.4278	22.3680
Functional transition	Crop production function	Average economic output value of cultivated land (x_{10})	yuan/hm^2	0.1132	+	Total output value of planting industry/cultivated land area.	873.6559	15,437.1585	4367.4795
		Average grain yield of cultivated land (x_{11})	kg/hm^2	0.0242	+	The total output of grain crops/the sown area of grain crops.	2164.4612	7918.1970	5869.9524
		Average cash crop yield of cultivated land (x_{12})	kg/hm^2	0.0644	+	The total output of cash crops/the sown area of cash crops.	3159.7650	8453.7768	4246.2841

Table 1. *Cont.*

Target	Factor	Indicators	Unit	Weights	Direction①	Description	Min	Max	Mean
	Life support function	Proportion of employees in planting industry (x_{13})	%	0.0423	+	Planting industry employees/total rural labor force.	0.1197	0.8532	0.5163
		Per capita food guarantee rate (x_{14})	%	0.0612	+	Total grain output/(resident population × 400 kg).	29.7647	280.0304	116.7594
		Per capita agricultural income ratio (x_{15})	%	0.1604	+	Farmers' per capita planting income/farmers' per capita income.	873.3799	22,038.8067	4694.1082
	Ecological maintenance function	Chemical load of cultivated land (x_{16})	t/hm^2	0.0107	−	Pesticides, fertilizers, and agricultural film usage/area of cultivated land.	0.1216	2.2480	0.6957
		Diversity of crop species (x_{17})	/	0.0239	+	Sim②.	0.2044	0.7695	0.5393
		Effective irrigation area ratio (x_{18})	%	0.0338	+	Effective irrigation area/total planting area of crops.	0.2076	0.8796	0.4055

Note: ① For the "Direction" column, the higher the value of the "+" direction index, the better the practical significance, and the lower the "−" direction index value, the better the practical significance. ② $Sim = 1 - \sum_{i=1}^{n} P_i^2$: Sim is an index of crop species diversity. Where P_i is the ratio of the sown area of the i-th type of crops to the total sown area of the crops, i is the type of crops, and n is the number of types of crops. Based on the aforementioned, the area, select food crops, vegetable crops, melon crops, and oil crops for calculation.

(3) The evaluation index of the cultivated land landscape pattern should represent the transition of its landscape morphology in agricultural production activities [43]. The landscape pattern mainly includes the degree of fragmentation and agglomeration, and the irregularities of farmland patches. This represents the concentration of cultivated land plots and the convenience of cultivators to plant. If cultivated land is relatively fragmented and has an irregular shape, then this will hinder large-scale development and mechanization. Therefore, the transition of the cultivated land landscape pattern is characterized by the fragmentation degree (x_7), aggregation degree (x_8), and landscape patterns index of cultivated land patches (x_9).

3.2.2. Selection of the Evaluation Index for the Functional Transition

(1) The crop production function evaluation indicators are considered from the two aspects of the economic value output capacity and the crop output capacity of cultivated land. The gross value of cultivation is an important manifestation of the productive capacity of the cultivated land economy. The output of food crops and cash crops can reflect the production value of agricultural products on cultivated land [37]. Therefore, the crop production function of cultivated land is calculated by using the average economic output value (x_{10}), average grain yield (x_{11}), and average cash crop yield of cultivated land (x_{12}).

(2) The life-support function evaluation index should show the ability of cultivated land to guarantee the rural population's employment, income, and social food security [42]. The proportion of planting in the rural population represents the employment absorption capacity of cultivated land for the rural population. The ratio of per capita agricultural income to total income represents the population's economic dependence on cultivated land. The per capita food guarantee rate represents the degree to which people depend on cultivated land for survival. Thus, the proportion of employees in the planting industry (x_{13}), the per capita food guarantee rate (x_{14}), and per capita agricultural income ratio (x_{15}) are selected to represent the life-support function of arable land.

(3) The evaluation indicators of the ecological maintenance function take into account the negative pressure of chemical substances on cultivated land ecosystems, the resilience of the ecosystems, and the resistance to natural disasters [46]. When the chemical load of cultivated land is heavier, the degree of ecological damage to cultivated soil is greater. When the crop species diversity index is higher, the resilience of cultivated land ecosystems is stronger. When the effective irrigation area is larger, the drought resistance of cultivated land is stronger. The reasonable and effective use of cultivated land will decrease ecological damage and maintain the balance of the ecosystem. Therefore, the ecological maintenance function is characterized by the chemical load of cultivated land (x_{16}), the diversity of crop species (x_{17}), and the effective irrigation area ratio (x_{18}).

3.3. Methodology

3.3.1. Entropy Weight Method

The entropy weight method is currently the most widely used objective weighting method, which can effectively eliminate the dimensional influence. This study uses the entropy weight method to determine the weight of each evaluation index, which reflects the contribution of each index to the CLUT. According to the evaluation index system of CLUT, the CLUT index is the sum of the products of each functional index and its respective weight. Since this method is relatively common, the specific formula refers to [47].

3.3.2. Exploratory Spatial Data Analysis

An exploratory spatial data analysis (ESDA) can perform a correlation and aggregation analysis of neighborhood spatial data, which can effectively verify the spatial clustering characteristics of regional CLUTs. Two types of autocorrelation coefficients are usually used for this measurement. The first is the global spatial autocorrelation coefficient: the distribution of the Moran scatter plot is used to show the spatial correlation of the CLUT in the study area. The expression is

$$I = \frac{n\sum_{i=1}^{n}\sum_{j=1}^{n} W_{ij}(x_i - \overline{x})(x_j - \overline{x})}{\sum_{i=1}^{n}\sum_{j=1}^{n} W_{ij}\sum_{i=1}^{n}(x_i - \overline{x})^2} \tag{1}$$

where I is the global Moran index, and x_i and x_j are the CLUT index in cities i and j, respectively, and $\overline{x}$ represents the average of the CLUT indices, and W_{ij} is the spatial weight matrix. In this study, a spatial adjacency matrix was used, which was constructed by GeoDa software. The value of I is $[-1,1]$. When $I = 0$, this indicates that the space is not autocorrelated; when $I > 0$, this means that there is a positive correlation, and when $I < 0$, this indicates that there is a negative correlation. The closer the absolute value of I is to 1, the greater the degree of clustering and the spatial correlation.

The second type is the local spatial autocorrelation coefficient: it can use an LISA graph to check the heterogeneity of the data calculation and reveal the correlation degree of the attribute values between spatial units and adjacent units. The formula is as follows:

$$I_i = \frac{n(x_i - \overline{x})\sum_{j=1}^{n} Wi_j(x_j - \overline{x})}{\sum_{i=1}^{n}(x_i - \overline{x})^2} \tag{2}$$

When $I_i > 0$, high-high/low-low means that the spatial unit value is higher/lower than all the surrounding units and that the integrated spatial difference is smaller. When $I_i < 0$, then low-high/high-low means that the lower/higher spatial unit value is higher/lower than the surrounding units and that the integrated spatial difference is smaller.

3.3.3. Standard Deviation Ellipse

The standard deviation ellipse (SDE) is used to quantitatively describe the spatial characteristics of the elements. The azimuth of the ellipse represents the main trend direction, the long axis represents the dispersion of the geospatial elements in its direction, and the center of gravity represents the relative position. The results of the SDE calculation can reflect the spatial change in the CLUT (Equations (3)–(5)).

$$tan\theta = \frac{\left(\sum_{i=1}^{n} w_i^2 x_i'^2 - \sum_{i=1}^{n} w_i^2 y_i'^2\right) + \sqrt{\left(\sum_{i=1}^{n} w_i^2 x_i'^2 - \sum_{i=1}^{n} w_i^2 y_i'^2\right)^2 + 4\sum_{i=1}^{n}(w_i^2 x_i'^2 y_i'^2)}}{2\sum_{i=1}^{n} w_i^2 x_i' y_i'} \tag{3}$$

$$\overline{X}_w = \frac{\sum_{i=1}^{n} w_i x_i}{\sum_{i=1}^{n} w_i}; \overline{Y}_w = \frac{\sum_{i=1}^{n} w_i y_i}{\sum_{i=1}^{n} w_i} \tag{4}$$

$$\sigma_x = \sqrt{\frac{\sum_{i=1}^{n}(w_i x_i' cos\theta - w_i y_i' sin\theta)^2}{\sum_{i=1}^{n} w_i^2}}; \sigma_y = \sqrt{\frac{\sum_{i=1}^{n}(w_i x_i' sin\theta - w_i y_i' cos\theta)^2}{\sum_{i=1}^{n} w_i^2}} \tag{5}$$

where $\tan\theta$ is the azimuth angle of the ellipse, i.e., the angle formed by the clockwise rotation from due north to the long axis of the ellipse; $(\overline{X}_w, \overline{Y}_w)$ are the center of gravity coordinates, and X_i and Y_i are the spatial location elements, W_i represents the weight, x_i', y_i' represents the deviation of the coordinates of the elements at different points from the mean center, and σ_x and σ_y are the standard deviations along the x- and y-axes, respectively.

3.3.4. Data Collection

The socioeconomic data come from the provincial and municipal statistical yearbooks of Hubei, Hunan, Jiangxi, Anhui, and Jiangsu in 2002, 2008, 2014, and 2020. Since ESA's land-use cover data has the advantages of authority, continuity, openness, and includes research time-point data, the land-use data in this research are derived from the ESA's global 300-m land cover data in 2001, 2007, 2013, and 2019. The administrative zoning data were obtained from the 1:4 million dataset of the China National Basic Geographic Information Center. To preprocess the land-use data, we first classified the land-use types into six types: namely, cultivated land, forestland, grassland, water bodies, construction land, and unused land. Afterward, the different land-use type data were recoded and assigned CLT1 to CLT6. On this basis, Fragstats software was then used to measure the fragmentation, aggregation, and landscape morphology indices of the cultivated land.

4. Results

4.1. Analysis of the CLUT

4.1.1. Measurement of Comprehensive Index of the CLUT

Overall, the comprehensive index of the CLUT in the study area increased from 0.0480 to 0.0711, which is an increase of 48.13%. The comprehensive CLUT index in the five provinces also increased (Figure 3). However, there was an imbalance in the CLUT between provinces. The order of the average value of the comprehensive transition index was: Hunan Province (0.0645) > Jiangsu Province (0.0617) > Anhui Province (0.0589) > Jiangxi Province (0.0575) > Hubei Province (0.0540). The order of the growth rate of the comprehensive transition index was: Jiangsu Province (62.26%) > Jiangxi Province (52.41%) > Hunan Province (46.65%) > Hubei Province (44.47%) > Anhui Province (34.84%). The extreme differences in the comprehensive transition index in the study area in 2001, 2007, 2013, and 2019 were 0.0345, 0.0438, 0.0587, and 0.0641, respectively. These results indicate that the morphology of cultivated land use in this region is undergoing rapid changes. Due to the heterogeneity of natural and socioeconomic characteristics, the differences in the CLUT between regions are expanding.

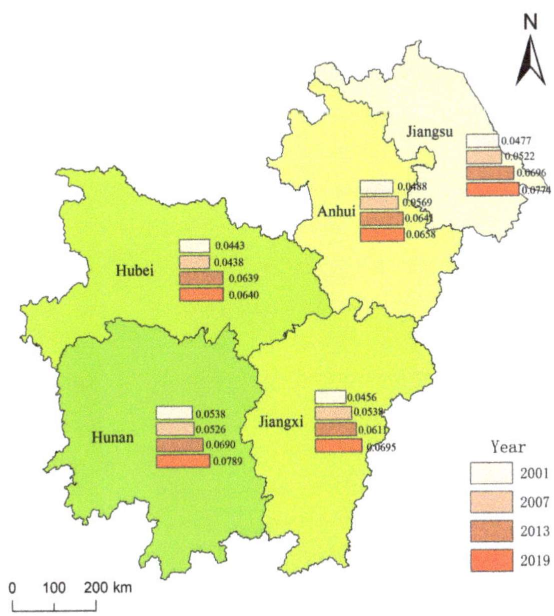

Figure 3. Change trend of the CLUT.

4.1.2. Variations in the Cultivated Land-Use Spatial Transition

Based on ArcGIS software, this study uses the Natural Breaks method to divide the CLUT index and change rate into five grades from high to low and visualize them.

From 2001 to 2019, the spatial transition of the high-value areas of cultivated land use was mainly concentrated in Jiangsu and Anhui Provinces in the northeast (Figure 4). There was no obvious law in the distribution of low-value areas, which were distributed in all provinces, and the central region was relatively concentrated in 2019. These results suggest that the change rate of the spatial transition index was higher in the northern and southern regions of the study area than that in the central region from 2001 to 2007 (Figure 5). Meanwhile, the overall difference in the change rate of the spatial transition index narrowed, and the regions with high change rates decreased significantly from 2007 to 2013. These results demonstrate that the regions with a high rate of change were clustered in the south and north from 2013 to 2019. Overall, the spatial transition index of cultivated land use in the study area increased from 0.0231 to 0.0288, and the spatial transition index change rate was 24.68% in the past 20 years. The distribution of the transition index and change rate are similar to the economic development and topographic differences in the study area. Areas with a higher CLUT index have the characteristics of a high economic development level, low altitude, and flat terrain, while regions with a lower CLUT index have the opposite characteristics.

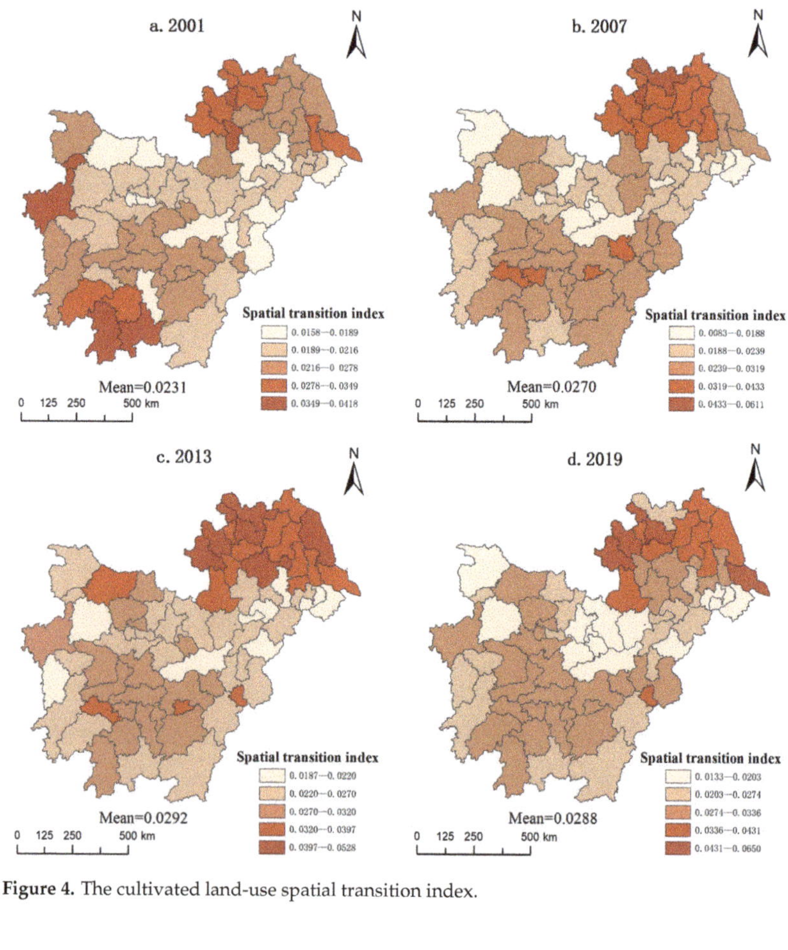

Figure 4. The cultivated land-use spatial transition index.

The Jiangsu and Anhui Provinces are both within the Yangtze River Delta, with flat terrain. In 1985, the Chinese government opened this region as a coastal economic open zone, with a high degree of economic development and advanced agricultural technology. From 2001 to 2019, the per capita GDP increased in Jiangsu and Anhui Provinces from \$1571 and \$725 to \$17,698 and \$8410, respectively, and the proportion of the grain and cash crop planting areas decreased from 1.052 and 1.271 to 0.977 and 1.166, respectively. This shows that people's demand for agricultural products continues to diversify with the growth of the economy, which results in an increase in the economic crop sowing area and changes in the cultivated land planting pattern. Furthermore, the social awareness of cultivated land protection in this area is stronger, and the response to national policies is also more active. From 2001 to 2019, the per capita cultivated land area in Jiangsu and Anhui Provinces increased from 0.099 hm^2 and 0.082 hm^2 to 0.125 hm^2 and 0.112 hm^2, respectively, and the patch aggregation degree increased from 93.384 and 93.320 to 94.659 and 96.660, respectively, which promoted the optimization and transition of the cultivated land spatial morphology. The rate of change in the spatial transition index decreased from 11.43% to 0.85%, which indicates that the landscape pattern has slowed in recent years.

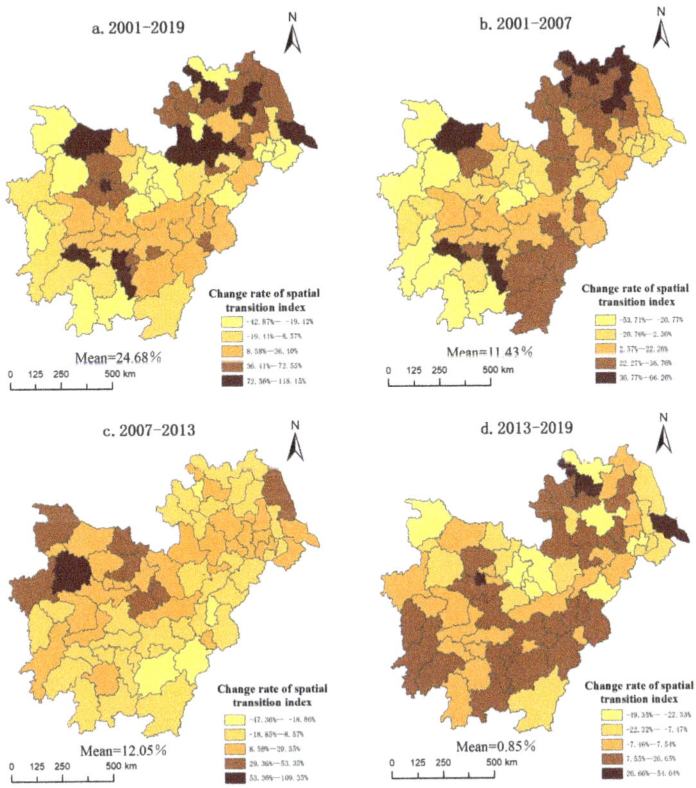

Figure 5. Change rate of cultivated land-use spatial transition index.

4.1.3. Variations in the Cultivated Land-Use Functional Transition

Figure 6 shows that the transition of the cultivated land-use function is generally strong. The high-value areas were mostly distributed in Hunan, Hubei, Jiangsu, and central Jiangxi from 2001 to 2013. Correspondingly, the number of high-value areas declined significantly, and the low-value areas were mainly concentrated in Anhui Province from 2013 to 2019. Meanwhile, Figure 7 demonstrates that the areas with a higher rate of change in the functional transition index were mainly in the central and southern regions from 2001 to 2007. This transferred to the western and eastern regions from 2007 to 2013 and moved to the south and northeast from 2013 to 2019. Overall, the transition index of the cultivated land-use function also increased from 0.0242 to 0.0407 during these periods. The change rate of the transition index is 68.18%, which is higher than that of the spatial transition index, indicating that the influence of the socioeconomic development process is stronger on the functional morphology of cultivated land than on the spatial morphology. The transition index and rate of change in the southwestern and northeastern regions are significantly higher, which is also similar to the difference in the local economic levels.

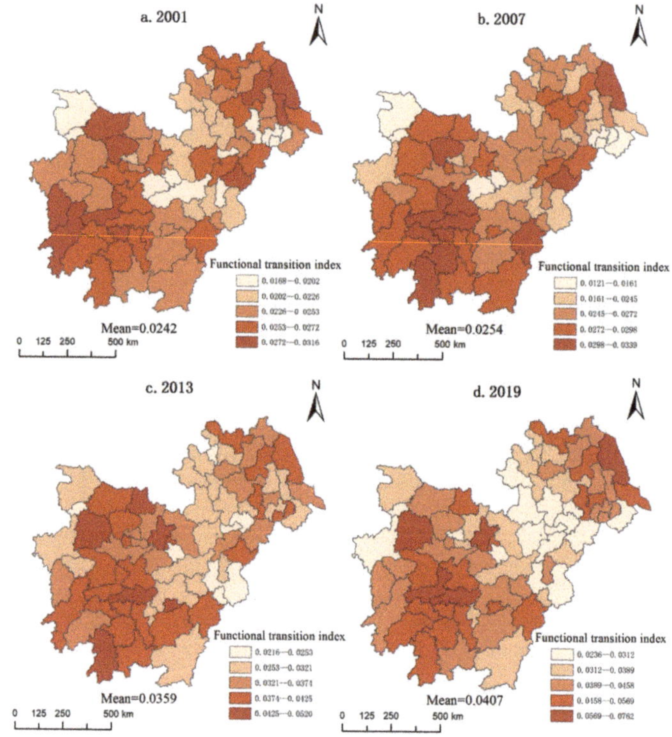

Figure 6. Cultivated land-use functional transition index.

From 2001 to 2019, the production, living, and ecological functional transition index of cultivated land in the study area increased from 0.0431, 0.0488, and 0.0329 to 0.0645, 0.0983, and 0.0333, respectively (Figure 8). During this period, the total number of plantation employees in the study area decreased from 75.44 million to 53.77 million. Although the average agricultural output value increased from 19,858.86 RMB/hm^2 to 64,764.13 RMB/hm^2, correspondingly, the average grain output increased from 5545.15 t/hm^2 to 6205.30 t/hm^2, and the farmers' per capita agricultural income rose from 1672.26 RMB to 8608.71 RMB; all of this resulted in a significant improvement in the production and living functions. From the perspective of the ecological function of cultivated land, the effective irrigation rate increased from 38.06% to 45.85%, the chemical load of cultivated land increased from 12.17 million tons to 25.85 million tons, and the crop species diversity index decreased from 0.581 to 0.508; therefore, the growth rate of the ecological function of cultivated land was not obvious.

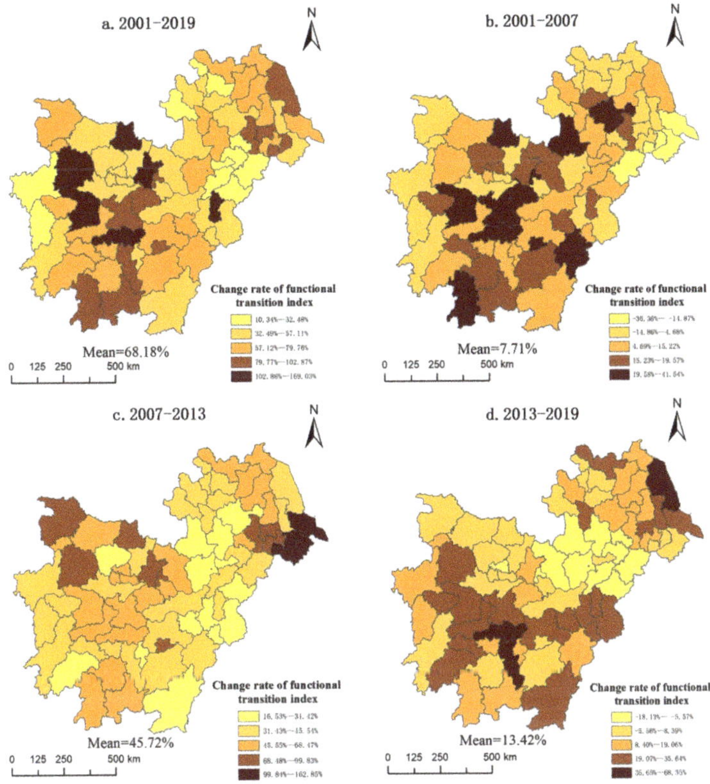

Figure 7. Change rate of the cultivated land-use functional transition index.

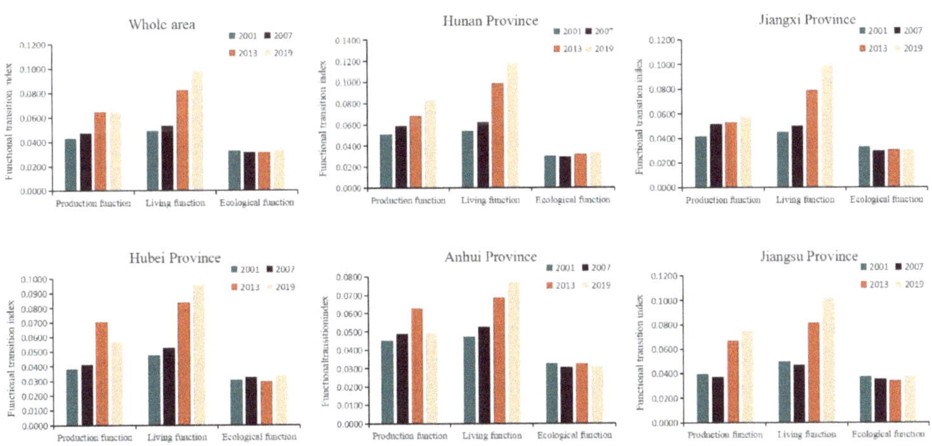

Figure 8. Change trend of the regional cultivated land functional transition index.

The production and living functions of cultivated land in Hunan, Hubei, Jiangxi, Anhui, and Jiangsu Provinces also showed an upward trend, while the ecological functions differed significantly. Hunan and Hubei showed an increasing trend, Jiangxi and Anhui showed a decreasing trend, and Jiangsu did not change significantly. The main reason is that, in 2019, the proportion of the agricultural industry in Hunan and Hubei reached 10.2% and 9.5%, respectively, while in Jiangxi it was 8.7%, and in Anhui it was 8.2%. The proportion of agriculture was relatively low, and the emphasis on agriculture was relatively insufficient, which resulted in the extensive use of cultivated land and biodiversity. Sexual decline limits the ecological maintenance of cultivated land. Similarly, the rate of change of the functional transition index also showed a slowdown, which increased from 7.71% to 45.72 and then decreased to 13.42%. This could be linked to a slowdown of China's economic growth.

4.2. Spatial Agglomeration Characteristics of the CLUT

4.2.1. Global Spatial Autocorrelation Analysis

Using GeoDa software, the global Moran's I index of the CLUT was calculated. Between 2001 and 2019, a significance test of the integrated, spatial, and functional transition of cultivated land-use was passed by 5% (Table 2), which indicates that the three transitions had a positive agglomeration effect on the geospatial distribution. Among the fluctuating changes in the agglomeration of the cultivated land spatial morphology transition, the agglomeration degree was the strongest in 2007, and Moran's I index reached 0.5698. The agglomeration degree decreased from 2007 to 2013. The reason is closely related to the implementation of the policy of "balance of compensation for land requisition" promulgated by China in 2006, which ensures the stability of cultivated land quantity and realizes spatial agglomeration changes. The agglomeration of the cultivated land functional morphology transformation shows an increasing trend of fluctuation. The agglomeration degree was the strongest in 2019, and the Moran's I index reached 0.2952.

Table 2. The results of Global Moran's I.

Type \ Year	2001			2007			2013			2019		
	I Value	Z Value	p Value	I Value	Z Value	p Value	I Value	Z Value	p Value	I Value	Z Value	p Value
Comprehensive transition	0.4170	5.7155	0.001	0.5360	7.3430	0.001	0.3845	5.2675	0.001	0.4018	5.5045	0.001
Spatial transition	0.3776	5.3576	0.001	0.5698	8.0846	0.001	0.3481	4.9390	0.001	0.4673	6.303	0.001
Functional transition	0.2127	3.1693	0.002	0.2679	3.7717	0.001	0.1683	2.5077	0.021	0.2952	4.3985	0.001

4.2.2. Local Spatial Autocorrelation Analysis

The LISA agglomeration shows the spatiotemporal evolution characteristics of the agglomeration of the CLUT (Figure 9). From 2001 to 2019, the high-high agglomerations of the comprehensive transition of cultivated land use were concentrated in the northeast and southwest of the region, the high-high agglomerations of the spatial transition were concentrated in the northeast, and the high-high agglomerations of the functional transition were distributed in the southwest. This is mainly affected by the differences in location, national policies, and economic development stages, which are analyzed in Sections 4.1.2 and 4.1.3 of this study. The low-low agglomerations of the cultivated land-use comprehensive, spatial, and functional transitions were distributed in the central region. Combined with Figure 2, it is found that most of these areas are in the transitional zone from mountainous and hilly to plains, with low economic development levels and poor agricultural production techniques. There are few contiguous and flat cultivated lands that are convenient for cultivation, and fragmentation is serious. Thus, the optimization of the cultivated land spatial and functional morphology is restricted to a certain extent.

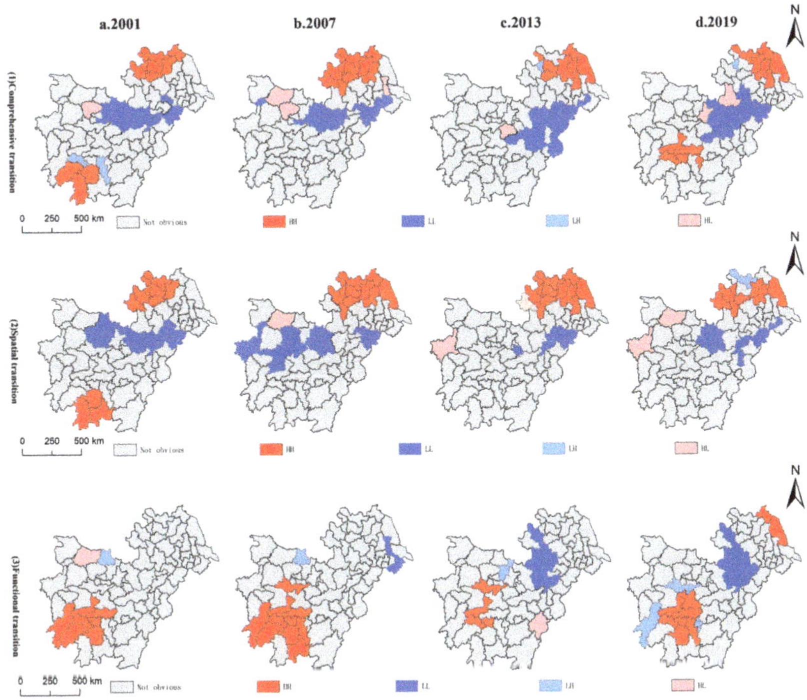

Figure 9. Local spatial correlation map of CLUT.

4.3. Changes in the Gravity Center of the CLUT

Figure 10 shows that the overall direction of the gravity center of the comprehensive transition moved from the southwest to the northeast from 2001 to 2019. The migration direction of the gravity center of the spatial transition is basically the same as that of the comprehensive transition, while the functional center of gravity moves in the opposite direction from northeast to southwest. However, since 2013, the movement direction of the gravity center of the spatial transition and functional transition has converged, and the distance between the centers of gravity has been reduced, which is due to the narrowing of the economic level gap within the region. The development of society is a staged process. Therefore, it is speculated that in the short term, the northeastern part of the study area will still be dominated by spatial transitions, while the southwestern part will still be dominated by functional transitions.

From the migration speed of the transition center of gravity, the comprehensive, spatial, and functional transitions all show a slowdown (Table 3). Among them, the center of gravity of the comprehensive transition moved 31.0182 km, and the moving speed dropped from 2.9401 km/year to 1.2370 km/year. The center of gravity of the space transition moved 69.0618 km, and the moving speed decreased from 8.3573 km/year to 1.0814 km/year. The functional transition center of gravity moved 38.5782 km, and the moving speed decreased from 3.2398 km/year to 1.0254 km/year. This was mainly the result of being influenced by the driving forces of socioeconomic transition. First, the central government has focused on high-quality regional development rather than on high-speed development. Therefore, the economic growth rate has slowed, which has led to a slowdown in the migration speed of the center of the cultivated land transition.

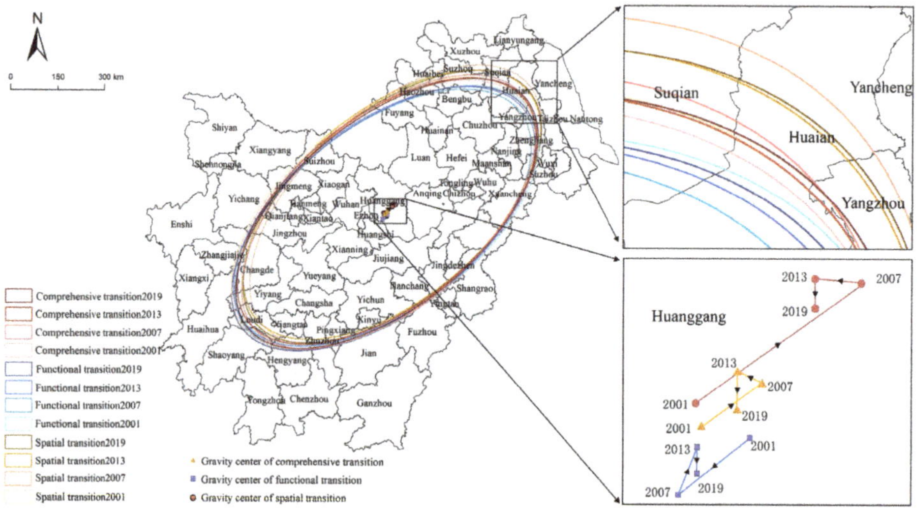

Figure 10. Path of gravity center of the CLUT.

Table 3. Speed of gravity center of the CLUT (km/year).

Type	2001–2007	2007–2013	2013–2019
Comprehensive transition	2.9401	0.9926	1.2370
Spatial transition	8.3573	2.0616	1.0814
Functional transition	3.2398	2.1645	1.0254

5. Discussion

5.1. Comparison of the Hidden Causes of the CLUT

Combined with the research results, this paper analyzes the driving factors of the CLUT. At present, some scholars have carried out the related research from an empirical perspective. Here, we selected several representative studies that measured and calculated the influencing factors of the cultivated land spatial transition or functional transition (Table 4). Through the comparison results, it is found that socioeconomic factors such as the decline in the agricultural population and regional economic development; increasing social demand; physical geographical factors such as the location, topography, hydrogeology, and soil types of cultivated land; and national-led policy factors all promote the CLUT.

The results show that the changes in the spatial morphology of cultivated land are mainly reflected in the reduction in the amount of cultivated land caused by the construction of urban and rural housing sites. Cultivated land generally changes in scale and intensification. However, there is still a small amount of cultivated land that moves toward intensification, because the improvement of living standards of urban residents has created high-end demand for "slow growing" crops. The country's greening policy also affects the area distribution of cultivated land crops. It can be seen that policies and social needs strongly influence the cultivated land-use direction. Furthermore, with the improvement of the regional socioeconomic level and the implementation of national culture and tourism policies, the functions of cultivated land have developed in a diversified way, such as in cultivated land-cultural functions and landscape functions. There are few studies on the empirical verification of the impact of engineering and technological factors on cultivated land. Nevertheless, engineering technology assists in the leveling and restoration of cultivated land.

Table 4. Comparison of hidden causes of CLUT.

Representative Example	Cultivated Land Morphology	Research Scale	Data Type	Conclusion
The slope CLUT characteristics and driving mechanism [48].	Spatial	Micro	Spatial, physical geography, and socioeconomic	Natural, socioeconomic, and humanistic policies are the main influencing factors.
The influence of state LED grain localization on the CLUT in suburban areas [49].	Spatial	Meso	Spatial, socioeconomic, policy	National leading food localization promotes the development of greenhouse agriculture.
The change of CLUT after the CAP greening [50].	Dominant	Micro	Spatial, policy	The CLUT are subject to greening rules.
The CLUT drivers in Tanzania [51].	Spatial	Meso	Spatial, biophysical, demographic, and socioeconomic	Demographic and socioeconomic factors are more significant.
Transitions in the function of cultivated land [33].	Functional	Macro	Socioeconomic	Cultivated land functions change with policy development.
Characteristics and influencing factors of the CLUT in Urban–Rural Coordination Area [52].	Dominant and recessive	Meso	Spatial and socioeconomic	Rural employment, per capita GDP, and proximity to central cities are the main factors affecting the CLUT.

5.2. Policy Implications

It can be seen from the above evaluation results that the degree of the CLUT between regions is imbalanced. If the CLUT meets the socioeconomic stage, then it can promote social development, but if it does not meet the socioeconomic stage, then it may hinder socioeconomic development. Therefore, policymakers need to clarify the leading factors of the CLUT, take social development and people's needs as guidance, and fully consider the regional differences in the CLUT to formulate cultivated land utilization policies. Based on the above analysis, we found that the relatively developed northeast region of the study area has a strong spatial transition of cultivated land, while the relatively underdeveloped economy of the southwest region has a strong functional transition. For areas with a strong spatial transition, urban construction has a greater demand for land, and people's demands are more abundant. The main goal should be to clarify the relationship between cultivated land use and social demand, and then promote the implementation of measures to compensate for the external cost of the loss of the cultivated land quantity in the following ways. First, it is necessary to control the amount of urban land supply, increase the level of land intensity, and strengthen the protection of basic farmland. Second, this study has demonstrated that the planting structure and type of cultivated land must not only match the needs of society but also meet sustainability requirements. Finally, a cultivated land landscape improvement plan should be established to ensure the degree of concentration and contiguous farmland.

For areas with strong arable land function transitions, the main goal should be to achieve the balanced development of various functions of cultivated land as follows. First, the reform of agricultural modernization should be deepened, land should be consolidated, the soil fertility should be improved, and farmland production efficiency should be increased. In addition, it is necessary to increase agricultural subsidies and absorb agricultural labor. Most importantly, it is essential to develop green agriculture from the perspective of ecological security and innovate agricultural ecological compensation mechanisms. Through these strategies, we can stimulate the endogenous development momentum of agriculture and realize the saving, green, and efficient use of cultivated land. In addition, both the overall and local CLUTs in the study area have positive aggregation effects, which indicates that the CLUT is a systematic problem, and there is a connection between the CLUTs in neighboring areas. Thus, it is necessary to strengthen regional cooperation, facilitate agricultural marketization and information construction, and invigorate agricultural operation mechanisms.

5.3. Limitations and Prospects

The CLUT is an inevitable consequence of socioeconomic transition. In previous studies, scholars have given more attention to changes in the amount or functional structure of cultivated land, while the comprehensive transition process of the cultivated land spatial and functional morphology has not received enough attention. Therefore, this study serves as a useful exploration to evaluate the CLUT and provides a reference for future research. The advantages of this research are as follows. First, this study combined quantitative and qualitative considerations to construct a "spatial–functional" integrated framework to identify the law of the CLUT. The transition of the cultivated land-use morphology is the dominant and recessive manifestation of CLUTs. This parameter is a comprehensive identification of the CLUT. Thus, compared with other regions, the main grain-producing areas have more important tasks for cultivated land protection and food security.

This research provides a novel perspective and content, but it still has some shortcomings. The evaluation of the CLUT is a complicated task, and a unified and standardized evaluation system needs to be established. In this study, natural factors are not considered in the "spatial–functional" comprehensive framework and should be included in subsequent studies. In addition, the defects of China's current farmland management regulations and property rights system have led to a decline in the quality of arable land. Moreover, the internal changes in cultivated land are more difficult to detect than the external changes. Therefore, research on the scientific and effective comprehensive evaluation systems for the transformation of cultivated land function forms should be the focus of future CLUT research. Subsequent research objects could be further enriched, such as the city scale, village scale, or field scale, especially in semi-urbanized areas where land conflict is more prominent and cultivated land protection is under great pressure. In addition, it is also a key consideration to put forward regulatory policies for the CLUTs in typical regions according to local conditions. In the future plan of this study, the spatial econometric model can be used to quantitatively analyze the core driving factors of CLUT.

6. Conclusions

This paper analyzes the mechanism of CLUT and evaluates the CLUT characteristics in the main grain-producing areas in the middle and lower reaches of the Yangtze River from 2001 to 2019 based on the "spatial–functional" integrated framework. The CLUT index of 71 prefecture-level cities was calculated, and then the magnitude, spatial correlation, direction, and speed of the inter-regional CLUT were analyzed.

The results show that the comprehensive transition index of cultivated land use continued to increase from 0.0480 to 0.0711 during the period. It can be said that social and economic factors have driven the acceleration of CLUT. In particular, the speed of functional transition is stronger than that of spatial transition, because the functional changes of cultivated land are directly intervened by human needs. However, the difference in the transition index between regions significantly expanded, and the transition range increased from 0.0345 to 0.0641. Such evidence suggests that the CLUT in the flat terrain area is larger than that in the mountainous area. Due to the advantages of terrain, the economic development, urbanization, and population growth of the plain area are strong, and the CLUT is fast.

The CLUT in the study area exhibited a positive aggregation effect and strengthened in the period. The degree of aggregation of the spatial and functional transitions increased from 0.3776 to 0.4673 and from 0.2127 to 0.2952, respectively. The CLUT of high-high agglomeration areas was concentrated in the northeast and southwest of the study area, and the low-low agglomerations areas were essentially distributed in the central region. This means that there is a spatial relationship between the CLUT in a region and its surrounding areas, and the activities of cultivated land used by humans are not separated.

The center of gravity of the CLUT has shifted from the southwest to the more economically developed northeast region. and the center of gravity migration speed of comprehensive, functional, and spatial transitions decreased from 2.9401 km/year to 1.2370 km/year, 8.3573 km/year to 1.0814 km/year, and 3.2398 km/year to 1.0254 km/year, respectively. Indeed, socioeconomic factors are often the main driving force behind the CLUT. In the current stage of China's economic slowdown, optimizing the morphology of cultivated land may be particularly important when adjusting the existing cultivated land management policy.

Author Contributions: Conceptualization, T.L. and S.F.; methodology, T.L., X.Z., G.W. and J.T.; software, S.F. and H.H; validation, T.L. and S.F.; formal analysis, T.L.; investigation, S.F.; resources, X.Z.; data curation, H.H.; writing—original draft preparation, T.L. and S.F.; writing—review and editing, T.L., G.W. and J.T.; visualization, S.F.; supervision, X.Z.; project administration, T.L. and X.Z.; funding acquisition, H.H. All authors have read and agreed to the published version of the manuscript.

Funding: This study was supported by the National Natural Science Foundation of China (No. 71864016 & 42261049), the Postdoctoral Science Foundation of China (No. 2017M622098), the Major Project of Social Science Foundation of China (21&ZD185), the Jiangxi Postdoctoral Science Foundation (No. 2017KY55), the Postdoctoral Daily Funding of Jiangxi Province (No. 2017RC036), the Science and Technology Project of Jiangxi Education Department (No. GJJ200509 & GJJ200542), the Humanities and Social Sciences Project of Jiangxi Education Department (No. JC20201), and the Student Research Project of Jiangxi University of Finance and Economics (No. 20210913221051660).

Institutional Review Board Statement: Not applicable.

Data Availability Statement: Not applicable.

Acknowledgments: We thank the editors and anonymous reviewers for their very constructive remarks in preparing the final version of the paper. We also thank all authors for their contributions and joint work to complete the paper. We are solely responsible for the opinions expressed in this article.

Conflicts of Interest: The authors declare no conflict of interest.

References

1. Foley, J.; DeFries, R.; Asner, G.; Barford, C.; Bonan, G.; Carpenter, S.; Stuart, F.; Coe, M.; Daily, G.; Gibbs, H.; et al. Global consequences of land use. *Science* **2005**, *309*, 570–574. [CrossRef] [PubMed]
2. Bai, X.; Shi, P.; Liu, Y. Realizing China's urban dream. *Nature* **2014**, *509*, 158–160. [CrossRef] [PubMed]
3. Jiang, G.; Wang, M.; Qu, Y.; Zhou, D.; Ma, W. Towards cultivated land multifunction assessment in china: Applying the "influencing factors-functions-products-demands" integrated framework. *Land Use Policy* **2020**, *99*, 104982. [CrossRef]
4. Liu, Y.S.; Wang, L.; Long, H. Spatio-temporal analysis of land-use conversion in the eastern coastal China during 1996–2005. *J. Geogr. Sci.* **2008**, *18*, 274–282. [CrossRef]
5. Long, H.; Zhang, Y.; Ma, L.; Tu, S. Land Use Transitions: Progress, Challenges and Prospects. *Land* **2021**, *10*, 903. [CrossRef]
6. Hong, T.; Yu, N.; Mao, Z.; Zhang, S. Government-driven urbanisation and its impact on regional economic growth in China. *Cities* **2021**, *117*, 103299. [CrossRef]
7. Lu, X.; Shi, Y.; Chen, C.; Yu, M. Monitoring cropland transition and its impact on ecosystem services value in developed regions of China: A case study of Jiangsu province. *Land Use Policy* **2017**, *69*, 25–40. [CrossRef]
8. Long, H.; Li, Y.; Liu, Y. Analysis of evolutive characteristics and their driving mechanism of hollowing villages in China. *Acta Geogr. Sin.* **2009**, *64*, 1203–1213.
9. Liu, Y.; Li, Y. Revitalize the world's countryside. *Nature* **2018**, *548*, 275–277. [CrossRef]
10. Li, Y.; Chen, C.; Wang, Y.; Liu, Y. Urban-rural transition and farmland conversion in China: The application of the environmental Kuznets Curve. *J. Rural Stud.* **2014**, *36*, 311–317. [CrossRef]
11. Mather, A. The forest transition. *Area* **1992**, *24*, 367–379.
12. Lu, X.; Qu, Y.; Sun, P.; Yu, W.; Peng, W. Green Transition of Cultivated Land Use in the Yellow River Basin: A Perspective of Green Utilization Efficiency Evaluation. *Land* **2020**, *9*, 475. [CrossRef]
13. Lambin, E.; Meyfroidt, P. Land use transitions: Socio-ecological feedback versus socio-economic change. *Land Use Policy* **2010**, *27*, 108–118. [CrossRef]

14. Long, H.L. Land use transition and rural transition development. *Prog.Geogr.* **2012**, *31*, 131–138.
15. Ma, L.; Long, H.L.; Tu, S.S. Farmland transition in China and its policy implications. *Land Use Policy* **2020**, *92*, 104470. [CrossRef]
16. Song, X.; Wu, Z.; Ouyang, Z. Changes of cultivated land function in China since 1949. *Acta Geograph. Sin.* **2014**, *4*, 435–447.
17. Qiang, W.; Liu, A.; Cheng, S.; Kastner, T.; Xie, G. Agricultural trade and virtual land use: The case of China's crop trade. *Land Use Policy* **2013**, *33*, 141–150. [CrossRef]
18. Bren d'Amour, C.; Reitsma, F.; Baiocchi, G.; Barthel, S.; Güneralp, B.; Erb, K.H.; Haberl, H.; Creutzig, F.; Seto, K. Future urban land expansion and implications for global croplands. *Proc. Natl. Acad. Sci. USA* **2017**, *114*, 8939–8944. [CrossRef]
19. Deng, X.; Huang, J.; Rozelle, S.; Zhang, J.; Li, Z. Impact of urbanization on cultivated land changes in China. *Land Use Policy* **2015**, *45*, 1–7. [CrossRef]
20. Bian, Z.; Kang, M.; Liu, L.; Zhu, R.; Yang, Z. Analysis on farmland multifunction in urban fringe area of Shenyang. *Chin. J. Soil Sci* **2015**, *3*, 533–538.
21. Swinton, S.; Lupi, F.; Robertson, G.; Hamiltonde, S. Ecosystem services and agriculture: Cultivating agricultural ecosystems for diverse benefits. *Ecol. Econ.* **2007**, *64*, 245–252. [CrossRef]
22. Eklund, L.; Persson, A.; Pilesj, P. Cropland changes in times of conflict, reconstruction, and economic development in iraqi kurdistan. *Ambio* **2016**, *45*, 78–88. [CrossRef] [PubMed]
23. Ma, W.; Jiang, G.; Chen, Y.; Qu, Y.; Zhou, T.; Li, W. How feasible is regional integration for reconciling land use conflicts across the urban–rural interface? Evidence from Beijing–Tianjin–Hebei metropolitan region in China. *Land Use Policy* **2020**, *92*, 104433. [CrossRef]
24. Li, T.; Long, H.; Zhang, Y.; Tu, S.; Ge, D.; Li, Y.; Hu, B. Analysis of the spatial mismatch of grain production and farmland resources in China based on the potential crop rotation system. *Land Use Policy* **2017**, *60*, 26–36. [CrossRef]
25. Long, H.; Liu, Y.; Hou, X.; Li, T.; Li, Y. Effects of land use transitions due to rapid urbanization on ecosystem services: Implications for urban planning in the new developing area of China. *Habitat Int.* **2014**, *44*, 536–544. [CrossRef]
26. Jiang, L.; Deng, X.; Seto, K. Multi-level modeling of urban expansion and cultivated land conversion for urban hotspot counties in China. *Landscape Urban Plann.* **2012**, *108*, 131–139. [CrossRef]
27. Yue, D.; Du, J.; Gong, J.; Jiang, T.; Zhang, J.; Guo, J.; Xiong, Y. Dynamic analysis of farmland ecosystem service value and multiple regression Analysis of the influence factors in Minqin Oasis. *Acta Ecol. Sin.* **2011**, *31*, 2567–2575.
28. Niu, S.; Lyu, X.; Gu, G. A New Framework of Green Transition of Cultivated Land-Use for the Coordination among the Water-Land-Food-Carbon Nexus in China. *Land* **2022**, *11*, 933. [CrossRef]
29. Chao, L.; Xu, Y.; An, H.; Liu, Y.; Wang, H.; Lu, L.; Sun, P.; Zheng, W. Spatial identification of land use multi-functionality at grid scale in farming-pastoral area: A case study of Zhangjiakou city, China. *Habitat Int.* **2018**, *76*, 48–61.
30. Vadrevu, K.; Ohara, T.; Justice, C. Land cover, land use changes and air pollution in Asia: A synthesis. *Environ. Res. Lett.* **2017**, *12*, 120201. [CrossRef]
31. Su, S.; Zhou, X.; Wan, C.; Li, Y.; Kong, W. Land use changes to cash crop plantations: Crop types, multilevel determinants and policy implications. *Land Use Policy* **2016**, *50*, 379–389. [CrossRef]
32. He, L.; Min, Q.; Zhang, D. Assessment models for multifunctionality of agriculture and their applications: A case study on Qingtian County in Zhejiang Province. China. *Resour. Sci.* **2010**, *32*, 1057–1064.
33. Tian, J.; Wang, B.; Zhang, C.; Li, W.; Wang, S. Mechanism of regional land use transition in underdeveloped areas of China: A case study of northeast China. *Land Use Policy* **2020**, *94*, 104538. [CrossRef]
34. Song, X.; Huang, Y.; Wu, Z.; Ouyang, Z. Does cultivated land function transition occur in china? *J. Geogr. Sci.* **2015**, *25*, 19. [CrossRef]
35. Daily, G.; Soderquist, T.; Aniyar, S.; Arrow, K.; Dasgupta, P.; Ehrlich, P.; Folke, C.; Jansson, A.; Jansson, B.; Kautsky, N. The value of nature and the nature of value. *Science* **2000**, *289*, 395–396. [CrossRef]
36. Xiang, J.; Liao, X.; Song, X.; Xiong, J.; Ma, W.; Huang, J. Regional convergence of cultivated land multifunctions in China. *Resour. Sci.* **2019**, *41*, 1959–1971. [CrossRef]
37. Long, H.; Li, T. The coupling characteristics and mechanism of farmland and rural housing land transition in china. *J. Geogr. Sci.* **2012**, *22*, 548–562. [CrossRef]
38. Meiyappan, P.; Roy, P.; Sharma, Y.; Ramachandran, R.; Joshi, P.K.; Defries, R.; Jain, A. Dynamics and determinants of land change in India: Integrating satellite data with village socioeconomics. *Reg. Environ. Chang.* **2017**, *17*, 753–766. [CrossRef]
39. Xu, Z.; Xu, J.; Deng, X.; Huang, J.; Uchida, E.; Rozelle, S. Grain for green versus grain: Conflict between food security and conservation set-aside in China. *World Dev.* **2006**, *34*, 130–148. [CrossRef]
40. Siciliano, G. Urbanization strategies, rural development and land use changes in China: A multiple-level integrated assessment. *Land Use Policy* **2012**, *29*, 165–178. [CrossRef]
41. Ma, W.; Jiang, G.; Li, W.; Zhou, T.; Zhang, R. Multifunctionality assessment of the land use system in rural residential areas: Confronting land use supply with rural sustainability demand. *J. Environ. Manag.* **2019**, *231*, 73–85. [CrossRef]
42. Jiang, G.; Ma, W.; Qu, Y.; Zhang, R.; Zhou, D. How does sprawl differ across urban built-up land types in China? A spatial-temporal analysis of the Beijing metropolitan area using granted land parcel data. *Cities* **2016**, *58*, 1–9. [CrossRef]
43. Rodríguez Sousa, A.; Parra-López, C.; Sayadi-Gmada, S.; Barandica, J.; Rescia, A. A multifunctional assessment of integrated and ecological farming in olive agroecosystems in southwestern Spain using the Analytic Hierarchy Process. *Ecol. Econ.* **2020**, *173*, 106658. [CrossRef]

44. Liu, Y.; Liu, Y.; Chen, Y.; Long, H. The process and driving forces of rural hollowing in China under rapid urbanization. *J. Geogr. Sci.* **2012**, *20*, 876–888. [CrossRef]
45. Jiang, P.; Chen, D.; Li, M. Farmland landscape fragmentation evolution and its driving mechanism from rural to urban: A case study of changzhou city. *J. Rural. Stud.* **2021**, *82*, 1–18.
46. Laliberte, E.; Tylianakis, J. Cascading effects of long-term land-use changes on plant traits and ecosystem functioning. *Ecology* **2011**, *93*, 145–155. [CrossRef]
47. Chen, C.-H. A Novel Multi-Criteria Decision-Making Model for Building Material Supplier Selection Based on Entropy-AHP Weighted TOPSIS. *Entropy* **2020**, *22*, 259. [CrossRef] [PubMed]
48. Wang, Q.; Li, Y.; Luo, G. Spatiotemporal change characteristics and driving mechanism of slope cultivated land transition in karst trough valley area of Guizhou province, China. *Environ. Earth. Sci.* **2020**, *79*, 284. [CrossRef]
49. Zhong, T.; Si, Z.; Shi, L.; Liu, S. Impact of state-led food localization on suburban districts'farmland use transformation: Greenhouse farming expansion in Nanjing city region, China. Landscape. *Urban Plan.* **2020**, *202*, 103872. [CrossRef]
50. Bertoni, D.; Aletti, G.; Ferrandi, G.; Micheletti, A.; Cavicchioli, D.; Pretolani, R. Farmland use transitions after the CAP greening: A preliminary analysis using markov chains approach. *Land Use Policy* **2018**, *79*, 789–800. [CrossRef]
51. Michael Uisso, A.; Tanrıvermis, H. Driving factors and assessment of changes in the use of arable land in Tanzania. *Land Use Policy* **2021**, *104*, 105359. [CrossRef]
52. Li, L.; Wang, L.; Qi, Z. The spatiotemporal variation of farmland use transition and its critical influential factors in coordinated urban-rural regions: A case of Chongqing in western china. *Sustain. Cities Soc.* **2021**, *70*, 102921. [CrossRef]

land

Article

The Effect of High-Speed Rail on Cropland Abandonment in China

Junhui Shi and Fang Wang *

College of Management, Sichuan Agricultural University, Chengdu 611130, China; 20147396@stu.sicau.edu.cn
* Correspondence: 11989@sicau.edu.cn

Abstract: This study analyzed Chinese labor-force survey data to determine the relationship between cropland abandonment and high-speed rail (HSR) infrastructure. A PSM-DID approach was employed to examine 2014 and 2016 data from the China Labor-force Dynamics Survey and estimate the impact of HSR, from which it was found that HSR accessibility promoted cropland abandonment in local farm households with a coefficient of 0.206, that is, HSR projects led to a 20.6% increase in area of cropland abandonment and these impacts were found to be greater in hilly areas and lower in plain areas. The results also suggested that HSR accessibility could have a "pull" effect, which resulted in rural labor force shifts to non-agricultural sectors in the local region. Countermeasures and policy suggestions are given to reduce cropland abandonment.

Keywords: high-speed rail; cropland abandonment; off-farm employment; migration; urbanization; land use

Citation: Shi, J.; Wang, F. The Effect of High-Speed Rail on Cropland Abandonment in China. *Land* **2022**, *11*, 1002. https://doi.org/10.3390/land11071002

Academic Editors: Li Ma, Yingnan Zhang, Muye Gan and Zhengying Shan

Received: 30 May 2022
Accepted: 27 June 2022
Published: 1 July 2022

1. Introduction

The relationship between high-speed rail (HSR) and land use has received significant global research attention. The construction of China's HSR started in 2005 and began operations in 2008, after which the network developed rapidly [1]. By the end of 2020, China's HSR network had reached 38,000 km, ranking it first in the world, and had connected administrative divisions in 31 provinces. China's "eight vertical eight horizontal" railway plan estimated that the HSR network would reach 45,000 km and connect 250 cities by 2030 (China's "eight vertical eight horizontal" railway plan was proposed in the *medium and long term railway network plan of China* published in 2016). This plan will come true in the near future. Evidence from Asia and Europe shows that the large-scale HSR cross-regional transportation network has shortened the space-time distance between regions, improved city accessibility and inter-regional and inter-city population, information, and technology mobility [2–4], and made labor migration easier by decreasing long-distance information decay and market adjustment costs [5]. Furthermore, the reduction in transport costs and the increase in information and labor flows has benefited local economies and spurred urbanization growth [6], which in turn has resulted in greater rural labor force migration to the cities.

China's HSR provides 2-h travel time services to approximately 74% of the population [7], which has increased China's internal migration, especially short-term (one or two years) rural-to-urban migration [8,9]. Because of this decrease in the agricultural labor force, significant village cropland has been abandoned, which has resulted in rural shrinkage [10–12].

Since the beginning of the 20th century, 385–472 million km^2 of global cropland has been abandoned [11,13–15], which has brought new challenges to food security [16–18]. Because cropland preservation is essential to guaranteeing food security [19,20], there have been many agricultural economics, human geography, and land management studies exploring the driving factors associated with cropland abandonment, such as environmental conditions [21], farm stability and viability [22], the rural labor force [23], agricultural production costs [22], and agricultural technology [24]. However, few studies have examined the effects of transport infrastructure on cropland abandonment.

The rapid construction of HSR in China has indeed promoted the development of the regional economy, but it has led to changes in land uses, which may have negative impacts. Therefore, these adverse impacts need to be fully considered when planning the construction of HSR, so that the construction of HSR can promote regional development more comprehensively. This study analyzed the causal effects of HSR projects on cropland abandonment to deepen the understanding of China's transportation improvement effects on land use. By revealing the heterogeneous impacts of HSR projects in the different regions, it is hoped that the results of this study can promote rural revitalization and sustainability.

The remainder of this article is organized as follows. Section 2 details the theoretical analytical framework and discusses the possible HSR accessibility effect mechanisms on cropland abandonment, Section 3 describes the data and the research methods and variables used in the empirical analyses, Section 4 presents and discusses the results, and Section 5 provides the research conclusions and some policy implications.

2. Conceptual Background

The HSR accessibility effect mechanisms affecting cropland abandonment are complex. First, the HSR has optimized accessibility between cities and reduced the cost of long-distance travel between cities [2,25], which has increased population migration [26], especially rural-to-urban population migration [9], which has contributed to the reduction in rural labor and led to significant cropland abandonment [12,27,28], that is, the HSR developments have had a possible "push" effect. Second, HSR transport infrastructure has enhanced economic development, regional consumption growth [29–31], and urbanization, which in turn has motivated a labor force shift to non-agricultura off-farm employment and cropland abandonment [22,32,33], that is, it has had a possible "pull" effect. This paper focused on these "push" and "pull" effects to assess the effect of the HSR on cropland abandonment. Figure 1 shows the HSR project impact mechanism on farm household cropland abandonment.

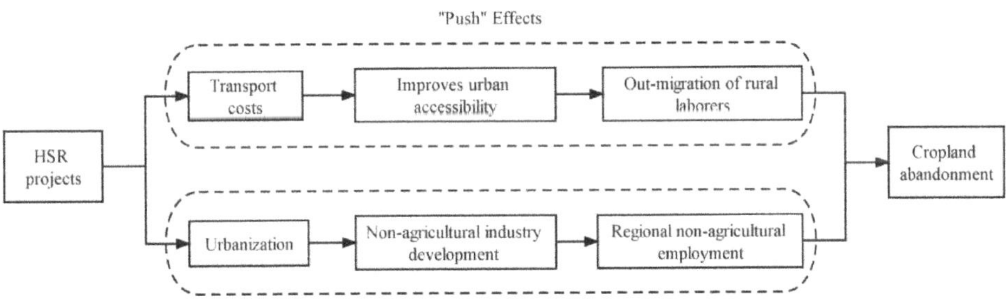

Figure 1. HSR project impact mechanism on cropland abandonment.

As shown in Figure 1, the HSR "push" effect improves urban accessibility by decreasing transport costs, shortening intercity and long-haul travel times [34], and improving regional transportation network efficiency [35–37]. The increased speed of the HSR can be converted into idle time and activity space [38], which is called the 'space-time compression' theory [39].The time and space compression reduces the time and space cost for rural workers to go to cities to obtain jobs. The decrease in transport costs has improved urban accessibility, making labor migration more frequent [5], that is, the opening of HSR stations directly impacts the abandonment of local agricultural activities and cropland because the rural labor force can more easily seek better urban employment [12,17,40]. The greater financial support the migrant workers provide reduces the need for other family members to continue farming [41].

As shown in Figure 1, the "pull" effect of the HSR projects improves non-agricultural industrial development and promotes urbanization in HSR-linked regions. Specifically, urbanization is the process of transforming rural agricultural communities into urban communities based on industry [42] by occupying farmland for urban development or encouraging rural workers to abandon their farmland to seek better non-agricultural opportunities [43,44]. As the HSR interconnected regional network structure promotes logistics, there are also additional opportunities for industrial development in small or medium-sized cities, which again encourages increased urbanization [45]. Therefore, as the HSR network has led to significant changes in regional industrial and factor input structures, there has been a commensurate increase in the labor and service markets in HSR connected cities [46], which has motivated rural workers to abandon cropland and seek better lives in non-agricultural industries [47]. Because of the constraints in the existing Chinese land system, the agricultural land property rights trading market is not perfect, which has also led to out-migration by rural workers to non-agricultural industries [48,49].

In summary, as the global food security situation becomes increasingly tense, especially in developing countries, reducing cropland abandonment is vital; however, as rural labor force availability falls, cropland abandonment has been increasing [12,43]. Several studies have examined the relationship between food security and transport infrastructure [50], with most finding that transport infrastructure had a positive effect on food security [51,52].

While previous studies have analyzed the impact of the HSR on population mobility and urbanization, few have comprehensively considered the HSR impacts on abandoned farmland. Many studies have only focused on the effects of HSR on urban land use [1,53] but not on the effects on rural land use. Therefore, this study considered the cropland abandonment associated with HSR projects and explored the heterogeneous HSR effects on cropland abandonment. Furthermore, as HSR has been found to affect cropland abandonment through labor force out-migration and urbanization, the empirical analyses also examined these potential mechanisms.

3. Data, Variables, and Method

3.1. Data

3.1.1. High-Speed Rail in China

HSR is a railway system that operates at speeds faster than 250 km/h (according to the definition presented by the *National Railway Administration of the People's Republic of China*). The first HSR went into operation in China in 2008, and by the end of 2020, had a 38,000-km network, which accounted for two-thirds of the world's HSR networks [54]. HSR infrastructure construction has strengthened social and economic activities between connected cities [55]. This study collected 2012 to 2014 HSR operations data (including site and opening time) from the National Railway Timetable released by the *Ministry of Railways of the People's Republic of China*. The HSR projects were treated as a quasi-natural experiment in this paper, with the linked HSR network cities being the treated regions, and the non-HSR-linked cities being the control regions. Because the household data collected in 2014 and 2016 reflected the farmer household situations in 2013 and 2015, 2012 and 2014 HSR data in the prefecture-level cities were collected. Figure 2 shows China's HSR network and the cities that had HSR services at the end of 2014.

3.1.2. Household Labor Transfer and Cropland Abandonment

To estimate the HSR impact on cropland abandonment, farm household data were collected from the China Labor-force Dynamics Survey (CLDS) organized by the Center for Social Science Survey at Sun Yat-sen University, which focused on household social-economic status and changes in China, such as urbanization, labor force migration, and land use. To ensure that the sample was representative, the CLDS collected data in 2014 and 2016 from 14,000 households in approximately 400 villages in 29 provinces or municipalities directly under the central government. The CLDS used a multistage cluster, stratified, probability proportional to size (PPS) sampling method, and was the most recent continuous data available for this study. As this study was focused on rural cropland abandonment, urban family samples were not considered. After data cleaning, 4880 valid household questionnaire data from 27 provinces in 2014 and 2016 were used in the analysis.

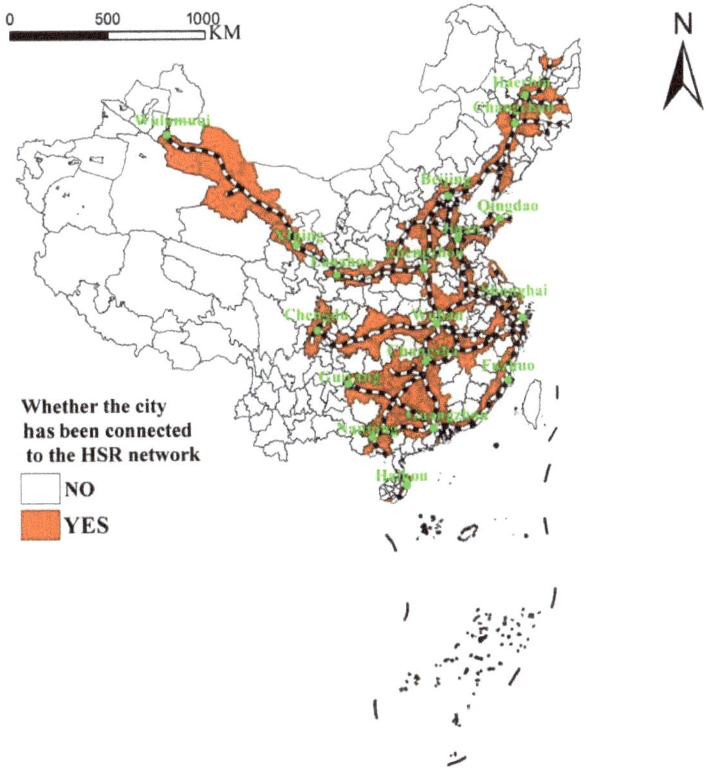

Figure 2. HSR lines and stations in the prefecture-level cities.

3.2. *Variables*

3.2.1. Dependent Variable

The core dependent variable was the total abandoned farm household cropland area in 2013 or 2015 [12,23], which was defined as cropland plots in which no crops were cultivated throughout an entire year. To cater to the differences in land size in the studied regions, the cropland abandonment ratio was used to test the robustness of the benchmark estimate.

3.2.2. Independent Variable

The focus dummy variable was whether the city in which the households lived had been linked to the HSR network or not [56]; when the city had been linked to the HSR network, HSR = 1; otherwise, HSR = 0.

3.2.3. Control Variables

Consistent with the existing studies, the main control variables considered to influence cropland abandonment were householder, household-level, and village-level characteristics [12,23,57]. The householder characteristics included householder age [17] and education level [58], the household characteristics included off-farm employment ratio [23], cropland area [59], total rural household members [60], and the number of family members over 64 years old [61], and the village-level characteristics included soil quality, and water availability [10,62]. The variables are detailed in Table 1.

Table 1. Model variables.

Variable	Definition and Assignment
Abandonment	Abandoned rural household cropland area (mu)
Abandonment Ratio	Share of abandoned cropland in total cropland (%)
Age	Age of the household head in the survey year
Education	Does the household head have a high school diploma or above (0 = no; 1 = yes)
Family	Total number of rural family members (number)
Elder	Number of household members over 64 years old (number)
Fixed assets	Current market value of per capita fixed assets owned by households (10^4 CNY/person)
Soil quality	Whether the quality of soil in the village is good (0 = no; 1= yes)
Water availability	Whether there is an irrigation facility in the village (0 = no; 1 = yes)
Urbanization	Share of urban households in total households with the same county in the sample (%)
Land size	Land area the rural households managed (mu)
Off-farm work	Share of off-farm employment labor in total labor (%)
Out-migration	Number of migrant workers in the household (number)

3.3. Methods

High-speed railway construction is part of the national strategic plan and is not regionally controlled. Therefore, by comparing the abandoned cropland area before and after the cities became connected to the HSR network, the impact of HSR accessibility was regarded as a quasi-natural experiment. Because of the differences in the railway foundations and natural conditions, the HSR accessibility completion was different between the cities; therefore, a PSM-DID analysis method was employed to estimate the HSR impact on cropland abandonment. The DID method is widely acknowledged as the best method to study quasi-natural experiments, or to evaluate the influences of external shocks. The DID method requires a completely random selection between the treatment group and control group, otherwise the result can be largely biased. To relieve the endogenous problems caused by the selection bias, we adopt the propensity score matching (PSM) method before applying DID, so as to ensure the accurate estimation of the HSR's influence.

3.3.1. Difference in Differences Method

As the main objective of this study was to investigate the effects of HSR accessibility on cropland abandonment, difference-in-difference (DID) models were used to estimate the periods and the treatment group's effects from the household panel data. The measured dependent variable was the rural household abandoned cropland area, with the treatment variable being the households linked to the HSR network.

To allow for an examination of the time-invariant unobserved heterogeneity and choices by changing the unbiased outcomes, the DID method was employed to estimate the policy effect by separating the time effect from the policy-treated implications [63,64]. To distinguish the HSR impact, the households living in towns with a linked HSR network were the treated group and the households in towns not linked to the HSR network were the control group. Therefore, all household samples were divided into four different subsamples, including the treated group households linked to the HSR network in 2013; the treated group households linked to the HSR network in 2014; the control group households not linked to the HSR network in 2013; and the control group households not linked to the HSR network in 2014. The significant HSR impact on the cropland abandonment was analyzed based on the following DID model:

$$Y_{it} = \alpha_0 + \alpha_1 HSR_{it} * After_{it} + \varepsilon_{it} \quad (1)$$

where Y_{it} was the size of rural household cropland abandonment, in which subscript i represented farm household i and subscript t represented year t, HSR_{it} denoted whether or not the households were linked to the HSR network (HSR = 1 meant the city was linked to the HSR network; otherwise, HSR = 0), $After_{it}$ = 1 indicated 2016 (after the city was linked to the HSR network), $After_{it} = 0$ indicated 2014, α_1 signified the HSR accessibility effect on the rural household abandoned cropland areas; ε_{it} was the residual.

3.3.2. PSM-DID Methods

An important underlying assumption of the DID method was that if the external effect (HSR connection) did not exist, the development trends for the dependent variable (cropland abandonment) in the treatment and control groups would be parallel, that is, there would be no apparent systematic differences in the cropland abandonment trends between the households linked to the HSR network and the other households. However, as the HSR was first built in provincial capital cities and cities that had railway foundations, the probability of the HSR network being in these cities was significantly higher. As only two panel data periods were consulted, it was difficult to ascertain the common trend hypothesis. To avoid any estimation biases resulting from the above assumptions, the PSM-DID method was applied [65,66].

The PSM method can solve matching problems by placing the pre-treatment features of any subject into a single index variable and matching the treatment groups to the control groups based on the propensity scores. Therefore, the PSM method was used before the DID method to allow for the selection of the appropriate control groups from the cities without a HSR connection, which helped solve the possible endogenous problems and ensured that the DID estimation results were unbiased [67,68]. The probability of assigning treatment depended on the following PSM model preprocessing variables:

$$p(X) = Pr\,[D = 1|X] \tag{2}$$

where D represents all the cities including both treatment and control groups; X represents the control variables.

While the PSM method can address sample selection bias problems, it cannot solve the endogeneity problems caused by missing variables; however, the DID model is unable to explain the sample deviation problem but can determine the endogeneity problem and the policy-treatment effect. Therefore, the following PSM-DID model (Equation (3) was the baseline model that was used to accurately evaluate the HSR accessibility impact:

$$Y_{it} = \alpha_0 + \alpha_1 HSR_{it} * After_{it} + \beta_1 X_{it} + \varepsilon_{it} \tag{3}$$

where X_{it} was the control variable set affecting the cropland abandonment factor.

4. Results

4.1. Summary Statistics

The summary statistics for the key variables are shown in Table 2. In 2014, most sample areas were not linked to the HSR network; however, in 2016, more than half of the sample areas were linked. The broad coverage of the HSR-linked cities also allowed for the use of the DID model. It was found that 10.3% of the rural households in the total sample had abandoned their cropland, with the average abandonment size being 0.42 mu in 2014 and 0.38 mu in 2016. In 2014, the average cropland abandonment area in the HSR accessible samples was 0.56 mu, which was over twice as high as in the HSR inaccessible samples (0.24 mu). In 2016, the average cropland abandonment area in the HSR accessible samples was also higher than in the HSR inaccessible samples. For the control variables, the average age for the total sample was 54 years, and only about 10% of householders had a high school education and above. In both 2014 and 2016, the average number of household members in the HSR accessible samples was lower than in the HSR inaccessible samples, the average share of off-farm employment labor in the total labor in the HSR accessible samples was lower than in the HSR inaccessible samples in all window periods, and the degree of urbanization in the HSR accessible areas was 13%, higher than in the HSR inaccessible regions in 2014; however, in 2016, the situation was reversed.

Table 2. Summary statistics for the crucial variables.

Variable	2014				2016			
	HSR		Non-HSR		HSR		Non-HSR	
	Mean	Std. Dev.	Mean	Std. Dev.	Mean	Std. Dev.	Mean	Std. Dev.
Abandonment	0.56	1.25	0.24	2.30	0.39	2.52	0.37	1.33
Abandonment Ratio	0.06	0.23	0.10	0.23	0.06	0.21	0.05	0.19
Age	54.96	12.63	53.79	12.69	55.83	12.53	53.77	12.64
Education	0.10	0.29	0.10	0.29	0.12	0.32	0.10	0.30
Family	4.28	2.07	4.85	2.14	6.57	3.87	6.91	3.79
Elder	0.13	0.41	0.18	0.47	0.15	0.44	0.17	0.45
Fixed assets	4.99	15.30	2.86	4.27	3.33	12.42	2.12	5.01
Soil quality	0.91	0.29	0.92	0.27	0.91	0.29	0.91	0.29
Water availability	0.52	0.50	0.54	0.50	0.56	0.50	0.52	0.50
Urbanization	0.13	0.20	0.12	0.20	0.08	0.19	0.09	0.18
Land size	5.65	10.62	5.60	10.65	5.88	25.79	8.31	94.63
Off-farm work	0.70	0.34	0.66	0.33	0.74	0.32	0.64	0.34
Out-migration	0.77	1.30	1.10	1.46	0.69	1.12	0.99	1.31
Number of observations	2005		2875		2882		1998	

4.2. PSM Results

First, the PSM method was used to process the variables that affected cropland abandonment. To test the reliability of the matching results, a logit model was employed to estimate a set of control variables. The nearest neighbor matching method was used, for which the caliper range was set to 0.05 to match the samples, after which a matching check was conducted to determine the covariate distributions between the processing group and the control group. The results are shown in Table 3.

Table 3 shows that after the propensity score matching, the standardized deviation for most covariates was reduced, the bias values of the matched samples were significantly less than 10%, and the *p*-values were all greater than 0.1, which indicated that the matched treatment and control groups had no systematic differences and satisfied the balance test. As can be observed in Figure 3, there were few out-of-support untreated samples and most observed values for the samples were supported. Therefore, it was appropriate to use the PSM method because the results were reliable. After deleting the samples that were off the common support domain, the DID model was then applied to precisely estimate the HSR impact.

4.3. DID Model Results

The estimated OLS, DID and PSM-DID cropland abandonment results for the HSR accessibility sample are shown in Table 4. Columns (1)–(6) show the comparative benchmark analysis for the different approaches with no control variables and when the individual, household, and village characteristics were respectively added. The results in column (1) show the correlations between HSR accessibility and cropland abandonment using the OLS method after controlling for the individual, household, and village characteristics. While column (2) shows a similar result for the effect of the HSR accessibility on cropland abandonment using the OLS approach, no significant correlation between HSR accessibility and cropland abandonment size was observed. The results in columns (2) and (3), which respectively show the correlations between HSR accessibility and cropland abandonment with and without the household and village characteristic controls, indicate that the HSR accessibility impact on cropland abandonment was significantly positive. Columns (5) and (6) show the PSM-DID method results, and the benchmark estimate for Equation (3) is shown in column (6). With a coefficient of 0.206, the benchmark regression results indicated that HSR accessibility had a significant effect on cropland abandonment size and was statistically significant at the 5% level, that is, compared with the cropland abandonment size of the HSR inaccessible sample, the cropland abandonment size in the HSR accessible sample was higher by about 20.6%, on average. The results in columns (5) and (6) show that the estimated results remained robust, regardless of whether or not there were control variables. Compared with the OLS and DID method results, column (6) indicates that the coefficient for HSR*After had a slight increase, which proved that the sample selection bias was reduced after the treatment group and control group were matched. All the above results suggested that HSR accessibility increased cropland abandonment likelihood and size.

Table 3. PSM validity test.

Variable	Unmatched/ Matched	Mean		%Bias	%Reduct ∣Bias∣	*t*-Test	
		Treated	Control			*t*	*p* > t
Age	U M	55.71 55.74	53.97 56.39	14.40 −5.40	 62.40	4.99 −1.88	0.00 0.16
Education	U M	0.11 0.10	0.09 0.11	5.10 −3.00	 41.40	1.77 −0.98	0.08 0.33
Family	U M	5.55 5.55	5.49 5.62	1.90 −2.20	 −19.90	0.65 −0.76	0.52 0.45
Elder	U M	0.17 0.17	0.18 0.18	−2.40 −1.50	 36.50	−0.83 −0.53	0.41 0.60
Land rent-out	U M	1.07 0.73	0.33 1.00	7.90 −3.00	 62.30	2.73 −1.06	0.01 0.29
Fixed assets	U M	3.65 3.14	2.65 3.0	9.60 0.90	 90.30	3.33 0.52	0.00 0.61
Soil quality	U M	0.92 0.92	0.91 0.93	5.00 −3.30	 32.80	1.72 1.23	0.09 0.22
Water availability	U M	0.56 0.56	0.55 0.57	3.70 −1.30	 65.40	1.27 −0.44	0.20 0.66
Urbanization	U M	0.07 0.07	0.12 0.08	−25.00 −2.20	 91.00	−8.66 −0.85	0.00 0.40
Land size	U M	7.43 6.70	5.63 5.97	8.60 3.50	 59.70	2.99 2.87	0.00 0.56
Off-farm work	U M	0.66 0.66	0.64 0.65	6.50 1.00	 85.20	2.27 0.33	0.02 0.74
Out migration	U M	0.72 0.72	1.12 0.70	−29.80 1.30	 95.50	−10.32 0.55	0.00 0.59

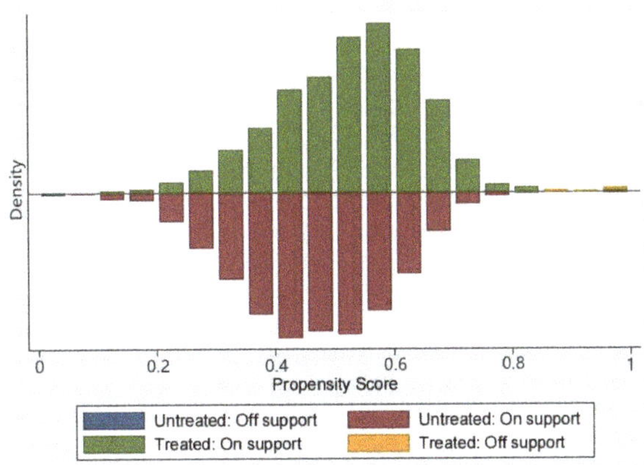

Figure 3. PSM validity test.

Table 4. OLS, DID, and PSM-DID estimation results for the HSR effect on cropland abandonment.

	(1) OLS	(2) OLS	(3) DID	(4) DID	(5) PSM-DID	(6) PSM-DID
HSR	0.028 (0.103)	−0.016 (0.109)				
HSR*After			0.132 ** (0.061)	0.183 ** (0.075)	0.162 ** (0.075)	0.206 ** (0.096)
Age		0.001 (0.006)		−0.000 (0.006)		−0.001 (0.008)
Education		−0.021 (0.172)		−0.035 (0.171)		0.064 (0.230)
Family		−0.029 (0.019)		−0.058 *** (0.021)		−0.067 ** (0.027)
Elder		0.207 ** (0.100)		0.216 ** (0.100)		0.155 (0.127)
Land rent-out		0.004 (0.004)		0.004 (0.004)		0.002 (0.009)
Fixed assets		−0.005 (0.005)		−0.004 (0.005)		−0.026 (0.017)
Soil quality		−0.155 (0.123)		−0.157 (0.123)		−0.268 * (0.155)
Water availability		−0.061 (0.082)		−0.065 (0.082)		−0.073 (0.101)
Urbanization		−1.291 *** (0.474)		−0.944 ** (0.470)		−1.629 ** (0.661)
Cropland size		−0.004 ** (0.002)		−0.004 ** (0.002)		0.004 (0.005)
Off-farm work		0.329 ** (0.140)		0.325 ** (0.140)		0.314 * (0.170)
Out-migration		−0.003 (0.029)		−0.009 (0.029)		−0.005 (0.036)
Constant	0.398 *** (0.055)	0.637 * (0.374)	0.380 *** (0.025)	0.767 ** (0.377)	0.424 *** (0.029)	1.124 ** (0.512)
Wave FEs	Y	Y	Y	Y	Y	Y
City FEs	Y	Y	Y	Y	Y	Y
Observations	7668	5907	7668	5907	4819	4819
R-squared	0.000	0.015	0.002	0.018	0.004	0.023
Number of no	4828	4219	4828	4219	3639	3639

Remarks: standard errors in parentheses; *** $p < 0.01$, ** $p < 0.05$, * $p < 0.1$.

The results in column (6) in Table 4 also show that some control variables had a significant impact on cropland abandonment. Specifically, total rural household members and cropland abandonment were closely related, which was consistent with the conclusions in previous research on the effects of the household labor force on cropland abandonment [12,28]. The number of family household members indicates the richness of the family labor force to a certain extent, with fewer family members indicating that the family does not have enough labor to farm effectively [69]. It was also found that the household off-farm work ratio affected the abandoned cropland area, which was also consistent with previous research [23]. The village-level characteristics, soil quality and urbanization, were found to be negative and significant, suggesting that improving soil quality could reduce cropland abandonment [70,71], that is, high-quality cropland was less likely to be abandoned than low-quality cropland.

4.4. Robustness Test

To further verify the positive effect of HSR accessibility on cropland abandonment, a robustness test was conducted. First, the robustness of the results was tested by using different dependent variables. However, even when the dependent variable measuring methods were changed, the cropland abandoned ratio was also an essential indicator for land quantity protection. Therefore, the cropland abandoned ratio was used to replace the cropland abandoned area [56]. The results in column (1) in Table 5 show that the regression coefficients for HSR*After were still significantly positive, which suggested that HSR accessibility had a significant positive effect on cropland abandonment.

Table 5. Robustness checks for different indicators and the matching method.

	(1) Abandonment Ratio	(2) Radius Matching	(3) Kernel Matching
HSR*After	0.030 ** (0.015)	0.209 ** (0.096)	0.207 ** (0.092)
Constant	0.114 (0.082)	1.067 ** (0.511)	1.081 ** (0.503)
Control variables	Y	Y	Y
Wave FEs	Y	Y	Y
City FEs	Y	Y	Y
Observations	4796	4803	4806
R-squared	0.037	0.023	0.023
Number of no	3628	3629	3629

Remarks: standard errors in parentheses; ** $p < 0.05$.

Another concern was the first-order nearest-neighbor matching that was used in the PSM method; therefore, to test the robustness of the nearest neighbor matching DID results, kernel and radius matching were also used to process the experimental and control groups, after which a DID analysis was conducted. As columns (2) and (3) in Table 5 show, regardless of which matching method was used, the HSR*After regression coefficients remained significantly positive, that is, the regression results still showed that HSR accessibility significantly promoted cropland abandonment, which confirmed that the PSM-DID estimation was robust.

4.5. Placebo Test

To further verify whether there were any unobservable factor biases, a virtual HSR access time was randomly assigned to households to conduct the placebo test, which was equivalent to constructing a "false" HSR accessibility variable. Then, the benchmark model was repeatedly estimated 500 times using the "false" HSR accessibility variable and the estimated coefficients stored.

Figure 4 shows the empirical cumulative distribution function and density for the estimated HSR accessibility coefficients. The estimated coefficients for the HSR accessibility placebo variable were distributed around zero. However, the real estimated value of the benchmark was 0.206, as shown in column (6) in Table 4. The benchmark HSR accessibility estimate coefficients clearly lay outside the range of the coefficients estimated for the virtual HSR accessibility. This placebo test method has been widely used in DID processing studies [72,73].

4.6. Regional Heterogeneity Analysis

Because of China's regional heterogeneity, the HSR accessibility effect on cropland abandonment could differ because of regional topographic and economic development differences [74,75]. For example, because towns and cities on a plain have relatively good locational and topographical conditions, the cost of building the HSR infrastructure is lower, the land mechanization rates relatively higher, the land scale operations easier, and the cropland abandonment probability lower [23]. Therefore, the topographic effects of HSR accessibility on cropland abandonment were examined.

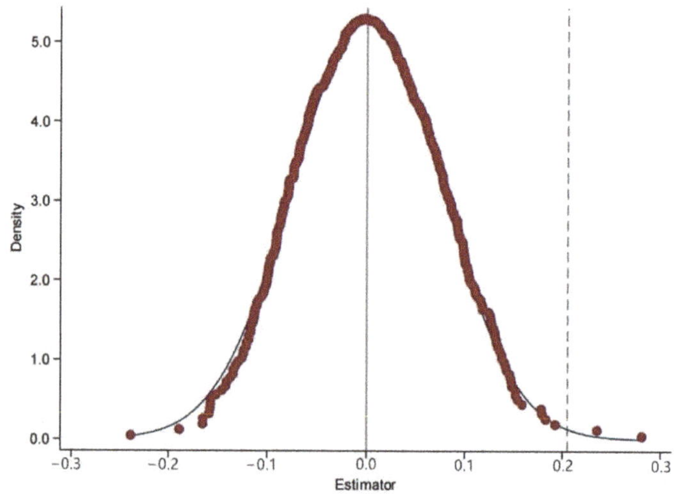

Figure 4. Land HSR and cropland abandonment: placebo test.

As Table 6 shows, in hilly and plain areas, the impact of HSR accessibility on cropland abandonment was significantly positive; however, in mountainous regions, the HSR accessibility coefficients were not significant. This could have been because HSR construction and operations in mountainous areas are costly, which means that many cities in mountainous areas are not linked to the HSR network and HSR accessibility was not the main influencing factor for cropland abandonment. Higher HSR infrastructure building costs in mountainous regions also mean higher HSR ticket prices. The marked development of low-cost air and coach services over the past decades suggests that not everyone can afford HSR travel [76]. Therefore, the high HSR ticket price also meant that HSR accessibility was not the main factor for the population migration in these areas [77] and was not the main reason for cropland abandonment. However, because the HSR construction and operating costs in hilly and plain areas are lower than in mountainous areas, HSR accessibility is higher and labor force mobility easier, which could increase the probability of cropland abandonment.

Table 6. HSR and cropland abandonment: regional heterogeneity.

	(1) Mountain	(2) Hill	(3) Plain
HSR*After	0.363 (0.288)	0.355 ** (0.184)	0.128 * (0.081)
Constant	0.902 (0.986)	0.721 (0.845)	0.667 (0.451)
Control variables	Y	Y	Y
Wave FEs	Y	Y	Y
City FEs	Y	Y	Y
Observations	1317	1684	2906
Number of no	984	1236	1999
R-squared	0.139	0.047	0.028

Remarks: standard errors in parentheses, ** $p < 0.05$, * $p < 0.1$.

The coefficient estimates in the hilly areas were substantially higher than in the plain areas. Compared with hilly areas, plain areas have flatter terrain and higher mechanization utilization rates, which could weaken the impact of labor migration and cropland abandonment [23] and the HSR accessibility effects.

4.7. Mechanism Analysis

The benchmark model results indicated that HSR accessibility influenced cropland abandonment and the out-migration of the rural labor force, that is, there was a possible "push" effect. Conversely, it is possible that HSR accessibility affected the probability of off-farm employment and there was a "pull" effect. These potential mechanisms were, therefore, examined.

First, the "push" effect mechanism that HSR accessibility affected the out-migration of the rural labor force was explored. Labor force migration is closely related to cropland abandonment, as the out-migration of the rural labor force decreases the number of people available for farm work and could result in households having insufficient labor to look after the farm. Therefore, the number of people working outside the rural household was evaluated to determine the effects of the out-migration on the rural labor force. The regression results are summarized in column (1) in Table 7. It was found that HSR accessibility decreased the rural labor force out-migration. This result may have been because HSR projects promote local economic development and urbanization, which provide more local employment opportunities.

Table 7. Mechanism analysis: impact of the HSR on rural labor force out-migration and off-farm employment.

	(1) Out-Migration	(2) Off-Farm Employment
HSR*After	−0.226 *** (−0.049)	0.019 *** (0.010)
Constant	0.954 *** (0.237)	0.682 *** (0.046)
Control variables	Y	Y
Wave FEs	Y	Y
City FEs	Y	Y
Observations	9760	9760
R-squared	0.009	0.002
Number of no	4880	4880

Remarks: standard errors in parentheses; *** $p < 0.01$.

Then, the "pull" effect mechanism on off-farm employment was examined. To determine whether this HSR accessibility effect mechanism improved cropland abandonment, the effects of HSR accessibility on off-farm employment were explored. Off-farm employment was defined as the share of off-farm employment labor in total labor. If HSR accessibility promoted local economic development and urbanization, then rural laborers were more likely to find non-agricultural jobs [28]. The results are shown in column (2) of Table 7. It was found that HSR accessibility significantly affected the share of off-farm employment labor in total labor, with all estimated coefficients being positive, that is, HSR accessibility promotes off-farm employment. Previous research has also shown that HSR accessibility can speed up the urbanization process, which in turn results in a labor shift to the non-agricultural sector [32].

Therefore, the results confirmed that HSR projects promoted the abandonment of croplands by increasing off-farm employment, which supported the "pull effect" hypothesis for non-farm sector employment.

5. Conclusions and Policy Implications

5.1. Conclusions

HSR promotes population migration and leads to economic and urbanization growth. Because HSR accessibility increases the ability of the labor force to move between industries, it has a positive effect on cropland abandonment. Taking HSR accessibility as a quasi-natural experiment, the DID method was used to investigate the effects of HSR accessibility on cropland abandonment. To avoid the traditional limitations in DID models, a PSM approach was first applied to ensure the DID estimation results would be unbiased. Using CLDS data from waves 2014 and 2016, the HSR accessibility cropland abandonment coefficient was estimated to be 0.205. When the ratio of cropland abandonment was taken as the dependent variable to estimate the HSR accessibility effect, the robustness of the conclusion improved. The heterogeneity analysis found that HSR accessibility in both hilly and plain areas had a significant effect on cropland abandonment. Several other potential variables were also examined to understand the underlying mechanisms for the impact of HSR accessibility on cropland abandonment. It was found that HSR accessibility decreased the out-migration of the rural labor force and that HSR accessibility promoted off-farm employment. The mechanistic analysis indicated that HSR accessibility promoted local economic development and urbanization growth, which resulted in local farm labor moving to the non-agricultural sector and abandoning their cropland.

5.2. Policy Implications

This study quantified the effects of HSR accessibility on cropland abandonment. Based on the above findings, some policy implications are proposed. Safeguarding food security remains a significant challenge in developing countries. While the development of the HSR network has increased production factor flows, which is important to China's urbanization and industrialization goals, it has also resulted in a large-scale loss of agricultural labor forces to non-agricultural employment, which in turn has resulted in cropland abandonment. Therefore, local and regional governments need to consider the potential rural hollowing and cropland abandonment risks of HSR projects when formulating relevant policies to improve transportation facilities. Furthermore, as China's rural labor force is continuing to decrease, labor substitution technologies, such as mechanization and no-tillage technologies, should be gradually employed in HSR-accessible areas. Local governments could also formulate relevant support policies to encourage land leasing to balance the withdrawal of agricultural personnel and cropland abandonment in HSR-accessible areas. For example, this could include building an efficient information platform for the cropland leasing market, or providing corresponding financial subsidies for infrastructure improvement to the cropland leaseholders. Encouraging non-farm personnel to contract land and invest in characteristic agriculture in these HSR-accessible areas could also inhibit cropland abandonment. For example, this could include providing corresponding financial subsidies to characteristic agricultural operators to help them improve agricultural infrastructure, etc.

Author Contributions: J.S. wrote the manuscript text and processed the data; F.W. provided the methodology and useful suggestions. All authors have read and agreed to the published version of the manuscript.

Funding: This research was funded by the National Social Science Fund of China (No. 19BGL152).

Institutional Review Board Statement: Not applicable.

Informed Consent Statement: Not applicable.

Data Availability Statement: Not applicable.

Conflicts of Interest: The authors declare no conflict of interest.

References

1. Huang, Z.; Du, X. How does high-speed rail affect land value? Evidence from China. *Land Use Policy* **2021**, *101*, 105068. [CrossRef]
2. Wang, L.; Liu, Y.; Sun, C.; Liu, Y. Accessibility impact of the present and future high-speed rail network: A case study of Jiangsu Province, China. *J. Transp. Geogr.* **2016**, *54*, 161–172. [CrossRef]
3. Shao, S.; Tian, Z.; Yang, L. High speed rail and urban service industry agglomeration: Evidence from China's Yangtze River Delta region. *J. Transp. Geogr.* **2017**, *64*, 174–183. [CrossRef]
4. Ravazzoli, E.; Streifeneder, T.; Cavallaro, F. The Effects of the Planned High-Speed Rail System on Travel Times and Spatial Development in the European Alps. *Mt. Res. Dev.* **2017**, *37*, 131–140. [CrossRef]
5. Haynes, K.E. Labor markets and regional transportation improvements: The case of high-speed trains. *Ann. Reg. Sci.* **1997**, *31*, 57–76. [CrossRef]
6. Motamed, M.J.; Florax, R.J.G.M.; Masters, W.A. Agriculture, transportation and the timing of urbanization: Global analysis at the grid cell level. *J. Econ. Growth* **2014**, *19*, 339–368. [CrossRef]
7. Wang, J.; Jiao, J.; Du, C.; Hu, H. Competition of spatial service hinterlands between high-speed rail and air transport in China: Present and future trends. *J. Geogr. Sci.* **2015**, *25*, 1137–1152. [CrossRef]
8. Wang, F.; Wei, X.; Liu, J.; He, L.; Gao, M. Impact of high-speed rail on population mobility and urbanisation: A case study on Yangtze River Delta urban agglomeration, China. *Transp. Res. Part A Policy Pract.* **2019**, *127*, 99–114. [CrossRef]
9. Mohabir, N.; Jiang, Y.; Ma, R. Chinese floating migrants: Rural-urban migrant labourers' intentions to stay or return. *Habitat Int.* **2017**, *60*, 101–110. [CrossRef]
10. van Leeuwen, C.C.; Cammeraat, E.L.; de Vente, J.; Boix-Fayos, C. The evolution of soil conservation policies targeting land abandonment and soil erosion in Spain: A review. *Land Use Policy* **2019**, *83*, 174–186. [CrossRef]
11. Li, Y.; Wu, W.; Liu, Y. Land consolidation for rural sustainability in China: Practical reflections and policy implications. *Land Use Policy* **2018**, *74*, 137–141. [CrossRef]
12. Xu, D.; Deng, X.; Huang, K.; Liu, Y.; Yong, Z.; Liu, S. Relationships between labor migration and cropland abandonment in rural China from the perspective of village types. *Land Use Policy* **2019**, *88*, 104164. [CrossRef]
13. Campbell, J.E.; Lobell, D.B.; Genova, R.C.; Field, C.B. The Global Potential of Bioenergy on Abandoned Agriculture Lands. *Environ. Sci. Technol.* **2008**, *42*, 5791–5794. [CrossRef]
14. Levers, C.; Schneider, M.; Prishchepov, A.V.; Estel, S.; Kuemmerle, T. Spatial variation in determinants of agricultural land abandonment in Europe. *Sci. Total Environ.* **2018**, *644*, 95–111. [CrossRef] [PubMed]
15. Long, H.; Tu, S.; Ge, D.; Li, T.; Liu, Y. The allocation and management of critical resources in rural China under restructuring: Problems and prospects. *J. Rural. Stud.* **2016**, *47*, 392–412. [CrossRef]
16. Liu, Y.; Zhou, Y. Reflections on China's food security and land use policy under rapid urbanization. *Land Use Policy* **2021**, *109*, 105699. [CrossRef]
17. Subedi, Y.R.; Kristiansen, P.; Cacho, O. Drivers and consequences of agricultural land abandonment and its reutilisation pathways: A systematic review. *Environ. Dev.* **2022**, *42*, 100681. [CrossRef]
18. Ye, L.; van Ranst, E. Production scenarios and the effect of soil degradation on long-term food security in China. *Glob. Environ. Chang.* **2009**, *19*, 464–481. [CrossRef]
19. Hervas, A. Mapping oil palm-related land use change in Guatemala, 2003–2019. Implications for food security. *Land Use Policy* **2021**, *109*, 105657. [CrossRef]
20. Jiang, L.; Guo, S.; Wang, G.; Kan, S.; Jiang, H. Changes in agricultural land requirements for food provision in China 2003–2011: A comparison between urban and rural residents. *Sci. Total Environ.* **2020**, *725*, 138293. [CrossRef]
21. Alonso-Sarría, F.; Martínez-Hernández, C.; Romero-Díaz, A.; Cánovas-García, F.; Gomariz-Castillo, F. Main Environmental Features Leading to Recent Land Abandonment in Murcia Region (Southeast Spain). *Land Degrad. Dev.* **2015**, *27*, 654–670. [CrossRef]
22. Lieskovský, J.; Bezák, P.; ŠPulerová, J.; Lieskovský, T.; Koleda, P.; Dobrovodská, M.; Bürgi, M.; Gimmi, U. The abandonment of traditional agricultural landscape in Slovakia—Analysis of extent and driving forces. *J. Rural. Stud.* **2015**, *37*, 75–84. [CrossRef]
23. Deng, X.; Xu, D.; Qi, Y.; Zeng, M. Labor Off-Farm Employment and Cropland Abandonment in Rural China: Spatial Distribution and Empirical Analysis. *Int. J. Environ. Res. Public Health* **2018**, *15*, 1808. [CrossRef] [PubMed]
24. Shively, G.; Martinez, E. Deforestation, irrigation, employment and cautious optimism in southern Palawan, the Philippines. *Agric. Technol. Trop. Deforest.* **2001**, 335–346. [CrossRef]
25. Abrate, G.; Viglia, G.; García, J.S.; Forgas-Coll, S. Price Competition within and between Airlines and High-Speed Trains: The Case of the Milan—Rome Route. *Tour. Econ.* **2016**, *22*, 311–323. [CrossRef]
26. Guirao, B.; Lara-Galera, A.; Campa, J.L. High Speed Rail commuting impacts on labour migration: The case of the concentration of metropolis in the Madrid functional area. *Land Use Policy* **2017**, *66*, 131–140. [CrossRef]
27. Shui, Y.; Xu, D.; Liu, S. Research on Role Cognition and Employment Strategy of Rural Female Laborers in Sichuan, China. *Sustainability* **2019**, *11*, 1708. [CrossRef]
28. Zhang, Y.; Li, X.; Song, W. Determinants of cropland abandonment at the parcel, household and village levels in mountain areas of China: A multi-level analysis. *Land Use Policy* **2014**, *41*, 186–192. [CrossRef]
29. Chen, Z.; Xue, J.; Rose, A.Z.; Haynes, K.E. The impact of high-speed rail investment on economic and environmental change in China: A dynamic CGE analysis. *Transp. Res. Part A Policy Pract.* **2016**, *92*, 232–245. [CrossRef]

30. Cheng, Y.S.; Loo, B.P.; Vickerman, R. High-speed rail networks, economic integration and regional specialisation in China and Europe. *Travel. Behav. Soc.* **2015**, *2*, 1–14. [CrossRef]
31. Rodrigue, J.P. The Economics and Politics of High-Speed Rail: Lessons from Experiences Abroad. *AAG Rev. Books* **2016**, *4*, 17–18. [CrossRef]
32. Emerick, K. Agricultural productivity and the sectoral reallocation of labor in rural India. *J. Dev. Econ.* **2018**, *135*, 488–503. [CrossRef]
33. Xie, Y.; Jiang, Q. Land arrangements for rural–urban migrant workers in China: Findings from Jiangsu Province. *Land Use Policy* **2016**, *50*, 262–267. [CrossRef]
34. Zhu, P. Does high-speed rail stimulate urban land growth? Experience from China. *Transp. Res. Part D Transp. Environ.* **2021**, *98*, 102974. [CrossRef]
35. Bentlage, M.; Lüthi, S.; Thierstein, A. Knowledge creation in German agglomerations and accessibility—An approach involving non-physical connectivity. *Cities* **2013**, *30*, 47–58. [CrossRef]
36. Cao, J.; Liu, X.C.; Wang, Y.; Li, Q. Accessibility impacts of China's high-speed rail network. *J. Transp. Geogr.* **2013**, *28*, 12–21. [CrossRef]
37. Shaw, S.L.; Fang, Z.; Lu, S.; Tao, R. Impacts of high speed rail on railroad network accessibility in China. *J. Transp. Geogr.* **2014**, *40*, 112–122. [CrossRef]
38. Spiekermann, K.; Wegener, M. The shrinking continent: New time-space maps of Europe. *Environ. Plan. B Plan. Des.* **1994**, *21*, 653–673. [CrossRef]
39. Harvey, D. The Condition of Postmodernity: An Enquiry into the Origins of Cultural Change. *Contemp. Sociol.* **1989**, *20*. [CrossRef]
40. Maharjan, A.; Kochhar, I.; Chitale, V.S.; Hussain, A.; Gioli, G. Understanding rural outmigration and agricultural land use change in the Gandaki Basin, Nepal. *Appl. Geogr.* **2020**, *124*, 102278. [CrossRef]
41. Ojha, H.R.; Shrestha, K.K.; Subedi, Y.R.; Shah, R.; Nuberg, I.; Heyojoo, B.; Cedamon, E.; Rigg, J.; Tamang, S.; Paudel, K.P.; et al. Agricultural land underutilisation in the hills of Nepal: Investigating socio-environmental pathways of change. *J. Rural. Stud.* **2017**, *53*, 156–172. [CrossRef]
42. Ahani, S.; Dadashpoor, H. A review of domains, approaches, methods and indicators in peri-urbanization literature. *Habitat Int.* **2021**, *114*, 102387. [CrossRef]
43. Hou, D.; Meng, F.; Prishchepov, A.V. How is urbanization shaping agricultural land-use? Unraveling the nexus between farmland abandonment and urbanization in China. *Landsc. Urban Plan.* **2021**, *214*, 104170. [CrossRef]
44. Rimal, B.; Sharma, R.; Kunwar, R.; Keshtkar, H.; Stork, N.E.; Rijal, S.; Rahman, S.A.; Baral, H. Effects of land use and land cover change on ecosystem services in the Koshi River Basin, Eastern Nepal. *Ecosyst. Serv.* **2019**, *38*, 100963. [CrossRef]
45. Verma, A.; Sudhira, H.S.; Rathi, S.; King, R.; Dash, N. Sustainable urbanization using high speed rail (HSR) in Karnataka, India. *Res. Transp. Econ.* **2013**, *38*, 67–77. [CrossRef]
46. Blum, U.; Haynes, K.E.; Karlsson, C. Introduction to the special issue the regional and urban effects of high-speed trains. *Ann. Reg. Sci.* **1997**, *31*, 1–20. [CrossRef]
47. MacDonald, D.; Crabtree, J.; Wiesinger, G.; Dax, T.; Stamou, N.; Fleury, P.; Gutierrez Lazpita, J.; Gibon, A. Agricultural abandonment in mountain areas of Europe: Environmental consequences and policy response. *J. Environ. Manag.* **2000**, *59*, 47–69. [CrossRef]
48. Cai, F. Has China's labor mobility exhausted its momentum. *Chin. Rural Econ.* **2018**, *9*, 2–13.
49. Tombe, T.; Zhu, X. Trade, Migration, and Productivity: A Quantitative Analysis of China. *Am. Econ. Rev.* **2019**, *109*, 1843–1872. [CrossRef]
50. Blimpo, M.P.; Harding, R.; Wantchekon, L. Public Investment in Rural Infrastructure: Some Political Economy Considerations. *J. Afr. Econ.* **2013**, *22* (Suppl. S2), ii57–ii83. [CrossRef]
51. Lang, T.; Barling, D. Food security and food sustainability: Reformulating the debate. *Geogr. J.* **2012**, *178*, 313–326. [CrossRef]
52. Yacamán Ochoa, C.; Ferrer Jiménez, D.; Mata Olmo, R. Green Infrastructure Planning in Metropolitan Regions to Improve the Connectivity of Agricultural Landscapes and Food Security. *Land* **2020**, *9*, 414. [CrossRef]
53. Niu, F.; Xin, Z.; Sun, D. Urban land use effects of high-speed railway network in China: A spatial spillover perspective. *Land Use Policy* **2021**, *105*, 105417. [CrossRef]
54. Guo, Y.; Cao, L.; Song, Y.; Wang, Y.; Li, Y. Understanding the formation of City-HSR network: A case study of Yangtze River Delta, China. *Transp. Policy* **2022**, *116*, 315–326. [CrossRef]
55. Jia, S.; Zhou, C.; Qin, C. No difference in effect of high-speed rail on regional economic growth based on match effect perspective? *Transp. Res. Part A Policy Pract.* **2017**, *106*, 144–157. [CrossRef]
56. Chen, F.; Hao, X.; Chen, Z. Can high-speed rail improve health and alleviate health inequality? Evidence from China. *Transp. Policy* **2021**, *114*, 266–279. [CrossRef]
57. Deng, X.; Xu, D.; Zeng, M.; Qi, Y. Does Internet use help reduce rural cropland abandonment? Evidence from China. *Land Use Policy* **2019**, *89*, 104243. [CrossRef]
58. Che, Y. Off-farm employments and land rental behavior: Evidence from rural China. *China Agric. Econ. Rev.* **2016**, *8*, 37–54. [CrossRef]
59. Zheng, L.; Qian, W. The impact of land certification on cropland abandonment: Evidence from rural China. *China Agric. Econ. Rev.* **2021**. *ahead of print*. [CrossRef]

60. He, Y.; Xie, H.; Peng, C. Analyzing the behavioural mechanism of farmland abandonment in the hilly mountainous areas in China from the perspective of farming household diversity. *Land Use Policy* **2020**, *99*, 104826. [CrossRef]
61. Lee, J.; Oh, Y.G.; Yoo, S.H.; Suh, K. Vulnerability assessment of rural aging community for abandoned farmlands in South Korea. *Land Use Policy* **2021**, *108*, 105544. [CrossRef]
62. Díaz, G.I.; Nahuelhual, L.; Echeverría, C.; Marín, S. Drivers of land abandonment in Southern Chile and implications for landscape planning. *Landsc. Urban Plan.* **2011**, *99*, 207–217. [CrossRef]
63. Meyer, B.D. Natural and Quasi-Experiments in Economics. *J. Bus. Econ. Stat.* **1995**, *13*, 151–161. [CrossRef]
64. Sunak, Y.; Madlener, R. The impact of wind farm visibility on property values: A spatial difference-in-differences analysis. *Energy Econ.* **2016**, *55*, 79–91. [CrossRef]
65. Heckman, J.J.; Ichimura, H.; Todd, P. Matching as an Econometric Evaluation Estimator. *Rev. Econ. Stud.* **1998**, *65*, 261–294. [CrossRef]
66. Heckman, J.J.; Ichimura, H.; Todd, P.E. Matching as an Econometric Evaluation Estimator: Evidence from Evaluating a Job Training Programme. *Rev. Econ. Stud.* **1997**, *64*, 605–654. [CrossRef]
67. Dehejia, R.H.; Wahba, S. Propensity Score-Matching Methods for Nonexperimental Causal Studies. *Rev. Econ. Stat.* **2002**, *84*, 151–161. [CrossRef]
68. Locke, C.M.; Butsic, V.; Rissman, A.R. Zoning effects on housing change vary with income, based on a four-decade panel model after propensity score matching. *Land Use Policy* **2017**, *64*, 353–362. [CrossRef]
69. Zimmerer, K.S. Soil erosion and labor shortages in the Andes with special reference to Bolivia, 1953–1991: Implications for "conservation-with-development". *World Dev.* **1993**, *21*, 1659–1675. [CrossRef]
70. Feranec, J.; Jaffrain, G.; Soukup, T.; Hazeu, G. Determining changes and flows in European landscapes 1990–2000 using CORINE land cover data. *Appl. Geogr.* **2010**, *30*, 19–35. [CrossRef]
71. Nadal, J.; Pèlachs, A.; Molina, D.; Soriano, J.M. Soil fertility evolution and landscape dynamics in a Mediterranean area: A case study in the Sant Llorenç Natural Park (Barcelona, NE Spain). *Area* **2009**, *41*, 129–138. [CrossRef]
72. Ferrara, E.L.; Chong, A.; Duryea, S. Soap Operas and Fertility: Evidence from Brazil. *Am. Econ. J. Appl. Econ.* **2012**, *4*, 1–31. [CrossRef]
73. Liu, Q.; Lu, Y. Firm investment and exporting: Evidence from China's value-added tax reform. *J. Int. Econ.* **2015**, *97*, 392–403. [CrossRef]
74. Campos, J.; de Rus, G. Some stylized facts about high-speed rail: A review of HSR experiences around the world. *Transp. Policy* **2009**, *16*, 19–28. [CrossRef]
75. van der Zanden, E.H.; Verburg, P.H.; Schulp, C.J.; Verkerk, P.J. Trade-offs of European agricultural abandonment. *Land Use Policy* **2017**, *62*, 290–301. [CrossRef]
76. Dobruszkes, F.; Chen, C.L.; Moyano, A.; Pagliara, F.; Endemann, P. Is high-speed rail socially exclusive? An evidence-based worldwide analysis. *Travel. Behav. Soc.* **2022**, *26*, 96–107. [CrossRef]
77. Garmendia, M.; Ureña, J.; Coronado, J. Long-distance trips in a sparsely populated region: The impact of high-speed infrastructures. *J. Transp. Geogr.* **2011**, *19*, 537–551. [CrossRef]

land

MDPI

Article

The Synchronous Development Pattern and Type Division of Functional Coupling Coordination and Human Activity Intensity Based on the "Production–Living–Ecological" Space Perspective: A Case Study of Wanzhou District

Tian Liang [1,2], Fei Yang [1,2], Yinchen Luo [2,3], Mengying Fang [1], Xi Huang [1], Zhiyong Zhang [2,4,5], Chuanhao Wen [6,*] and Xiaohong Ren [7,*]

1 College of Public Management, Chongqing Finance and Economics College, Chongqing 401320, China
2 Institute of Green Development, Chongqing Finance and Economics College, Chongqing 401320, China
3 College of Economics, Chongqing Finance and Economics College, Chongqing 401320, China
4 Research Center for Economic of Upper Reaches of Yangtze River, Chongqing Technology and Business University, Chongqing 400067, China
5 Guizhou Academy of Social Sciences, Guiyang 550002, China
6 College of Economics, Yunnan University, Kunming 650091, China
7 College of Economics and Management, Chongqing Jiaotong University, Chongqing 400074, China
* Correspondence: chhwen1972@ynu.edu.cn (C.W.); renxiaohong@cqjtu.edu.cn (X.R.); Tel.: +86-185-0234-0601 (C.W.); +86-152-2321-9001 (X.R.)

Citation: Liang, T.; Yang, F.; Luo, Y.; Fang, M.; Huang, X.; Zhang, Z.; Wen, C.; Ren, X. The Synchronous Development Pattern and Type Division of Functional Coupling Coordination and Human Activity Intensity Based on the "Production–Living–Ecological" Space Perspective: A Case Study of Wanzhou District. *Land* **2022**, *11*, 1897. https://doi.org/10.3390/land11111897

Academic Editors: Li Ma, Yingnan Zhang, Muye Gan and Zhengying Shan

Received: 16 September 2022
Accepted: 23 October 2022
Published: 26 October 2022

Abstract: The coupling and coordinated change characteristics of land-use production, living, and ecological functions (PLEFs) and their relationship with human activity intensity (*HAI*) in ecologically fragile areas are important to study, especially in promoting the sustainable development of regional land-use and revealing the evolution of the human–land relationship. In this paper, the coupling coordination degree (CCD) model was used to analyze the coordinated development level of PLEF in Wanzhou District from 2000 to 2020. The *HAI* was measured by the equivalent of construction land. The synchronous development model was introduced to analyze the relationship between them. The results showed that, in Wanzhou District, the PLEFs showed significant spatial distribution differences and evident spatial complementarity. The PLEFs of Wanzhou District were at a good coordination level, but exhibited a downward trend. A spatial pattern of "high in the west and low in the southeast" was presented. The CCD of the production–living function was poor, which is the critical direction of future optimization. The value of *HAI* in Wanzhou District showed an increasing trend and exhibited a high concentration in the central town and its surrounding regions. According to the synchronous development state of the *HAI* and the CCD of the PLEFs, Wanzhou District was divided into three development types. The development type of most areas of Wanzhou District was positive, but the area decreased over the past 20 years. Therefore, it is crucial to propose other regulatory strategies for regions with different development types. This research will provide a decision-making reference for promoting the coordination of the PLEFs and alleviating human–land relations in the reservoir area of central and western China, mountainous regions, and similar areas in developing countries.

Keywords: coupling coordination degree model; "production–living–ecological" functions; human activity intensity; the synchronous development model

1. Introduction

Land is the spatial carrier of national survival and development [1–3]. Land is multifunctional. It can provide various products and services to human beings through different land-use functions [4,5]. The classification of land-use functions from the perspective of "production–living–ecological" space (PLES) has been recognized by more and more scholars [6–8]. It includes production, living, and ecological functions (PLEFs) [9,10]. The production function (PF) refers to the social output function of using land as the labor carrier to produce products and services [11]. The living function (LF) refers to the function that various land types provide to ensure human survival and growth [12]. The ecological function (EF) is the natural basis and function for maintaining human survival and development, which occurs in the ecosystem and its processes [13]. With the rapid growth of urbanization, intensified human activities have caused severe disturbances to land-use. The function and pattern of land have undergone drastic changes [14,15]. The competition and conflict of the PLES are becoming increasingly fierce [16]. Additionally, the imbalance problems and contradictions of the PLEFs are becoming increasingly prominent [11,17,18] and many other issues frequently occur [19–24]. The sustainable utilization of territorial space and the relationship between humans and land are facing severe crises and challenges [11,25]. Accurately understanding the changes in the PLEFs and their relations with the intensity of human activities is the basis for carrying out territorial space planning and coordinating the relationship between humans and land [26–28].

The research on PLES functions has mainly focused on the following aspects: (1) the definition of concept and connotation [29], spatio-temporal pattern evolution research [30], classification and function evaluation [31], coupling coordination degree (CCD) research [6,32], spatial layout optimization of PLES and land-use policy, etc. The degree of interaction and coordination between multiple systems can be calculated by the CCD model [33], which has been widely used and achieved fruitful results [34,35]. Scholars generally use human activity intensity (*HAI*) to measure the degree of human interference with the natural environment. However, when related to research on the degree of land-use, single indicators, such as the land utilization rate and land development intensity, are often used [36], resulting in one-sided research results. In addition, the selection of indicators and the determination of weights in these studies are primarily targeted at specific regions, with the defects of intense subjectivity and poor universality. Based on this, this study adopts the *HAI* measurement model [37]. The percentage of the equivalent area of construction land to the total area of provincial land is used in this model. This method can reflect the comprehensive status of regional land-use degrees and has strong universality. These methods provide an essential reference for measuring the CCD and *HAI* in Wanzhou District. However, research on the relationship between them in ecologically fragile areas has not been systematically and deeply carried out. It is urgent that new research ideas are injected for further research.

Wanzhou District is located in the central area of the Three Gorges Reservoir Region (TGRR) and is the financial center of the TGRR. The ecological environment in this area is fragile and highly sensitive to human activities [38]. Therefore, we took ArcGIS 10.4 as the technology platform and adopted land-use raster data. The technical solution for this paper is shown in Figure 1. We aimed to study the evolution characteristics of the CCD and its relationship with the *HAI* in Wanzhou District from 2000 to 2020. This could reveal the evolution of the human–land relationship in Wanzhou District. The results of this study will contribute to the adjustment and optimization of various ecological policies in the hinterland of the TGRR. We provided a typical example for the study of sustainable development in the reservoir area of central and western China, mountainous areas, and similar areas in developing countries. The objectives of this paper include the following:

(1) Quantitatively evaluate the CCD of the PLEFs;
(2) Calculate the spatio-temporal evolution characteristics of the *HAI* in Wanzhou District;
(3) Use a synchronous development model to explore the response process of the *HAI* to the CCD of the PLEFs;
(4) Divide the development types and put forward different regulation strategies.

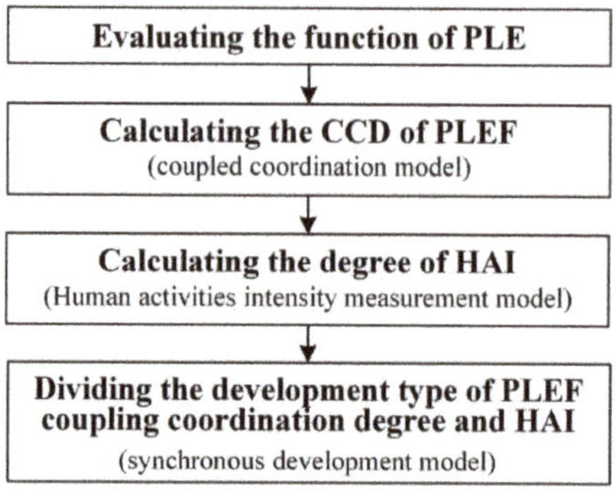

Figure 1. The primary steps of the technical solution.

2. Materials and Methods

2.1. Study Area

Wanzhou District is located in the upper reaches of the Yangtze River and the heart of the TGRR. It is the second-largest city in Chongqing. It covers an area of 3457 km^2, with 12 townships, 29 towns, and 11 subdistricts under its jurisdiction. It also has the dual characteristics of a mountain city and a river city (Figure 2). According to the Seventh National Census bulletin of Wanzhou District, by the end of 2020, the total population was 172.57 million, with an urbanization rate of 68.92%. It is the district with the largest population, the most significant urban volume, and the most management units in Chongqing. The terrain of Wanzhou District is mainly mountainous, and forest is the primary land cover type (53.5%). Purple soil, yellow soil, and lime soil are the main soil types. After three impoundments in the TGRR, the inundated area and the scale of the resettlement project in Wanzhou District are huge. During this period, land-use and the intensity of human activities have undergone significant changes. The ecological environment and landscape pattern change clearly in the area, which is a typical representative area of the hinterland of the TGRR. Therefore, taking Wanzhou District as the study area, this paper will be helpful in revealing the general law of the coupling and coordinated evolution of the PLEFs and their response to the human activity intensity in other ecologically vulnerable areas.

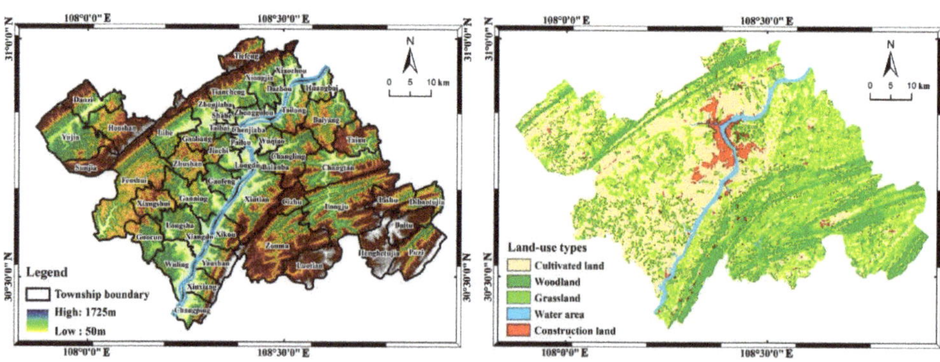

Figure 2. Administrative division of Wanzhou District.

2.2. Data

This paper adopted the land-use raster data of Chongqing from 2000, 2010, and 2020. The resolution is 30 m × 30 m. The data were downloaded from the Data Center for Resources and Environmental Sciences, Chinese Academy of Sciences (http://www.resdc.cn/ (accessed on 5 June 2022)). We used a mask extraction tool to extract the data from

Wanzhou District. These data include the following types: cultivated land, woodland, grassland, water area, urban and rural industrial and mining areas and residential land, and unused land. The social–economic data of the Wanzhou District primarily came from the Chongqing Statistical Yearbook (2021), including the total area, population, urban population, urbanization rate, forest coverage rate, etc.

2.3. Methods

2.3.1. Evaluation of PLEFs

Based on the current research results and guided by the functional theory of PLEF, the concepts of strong production land, semi-production land, and weak production land; strong living land, semi-living land, and weak living land; and strong ecological land, semi-ecological land, and weak ecological land were introduced. In addition, according to the existing studies [2,6,39,40], the hierarchical assignment method was used to assign the applicable scores of various functional lands (Table 1).

Table 1. Index system of PLEFs.

The PLES Types	Secondary Categories	Land-Use Type	Function Score
Production land	Strong production land	Industrial and mining construction land	5
	Semi-production land	Paddy fields, dry land, urban land, and rural settlements	3
	Weak production land	Woodland, grassland, rivers, reservoir ponds, and wetlands	1
Living land	Strong living land	Urban land and rural settlements	5
	Semi-living land	Industrial and mining construction land	3
	Weak living land	Paddy fields, dry land, rivers, reservoir ponds, and wetlands	1
Ecological land	Strong ecological land	Woodland, grassland, rivers, and wetlands	5
	Semi-ecological land	Paddy fields and dry land	3
	Weak ecological land	Reservoirs and ponds	1

Firstly, to more accurately identify the land types and evaluate the "production–living–ecological" function, based on the existing research [24] and the actual situation of the research area, the fishing net tool was used to determine the 300 m × 300 m grid as the evaluation unit. We divided Wanzhou District into 39,230 evaluation units. Secondly, the frequency statistics tool was used to calculate the area of each land-use type in each evaluation unit. Thirdly, the multi-factor weighted summation method was used to sum up the functional scores of each evaluation unit. Finally, the results were analyzed by visual mapping to reveal the spatial and temporal evolution characteristics of the PLEFs in Wanzhou District. The formula is as follows:

$$W_j = \sum_{i=1}^{n} S_i \times V_i \quad (1)$$

where W_j is the functional evaluation value of subsystem j in the study area, j = 1, 2, 3; W_1, W_2, and W_3 are the evaluation values of the production space functional system, living space functional system, and ecological space functional system, respectively; I is the land-use type; n is the number of land-use types included in each evaluation unit; S_i represents the area of each land-use type in each evaluation unit, and the unit is km^2; and V_i represents the function score corresponding to the i-th land-use type (Table 1).

2.3.2. PLEF Coupled Coordination Model

The CCD, introduced from physics, describes the degree of the interaction influence of two or more systems. The coupling effect and coordination degree determine the sustainable development of the system [33]. To explore the spatial distribution and evolution characteristics of the coupling coordination relationship between the PLEFs in Wanzhou District at different periods, considering relevant research results [41,42] and based on the actual study, a measuring model of the coupling degree of the PLEFs in Wanzhou District was constructed.

$$C = 3\left\{\frac{P_i \times R_i \times E_i}{(P_i + R_i + E_i)^3}\right\}^{1/3} \tag{2}$$

where C is the coupling degree of the PLEFs in Wanzhou District, and the value range is [0–1]. The value of C is determined by the evaluation value of the PLEFs. P_i, R_i, and E_i represent the evaluation scores of the production, living, and ecological functions in Wanzhou District, respectively. The higher the value of C, the stronger the interaction and influence within the PLEFs. Using the Natural Breakpoint classification method, the coupling degree of the PLEFs in the study area was divided into four types (Table 2).

Table 2. Stage division of the coupling degree of PLEF.

Coupling Degree	Specific Stage	Characteristic
C ∈ [0, 0.2]	Low coupling period	The game within the PLEFs began to be played, which were in a low-level coupling period. When C = 0, the PLEFs are unrelated and develop into disorder.
C ∈ [0.2, 0.5]	Antagonism period	The interaction between the PLEFs is strengthened. The phenomenon is that the dominant function becomes stronger and occupies the space of other parts, while other parts continue to weaken.
C ∈ [0.5, 0.8)	Run-in period	PLEF begins to balance and cooperate, showing a benign coupling feature.
C ∈ [0.8, 1)	Coordinated coupling period	The benign coupling between the PLEFs is stronger and gradually develops in an orderly direction during the period of high-level coordination coupling. When C = 1.0, the PLEFs achieve benign resonance coupling and tend toward a new orderly structure.

To further explore whether the PLEFs are mutually promoted to a high level or mutually restricted to a low level, we constructed the CCD model to further analyze the comprehensive coordination development degree between the PLEFs. The formula is as follows:

$$D = (C \times T)^{1/2},\ T = \alpha P_i + \beta R_i + \gamma E_i \tag{3}$$

where D is the CCD between the PLEFs in Wanzhou District. The higher the D value of the PLEFs, the better the coupling coordination between functions. T is the comprehensive coordination index among the PLEFs; and α, β, and γ are the undetermined coefficients of the PF, LF, and EF, respectively. Based on the existing research results, the undetermined coefficient was determined to be $\alpha = \beta = 0.3$, $\gamma = 0.4$.

To further explore the interaction degree and coordinated development of the production–living function, production–ecological function, and living–ecological function, Equations (2) and (3) can be further subdivided into:

$$C_1 = 2\left\{\frac{P_i \times R_i}{(P_i + R_i)^2}\right\}^{1/2},\ C_2 = 2\left\{\frac{P_i \times E_i}{(P_i + E_i)^2}\right\}^{1/2},\ C_3 = 2\left\{\frac{R_i \times E_i}{(R_i + E_i)^2}\right\}^{1/2} \tag{4}$$

$$D = (C \times T)^{1/2},\ T_1 = \alpha P_i + \beta R_i \ or\ T_2 = \alpha P_i + \gamma E_i \ or\ T_3 = \beta R_i + \gamma E_i \tag{5}$$

According to current research results [43], when measuring the CCD between the production and living functions, $\alpha = \beta = 0.5$. When calculating the CCD between the production and ecological functions, $\alpha = 0.45$, $\gamma = 0.55$. When calculating the CCD between the ecological and living functions, $\beta = 0.45$, $\gamma = 0.55$. By referring to relevant research results [36,37] and combining them with this study, the CCD of the PLEFs can be divided into three stages: good coordination, barely coordinated, and dysregulation recession (Table 3).

Table 3. Stage division of the CCD of the PLEFs.

Coupling Coordination Degree	Specific Stage	Characteristic
D ∈ [0, 0.15)	Dysregulation recession	PLEFs conflict with each other and are in the low level of coupling and coordination stage.
D ∈ [0.15, 0.35)	Barely coordinated	PLEFs check and balance each other and cooperate, and enter the transitional stage of coupling, coordination, or conflict.
D ∈ [0.35, 0.5]	Good coordination	Coordinated development of the PLEFs with a high level of coupling and coordination stage

2.3.3. Human Activities Intensity Measurement Model

The *HAI* is an effective index to calculate the intensity of the human transformation of nature. In this paper, we used the construction land equivalent to measure the *HAI* in Wanzhou District. The areas of different land-use types were converted into corresponding construction land equivalents according to their conversion coefficients of construction land equivalents. Then, *HAI* was calculated according to the equivalent sum of construction land of different land-use types in the region [37]. The calculation formulas are as follows:

$$HAI_i = \frac{S_{CLE-i}}{S_i} \times 100\% \tag{6}$$

$$S_{CLE-ij} = \sum_{j=1}^{n} SL_{ij} \times CI_{ij} \tag{7}$$

where HAI_i is the human activity intensity in the i-th region, S_{CLE-i} is the equivalent area of construction land in the i-th region, S_i is the total area of the i-th region, SL_{ij} is the area of the j-th land-use type in the i-th region, CI_{ij} is the equivalent conversion coefficient of construction land of the j-th land-use type in the i-th region, and n is the number of land-use types in the i-th region. Referring to existing studies [44], the equivalent conversion coefficient (CI_{ij}) of construction land for different land-use types is shown in Table 4.

Table 4. Equivalent conversion coefficient of construction land for different land-use types.

Land-use Types	Cultivated Land	Woodland	Grassland	Water Area	Urban and Rural Industrial and Mining Areas and Residential Land	Unused Land
CI_{ij}	0.20	0.13	0.10	0.60	1.00	0.00

2.3.4. The Synchronous Development Model

The relationship between the *HAI* and the CCD of the PLEFs is not only a state at a specific time point, but also a changing process at different time points. Therefore, a synchronous development model was constructed in this paper. We used this model to reveal the synchrony characteristics (lead or lag) of the changes in the CCD of the PLEFs and the *HAI* level in different years [45,46]. The results can provide a basis for the formulation of differentiated regulation strategies. The calculation formula is as follows:

$$T = D - HAI \quad (8)$$

where T is the synchronous development degree. If $T > 0.1$, the coordination degree of the PLEFs is relatively higher than the *HAI*, and the *HAI* is lagging, which is defined as the lagging type of development. In this type, *HAI* does not negatively affect the coupling and coordinated development of the PLEFs. If $-0.1 \leq T \leq 0.1$, their difference is slight. When there is more balance, it is defined as the synchronous type of development. In this type, the *HAI* and coupling coordination development of the PLEFs are in synchronous coordination; If $T < -0.1$, it means that the *HAI* is relatively higher than the CCD of the PLEFs. The *HAI* is leading development, defined as the leading type of development. This type of *HAI* negatively affects the coupling and coordinated development of the PLEFs.

3. Results

3.1. Spatial–Temporal Evolution Characteristics of the PLEFs

3.1.1. Production Function

From 2000 to 2020, the production function in Wanzhou District increased yearly, showing a spatial distribution characteristic of higher values in the west and lower values in the east (Figure 3). High values were mainly distributed in Gaofeng Town, Shuanghekou Street, Longdu Street, and other places. In the east of Wanzhou District, Zouma Town, Puzi Town, Lishu Town, Dibao Tujia Town, and the Yangtze River had lower values. Owing to the implementation of environmental protection policies along the Yangtze River basin, the land used for industrial and mining construction in the coastal areas was rectified and transferred. The area of land used for intensive production in the Yangtze River basin is decreasing. Tai'an Town, Puzi Township East, Dibao Tujia Township, and the other regions in the past 20 years have developed their strong production land area to vigorously develop the economy. Therefore, the production function scores of the coastal regions of the Yangtze River Basin, Tai'an Town, the eastern part of Puzi Town, Dibao Tujia Town, and other areas changed significantly.

3.1.2. Living Function

From 2000 to 2020, the area with a high living function value in Wanzhou District expanded year by year, showing the spatial distribution characteristics with higher elevation in the central area and lower around (Figure 3). The high value was mainly distributed in the central towns of Wanzhou District, including Pailou Street, Chenjiaba Street, and Wuqiao Street. In the rapid urbanization process in Wanzhou District, especially in the combination of urban and rural areas, the population flow was large, leading to increases in the urban land around the central towns. The living function score changed most significantly in these areas.

3.1.3. Ecological Function

From 2000 to 2020, the evaluation value of the ecological function in Wanzhou District showed a spatial distribution with higher values in the east and lower values in the west (Figure 3). The area with high weight was mainly distributed in Puzi Township, Lishu Township, and Zouma Town. The area with low weight was mainly distributed in the central towns of Wanzhou District and the coastal areas of the Yangtze River Basin, and the scope expanded year by year. Owing to the development concept of "clear water and green mountains are gold and silver mountains" and the implementation of the policies of "returning farmland to forest or grassland" and "returning farmland to the lake", the ecological environment along the Yangtze River basin was restored, and its ecological function was enhanced. Therefore, the ecological function scores in these areas changed dramatically.

In general, the evaluation values of the ecological function and the other two functions in Wanzhou District showed evident spatial complementarity in their distribution.

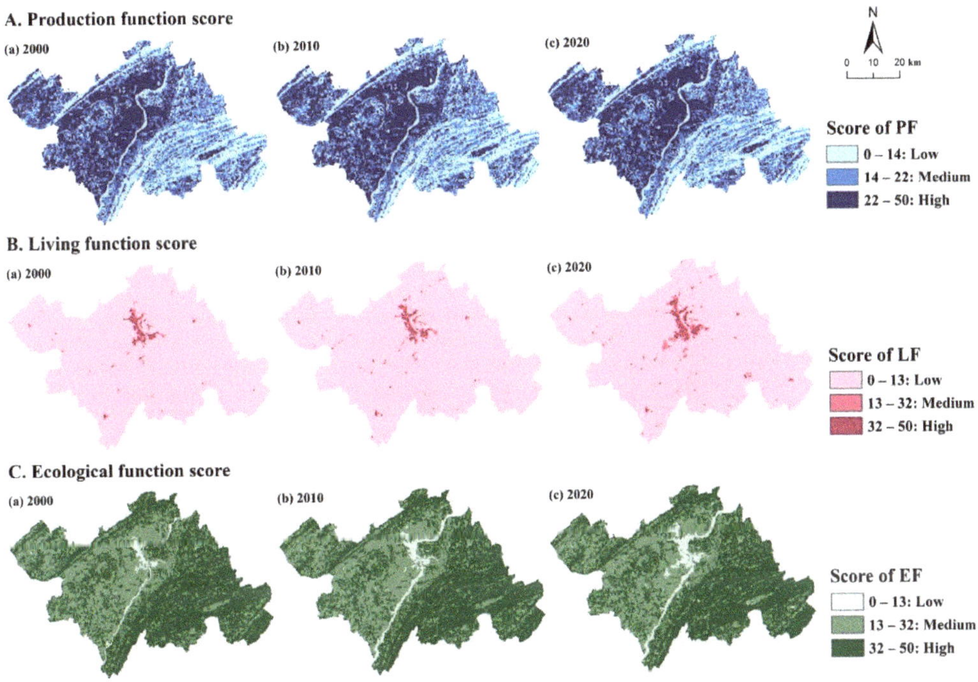

Figure 3. Spatial–temporal evolution of the PF(A), LF(B), and EF(C) from 2000 to 2020.

3.2. Coupling Coordination Degree of the PLEsF

3.2.1. Coupling Degree

In the spatial dimension, the coupling degree of the PLEFs in Wanzhou District showed the distribution characteristics of high in the west and low in the east. In Gaofeng Town, Jiuchi Town, and other areas, the coupling degree of the PLEFs reached the highest value of "1". This indicates that the PLEFs of these regions had a great degree of mutual influence. In the temporal dimension, the interaction within the PLEFs in Wanzhou District weakened from 2000 to 2020. Additionally, the weakening range gradually expanded in the center of Chenjiaba Street (Figure 4).

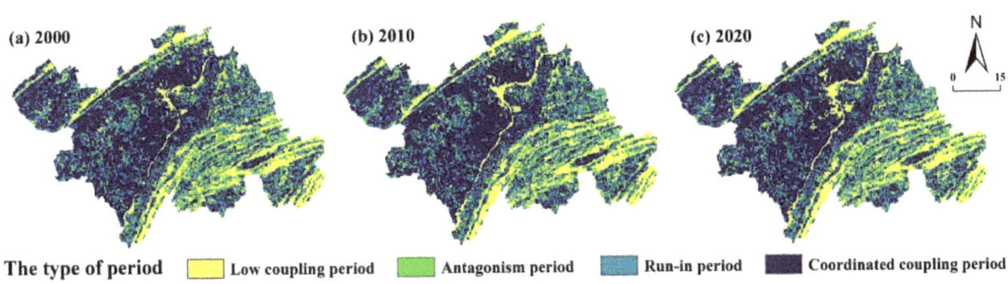

Figure 4. Coupling degree of the PLEFs in 2000 (**a**), 2010 (**b**), and 2020 (**c**).

Since the establishment of the TGRR and the direct jurisdiction of Chongqing Municipality, Wanzhou District, the second-largest city of Chongqing and the heart of the TGRR, has received counterpart support from China. In this stage, Wanzhou District vigorously developed its economy, and the areas of urban residential land, industrial and mining land, and land for transportation and water conservancy increased significantly. Moreover, it constantly encroached on ecological space, leading to the weakening of ecological functions and causing the coupling degree of the PLEFs to become lower and lower from 2000 to 2020.

As shown in Figure 5, the production–living function (PLF) coupling degree of central towns in Wanzhou District had a significant value, with the highest value reaching "1", the scope of which gradually expanded with Pailou Street as the center. This shows that the interaction of central towns in Wanzhou District continued to strengthen. With the continuous development of the economy, the infrastructure of Wanzhou District continuously improved. During the past 20 years, economic production developed vigorously, and the production function reached a dominant position. The production function provides the financial basis and material guarantee for the living function. Therefore, the living function of the central town also improved, and the degree of mutual influence between production and living function increased. The coupling degree of the living–ecological function (LEF) changed little, except in the central urban area. The spatial distribution of the production–ecological function (PEF) coupling was higher in the west than in the east. In 2015, Wanzhou District actively implemented the policy of "returning farmland to forest or grassland." The area of forest and grassland in Wanzhou District significantly increased, and the ecosystem was restored. Therefore, the coupling degree of PEF in Wanzhou District grew yearly. The interaction between the production and ecological functions has become more robust. However, due to the irreversible impact of economic construction on the ecological environment, the coupling degree of PEF in the central towns of Wanzhou District decreased.

3.2.2. Coupling Coordination Degree

In Wanzhou District, the CCD of the PLEFs was at the coordination level. The spatial distribution decreased from northwest to southeast, and the regional difference was noticeable. The western region of Wanzhou District exhibited good coordination, including Zhushan Township, Gaofeng Town, Lihe Town, and other areas. However, the area of this region decreased from 2198.46 km^2 in 2000 to 2142.99 km^2 in 2020. This shows that the CCD of the PLEFs decreased in the study area. The barely coordinated area had a wide distribution range, but was relatively scattered, and its area showed a trend of decreasing first and then increasing. The dysregulation recession area was mainly distributed in the eastern and southeastern regions of Wanzhou District, including Longju Town, Lishu Town, Dibao Tujia Town, and the east part of Puzi Town. Its area showed an increasing trend from 566.71 km^2 in 2000 to 625.84 km^2 in 2020. It can be concluded that, although the coordination area of the PLEFs was dominant, the coordination level decreased in Wanzhou District (Figure 6).

Regarding the CCD of PLF, Wanzhou District was generally in a moderate state of coordination. The spatial distribution included high values in the middle and low values on the two sides, and regional differences were apparent. Most of the western part of Wanzhou District was in a barely coordinated area, including Longsha Town, Ganning Town, Sunjia Town, etc. The moderate coordination area decreased from 2557.00 km^2 in 2000 to 2478.39 km^2 in 2020. The eastern and southeastern regions were in dysregulated recession, including Longju Town, Puzi Town, Zouma Town, etc. Its area showed a trend of first increasing and then decreasing. The good coordination area was mainly distributed in the central towns of Wanzhou District, such as Longdu Street, Bell and Drum Tower Street, Pailou Street, etc. The spatial distribution was more concentrated, and its area showed an increasing trend. This indicates that the CCD of PLF in Wanzhou District improved. This was mainly due to the construction and development of central towns in Wanzhou District and the continuous improvement of infrastructure, which provide space-bearing and material guarantees of living functions. The CCD between the production and life functions improved and was at the coordination level (Figure 7).

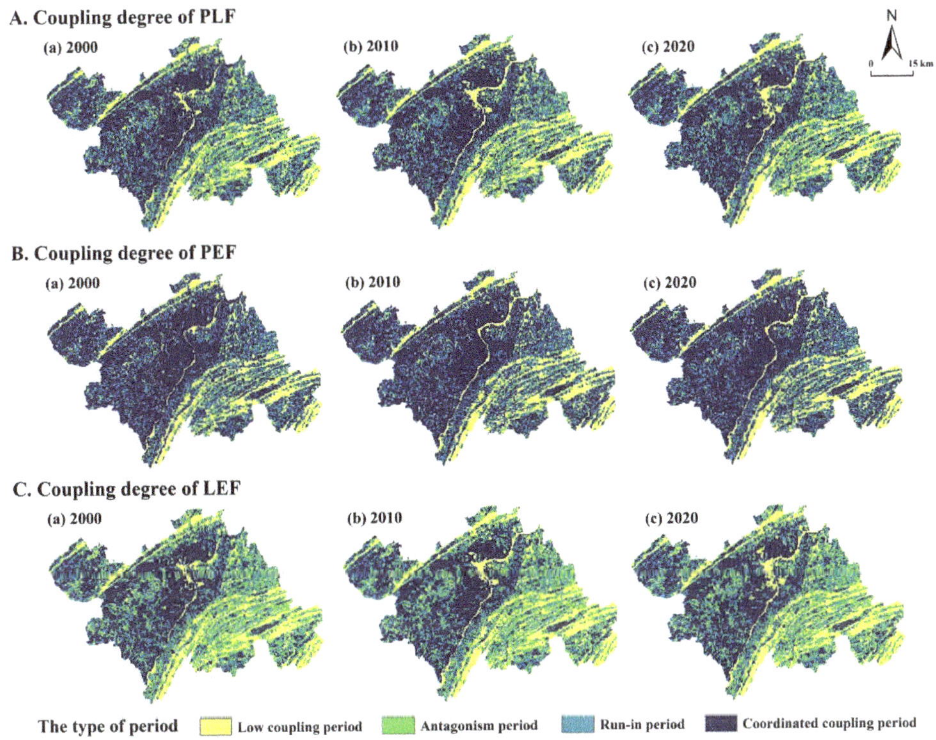

Figure 5. Coupling degree of PLF (**A**), PEF (**B**), and LEF (**C**).

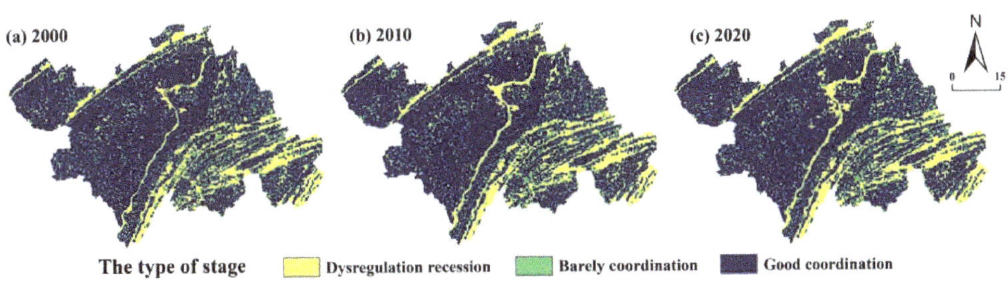

Figure 6. Coupling coordination degree of the PLEFs in 2000 (**a**), 2010 (**b**), and 2020 (**c**).

Regarding the CCD of PEF, Wanzhou District was in a moderate coordination state. The spatial distribution features included high values on both sides and low values in the middle, and the regional difference was pronounced. Most of the eastern and western areas of Wanzhou District were in good coordination, such as Zishui Town, Danzi Town, Puzi Town, etc. The coordination area decreased from 3358.24 km^2 in 2000 to 3300.50 km^2 in 2020, indicating that the CCD of production and life decreased. The barely coordinated area showed a point distribution, and the area showed an upward trend. The moderate coordination area increased by 12.48 km^2. The central region was in dysregulation recession, including Chenjiaba Street, Pailou Street, etc. Its area showed an increasing trend, and the area of the discordant area increased by 45.26 km^2 over the past 20 years. Owing to the excessive production development and human activities, ecological damage in the central region was caused. The CCD of PEF was at an uncoordinated level, and the scope gradually expanded (Figure 7).

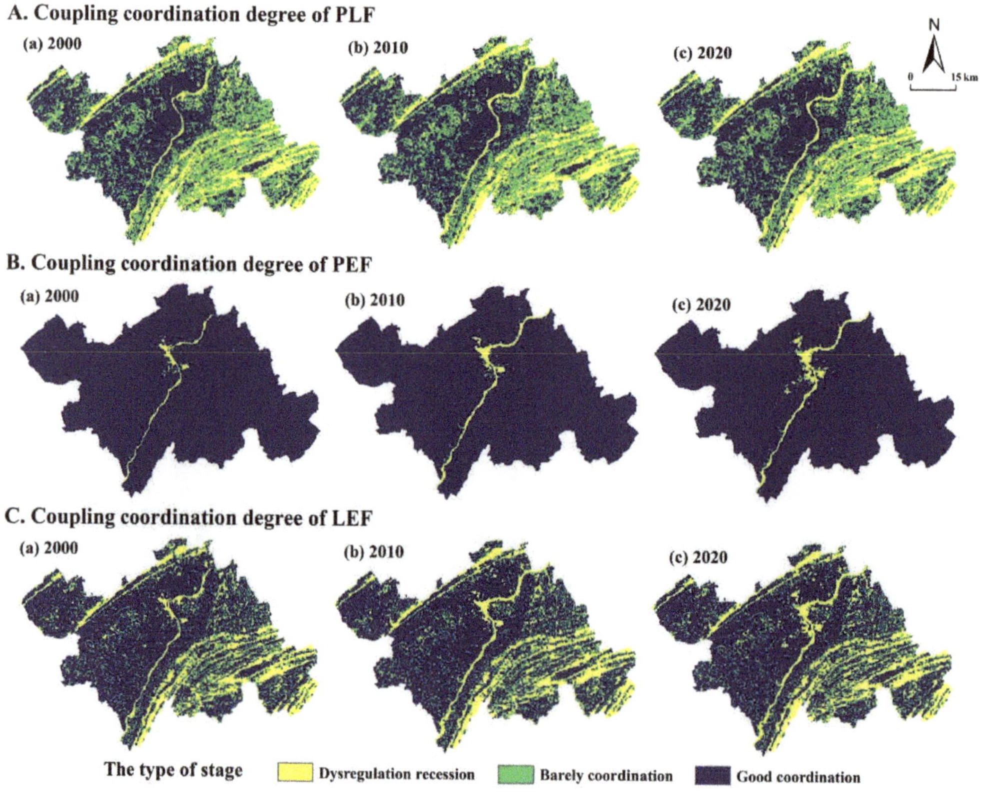

Figure 7. Coupling coordination degree of PLE (**A**), PEF (**B**), and LEF (**C**).

Regarding the CCD of LEF, Wanzhou District was in a state of coupling coordination. The spatial distribution exhibited a decrease from northwest to southeast. Most of the western and northwestern areas were in good coordination, including Fanshui Town, Yujia Town, Houshan Town, etc. The area of coordination decreased from 2325.32 km^2 in 2000 to 2264.59 km^2 in 2020. The CCD of LEF in the southeast region was in dysregulation recession, including Lishu township, the eastern part of Puzi township, and the eastern part of Yanshan township. During the past 20 years, the discordant area increased by 60.22 km^2. The barely coordinated area was scattered between the uncoordinated area and the communal area, and its area showed a trend of decreasing first and then increasing. The distribution of CCD between the production and ecological functions of central towns in Wanzhou District changed obviously. The uncoordinated area showed a trend of expansion year by year. This was mainly because domestic land continuously encroached on ecological land, resulting in the reduction of the area of ecological land and the reduction of the CCD between the living and ecological functions (Figure 7).

3.3. Spatial–Temporal Evolution Characteristics of HAI

From 2000 to 2020, the human activity intensity in Wanzhou District showed an increasing trend. It was 17.47% in 2000, 17.97% in 2010, and 18.81% in 2020. The *HAI* in

Wanzhou District during different periods was divided into five levels by the GIS natural break point method (Table 5).

Table 5. Human activity intensity at different times.

Level of *HAI*	Standards (%)	Area Ratio		
		2000	2010	2020
Lowest	<8	0.61%	0.59%	0.58%
Lower	8–15	38.49%	38.29%	38.03%
Medium	15–19	27.62%	27.61%	27.46%
Higher	19–38	30.12%	29.29%	28.46%
Highest	>38	3.16%	4.22%	5.47%

Only the highest level has expanded, while the other four scales reduced. The spatial distribution of the *HAI* in Wanzhou District differed. The distribution pattern was "high in the west and low in the east". The areas with the highest level were primarily distributed in the central towns of Wanzhou District, and the scope expanded yearly. The areas with lower and lowest levels were primarily distributed in the eastern part of Wanzhou District. This area had high elevation, and its land-use type was ecological land dominated by forest and grass, playing an ecological protection role (Figure 8).

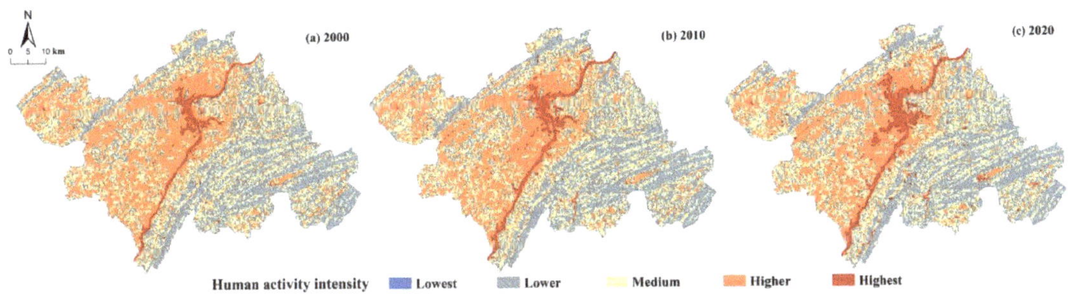

Figure 8. Degree of human activity intensity in 2000 (**a**), 2010 (**b**), and 2020 (**c**).

3.4. The Development Type of the PLEF Coupling Coordination Degree and HAI

The development types were preliminarily determined according to the synchronous development index of the *HAI* and the CCD of the PLEFs (Table 6). According to the coupling coordination index (D), Wanzhou District could be divided into three types: dysregulation recession, barely coordinated, and good coordination. According to the synchronous development index (T), Wanzhou District could be divided into three types: lagging development, synchronous development, and advanced development. The results of the two indices can be combined into nine classes, including dysregulation recession–advanced development, dysregulation recession–synchronous development, dysregulation recession–lagging development, barely coordinated–advanced development, barely coordinated–synchronous development, barely coordinated–lagging development, good coordination–advanced development, good coordination–synchronous development, and good coordination–lagging development. It is considered that regulating the synchronous development index is key to promoting the coordinated development of regional human–land relations. Therefore, we used the synchronous development index to classify the study area into three types: negative, positive, and synchronous.

Table 6. Preliminary results of the development types in Wanzhou District in 2000, 2010, and 2020.

Development Types	Specific Types	Area Ratio (%)			
		2000	2010	2020	Variation
Negative type	I: Dysregulation recession—Advanced development	13.84	15.65	17.44	3.6
	II: Barely coordinated—Advanced development	0.63	0.62	0.81	0.18
	III: Good coordination—Advanced development	0.75	0.87	1.23	0.48
	Subtotal	15.22	17.14	19.48	4.26
Synchronous type	IV: Dysregulation recession—Synchronous development	2.26	1.98	0.45	−1.81
	V: Barely coordinated—Synchronous development	1.53	1.78	1.68	0.15
	VI: Good coordination—Synchronous development	0.69	0.84	1.1	0.41
	Subtotal	4.48	4.6	3.23	−1.25
Positive type	VII: Dysregulation recession—Lagging development	0	0	0.01	0.01
	VIII: Barely coordinated—Lagging development	15.13	14.56	14.55	−0.58
	IX: Good coordination—Lagging development	65.17	63.70	62.73	−2.44
	Subtotal	80.3	78.26	77.29	−3.01

According to the preliminary results, the development type of most areas in Wanzhou District was positive. That is, the development of regional economic construction was not based on the sacrifice of the coordinated development of the PLEFs. However, the proportion of positive and synchronous types decreased over 20 years. Among them, the proportions of the "good coordination–lagging development" type and the "dysregulation recession–synchronous development" type fell the most obviously, which decreased by 2.44% and 1.81%, respectively. The ratio of the negative area increased by 4.26%, among which the proportion of the "dysregulation recession–advanced development" type increased the most, with a total increase of 3.6% over the past 20 years. The development of these regions was the sacrifice of the coordinated development of PLEFs. The relationship between humans and land is tense and is in urgent need of regulation. Therefore, according to the synchronous development state of the *HAI* and the CCD of PLEFs in Wanzhou District, it is essential to classify the study area and put forward different regulation strategies (Figure 9).

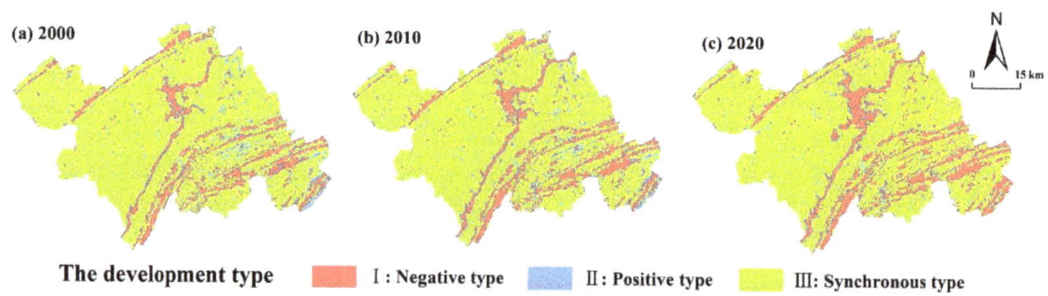

Figure 9. The development type of the PLEF coupling coordination degree and *HAI* in 2000 (**a**), 2010 (**b**), and 2020 (**c**).

4. Discussion

4.1. Policy Recommendations

The first recommendation is for the development of hostile regions. The scale of construction land should be reasonably determined and the *HAI* should be controlled. To avoid irreversible damage to the coupling coordinated development of the PLEFs in the process of *HAI* growth, it is also necessary to increase investment in environmental protection and environmental governance. The remediation and the closure of seriously polluting factories, and promoting the ecological transformation of the industry are important. The implementation of policies and striving to improve the maladjustment of human–land relationships should be strengthened.

The second is for the development of synchronous regions. It is necessary to maintain the synchronous and coordinated development state of the current *HAI* and the coupling coordination level of PLEFs. It is also important to upgrade and relocate industries that are energy-intensive and highly polluting, treating wastewater, waste gas, and waste residue. The government should also actively promote knowledge of agricultural pollution, and improve the awareness of farmers to use land and raise land in order to reduce agrarian non–point source pollution and soil pollution considering the synchronous coordination of economic development, ecological protection, and comfortable living. Thus, this will promote the deep integration of the PLEFs.

The third is for the development of positive regions. Under the guidance of the concept of "ecological priority and green development", The relationship between protecting the ecological environment, guaranteeing economic development, and preserving bare farmland should be resolved. Through land space planning, the layout of ecological lands, such as green heartlands and the green belt, should be increased. An appropriate amount of development of unused land, easing the central area of land tension, and undertaking industrial transfer should be conducted. Combined with local characteristics, ecological leisure tourism should be properly developed. The advanced production technology should be introduced and there should be scientific increases in various production input factors. A three-dimensional, circular, ecological, high–quality, and sustainable agricultural development model should be formed.

4.2. Contributions and Limitations

Promoting the coupling and coordinated development of the regional PLEF has become one of the critical issues for regional sustainable development. To clarify the relationship between the *HAI* and the CCD of the PLEFs, this paper has the following advantages. Firstly, compared with previous studies [47], we used a 300 m × 300 m grid as the evaluation unit to make the evaluation result more accurate. This provided more precise location information for land-use decisions and land management. Secondly, we referred to the existing research [37] and calculated the *HAI* according to the sum of the construction land equivalents of different regional land use types on the grid scale. The calculated results could not only reflect the comprehensive effect degree of human social and economic activities on land resources, but also enhance the comparability of the research results of the intensity of human activities in different regions. Thirdly, exploring the relationship between the *HAI* and the coupling coordination of PLEF is a point that was easily neglected in previous studies. In this study, we try to use the synchronous development model [45,46] to preliminarily discuss and classify the development relationship. Fourthly, based on analyzing the level of *HAI* and the CCD of PLEF, this paper scientifically discriminated the factors restricting synchronous development and divided the regions into different types. The differentiation promotion strategy was put forward, providing a theoretical and practical reference for regional sustainable development.

However, because of the complexity of the changing interaction between the *HAI* and the CCD of the PLEFs, and due to the limitations of data and data acquisition, this paper only tried to explore the interaction relationship between them from the perspective of synchronous development and the spatial grid scale. The scientific prediction of the evolution of the relationship and taking scientific measures in advance for accurate regulation are of significance to alleviate the contradiction between humans and land and achieve regional sustainable development. Therefore, in the future, it will be necessary to carry out relevant research on the evolution direction of the synchronous development relationship between the *HAI* and the CCD of PLEF from other perspectives and different spatial scales in order to obtain its internal operating mechanism and distribution law systematically.

5. Conclusions

The methods of the CCD model, human activity intensity model, and synchronous development model were used in this paper. This paper calculated and analyzed the evolution characteristics of the CCD of the PLEFs and the intensity of human activities in Wanzhou District from 2000 to 2020, as well as the types of their development relations. The following conclusions can be obtained:

(1) In Wanzhou District, the PLEFs showed significant spatial distribution differences and evident spatial complementarity. The spatial distribution of the production function was higher in the west than in the east. The living function showed the spatial distribution characteristics of being higher in the middle. The ecological function showed the spatial distribution characteristics of being higher in the east and lower in the west.

(2) From 2000 to 2020, the coupling degree of the PLEFs in the Wanzhou District decreased and the interaction became weak. The CCD of the PLEFs was at a good coordination level, but there was also a downward trend. The coupling coordination of the living–production function was poor, which is the critical direction for future optimization.

(3) The *HAI* in Wanzhou District showed an increasing trend and formed a high concentration in the central town and its surrounding areas. The development type of most regions in Wanzhou District was positive.

(4) In Wanzhou District, the "good coordination–lagging development" type was dominant, but the area ratio decreased, while the proportion of "dysregulation recession–advanced development" increased. We proposed different regulation strategies for further development types to improve the regional CCD of the PLEFs and guide the harmonious development of regional human–land relationships.

Author Contributions: The co-authors together contributed to the completion of this article. Specifically, their individual contributions are as follows: conceptualization, T.L., F.Y. and Y.L.; software, M.F., X.H. and F.Y.; methodology, T.L., X.H. and M.F.; validation, T.L., Y.L. and M.F.; formal analysis, T.L. and X.H.; resources, T.L.; data curation, X.H. and M.F.; writing—original draft preparation, T.L., X.H. and M.F.; writing—review and editing, T.L., F.Y. and Y.L.; visualization, T.L. and F.Y.; supervision, T.L., Z.Z., C.W. and X.R. All authors have read and agreed to the published version of the manuscript.

Funding: This research was funded by the Key Program of the National Social Science Foundation of China (Grant No. 20&ZD095), the National Social Science Foundation of China (Grant No. 21BJY223), the 2020 General Project of Water Transport Economics Research Center, Key Research Base of Humanities and Social Sciences of Colleges and Universities of Sichuan Province (Grant No. SYJJ2020C02), the National Social Science Youth Foundation of China (Grant No. 21CJY044), the National Social Science Youth Foundation of China (Grant No. 18CJL031), the Science and Technology Research project of Chongqing Education Commission (Grant No. KJQN202102103), the Humanities and Social Sciences Research Project of Chongqing Education Commission (Grant No. 21SKGH308), and the Chongqing Social Science Planning Social Organization Project (Grant No. 2021SZ26).

Institutional Review Board Statement: Not applicable.

Informed Consent Statement: Not applicable.

Data Availability Statement: The data presented in this study are available on request from the author.

Conflicts of Interest: The authors declare no conflict of interest.

References

1. Ahmad, M.; Rehman, A.; Shah, S.A.A.; Solangi, Y.A.; Chandiod, A.A.; Jabeen, G. Stylized heterogeneous dynamic links among healthcare expenditures, land urbanization, and CO_2 emissions across economic development levels. *Sci. Total Environ.* **2021**, *753*, 142228. [CrossRef]
2. Wang, D.; Jiang, D.; Fu, J.; Lin, G.; Zhang, J. Comprehensive assessment of production–living–ecological space based on the coupling coordination degree model. *Sustainability* **2020**, *12*, 2009. [CrossRef]
3. Lin, G.; Fu, J.; Jiang, D. Production–living–ecological conflict identification using a multiscale integration model based on spatial suitability analysis and sustainable development evaluation: A case study of Ningbo, China. *Land* **2021**, *10*, 383. [CrossRef]

4. Helming, K.; Perez-Soba, M.; Tabbush, P. (Eds.) *Sustainability Impact Assessment of Land Use Changes*; Springer: Berlin/Heidelberg, Germany, 2008.
5. Liu, Y.S. Introduction to land use and rural sustainability in China. *Land Use Policy* **2018**, *74*, 1–4. [CrossRef]
6. Yang, Y.Y.; Bao, W.K.; Liu, Y.S. Coupling coordination analysis of rural production-living-ecological space in the Beijing-Tianjin-Hebei region. *Ecol. Ind.* **2020**, *117*, 106512. [CrossRef]
7. Zhou, D.; Xu, J.C.; Lin, Z.L. Conflict or coordination? Assessing land use multi-functionalization using production-living- ecology analysis. *Sci. Total Environ.* **2017**, *577*, 136–147. [CrossRef] [PubMed]
8. Zou, L.L.; Liu, Y.S.; Yang, J.X.; Yang, S.F.; Wang, Y.S.; Cao, Z.; Hu, X.D. Quantitative identification and spatial analysis of land use ecological-production-living functions in rural areas on China's southeast coast. *Habitat Int.* **2020**, *100*, 102182. [CrossRef]
9. Fang, Y.G.; Liu, J.S. Diversified agriculture and rural development in China based on multi–function theory: Beyond modernization paradigm. *Acta Geogr. Sin.* **2015**, *70*, 257–270.
10. Long, H.L.; Tu, S.S.; Ge, D.Z.; Li, T.T.; Liu, Y.S. The allocation and management of critical resources in rural China under restructuring: Problems and prospects. *J. Rural Stud.* **2016**, *47*, 392–412. [CrossRef]
11. Zhang, Y.N.; Long, H.L.; Tu, S.S.; Ge, D.Z.; Ma, L.; Wang, L.Z. Spatial identification of land use functions and their tradeoffs/synergies in China: Implications for sustainable land management. *Ecol. Ind.* **2019**, *107*, 1100–1114. [CrossRef]
12. Geoghegan, J. The value of open spaces in residential land use. *Land Use Policy* **2002**, *19*, 91–98. [CrossRef]
13. Shan, Y.J.; Wei, S.K.; Yuan, W.L.; Miao, Y. Spatial-temporal differentiation and influencing factors of coupling coordination of "production-living-ecological" functions in Yangtze River Delta urban agglomeration. *Acta Ecol. Sin.* **2022**, *42*, 1–12.
14. Azadi, H.; Vanhaute, E.; Janekova, K.; Sklenick, P.; Teklemariamet, D.; Feng, L.; Witlox, F. Evolution of land distribution in the context of development theories. *Land Use Pol.* **2020**, *97*, 104730. [CrossRef]
15. Sumari, N.S.; Cobbinah, P.B.; Ujoh, F.; Xu, G. On the absurdity of rapid urbanization: Spatio-temporal analysis of land-use changes in Morogoro, Tanzania. *Cities* **2020**, *107*, 102876. [CrossRef]
16. Liang, T.; Du, P.; Yang, F.; Su, Y.; Luo, Y.; Wu, Y.; Wen, C. Potential land-use conflicts in the urban center of Chongqing based on the "production–living–ecological space" perspective. *Land* **2022**, *11*, 1415. [CrossRef]
17. Zhang, L.L.; Zheng, X.Q.; Meng, C.; Zhang, P.T. Space and temporal division of land multi-function coupling and coordination in Hunan Province. *China Land Sci.* **2019**, *33*, 85–94.
18. Bernues, A.; Rodríguez-Ortega, T.; Alfnes, F.; Clemetsen, M.; Eik, L.O. Quantifying the multifunctionality of fjord and mountain agriculture by means of sociocultural and economic valuation of ecosystem services. *Land Use Policy* **2015**, *48*, 170–178. [CrossRef]
19. Chen, L.D.; Sun, R.H.; Liu, H.L. Progress on the evolution of urban landscape pattern. *J. Ecol.* **2013**, *33*, 1042–1050.
20. Yang, Q.Y.; Luo, K.; Lao, X. Evolution and enlightenment of foreign spatial planning: Exploration from the perspective of geography. *Acta Geogr. Sin.* **2020**, *75*, 1223.
21. Liu, Y.S.; Fang, F.; Li, Y.H. Key issues of land use in China and implications for policy making. *Land Use Policy* **2014**, *40*, 6–12. [CrossRef]
22. Lu, D.; Wang, Y.; Yang, Q.; Su, K.; Zhang, H.; Li, Y. Modeling spatiotemporal population changes by integrating DMSP-OLS and NPP-VIIRS nighttime light data in Chongqing, China. *Remote Sens.* **2021**, *13*, 284. [CrossRef]
23. Chen, B.; Wu, C.; Huang, X.; Yang, X. Examining the relationship between urban land expansion and economic linkage using coupling analysis: A case study of the Yangtze River Economic Belt, China. *Sustainability* **2020**, *12*, 1227. [CrossRef]
24. Chen, H.; Yang, Q.; Su, K.; Zhang, H.; Lu, D.; Xiang, H.; Zhou, L. Identification and optimization of production-living- ecological space in an ecological foundation area in the upper reaches of the Yangtze River: A case study of Jiangjin district of Chongqing, China. *Land* **2021**, *10*, 863. [CrossRef]
25. Zou, L.L.; Liu, Y.S.; Wang, J.Y.; Yang, Y.Y. An analysis of land use conflict potentials based on ecological-production-living function in the southeast coastal area of China. *Ecol. Indic.* **2021**, *122*, 107297. [CrossRef]
26. Dullinger, I.; Essl, F.; Moser, D.; Erb, K.; Haberl, H.; Dullinger, S. Biodiversity models need to represent land-use intensity more comprehensively. *Glob. Ecol. Biogeogr.* **2021**, *30*, 924–932. [CrossRef]
27. Gkargkavouzi, A.; Halkos, G.; Matsiori, S. A multi-dimensional measure of environmental behavior: Exploring the predictive power of connectedness to nature, ecological worldview and environmental concern. *Soc. Indicat. Res.* **2019**, *143*, 859–879. [CrossRef]
28. Duan, Y.M.; Wang, H.; Huang, A.; Xu, Y.Q.; Lu, L.H.; Ji, Z.X. Identification and spatial-temporal evolution of rural "production-living-ecological" space from the perspective of villagers' behavior–A case study of Ertai Town, Zhangjiakou City. *Land Use Policy* **2021**, *106*, 105457. [CrossRef]
29. Huang, J.C.; Lin, H.X.; Qi, X.X. A literature review on optimization of spatial development pattern based on ecological-production-living space. *Prog. Geogr.* **2017**, *36*, 378–391.
30. Liang, T.; Yang, F.; Huang, D.; Luo, Y.; Wu, Y.; Wen, C. Land-Use Transformation and Landscape Ecological Risk Assessment in the Three Gorges Reservoir Region Based on the "Production–Living–Ecological Space" Perspective. *Land* **2022**, *11*, 1234. [CrossRef]
31. Gao, X.; Liu, Z.W.; Li, C.X.; Cha, L.S.; Song, Z.Y.; Zhang, X.R. Land use function transformation in the Xiong'an New Area based on ecological-production-living spaces and associated eco-environment effects. *Acta Ecol. Sin.* **2020**, *40*, 7113–7122.
32. Wang, C.; Tang, N. Spatio-temporal characteristics and evolution of rural production-living-ecological space function coupling coordination in Chongqing Municipality. *Geogr. Res.* **2018**, *37*, 1100–1114.

33. Zhang, Y.; Yang, Q.Y.; Min, J. An analysis of coupling between the bearing capacity of the ecological environment and the quality of new urbanization in Chongqing. *Acta Geogr. Sin.* **2016**, *71*, 817–828.
34. Cheng, X.; Long, R.; Chen, H.; Li, Q. Coupling coordination degree and spatial dynamic evolution of a regional green competitiveness system-A case study from China. *Ecol. Ind.* **2019**, *104*, 489–500. [CrossRef]
35. Li, W.; Yi, P. Assessment of city sustainability-Coupling coordinated development among economy, society and environment. *J. Clean. Prod.* **2020**, *256*, 120453. [CrossRef]
36. Fan, J.; Li, P.X. The scientific foundation of major function oriented zoning in China. *J. Geogr. Sci.* **2009**, *19*, 515–531. [CrossRef]
37. Xu, Y.; Sun, X.Y.; Tang, Q. Human activity intensity of land surface: Concept, method, and application in China. *Acta Geogr. Sin.* **2015**, *70*, 1068–1079. [CrossRef]
38. Schönbrodt-Stitt, S.; Behrens, T.; Schmidt, K.; Shi, X. Thomas Scholten. Degradation of cultivated bench terraces in the Three Gorges Area: Field mapping and data mining. *Ecol. Indic.* **2013**, *34*, 478–493. [CrossRef]
39. Wei, L.; Zhang, Y.; Wang, L.; Mi, X.; Wu, X.; Cheng, Z. Spatiotemporal Evolution Patterns of "Production- Living-Ecological" Spaces and the Coordination Level and Optimization of the Functions in Jilin Province. *Sustainability* **2021**, *13*, 13192. [CrossRef]
40. Wang, M.; Qin, K.; Jia, Y.; Yuan, X.; Yang, S. Land Use Transition and Eco-Environmental Effects in Karst Mountain Area Based on Production-Living-Ecological Space: A Case Study of Longlin Multinational Autonomous County, Southwest China. *Int. J. Environ. Res. Public Health* **2022**, *19*, 7587. [CrossRef]
41. Zhao, Y.; Hou, P.; Jiang, J.; Zhai, J.; Chen, Y.; Wang, Y.; Bai, J.; Zhang, B.; Xu, H. Coordination Study on Ecological and Economic Coupling of the Yellow River Basin. *Int. J. Environ. Res. Public Health* **2021**, *18*, 10664. [CrossRef]
42. Liao, G.; He, P.; Gao, X.; Deng, L.; Zhang, H.; Feng, N.; Zhou, W.; Deng, O. The Production–Living–Ecological Land Classification System and Its Characteristics in the Hilly Area of Sichuan Province, Southwest China Based on Identification of the Main Functions. *Sustainability* **2019**, *11*, 1600. [CrossRef]
43. Li, Q.Y.; Fang, C.L.; Wang, S.J. Quality evaluation of provincial territorial space utilization in China: Based on the spatial perspective of "production-living-ecological". *Reg. Res. Dev.* **2016**, *35*, 163–169.
44. Han, M.; Zhang, C.; Lu, G.; Liu, Y.B.; Yu, H.Z. Response of wetland landscape pattern gradient to human activity intensity in Yellow River Delta. *Trans. CSAE* **2017**, *33*, 265–274.
45. Sun, Y.S.; Tong, L.J. Spatio-temporal coupling relationship between development strength and eco-environment in the Restricted Development Zone of Northeast China. *Sci. Geogr. Sin.* **2021**, *41*, 684–694.
46. Wang, C.; Ji, M.Z.; Dai, R.L.; Chen, J.; Shen, Y. Coupling and coordination pattern and type division of village and town construction land expansion and ecological environment effect: A case study of Rongchang District, Chongqing Municipality. *Prog. Geogr.* **2022**, *41*, 409–422.
47. Huang, M.Q.; Li, Y.B.; Li, Y.B.; Li, M.Z.; Chen, S.; Zeng, C.C.; Zhang, B.; Xia, C.H. Coupling respinse of human activity intensity and ladscape pattern in the Three Gorges Reservoir Area. *Acta Ecol. Sin.* **2022**, *42*, 3959–3973.

land

MDPI

Article

Does Environmental Decentralization Affect the Supply of Urban Construction Land? Evidence from China

Lida Han [1], Xi Wu [1] and Peng Tang [2,*]

[1] School of Economics, Sichuan University, Chengdu 610065, China
[2] Department of Land Resource and Real Estate Management, School of Public Administration, Sichuan University, Chengdu 610065, China
* Correspondence: tp@scu.edu.cn

Abstract: Against the background of Chinese decentralization, the preferences and choices of local governments significantly affect the scale and structure of urban construction land supply. Due to the shortage of financial funds and the political performance pursuit of local governments, environmental decentralization gives local governments greater autonomy in environmental management, and increases the possibility for local governments relying on land transfer income to make up for the financial gap and provide public goods and services. This paper analyses the impact of environmental decentralization on the construction land supply scale of local government based on the panel data of 30 provinces in China from 2003 to 2015. The results indicate that: (1) environmental decentralization has a positive effect on the increase in urban construction land supply scale; (2) environmental decentralization affects urban construction land supply by strengthening land financial dependence and distorting land resources misallocation; (3) there are regional disparities in the effect of environmental decentralization on urban construction land supply. The impact is greater in regions with high financial pressure, high economic growth pressure, and low environmental protection pressure. In summary, some policy suggestions are put forward to reasonably supply urban construction land against the background of Chinese decentralization.

Keywords: environmental decentralization; urban construction land supply; fiscal decentralization; political centralization

Citation: Han, L.; Wu, X.; Tang, P. Does Environmental Decentralization Affect the Supply of Urban Construction Land? Evidence from China. *Land* **2023**, *12*, 601. https://doi.org/10.3390/land12030601

Academic Editors: Li Ma, Yingnan Zhang, Muye Gan and Zhengying Shan

Received: 10 February 2023
Revised: 26 February 2023
Accepted: 1 March 2023
Published: 3 March 2023

1. Introduction

Land resources are extremely important for ensuring food security, facilitating economic development, and protecting the ecological environment [1,2]. Urbanization has accelerated the conversion of agricultural land around cities to non-agricultural construction uses, which is a common problem faced by the whole world [3,4]. This phenomenon is particularly acute in developing countries, such as India and China [5]. Over the years, the important role of land in China's rapid economic development cannot be ignored. A great deal of construction land has been provided by the Chinese government to support urbanization and industrialization [6]. According to data from the "China urban construction statistical yearbook", the area of urban construction land in China increased from 6720 km^2 to 52,761.3 km^2, with an average annual growth of 6.06%, from 1981 to 2016. In contrast, this rate of urban population growth over the same period was only 2.98%. However, economic growth that heavily relies on land not only causes many social problems such as land finance, illegal land use, inefficiency of industrial land, and soaring housing prices [7], but also leads to environmental problems, such as declining biodiversity and environmental pollution [8]. Governments are working to curb urban sprawl through environmental regulation [9–13].

In response to environmental problems, including those caused by urban sprawl and land resources misallocation, the Chinese government has adopted a decentralized management model (environmental federalism) [14]. In other words, central and local governments share environmental management powers. Local governments are responsible for specific matters of environmental management, including pollution prevention, ecological protection, and environmental access management. The central government is responsible for supervising and coordinating local governments. According to Environmental Protection Law of the People's Republic of China, all factors affecting human survival and development are the objects of environmental management, including natural elements (atmosphere, water, land, etc.), as well as artificial transformation elements (nature reserves, urban and rural areas, etc.) [15]. In terms of land use management, the improvement of central environmental protection supervision has prompted local governments to rationally plan the scale and structure of construction land supply, and the supply of highly polluting industrial land has been restricted. Statistics show that the supply area of construction land gradually declined after reaching a peak of 374,804.03 hm^2 in 2013; the proportion of the land transferred by agreement plummeted from 50.08% in 2007 to 16.06% in 2008 (the agreed transfer can reflect the supply of industrial land) [16]. This suggests that changes in local government autonomy alter the allocation of resources. This phenomenon also exists in other locations worldwide, such as Montpellier (France), Rome (Italy), and Jakarta (Indonesia), where the expansion of construction use is also affected by decentralization [17,18].

Construction land expansion is a complex issue. In addition to environmental regulations, there are many factors that affect the expansion of urban construction land, including land use planning, land systems, public policy, etc. [19,20]. The factors influencing the supply of construction land under China's unique institutional background are particularly complex. First, the Chinese government is the sole subject of land acquisition and transfer. The government can obtain most of the land appreciation profits by acquiring land at a low price and selling it at a high price [21]. Therefore, they intervene in the scale and structure of urban construction land supply according to their own interests [22–24]. Second, fiscal decentralization and political centralization increase incentives for local government officials to sell land for fiscal revenue and political performance [25–29], resulting in an excessive supply of urban construction land [30]. In addition, some factors, such as economic development level, population size, and industrial structure, can also affect the government's land supply [31–33].

With the advancement of urbanization around the world, curbing the disorderly expansion of cities and determining the reasonable scale of urban construction land supply are common concerns for all countries [34,35]. The government plays an important role in controlling the scale and structure of construction land supply. The scale of construction land supply is the result of the government's resource allocation after comprehensive consideration of various factors, such as environmental protection and economic development. Decentralization affects the allocation of resources by changing the government's autonomy in the choice of economic, political, environmental, and other goals. How to design the administrative management system to achieve better allocation of resources by the government has attracted much attention. Our research provides Chinese experience for this. Some of the literature has analyzed the logic of massive land supply by local governments under the background of fiscal decentralization and political centralization. However, the impact of the environmental protection assessment requirements of the central government and the evolution of the environmental management system on construction land supply by local governments has not been paid attention to. Therefore, this paper incorporates environmental decentralization into the framework of Chinese decentralization and analyzes the influence of environmental decentralization on local government land supply. The main issues addressed in this paper include: (1) Does environmental decentralization affect local government urban construction land supply? (2) How does environmental decentralization affect the supply of urban construction land by local government? (3) Is there regional difference in the influence of environmental decentralization on urban construction land supply? Compared with the existing research of scholars, this paper proposes innovations in the following aspects: (1) environmental decentralization is incorporated into the Chinese decentralization framework, which expands the Chinese decentralization system. (2) Based on the Chinese decentralization framework, the urban construction land supply behaviors of local governments are analyzed. The impact of environmental decentralization on urban construction land supply is emphatically analyzed, which provides a new concept for studying the influencing factors of local government land supply. (3) Combined with the incentive mechanism of urban construction land supply under Chinese decentralization, the impact of environmental decentralization on land supply in regions with different financial pressure, economic growth pressure, and environmental protection pressure is considered, which provides policy directions for coordinating economic development and environmental protection.

2. Institutional Background and Theoretical Hypothesis

2.1. Institutional Background

China has a unique institutional background of fiscal decentralization and political centralization, which is different from the fiscal decentralization and federal political system of Western countries [36]. In terms of finance, China implements a tax-sharing system, and the fiscal budgets of governments at all levels are relatively independent. This means that local governments have certain economic decision-making power and can govern according to their own preferences [37]. In terms of politics and bureaucracy, China is in power with one party and implements vertical management with upward responsibility, which forms a multi-departmental M-shaped hierarchy [38,39]. The superior (or central) government has the right to appraise the performance of local government officials and decide on their promotion [40].

Corresponding to the delegated powers, China implements a decentralized environmental management system [41]. That is, environmental decentralization. Environmental decentralization originates from environmental federalism, which studies the environmental management functions of all levels of government [42]. In the "pyramid" system, where the central government is located above and the local governments are located below, local governments are authorized to participate in environmental management affairs [43]. It is generally believed that environmental centralization can effectively control cross-regional pollution, and that environmental decentralization is more flexible and targeted [44–47]. Since the reform and opening up in 1978, the evolution of environmental decentralization in China can be divided into three stages [48]. The first stage was from 1978 to 1993. During this stage, financial and administrative powers were highly decentralized, and the degree of environmental decentralization was high. The central government lacked supervision over local environmental affairs. The second stage was from 1994 to 2007. During this stage, with the increase in central government's fiscal revenue, the environmental management powers of central government improved, and the degree of environmental decentralization decreased. The third stage is from 2008 to the present. With the adjustment of the government environmental protection institutions and the increased emphasis on environmental protection, the central government decentralized environmental administrative power and enhanced environmental supervision power and environmental governance incentives. Several instances in the literature have confirmed the beneficial effects of environmental decentralization in China, including promoting green development and pollution control [49,50]. However, some scholars believe that the institutional background of fiscal decentralization and political centralization in China makes the incentives and constraints of local government environmental protection mismatched. When decentralizing environmental rights, local governments will relax environmental control, which forms a race to the bottom and aggravates environmental pollution [51,52].

The fiscal decentralization, political centralization, and environmental decentralization implemented in China can be summarized as Chinese decentralization.

2.2. Theoretical Hypothesis

The institutional arrangement of Chinese decentralization affects the construction land supply by local government in three aspects: fiscal decentralization, political centralization, and environmental decentralization. In terms of fiscal decentralization, the 1994 tax-sharing reform led to a misalignment of the executive and fiscal powers of the government [53]. Fiscal revenues flow upwards to the central government, and the responsible affairs remain in the local government, resulting in a huge fiscal gap [54,55]. Decentralization has led local governments to provide land for construction to ease fiscal pressures and cover fiscal deficits. This is manifested in the expansion of fiscal revenue through "land rent" and "land tax" [56]. The "land rent" refers to the one-time income obtained by transferring land. Selling commercial and residential land at a high price can obtain more "land rent" [57,58]. The "land tax" refers to attracting investment by selling industrial land at low prices, and developing industries to obtain continuous tax revenue [59]. When it comes to political centralization, Chinese officials are usually accountable to the top. Local government officials are promoted through excellent performance appraisals [60]. Economic growth is the core of performance measurement, so there is fierce competition around GDP growth [61–63]. In order to win the GDP championship, local governments often sell industrial land at reserve prices or free of charge to attract investment and promote economic growth, which increases the supply of construction land [28,29]. The importance of environmental indicators in performance appraisals has increased in recent years, which limits the expansion of construction land [36]. Environmental decentralization affects the supply of construction land by relaxing or tightening environmental constraints after weighing the importance of economic growth and environmental protection. After incorporating environmental decentralization into the Chinese decentralization framework, the incentives and constraints of local government construction land supply are shown in Figure 1.

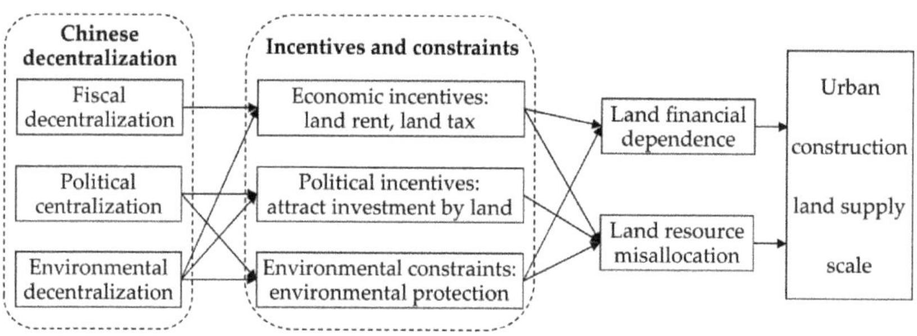

Figure 1. The influence of Chinese decentralization on construction land supply.

Under the specific institutional background of Chinese decentralization, environmental decentralization gives local governments greater autonomy in choosing economic development goals and environmental protection goals [64]. Local governments' economic incentives and political incentives do not match the environmental protection constraints, which causes local governments to prioritize economic growth over environmental welfare. Local governments ignore environmental issues and supply a great deal of construction land to generate revenue and boost the economy.

Hence, the first hypothesis is obtained: environmental decentralization will increase the urban construction land supply by local government.

When environmental management power is decentralized, the local government will compare the importance of economic incentives, political incentives, and environmental constraints. Against the background of the shortage of local government's fiscal revenue and the pursuit of political performance, environmental decentralization increases the motivation of local governments to pursue "land rent", "land tax", and "attracting investment from land", so that local governments reduce environmental protection requirements. This leads to the intensification of the local government's land financial dependence and land resources misallocation. Thus, the scale of construction land supply is increased.

Hence, the second hypothesis is obtained: environmental decentralization increases the urban construction land supply scale of local governments by increasing land financial dependence and land resources misallocation.

Environmental decentralization affects construction land supply by changing the economic incentives, political incentives, and environmental constraints of local governments. Therefore, in regions with different financial pressures, economic growth pressures, and environmental protection pressures, the incentives and constraints of land supply are also distinct. In regions with high financial pressure, local governments have stronger economic incentives. When environmental power is decentralized, there is a stronger incentive to provide a great deal of land at the expense of the environment [65]. In regions with high economic growth pressure, local governments have a stronger motivation to attract investment by land [66,67]. Environmental decentralization will increase the supply of industrial land, which expands the scale of construction land supply. In regions with high environmental protection pressures, local governments have stricter constraints on the preservation of the environment. When environmental power is devolved, local governments will optimize local environmental management and limit the scale of land supply [68].

Hence, the third hypothesis is obtained: there are regional disparities in the effects of environmental decentralization on urban construction land supply. In the regions with high financial pressure, high economic growth pressure, and low environmental protection pressure, the impact of environmental decentralization on construction land supply is more significant.

3. Research Design

3.1. Model Settings

In the context of Chinese decentralization, environmental, fiscal decentralization, and political centralization jointly affect the urban construction land supply by local governments. Thus, the following fixed-effect model is established [50].

$$LS_{it} = \beta_0 + \beta_1 ED_{it} + \beta_2 FD_{it} + \beta_3 FAI_{it} + \beta_4 IPC_{it} + \beta_i X_{it} + \theta_i + \mu_t + \varepsilon_{it} \quad (1)$$

where LS_{it} is the explained variable, representing construction land supply; ED_{it} is the explanatory variable, representing environmental decentralization; FD_{it} is fiscal decentralization; FAI_{it} and IPC_{it} are fixed asset investment and environmental protection investment, respectively, representing political centralization; X_{it} are other control variables, including per capita GDP, population, urbanization rate of land, the proportion of secondary and tertiary industries, financial pressure; i and t represent the provinces and periods under consideration, respectively; β_0 is the constant term; β_1–β_i are coefficients; θ_i and μ_t represent the province and time effects, respectively; ε_{it} is the random error term. In order to make the data smoother and reduce heteroscedasticity, the non-proportional data are dealt with as logarithms.

3.2. Variable Selection

(1) Explained variable. The construction land supply scale is used to describe the construction land supply behavior of local government. The data sources of urban construction land supply in the existing literature mainly include the "China Land and Resources Statistical Yearbook" [33], the land survey results sharing application service platform, and the network of "landchina" [69,70]. Among them, the time span of the data in the "China Land and Resources Statistical Yearbook" is relatively long. Therefore, this paper uses the construction land transfer area in the "China Land and Resources Statistical Yearbook" to express the scale of land supply.

(2) Explanatory variable. Environmental decentralization reflects the functional division of central and local governments in environmental management affairs. The higher the

degree of environmental decentralization, the greater the authority of local governments in environmental affairs. Referring to the existing literature, the relative personnel number of environmental protection agencies at all levels is used to calculate environmental decentralization [71,72]. The calculation formula for environmental decentralization is as follows.

$$ED_{it} = \left[\frac{LEPP_{it}/POP_{it}}{NEPP_t/POP_t}\right] \times \left[1 - \left(\frac{GDP_{it}}{GDP_t}\right)\right] \quad (2)$$

where ED_{it} is environmental decentralization; $LEPP_{it}$, POP_{it}, GDP_{it} are the personnel number of environmental protection agencies, population, and gross domestic product of province i in year t; $LEPP_t$, POP_t, GDP_t are the number of personnel in environmental protection agencies, population, and gross domestic product of whole country in year t; $(1 - GDP_{it}/GDP_t)$ is the economic scaling factor for reducing endogenous interference [73].

(3) Control variables. Under the framework of Chinese decentralization, the impact of fiscal decentralization and political centralization on land supply cannot be ignored. Therefore, three special control variables were selected, including fiscal decentralization, fixed asset investment per unit of GDP, and industrial pollution control completed investment. Fiscal decentralization typically includes revenue decentralization and expenditure decentralization [74]. Fiscal revenue incentives can increase local government construction land supply. Thus, this paper uses fiscal revenue decentralization. Fixed assets investment is used to represent the promotion incentives of political centralization [75], and industrial pollution control completed investment is used to represent the environmental constraints of political centralization [76]. For the control of other social and economic factors, referring to the research of Wang (2015), Zhou (2019), and Li (2021) [32,33,50], six control variables are selected from the five aspects: economic development, population size, urban expansion, industrial structure, and financial pressure. Specific indicators include the per capita GDP and its square, population, land urbanization rate, the proportion of secondary and tertiary industries, and financial pressure.

The contents of the variables are shown in Table 1.

Table 1. Variable selection.

Variable Type	Variable Name	Variable Connotation
Explained variable	Supply scale of construction land (LS)	Total area of construction land supply
Explanatory variable	Environmental decentralization (ED)	Local government environmental protection employees/national environmental protection employees
Special control variables	Fiscal decentralization (FD)	Local government per capita fiscal revenue/central government per capita fiscal revenue
	Promotion incentive (FAI)	Fixed asset investment/GDP
	Environmental constraints (IPC)	Industrial pollution control completed investment
Other control variables	Economic development level (PGDP)	Per capita GDP
	Economic development level square ($PGDP^2$)	Square of per capita GDP
	Population (POP)	Total population
	Land urbanization rate (URL)	Built up area/land area
	Proportion of secondary and tertiary industries (STI)	(Secondary industry added value + tertiary industry added value)/GDP
	Financial pressure (FP)	(Local government fiscal expenditure—Local government fiscal revenue)/GDP

3.3. Data Sources

Since the personnel number of environmental protection agencies at provincial level has only been updated to 2015, the panel data of 30 provinces (Hong Kong, Macao, Taiwan, and Tibet are not included) from 2003 to 2015 are used for analysis. The data on urban construction land supply area and land transfer income are from the "China Land and Resources Statistical Yearbook (2004–2016)". The data on the number of employees in environmental protection agencies are from the "China Environmental Yearbook (2004–2016)". Other data are from the provincial annual database on the website of the "National Bureau of Statistics of China". The descriptive statistics of the variables are shown in Table 2.

Table 2. Descriptive statistics of variables.

Variable Name	Unit	Mean	Std. Dev	Min	Max
lnLS	hectare	8.662	0.953	4.797	12.445
ED	-	0.971	0.353	0.438	2.290
FD	-	1.132	1.008	0.343	5.926
FAI	-	0.662	0.255	0.131	1.564
lnIPC	10^8 yuan	2.257	1.038	−1.693	4.658
lnPGDP	10^4 yuan/person	0.673	0.611	−0.996	2.158
$lnPGDP^2$	10^4 yuan/person	0.827	0.973	0.000	4.657
lnPOP	10^4 persons	8.161	0.754	6.280	9.365
URL	-	0.023	0.129	0.000	2.500
STI	-	0.881	0.062	0.658	0.995
FP	-	0.115	0.095	0.008	0.636

The spatiotemporal evolution of urban construction land supply and environmental decentralization is plotted in Figure 2.

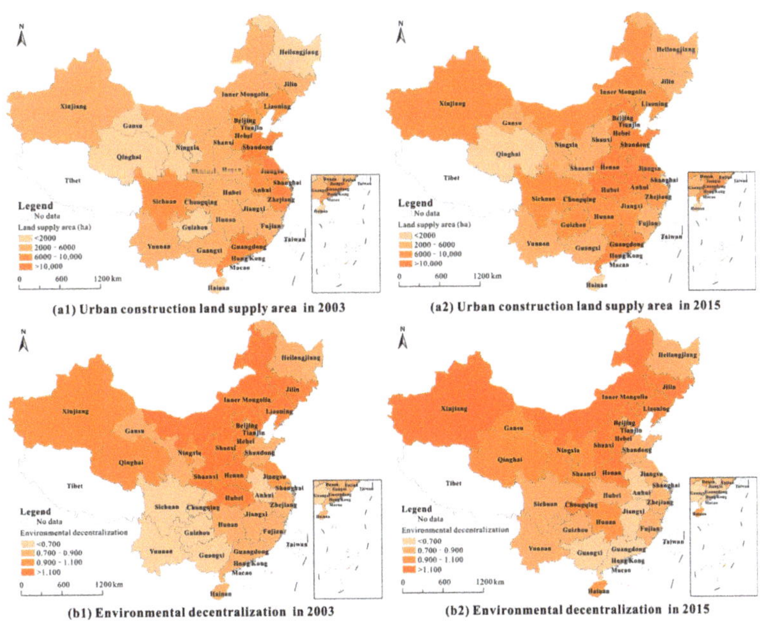

Figure 2. Temporal and spatial evolution of urban construction land supply and environmental decentralization.

Figure 2(a1,a2) shows that there is more construction land supply in the east than in the west, and the land supply in the Yangtze River Delta and Pearl River Delta is the largest. From 2003 to 2015, land supply in some provinces in the central and western regions of China increased. Figure 2(b1,b2) shows that the provinces closer to the capital (Beijing) have a higher degree of environmental decentralization. From 2003 to 2015, environmental decentralization in the central region decreased, while it marginally increased in the western region.

4. Empirical Results

4.1. Basic Estimation Results

For fear of unrealistic regression results, it is necessary to check that the data is stable before regression. Since this paper uses short panel data, the LLC unit root test is used. The test results indicate that the data used for regression are all stationary. The basic estimation results are shown in Table 3. FE1 is the estimation result of adding no control variables, and FE2 and FE3 are the estimation results of gradually adding special control variables and other control variables.

Table 3. Basic estimation results.

Variables	FE1	FE2	FE3
ED	1.035 *** (4.498)	1.010 *** (4.967)	0.543 ** (2.533)
FD	—	0.394 *** (3.614)	0.360 *** (3.059)
FAI	—	1.427 *** (7.483)	0.849 *** (4.033)
lnIPC	—	−0.062 (−1.334)	−0.115 ** (−2.530)
lnPGDP	—	—	1.196 *** (3.223)
$lnPGDP^2$	—	—	−0.327 *** (−3.184)
lnPOP	—	—	0.415 (0.584)
URL	—	—	0.176 (1.135)
STI	—	—	−4.661 *** (−2.955)
FP	—	—	1.839 ** (2.207)
Cons	7.242 *** (30.553)	6.350 *** (27.700)	7.729 (1.378)
Time-fixed effect	Yes	Yes	Yes
Province-fixed effect	Yes	Yes	Yes
R^2	0.403	0.546	0.600
Obs	390	390	390

Note: values in parentheses are *t* statistics; **, and *** represent that the coefficients are significant at the levels of 5%, and 1%, respectively.

The regression results in Table 3 show that after adding all control variables, the coefficient of environmental decentralization is significant, at the level of 5%, which indicates that environmental decentralization increases the scale of construction land supply. The effect of fiscal decentralization is positive, at a 1% significance level, suggesting that the greater financial autonomy of local governments, the stronger the motivation to provide a great deal of construction land. The coefficient of fixed asset investment per unit of GDP is positive, at a 1% significance level, suggesting that the scale of construction land supply is expanding due to the competition of local governments to attract investment. The effect of industrial pollution control investment is negative, at a 5% significance level, indicating that environmental protection constraints limit the expansion of construction land supply. The coefficient of square per GDP is markedly negative, at the level of 1%, indicating that there is an inverted U-shaped Kuznets curve relationship between economic development and construction land supply. The scale of construction land supply first increases and then decreases with economic growth. The proportion of secondary and tertiary industries is significantly negative, at the level of 1%, indicating that industrial structure upgrading can reduce construction land supply. The effect of local government financial pressure on land supply is markedly positive, at the level of 5%, indicating that the greater financial pressure, the more construction land supply. Population size and land urbanization have a positive influence on construction land supply, but they are not significant enough.

4.2. Robustness Test

To make the regression results more robust and reliable, the explained and explanatory variables are replaced, and the estimation method is changed. The robustness test results are shown in Table 4. In Table 4, FE4 is the regression result of replacing the explained variable with land transfer income (LTI). FE5 is the regression result of replacing the explanatory variable by environmental decentralization with a lag of one period. In addition, the model may suffer from heteroscedasticity and autocorrelation, and the feasible generalized least squares (FGLS) method is used to mitigate these problems [37,77]. The third column in Table 4 are the regression results with FGLS. The main work of environmental management includes administration, supervision, and monitoring. Therefore, subdivided environmental decentralization can be used for robustness testing. FE6, FE7, and FE8 are the estimated results of environmental administrative decentralization, environmental supervision decentralization, and environmental monitoring decentralization, respectively.

Table 4. Robustness test results.

Variables	FE4	FE5	FGLS	FE6	FE7	FE8
ED	0.720 *** (3.836)	—	0.290 ** (2.072)	—	—	—
L.ED	—	0.767 *** (3.247)	—	—	—	—
EAD	—	—	—	0.013 (0.317)	—	—
ESD	—	—	—	—	0.082 ** (2.481)	—
EMD	—	—	—	—	—	0.038 (0.857)
FD	0.174 * (1.688)	0.360 *** (2.872)	0.258 * (1.785)	0.351 *** (2.907)	0.392 *** (3.291)	0.373 *** (3.026)
FAI	0.361 * (1.958)	0.833 *** (3.759)	0.419 *** (3.470)	0.743 *** (3.567)	0.756 *** (3.658)	0.760 *** (3.635)
lnIPC	−0.045 (−1.126)	−0.135 *** (−2.840)	−0.086 *** (−2.994)	−0.112 ** (−2.454)	−0.112 ** (−2.468)	−0.103 ** (−2.190)
Cons	−2.067 (−0.421)	8.909 (1.471)	10.392 ** (2.106)	8.066 (1.379)	1.827 (0.294)	6.324 (1.018)
Other control variables	Yes	Yes	Yes	Yes	Yes	Yes
R^2	0.867	0.579	—	0.593	0.600	0.593
Obs	390	360	390	390	390	390

Note: values in parentheses are t statistics; *, **, and *** represent that the coefficients are significant at the levels of 10%, 5%, and 1%, respectively.

Table 4 indicates that environmental decentralization has a remarkable positive influence on land transfer income. The impacts of fiscal decentralization and competition for attracting investment on land transfer income are also significantly positive. The impact of environmental decentralization lagging one period on land supply is significant, at a level of 1%, indicating that the current land supply is affected by environmental decentralization in the previous period. The regression results with FGLS are not much different from the basic estimation. From the perspective of subdivided environmental decentralization, the influences of environmental administrative and monitoring decentralization are not significant, while the impact of environmental supervision decentralization is significant, at a level of 5%. The reason for this may be that administrative and monitoring decentralization increases the flexibility and positiveness of environmental management and helps to improve environmental performance. Supervision decentralization weakens the importance of environmental constraints and reduces environmental performance. Robustness test results are consistent with basic estimation results. It can be considered that the impact of environmental decentralization on urban construction land supply is stable.

4.3. Endogenetic Test

There are many factors that affect urban construction land supply. Although some variables affecting land supply are controlled in this paper, there may still be some factors that are not considered by the model. In addition, the bidirectional causality relationship between explanatory variables and explained variables may also produce endogeneity. In this paper, the first-order and second-order lags of environmental decentralization are used as instrumental variables, and two-stage least square (2SLS) regression is used to eliminate endogenous effects [78]. The generalized method of moments (GMM) can deal with endogenous issues [79]. In this paper, the system generalized method of moments (sys-GMM) is used to further avoid the endogenous and weak instrumental variables [50]. The endogeneity test results are shown in Table 5.

Table 5. Endogenous test results.

Variables	2SLS		Sys-GMM
	1st Step (ED)	2nd Step (LS)	
L.LS	—	—	0.025 (0.323)
L.ED	0.854 *** (14.334)	—	—
L2.ED	−0.045 (−0.723)	—	—
ED	—	0.869 *** (2.671)	0.414 * (1.701)
FD	−0.012 (−0.672)	0.297 ** (2.153)	0.459 ** (2.524)
FAI	−0.044 (−1.367)	0.885 *** (3.677)	0.404 (1.611)
lnIPC	−0.005 (−0.784)	−0.147 *** (−2.874)	−0.150 (−1.338)
Cons	0.101 (0.108)	7.290 (1.051)	7.467 (1.445)
Other control variables	Yes	Yes	Yes
Underidentification test	22.274 (0.000)		—
Weak identification test	137.478 (19.93)		—
Hansen J statistic	1.664 (0.197)		—
AR(2)	—	—	−0.83 (0.405)
Sargan test	—	—	9.22 (0.324)
R^2	0.727	0.546	—
Obs	330	330	330

Note: values in parentheses are *t* statistics; *, **, and *** represent that the coefficients are significant at the levels of 10%, 5%, and 1%, respectively; the *p* values of Underidentification test, Hansen J statistic, AR(2) and Sargan test are in parentheses; the critical value of 10% significance is shown in the brackets of Weak identification test.

Table 5 shows that the impact of environmental decentralization on urban construction land supply is significantly positive, at a 10% level, whether estimated using the 2SLS model or the sys-GMM model. After accounting for endogeneity, the impact of environmental decentralization on urban construction land supply is still significantly positive.

5. Further Analysis

5.1. Intermediary Mechanism

Theoretical analysis has shown that land finance dependence and land resources misallocation increase the scale of local government construction land supply. Land financial dependence can be measured by indicators such as the proportion of land transfer income to GDP [80], the proportion of land transfer income to local government fiscal revenue [81,82], and per capita land transfer income [83]. In this paper, the proportion of land transfer revenue in the general public budget revenue of the local government is used to represent the land financial dependence of local government. Land resources misallocation in this paper refers to the unreasonable allocation between different uses of construction land. This is because the government supplies a great deal of industrial land at low prices, resulting in a high proportion of industrial land. Scholars believe that agreement transfers go hand in hand with the supply of industrial land at low prices. Thus, the proportion of land transferred by agreement to the total land transfer area can be used to measure the degree of land resource misallocation [16]. The estimated results of the intermediary effect are shown in Table 6.

Table 6 shows that the impact of environmental decentralization on land financial dependence and land resource misallocation is significantly positive, at a level of 1%, which indicates that environmental decentralization exacerbates land financial dependence and land resource misallocation. Comparing the columns (3)–(5) in Table 6, it can be found that the impact of environmental decentralization on land supply is significantly reduced after controlling land financial dependence and land resource misallocation. It shows that the two intermediary variables have a remarkable influence on urban construction land supply [84].

Table 6. Intermediary effect estimation results.

Variables	LF	LRM	LS (ED)	LS (LF)	LS (LRM)
ED	0.341 *** (4.371)	0.221 *** (4.205)	1.035 *** (4.498)	0.643 *** (2.950)	0.904 *** (3.862)
LF	—	—	—	1.148 *** (7.866)	—
LRM	—	—	—	—	0.591 ** (2.539)
Cons	0.122 (1.523)	0.452 *** (8.346)	7.242 *** (30.553)	7.101 *** (32.373)	6.974 *** (27.062)
Control variables	No	No	No	No	No
R^2	0.324	0.870	0.403	0.493	0.414
Obs	390	390	390	390	390

Note: values in parentheses are *t* statistics; **, and *** represent that the coefficients are significant at the levels of 5%, and 1%, respectively.

5.2. Heterogeneity Analysis

After environmental decentralization, urban construction land supply by local governments should be balanced and chosen among fiscal revenue, economic growth, and environmental protection. Local governments with high pressure on fiscal revenue and economic growth may loosen environmental control and transfer a great deal of construction land for obtaining more land transfer income and better political performance. Local governments with high environmental protection pressures are more inclined to protect the environment and will not supply a great deal of construction land. In order to contrast the different influences in the regions with different financial pressure, economic growth pressure, and environmental protection pressure, this paper uses the median of pressure to divide the provinces into two types of regions. Referring to Li (2015), the calculation of financial pressure is attained by the proportion of the disparity between fiscal expenditure and fiscal revenue to GDP [50]. The provinces are divided into high and low financial pressure regions based on the median of the average financial pressure in each province from 2003 to 2015. Providing more land can boost economic growth [85]. The regions with high economic growth pressure have a stronger willingness to supply more construction land. Referring to Yang (2016), regional economic growth pressure is represented by the divergence between the local GDP growth rate of the current year and the last year [86]. Similarly, the regional division standard is the median of the average economic growth pressure. Chemical oxygen demand (COD) and SO_2 are often used to indicate pollution levels [87,88]. COD is mainly produced by industrial enterprises, and is often used as an important indicator to measure environmental pollution. Therefore, COD emissions per unit of GDP are used to measure environmental protection pressure. The regional division of environmental pressure is in accordance with financial pressure and economic growth pressure. The estimated results for regions with distinct financial pressures, economic growth pressure, and environmental protection pressures are shown in Table 7.

From the perspective of financial pressure, the effect of environmental decentralization on urban construction land supply is significantly positive, at a level of 10%, in the regions with high financial pressure, but it is not significant in the regions with low financial pressure. Moreover, the regression coefficient of environmental decentralization in regions with high financial pressure is larger (0.609 > 0.396). The results show that environmental decentralization prompts local governments with high financial pressure to increase construction land supply to ease financial pressure.

From the perspective of economic growth pressure, the effect of environmental decentralization on urban construction land supply is markedly positive, at a level of 10%, in regions with high economic growth pressure. The influence is negative in regions with low economic growth pressure, and it is not significant. The regression coefficient of environmental decentralization in regions with high economic growth pressure is larger

(0.845 > −0.007). It shows that the motivation of local governments, which relies on construction land supply to boost economy, distorts the impact of environmental decentralization.

Table 7. Subregional estimation results.

Variables	Financial Pressure		Economic Growth Pressure		Environmental Protection Pressure	
	High	Low	High	Low	High	Low
ED	0.609 * (1.666)	0.396 (1.364)	0.845 * (1.731)	−0.007 (−0.033)	0.245 (0.749)	0.760 ** (2.325)
FD	0.857 ** (2.108)	0.263 * (1.753)	0.300 * (1.783)	−0.160 (−0.486)	0.684 ** (2.342)	0.195 (1.173)
FAI	0.750 *** (2.709)	0.661 ** (2.045)	0.852 *** (2.692)	0.745 ** (2.605)	0.516 ** (2.580)	1.567 *** (3.030)
lnIPC	−0.041 (−0.782)	−0.348 *** (−4.008)	−0.182 ** (−2.064)	−0.091 * (−1.905)	−0.018 (−0.371)	−0.349 *** (−4.123)
Cons	−1.422 (−0.198)	19.834 * (1.893)	10.897 (1.144)	−18.501 * (−1.819)	−5.990 (−0.813)	23.565 * (1.905)
Other control variables	Yes	Yes	Yes	Yes	Yes	Yes
R^2	0.771	0.506	0.551	0.780	0.772	0.522
Obs	195	195	195	195	195	195

Note: values in parentheses are t statistics; *, **, and *** represent that the coefficients are significant at the levels of 10%, 5%, and 1%, respectively.

From the perspective of environmental protection pressure, the impact of environmental decentralization on urban construction land supply is significant, at a level of 5%, in the regions with low environmental protection pressure, and it is not significant in the regions with high environmental protection pressure. Moreover, the regression coefficient of environmental decentralization in the regions with low environmental protection pressure is larger (0.760 > 0.245), which indicates that environmental decentralization has a greater influence on construction land supply in regions with low environmental protection pressure. The cause for such a phenomenon is that the environmental constraints in regions with low environmental protection pressure are relatively weak. When the degree of environmental decentralization increases, local governments tend to supply more construction land to improve economic and political performance.

6. Policy Suggestions

The research of this paper provides some policy ideas for local governments to reasonably supply urban construction land against the background of Chinese decentralization.

(1) The environmental management power and responsibilities of the central and local governments should be reasonably divided, and the incentive and restraint mechanisms for construction land supply should be adjusted. The original intention of environmental decentralization is to realize the localization of environmental management and improve the efficiency of environmental management. However, in practice, local governments are encouraged by fiscal revenue and political performance to loosen environmental control and increase construction land supply in the background of environmental decentralization. By decentralizing the environmental administrative power and centralizing the environmental supervision power, local government's target choice and land supply behavior can be efficaciously restrained. In addition, strengthening the central environmental protection constraints can increase the environmental preferences of local governments and encourage them to reasonably supply construction land. Some measures should be used, such as strictly implementing the environmental protection supervision system and improving the importance of environmental protection indicators in performance appraisal.

(2) Differentiated environmental decentralization should be implemented to form differentiated construction land supply incentives. The strong economic and political incentives brought by urban construction land to local governments make environmental decentralization ineffective in regions with high financial pressure, high economic growth pressure, and low environmental protection pressure. The central government ought to strengthen environmental centralization in these three types of regions. While empowering local governments with environmental administrative power, the central government's regulatory powers ought to be enhanced to restrict local governments' land supply behavior that damages the environment. In regions with low financial pressure, low economic

growth pressure, and high environmental protection pressure, the central government should increase the degree of environmental decentralization to improve environmental management efficiency.

(3) The land financial dependence of local governments ought to be weakened. Looking for new sources of income and changing financing methods can alleviate the financial pressure of local governments and weaken their land financial dependence. Specific measures can be implemented from two aspects: open source and throttling. Firstly, the fiscal and taxation systems ought to be innovated to increase the tax revenue of local governments, such as levying property tax and inheritance tax, and so on [89]. Secondly, BOT (build-operate-transfer) and other financing modes could be adopted in infrastructure construction to alleviate the financial pressure of the government [90]. Moreover, the land lease can be changed to annual lease, and the land transfer income can be changed from one-time income to continuous income. While ensuring the government's stable income, the urban construction land supply for obtaining short-term economic profits can be controlled.

In addition, urban planning is an important means to control the expansion of construction land scale. The government's environmental policy objectives should be fully considered when formulating land use plans [91]. In the use of land, it is necessary to strengthen the guiding and standardizing roles of planning, and realize the scientific and rational allocation of land for different purposes, such as urban construction land and agricultural land.

7. Discussion and Conclusions

7.1. Discussion

In recent years, China's environmental problems have gradually come to light. Under the environmental decentralization system, the environmental management power of the central government has gradually been strengthened [92]. This shows that the central government is trying to find a balance of environmental decentralization to coordinate the economic growth goals and the environmental protection goals of local government. We integrated environmental decentralization into the Chinese decentralization framework. Then, under the framework of Chinese decentralization, the incentives and constraints of environmental decentralization, fiscal decentralization, and political centralization on the supply of construction land by local governments were studied.

Our research enriches the Chinese decentralization framework. Previous studies have only emphasized the incentives of fiscal decentralization and political centralization on local government land supply [64,85,93]. Tang (2019) focused on the impact of environmental politically binding indicators on local governments' land violations [36]. However, there has been no in-depth analysis of the impact of environmental management systems on local government land supply. More importantly, the expansion of urban construction land against the background of resource constraints and environmental protection has been more strictly regulated [94]. Clarifying the incentives and constraints of government land supply is conducive to achieving the rational allocation of urban land.

Our research results suggest that environmental decentralization reduces the performance of land environmental management in China. This differs from Laskowski (2005) and Blundell (2021), in that environmental decentralization improves performance by improving environmental policy adaptability [44,45]. As Ulph (1998) argued, information asymmetry between local and central governments in the context of Chinese decentralization distorts the performance of environmental decentralization [46]. Among the three environmental management affairs of administration, supervision, and monitoring, supervision decentralization has the most obvious promotion of the expansion of construction land. Our research also shows that the impact of environmental decentralization on construction land supply regionally varies. Where economic, political, and environmental protection constraints differ, the performance of environmental decentralization is also different. Fredriksson's study of 110 countries also concluded that there were regional differences in environmental decentralization performance [95].

Although our study is based on the specific situation in China, it is also meaningful to analyze government actions in the supply of urban construction land in other countries. The impact of local government incentives and constraints on urban construction land supply also exists in Western countries. Götze (2021) found that municipalities in Germany and the Netherlands also provide a loose supply of urban land due to financial incentives [96]. Perrin (2018) and Kurnia (2021) studied the impact of decentralization on the expansion of urban construction land in France, Italy, and Indonesia [17,18].

7.2. Conclusions

The economical and intensive use of land resources is an important measure to push ecological civilization construction. Practice in western countries has proved that decentralization can reduce information asymmetry and principal-agent risks, and improve policy efficiency. However, decentralization may also allow local governments to loosen environmental regulations to attract investment, which could boost economic development. Under the Chinese decentralization framework, fiscal decentralization and political centralization provide incentives for local governments to supply a great deal of urban construction land to make up for fiscal gaps and support economic growth. Environmental decentralization increases the possibility of local governments to loosen environmental regulation, which promotes the expansion of construction land supply. The efficiency of environmental decentralization is further reduced. Based on the theoretical analysis of the effect of environmental decentralization on construction land supply, the interprovincial panel data from 2003 to 2015 were used for analysis. The main conclusions are as follows:

(1) Environmental decentralization promotes the expansion of urban construction land supply scale. Environmental decentralization intensifies the positive incentives of economic and political incentives to the supply of construction land, and weakens environmental constraints. Due to the important role of construction land in local economic development, local governments usually choose to relax environmental control and supply a great deal of construction land.

(2) Environmental decentralization promotes the expansion of urban construction land supply by strengthening land financial dependence and distorting land resources misallocation. Environmental decentralization increases the possibility of "land rent", "land tax", and "attract investment by land", while weakening environmental constraints. This raises land financial dependence and the misallocation of land resources, which accelerates the swell of construction land supply scale.

(3) Environmental decentralization has diverse incentives and constraints in regions with distinct financial pressure, economic growth pressure, and environmental protection pressure, and has different impacts on construction land supply. In regions with high financial and economic growth pressure, local governments, driven by economic and political incentives, choose to relax environmental constraints and supply a great deal of construction land. In regions with low environmental protection pressure, local governments have lower environmental constraints. The impact of environmental decentralization on urban construction land supply is greater than that in regions with high environmental protection pressure.

7.3. Limitations and Future Research

There are some deficiencies in this study. First, this paper only pays attention to the effect of environmental decentralization on the urban construction land supply scale of local government. However, the construction land supply structure of different uses and industries is also affected by the preferences and behaviors of local government, which is a content worthy of study. Second, due to data limitations, this paper only focuses on the impact of decentralization and centralization between the central government and provincial governments on urban construction land supply. However, Chinese governments at or above the county level are empowered to supply land. The impact of decentralization and centralization between the upper and lower governments below the provincial level on land supply is also worth studying. Third, this paper only studies the impact of environmental decentralization on construction land supply in the context of China's specific system. However, there may be differences in the performance of environmental decentralization in different institutional contexts. In the future, comparative analysis with other countries could also be carried out.

Author Contributions: Conceptualization, L.H., P.T. and X.W.; methodology, P.T. and X.W.; software, X.W.; validation, P.T. and X.W.; formal analysis, P.T. and X.W.; investigation, X.W.; resources, P.T. and L.H.; data curation, X.W.; writing—original draft preparation, P.T. and X.W.; writing—review and editing, P.T., L.H. and X.W.; visualization, X.W.; supervision, P.T. and L.H.; project administration, P.T. and X.W.; funding acquisition, P.T. and L.H. All authors have read and agreed to the published version of the manuscript.

Funding: This research was funded by the National Social Science Fund (19BJY110) and the Project of Sichuan University (2021CXC03).

Institutional Review Board Statement: Not applicable.

Informed Consent Statement: Not applicable.

Data Availability Statement: Publicly available datasets were analyzed in this study. These data can be found at: https://data.stats.gov.cn/, and https://data.cnki.net/Yearbook/Navi?type=type&code=A, accessed on 14 June 2022.

Acknowledgments: The authors are particularly grateful to the anonymous reviewers for their comments and suggestions which contributed to the further improvement of this paper.

Conflicts of Interest: The authors declare no conflict of interest.

References

1. Liu, J.; Jin, X.B.; Xu, W.Y.; Gu, Z.M.; Yang, X.H.; Ren, J.; Fan, Y.T.; Zhou, Y.K. A New Framework of Land Use Efficiency for the Coordination among Food, Economy and Ecology in Regional Development. *Sci. Total Environ.* **2020**, *710*, 135670. [CrossRef] [PubMed]
2. Fleming, A.; Anthony, P.O.; Cara, S.; Sue, O.; Daniel, M.; Matthew, T.H. Improving Acceptance of Natural Capital Accounting in Land Use Decision Making: Barriers and Opportunities. *Ecol. Econ.* **2022**, *200*, 107510. [CrossRef]
3. Imbrenda, V.; Giovanni, Q.; Rosanna, S.; Gianluca, E.; Luca, S.; Marcela, P.; Rosa, C.; Maria, L. Land Degradation and Metropolitan Expansion in a Peri-urban Environment. *Geomat. Nat. Hazards Risk* **2021**, *12*, 1797–1818. [CrossRef]
4. Dolley, J.; Fiona, M.; Bradley, B.; Jeremy, R.; James, A.R.; Baris, E.; Novi, Q. Analysing Trade-offs and Synergies between SDGs for Urban Development, Food Security and Poverty Alleviation in Rapidly Changing Peri-urban Areas: A Tool to Support Inclusive Urban Planning. *Sustain. Sci.* **2020**, *15*, 1601–1619. [CrossRef]
5. Simon, D. Urban Environments: Issues on the Peri-urban Fringe. *Annu. Rev. Environ. Resour.* **2008**, *33*, 167–185. [CrossRef]
6. Ding, C.R.; Zhao, X.S. Land Market, Land Development and Urban Spatial Structure in Beijing. *Land Use Policy* **2014**, *40*, 83–90. [CrossRef]
7. Feng, L.; Yang, Y.L.; Yao, L. Decentralization and Incentives: Research on the Central-Local Relations within the Land Supply System. *China Land Sci.* **2013**, *27*, 4–10+23.
8. Hassan, M.M.; Nazem, M.N.I. Examination of Land Use/land Cover Changes, Urban Growth Dynamics, and Environmental Sustainability in Chittagong City, Bangladesh. *Environ. Dev. Sustain.* **2016**, *18*, 697–716. [CrossRef]
9. Huang, K.X.D. Growth and Cycles in China's Unbalanced Development: Resource Misallocation, Debt Overhang, Economic Inequality, and the Importance of Structural Reforms. *Front. Econ. China* **2019**, *14*, 53–71.
10. De Mello, K.; Taniwaki, R.H.; Paula, F.R.D.; Valente, R.V.; Randhir, T.O.; Macedo, D.R.; Leal, C.G.; Rodrigues, C.B.; Hughes, R.M. Multiscale Land Use Impacts on Water Quality: Assessment, Planning, and Future Perspectives in Brazil. *J. Environ. Manag.* **2020**, *270*, 110879. [CrossRef]
11. Vito, I.; Rosa, C.; Leonardo, B.; Valerio, D.S.; Luca, S. Urban sprawl: Theory and practice. *Adv. Chem. Pollut. Environ. Manag. Prot.* **2022**, *8*, 23–46.
12. Amponsah, O.; Blija, D.K.; Ayambire, R.A.; Takyi, S.A.; Mensah, H.; Braimah, I. Global Urban Sprawl Containment Strategies and Their Implications for Rapidly Urbanising Cities in Ghana. *Land Use Policy* **2022**, *114*, 105979. [CrossRef]
13. Pourtaherian, P.; Jaeger, J.A.G. How Effective Are Greenbelts at Mitigating Urban Sprawl? A Comparative Study of 60 European Cities. *Landsc. Urban Plan.* **2022**, *227*, 104532. [CrossRef]
14. Millimet, D.L. Environmental Federalism: A Survey of the Empirical Literature. *Case West. Reserve Law Rev.* **2014**, *64*, 1669. [CrossRef]

15. The National People's Congress of the People's Republic of China Home Page. Available online: http://www.npc.gov.cn/zgrdw/englishnpc/Special/CombatingClimateChange/2009-08/28/content_1516267.htm (accessed on 23 February 2023).
16. Du, W.J.; Li, M.J. The Impact of Land Resource Mismatch and Land Marketization on Pollution Emissions of Industrial Enterprises in China. *J. Environ. Manag.* **2021**, *299*, 113565. [CrossRef] [PubMed]
17. Perrin, C.; Nougarèdes, B.; Sini, L.; Branduini, P.; Salvati, L. Governance Changes in Peri-urban Farmland Protection following Decentralisation: A Comparison between Montpellier (France) and Rome (Italy). *Land Use Policy* **2018**, *70*, 535–546. [CrossRef]
18. Kurnia, A.A.; Rustiadi, E.; Fauzi, A.; Pravitasari, A.E.; Saizen, I.; Ženka, J. Understanding Industrial Land Development on Rural-Urban Land Transformation of Jakarta Megacity's Outer Suburb. *Land* **2022**, *11*, 670. [CrossRef]
19. Lawton, A.; Morrison, N. The Loss of Peri-urban Agricultural Land and the State-local Tensions in Managing Its Demise: The Case of Greater Western Sydney, Australia. *Land Use Policy* **2022**, *120*, 106265. [CrossRef]
20. Spyra, M.; Kleemann, J.; Calò, N.C.; Schürmann, A.; Fürst, C. Protection of Peri-urban Open Spaces at the Level of Regional Policy-making: Examples from Six European Regions. *Land Use Policy* **2021**, *107*, 105480. [CrossRef]
21. Landesa Home Page. Available online: http://www.landesa.org/china-survey-6/ (accessed on 23 February 2023).
22. Zhang, J.; Fan, J.; Mo, J. Government Intervention, Land Market, and Urban Development: Evidence from Chinese Cities. *Econ. Inq.* **2017**, *55*, 115–136. [CrossRef]
23. Jiang, N.P.; Xu, H. Research on the effect of local government to urban land supply in China. *Econ. Theory Bus. Manag.* **2015**, *01*, 55–66.
24. Ran, R. Perverse Incentive Structure and Policy Implementation Gap in China's Local Environmental Politics. *J. Environ. Policy Plan.* **2013**, *15*, 17–39. [CrossRef]
25. Lichtenberg, E.; Ding, C. Local Officials as Land Developers: Urban Spatial Expansion in China. *J. Urban Econ.* **2009**, *66*, 57–64. [CrossRef]
26. Head, K.; Ries, J. Inter-City Competition for Foreign Investment: Static and Dynamic Effects of China's Incentive Areas. *J. Urban Econ.* **1996**, *40*, 38–60. [CrossRef]
27. Cai, M.N.; Murtazashvili, J.B.; Murtazashvili, I.; Wang, H. Sugarcoating the Bitter Pill: Compensation, Land Governance, and Opposition to Land Expropriation in China. *J. Peasant. Stud.* **2020**, *47*, 1371–1392. [CrossRef]
28. Tao, R.; Su, F.B.; Liu, M.X.; Cao, G.Z. Land Leasing and Local Public Finance in China's Regional Development: Evidence from Prefecture-level Cities. *Urban Stud.* **2010**, *47*, 2217–2236.
29. Wang, R.; Hou, J. Land Finance, Land Attracting Investment and Housing Price Fluctuations in China. *Int. Rev. Econ. Financ.* **2021**, *72*, 690–699. [CrossRef]
30. Liu, S.Y.; Wang, Z.F.; Zhang, W.F.; Xiong, X.F. The Exhaustion of China's "Land-Driven Development" Mode: An Analysis Based on Threshold Regression. *J. Manag. World* **2020**, *36*, 80–92+119+246.
31. Jiang, M.; Xin, L.J.; Li, X.B.; Tan, M.H. Spatiotemporal Variation of China's State-Owned Construction Land Supply from 2003 to 2014. *Sustainability* **2016**, *8*, 1137. [CrossRef]
32. Zhou, C.S.; Jin, W.F.; Zhang, G.J.; Li, M.; Wang, S.J. Spatio-temporal characteristics and influencing factors of state owned construction land supply in China. *Acta Geogr. Sin.* **2019**, *74*, 16–31.
33. Wang, X.B.; Shao, Y.F.; Zhang, Q.H. Influential factors and regional difference of local government land transfer: Based on 1999–2011 provincial panel data. *J. Wuhan Univ. Technol. (Soc. Sci. Ed.)* **2015**, *28*, 627–634.
34. Bidandi, F.; Williams, J.J. Understanding Urban Land, Politics, and Planning: A Critical Appraisal of Kampala's Urban Sprawl. *Cities* **2020**, *106*, 102858. [CrossRef]
35. Banzhaf, H.S.; Lavery, N. Can the Land Tax Help Curb Urban Sprawl? Evidence from Growth Patterns in Pennsylvania. *J. Urban Econ.* **2010**, *67*, 169–179. [CrossRef]
36. Tang, P.; Feng, Y.; Li, M.; Zhang, Y.Y. Can the Performance Evaluation Change from Central Government Suppress Illegal Land Use in Local Governments? A New Interpretation of Chinese Decentralisation. *Land Use Policy* **2021**, *108*, 105578. [CrossRef]
37. Xie, J. Why Natural Resources Become A Curse: "Grabbing Hand" under Market Distortion—An Empirical Test from the Perspective of Officials 'Promotion Incentive Caused by "Chinese Decentralization". *Commer. Res.* **2019**, *11*, 23–31.
38. Lin, N. Capitalism in China: A Centrally Managed Capitalism (CMC) and Its Future. *Manag. Organ. Rev.* **2011**, *7*, 63–96. [CrossRef]
39. Maskin, E.; Qian, Y.Y.; Xu, C.G. Incentives, Information, and Organizational Form. *Rev. Econ. Stud.* **2000**, *67*, 359–378. [CrossRef]
40. Zhang, X.B. Fiscal Decentralization and Political Centralization in China: Implications for Growth and Inequality. *J. Comp. Econ.* **2006**, *34*, 713–726. [CrossRef]
41. Lo, K. How Authoritarian Is the Environmental Governance of China? *Environ. Sci. Policy* **2015**, *54*, 152–159. [CrossRef]
42. Loomis, J.J.; Oliveira, C.M.R.D.; Dziedzic, M. Environmental Federalism in EIA Policy: A Comparative Case Study of Paraná, Brazil and California, US. *Environ. Sci. Policy* **2021**, *122*, 75–82. [CrossRef]
43. Feng, S.L.; Sui, B.; Liu, H.M.; Li, G.X. Environmental Decentralization and Innovation in China. *Econ. Model.* **2020**, *93*, 660–674. [CrossRef]
44. Laskowski, S.; Morgenstern, R.D.; Blackman, A. Environmental Decentralization in the United States: Seeking the Proper Balance between National and State Authority. *IDEAS Work. Pap. Ser. RePEc* **2005**. [CrossRef]
45. Blundell, W.; Evans, M.F.; Stafford, S.L. Regulating Hazardous Wastes under U.S. Environmental Federalism: The Role of State Resources. *J. Environ. Econ. Manag.* **2021**, *108*, 102464. [CrossRef]

46. Ulph, A. Political Institutions and the Design of Environmental Policy in a Federal System with Asymmetric Information. *Eur. Econ. Rev.* **1998**, *42*, 583–592. [CrossRef]
47. Gray, W.B.; Shadbegian, R.J. 'Optimal' Pollution Abatement—Whose Benefits Matter, and How Much? *J. Environ. Econ. Manag.* **2004**, *47*, 510–534. [CrossRef]
48. Qi, Y.; Lu, H.Y.; Xu, Y.K. Research on the Reformation of China's Environmental Decentralization System: Institutional Change, Numerical Estimates and Effects Assessment. *China Ind. Econ.* **2014**, *1*, 31–43.
49. Wu, H.T.; Li, Y.W.; Hao, Y.; Ren, S.Y.; Zhang, P.F. Environmental Decentralization, Local Government Competition, and Regional Green Development: Evidence from China. *Sci. Total Environ.* **2020**, *708*, 135085. [CrossRef]
50. Li, G.X.; Guo, F.Y.; Di, D.Y. Regional Coial factors aironmental Decentralization, and Target Selection of Local Governments. *Sci. Total Environ.* **2021**, *755*, 142536. [CrossRef]
51. Ran, Q.Y.; Zhang, J.N.; Hao, Y. Does Environmental Decentralization Exacerbate China's Carbon Emissions? Evidence Based on Dynamic Threshold Effect Analysis. *Sci. Total Environ.* **2020**, *721*, 137656. [CrossRef]
52. Cheng, S.L.; Fan, W.; Chen, J.D.; Meng, F.X.; Liu, G.Y.; Song, M.L.; Yang, Z.F. The Impact of Fiscal Decentralization on CO_2 Emissions in China. *Energy* **2020**, *192*, 116685. [CrossRef]
53. Ding, Y.; McQuoid, A.; Karayalcin, C. Fiscal Decentralization, Fiscal Reform, and Economic Growth in China. *China Econ. Rev.* **2019**, *53*, 152–167. [CrossRef]
54. Tu, L.H.; Padovani, E. A Research on the Debt Sustainability of China's Major City Governments in Post-Land Finance Era. *Sustainability* **2018**, *10*, 1606. [CrossRef]
55. Ahmad, E.; Singh, R.; Fortuna, M. *Toward More Effective Redistribution: Reform Options for Intergovernmental Transfers in China*; International Monetary Fund: Washington, DC, USA, 2004.
56. Huang, Z.H.; Du, X.J. Government intervention and land misallocation: Evidence from China. *Cities* **2017**, *60*, 323–332. [CrossRef]
57. Geng, B.; Zhang, X.L.; Liang, Y.; Bao, H.J.; Skitmore, M. Sustainable Land Financing in a New Urbanization Context: Theoretical Connotations, Empirical Tests and Policy Recommendations. *Resour. Conserv. Recycl.* **2018**, *128*, 336–344. [CrossRef]
58. Chiang, T.F.; Hou, J.; Tsai, P.H. Fiscal Incentives and Land Finance Cycles of Prefectures in China. *World Econ.* **2022**, *45*, 1262–1293. [CrossRef]
59. Cai, M.N. Revenue, Time Horizon, and Land Allocation in China. *Land Use Policy* **2017**, *62*, 101–112. [CrossRef]
60. Chen, T.; Kung, J.K.S. Do Land and Revenue Windfalls Create a Political Resource Curse? Evidence from China. *J. Dev. Econ.* **2016**, *123*, 86–106. [CrossRef]
61. Zhang, B.B.; Li, J.X.; Tian, W.M.; Chen, H.B.; Kong, X.B.; Chen, W.; Zhao, M.J.; Xia, X.L. Spatio-temporal Variances and Risk Evaluation of Land Finance in China at the Provincial Level from 1998 to 2017. *Land Use Policy* **2020**, *99*, 104804. [CrossRef]
62. Li, H.B.; Zhou, L.A. Political Turnover and Economic Performance: The Incentive Role of Personnel Control in China. *J. Public Econ.* **2005**, *89*, 1743–1762. [CrossRef]
63. Landry, P.F. The Political Management of Mayors in Post-Deng China The Political Management of Mayors in Post-Deng China. *Cph. J. Asian Stud.* **2003**, *17*, 31–58.
64. Van, D.K.D.; Lorentzen, P.; Mattingly, D. Racing to the Bottom or to the Top? Decentralization, Revenue Pressures, and Governance Reform in China. *World Dev.* **2017**, *95*, 164–176.
65. Hong, T.; Yu, N.N.; Mao, Z.G. Does Environment Centralization Prevent Local Governments from Racing to the Bottom?—Evidence from China. *J. Clean. Prod.* **2019**, *231*, 649–659. [CrossRef]
66. Qi, S.W.; Mao, H.; Zhang, J.D. Financial Pressure, Economic Stimulus, and Land Transfer for Investment Attraction—Empirical Evidence Based on Micro-Data of Industrial Land. *Financ. Trade Econ.* **2020**, *41*, 20–34.
67. Wu, Y.; Heerink, N. Foreign Direct Investment, Fiscal Decentralization and Land Conflicts in China. *China Econ. Rev.* **2016**, *38*, 92–107. [CrossRef]
68. Millimet, D.L. Assessing the Empirical Impact of Environmental Federalism. *J. Reg. Sci.* **2003**, *43*, 711–733. [CrossRef]
69. Wang, L.; Wang, K.; Zhang, J.J.; Zhang, D.; Wu, X.; Zhang, L.J. Multiple Objective-oriented Land Supply for Sustainable Transportation: A Perspective from Industrial Dependence, Dominance and Restrictions of 127 Cities in the Yangtze River Economic Belt of China. *Land Use Policy* **2020**, *99*, 105069. [CrossRef]
70. Jin, W.F.; Zhou, C.S.; Zhang, G.J. Characteristics of State-owned Construction Land Supply in Chinese Cities by Development Stage and Industry. *Land Use Policy* **2020**, *96*, 104630. [CrossRef]
71. Guo, L.; Jiang, X.N. Decentralization of Environmental Management and Enterprises' Environmental Technology Innovation: Evidence from China. *Appl. Econ.* **2022**, *54*, 4170–4186. [CrossRef]
72. Hao, Y.; Xu, L.; Guo, Y.X.; Wu, H.T. The Inducing Factors of Environmental Emergencies: Do Environmental Decentralization and Regional Corruption Matter? *J. Environ. Manag.* **2022**, *302*, 114098. [CrossRef]
73. Li, G.L.; Zhou, Y.L. Environmental Decentralization, Local Government Competition and Green Development. *Public Financ. Res.* **2019**, *10*, 73–86.
74. Ahmad, M.; Satrovic, E. Relating Fiscal Decentralization and Financial Inclusion to Environmental Sustainability: Criticality of Natural Resources. *J. Environ. Manag.* **2023**, *325*, 116633. [CrossRef] [PubMed]
75. Pang, W.; Sun, Y.D. Research on the Spatial Effect of Fiscal Decentralization and Promotion Incentive on Income Distribution. *J. Shanxi Univ. Financ. Econ.* **2018**, *40*, 1–14.

76. Li, Q.; Li, X.H. Local Government Competition and Environmental Governance-the Regulation of Environmental Decentralization. *J. Guizhou Univ. Financ. Econ.* **2020**, *3*, 101–110.
77. Reed, W.R.; Ye, H. Which Panel Data Estimator Should I Use? *Appl. Econ.* **2011**, *43*, 985–1000. [CrossRef]
78. Aston, L.; Currie, G.; Kamruzzaman, M.; Delbosc, A.; Brands, T.; Oort, N.V.; Teller, D. Multi-city Exploration of Built Environment and Transit Mode Use: Comparison of Melbourne, Amsterdam and Boston. *J. Transp. Geogr.* **2021**, *95*, 103136. [CrossRef]
79. Majumder, S.C.; Rahman, M.H.; Martial, A.A.A. The Effects of Foreign Direct Investment on Export Processing Zones in Bangladesh Using Generalized Method of Moments Approach. *Soc. Sci. Humanit. Open* **2022**, *6*, 100277. [CrossRef]
80. Cheng, Y.D.; Jia, S.H.; Meng, H. Fiscal Policy Choices of Local Governments in China: Land Finance or Local Government Debt? *Int. Rev. Econ. Financ.* **2022**, *80*, 294–308. [CrossRef]
81. Lu, J.; Li, B.; Li, H. The Influence of Land Finance and Public Service Supply on Peri-urbanization: Evidence from the Counties in China. *Habitat Int.* **2019**, *92*, 102039. [CrossRef]
82. Tang, P.; Shi, X.P.; Gao, J.L.; Feng, S.Y.; Qu, F.T. Demystifying the Key for Intoxicating Land Finance in China: An Empirical Study through the Lens of Government Expenditure. *Land Use Policy* **2019**, *85*, 302–309. [CrossRef]
83. Wang, L.O.; Wu, H.T.; Hao, Y. How Does China's Land Finance Affect Its Carbon Emissions? *Struct. Chang. Econ. Dyn.* **2020**, *54*, 267–281. [CrossRef]
84. Jiang, T. Mediating Effects and Moderating Effects in Causal Inference. *China Ind. Econ.* **2022**, *410*, 100–120.
85. Li, J. Lan Impact of Fiscal Decentralization on Economic Growth Path: Evidence from China's Urban Land Market. *Habitat Int.* **2014**, *41*, 307–313. [CrossRef]
86. Yang, J.D.; Yang, Q.J. Growth Pressures, Stimulus Plan and Industrial Land Transfer. *Econ. Res. J.* **2016**, *51*, 99–113.
87. Fan, H.C.; Graff, Z.J.; Kou, Z.L.; Liu, X.Y.; Wang, H.H. *Going Green in China: Firms' Responses to Stricter Environmental Regulations*; NBER Working Paper No. w26540; NBER: Cambridge, MA, USA, 2019.
88. Justyna, L.C.; Socha, J.; Gruba, P.; Pietrzykowski, M. The Current State of Environmental Pollution with Sulfur Dioxide (SO_2) in Poland Based on Sulfur Concentration in Scots Pine Needles. *Environ. Pollut.* **2020**, *258*, 113559.
89. Wu, W.Z.; Zhou, A.L. The Causes for House Price Fluctuation: Rigid Demand, or Finance? *Shandong Soc. Sci.* **2021**, *3*, 126–132.
90. Ran, Q.Y.; Wang, J.L.; Wang, W.L.; Yang, X.D. Land Finance, Technological Innovation and Green Development in China. *J. Nanjing Univ. Financ. Econ.* **2021**, *04*, 64–74.
91. OECD. *The Governance of Land Use in OECD Countries: Policy Analysis and Recommendations*; OECD Publishing: Paris, France, 2020; Volume 24.
92. Lin, J.Y.; Long, C.H.; Yi, C.Z. Has Central Environmental Protection Inspection Improved Air Quality? Evidence from 291 Chinese Cities. *Environ. Impact Assess. Rev.* **2021**, *90*, 106621. [CrossRef]
93. Henderson, J.V.; Su, D.L.; Zhang, Q.H.; Zheng, S.Q. Political Manipulation of Urban Land Markets: Evidence from China. *J. Public Econ.* **2022**, *214*, 104730. [CrossRef]
94. Bovet, J.; Reese, M.; Köck, W. Taming Expansive Land Use Dynamics–Sustainable Land Use Regulation and Urban Sprawl in a Comparative Perspective. *Land Use Policy* **2018**, *77*, 837–845. [CrossRef]
95. Fredriksson, P.G.; Wollscheid, J.R. Environmental Decentralization and Political Centralization. *Ecol. Econ.* **2014**, *107*, 402–410. [CrossRef]
96. Götze, V.; Hartmann, T. Why Municipalities Grow: The Influence of Fiscal Incentives on Municipal Land Policies in Germany and the Netherlands. *Land Use Policy* **2021**, *109*, 105681. [CrossRef]

land

Article

Coupling and Coordination Relationship between Urbanization Quality and Ecosystem Services in the Upper Yellow River: A Case Study of the Lanzhou–Xining Urban Agglomeration, China

Wusheng Zhao [1], Peiji Shi [1,*], Ya Wan [1] and Yan Yao [2]

[1] College of Geography and Environment Science, Northwest Normal University, Lanzhou 730070, China; 2017212169@nwnu.edu.cn (W.Z.)

[2] College of Economics, Northwest Normal University, Lanzhou 730070, China

* Correspondence: shipj@nwnu.edu.cn; Tel.: +86-138-9366-5158

Abstract: The study of the man–land relationship in the urbanization process is the current frontier and focus of international research. How to balance urban development and ecosystem conservation in the Upper Yellow River is a key issue for sustainable development in China. In this study, we evaluated the Lanzhou–Xining urban agglomeration (LXUA) by constructing a multi-dimensional assessment system for urbanization quality and ecosystem services. The efficacy function model, entropy weight method, and Integrated Valuation of Ecosystem Services and Tradeoffs (InVEST) model were used to quantitatively assess the subsystems' state of development. Then, the coupling model (CD) and the coordination degree (CCD) model were used to explore the coupling coordination relationship and spatial-temporal change characteristics of the composite system. The findings indicate that: (1) In 2020, the quality of urbanization in the LXUA showed the pattern of "double core". The development of urban centers in each city is insufficient, and the proportion of counties with a low level is too high. (2) Integrated ecosystem services showed an increasing distribution pattern from the northeast to the southwest. Water provision services, soil conservation services and carbon fixation services all showed growth trends. (3) Each county's composite system was in the run-in stage or highly coupled stage. The subsystems were closely related to each other. (4) The CCD was decreased by 6% between two decades. The number of counties on the verge of disorder was the highest. About 80% of the counties were relatively lagging behind in ecosystem services.

Keywords: urbanization quality; ecosystem services; coupling coordination; spatial-temporal variations; Lanzhou–Xining urban agglomeration

Citation: Zhao, W.; Shi, P.; Wan, Y.; Yao, Y. Coupling and Coordination Relationship between Urbanization Quality and Ecosystem Services in the Upper Yellow River: A Case Study of the Lanzhou–Xining Urban Agglomeration, China. *Land* **2023**, *12*, 1085. https://doi.org/10.3390/land12051085

Academic Editors: Li Ma, Yingnan Zhang, Muye Gan and Zhengying Shan

Received: 1 April 2023
Revised: 12 May 2023
Accepted: 17 May 2023
Published: 18 May 2023

1. Introduction

The first explosion of industry in the late 18th century required a greater concentration and continuity of production. Factors of production such as capital, manpower and resources are highly combined in a limited space, driving the formation and development of cities. Currently, more than half of the world's people live in urban areas. Although urbanization is of great significance in promoting population transformation, industrial development, scientific and technological progress, and cultural exchange, it has also produced some negative effects, such as widening the urban–rural gap, tightening resources and energy, intensifying environmental pollution, and overwhelming the ecosystem [1]. In this context, the Third United Nations Conference on Housing and Sustainable Urban Development (Habitat III) and the "Future Earth" (FE) all emphasized that the regional urbanization process should be coordinated with the state of the ecosystem and matched with the carrying capacity of the resource and environment. The goals of building inclusive, safe, disaster-resilient and sustainable cities and human settlements, and protecting, restoring and promoting the sustainable use of terrestrial ecosystems were included in the 2023 Agenda for Sustainable Development as part of the next 17 global sustainable development priorities. How to reduce the negative impacts of rapid urbanization on ecosystems and promote the synergistic development of urbanization and ecosystem services has become a hot topic of widespread concern worldwide. At present, China's urbanization development is in a critical transition period from the medium-term rapid growth stage to the later stage of quality improvement. The report of the 20th National Congress of the CPC and the Central Urbanization Working Conference pointed out that we should plan for development at the height of harmonious coexistence between human and nature; focus on improving the quality of urbanization development; improve the diversity, stability, and sustainability of the ecosystem; and take a green, intensive and efficient high-quality urbanization road. Therefore, from the perspective of system coupling, it is of great theoretical and practical significance to scientifically and accurately evaluate the current situation of urbanization quality and ecosystem services, as well as the coupling and coordination relationship between them.

International research is abundant regarding the relationship between urbanization and ecosystems. The Organization for Economic Cooperation and Development (OECD) and the United Nations Environment Programme (UNEP) pioneered the "Pressure-State-Response" model in the 1980s. This model fostered a two-way perspective for studying the interplay between urbanization and ecosystems. Through the application of econometric methodologies, Grossman and Krueger discovered the renowned environmental Kuznets curve based on panel data from 42 developed countries in 1995. This curve uncovers an inverted "U"-shaped evolution law, correlating urban economic development with the quality of urban ecological environments, thus providing a basis for further research. Contemporary international research can be classified broadly into two categories from a research perspective. One category focuses on the quantitative relationship between urbanization and ecosystems on a global scale. For example, Li studied the coupling mechanism between urbanization systems and ecosystems [2]. Howard delved into the interaction mechanism between urbanization and environmental evolution [3]. On the other hand, Deosthali used simulation to assess the impact of urbanization on the local climates of cities [4]. Vester uncovered the mechanism linking urban economic growth and its environmental evolution [5]. Girmm explored the correlation between changes in urban landscape ecology and global change [6]. Berry identified the primary factors impacting urban ecology due to urbanization. As for research methods, current studies primarily employ disciplines such as economics, ecology, biology, and physics. Howard implemented a system dynamics model, Berry utilized ecological factor analysis, Deosthali leveraged bioclimatic indices, and Vester applied sensitivity models. Overall, the regional scale of foreign research is larger, focusing on exploring the general rules of urbanization and ecosystems in long-time serial variation.

The domestic research began in the 1980s. In 1979, Wu Chuanjun innovatively proposed the theory of regional system of the man–land relationship, underlining the importance of geographical studies focusing on the interaction and negative feedback between humanity and nature within the man–land system [7]. As articulated by Lu Daodao, studying regional human–earth systems requires an understanding of their dynamic changes across different stages of societal development, necessitating an integrated qualitative and quantitative approach and advocating for a harmonious relationship between humans and nature at varying scales [8,9]. Domestic research in this field has also yielded a wealth of findings. Concerning research content, scholars have deployed mathematical models to elucidate various relationships between urbanization and ecosystems, such as "positive", "negative", or "inverted-U" relationships [10–13]. Evaluations of urbanization are predominantly conducted from the perspectives of population movement [14,15], industrial agglomeration [16,17], expansion of construction land [18,19], and infrastructure development [20,21]. Assessments of ecosystem services typically rely on the value scale formulated by Costanza and others [22], with the value of ecosystem services determined by continuously refining the value equivalent factor [23]. As for research methodologies, most existing studies have employed mathematical and statistical models such as regression analysis [24,25], input-output models [26,27], system dynamics models [28,29], or spatial analysis models. Regarding study area selection, empirical studies have mainly concentrated on provincial or municipal levels, predominantly targeting economically advanced regions, such as the Yangtze River Delta, Pearl River Delta, Beijing-Tianjin-Hebei region, and Chengdu-Chongqing region. Overall, contemporary domestic research on the relationship between urbanization and ecosystems exhibits three key characteristics. Firstly, the assessment dimensions of urbanization tend to be somewhat singular, primarily encompassing population aggregation, economic growth, and construction land expansion, with inadequate attention given to the comprehensive benefits of urbanization. Secondly, an over-reliance on statistical data often overlooks the natural attributes of ecosystems, which may result in the assessment findings not accurately reflecting the actual regional ecosystem conditions. Thirdly, the limitations of one-way research remain unaddressed, lacking the analysis of factor relationships and subject behaviors guided by synergistic ideas. The research primarily centers around the ecological effects prompted by urbanization, with insufficient focus on the feedback mechanism and mode of action of the ecosystem.

Ecosystem services refer to the conditions and processes that ecosystems and their species can provide to humans to satisfy and sustain their needs [30]. It is a frontier area of research in ecology and geography, and a link and bridge that connects natural and hu-man processes [31]. For the purpose of identifying regional ecosystem service issues, preserving regional ecological balance, and advancing regional sustainable development, it is crucial to explore the intrinsic interaction mechanism between the external spatial and temporal evolution of ecosystem services and the economic society [32–34]. We introduced an exponential efficacy function model to quantitatively measure the development of economic and social systems based on pertinent studies. Meanwhile, we used multivariate data and the InVEST model to analyze the regional ecosystem services and explore the current development status and coupled coordination of the complex ecosystem of the LXUA. The purpose of this research is to clarify the synergistic evolution mechanism of the man–land relationship in the ecologically sensitive area of the Upper Yellow River and provide a reference for ecological protection and high-quality development.

2. Materials and Methods

2.1. Study Area

The LXUA, with coordinates of 34°26′ N–37°38′ N, 98°55′ E–105°55′ E, is the westernmost town-dense region in the Yellow River basin. It is situated below Longyangxia, in the basin of the Yellow River and Huangshui River valley (Figure 1). The LXUA, which spans 97,500 km^2, consists of 39 counties in 9 cities, including Lanzhou, Xining, and Haidong. Mountains and river valleys dominate the region's terrain, which is complicated and varied. With an average height of 2000 m or more, the elevation varies from 1258 to 5255 m. In 2020, the GDP of LXUA reached 61.4 billion RMB, accounting for 51% of the GDP in the two provinces. The population reached 12.19 million, accounting for 66.5% of the permanent population in the two provinces. The city group is rich in hydraulic resources; climate geographic distribution differences; thick soil; and complex and diverse vegetation types, among which the eastern agricultural area of Qinghai is located in the Huangshui and Yellow River basin triangle. It has fertile soil, a mild climate, and a wealth of natural resources that are advantageous for developing agriculture and animal husbandry. Lanzhou, Xining, Huangzhong, Datong, Xunhua, etc. are important towns on the ancient Silk Road transportation route. Lanzhou is known as the "heart of the land", Xining is the "Pearl City" on the Qinghai-Tibet Plateau, and the two cities are the "growth poles" to promote the population clustering and economic development of the urban agglomeration.

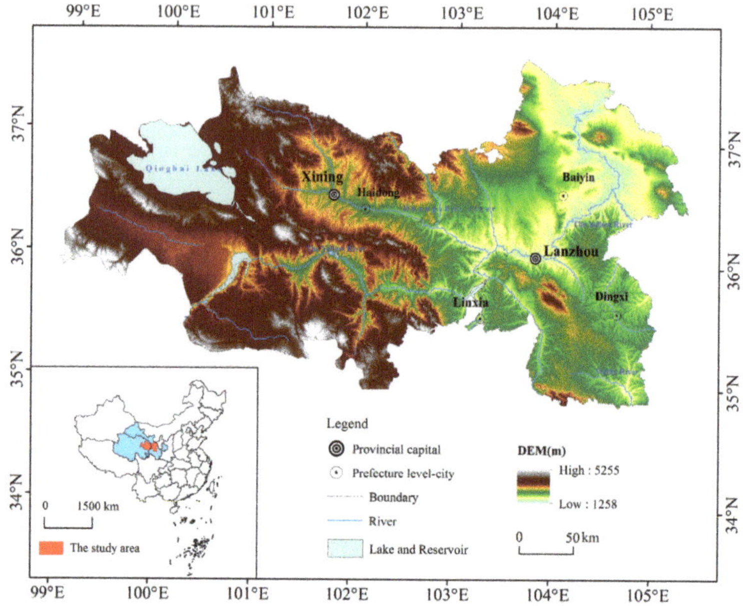

Figure 1. Diagram of the study area.

2.2. Analysis Framework

Urbanization and ecosystems together make up a complex system that is a synthesis of ecological processes brought about by the interaction of human social, economic, and cultural actions with the environment [35]. The two subsystems are connected and engage in interactions with one another in terms of amount, structure, order, quantity in space, and time. The quality of urbanization refers to a comprehensive concept with a rich connotation that reflects the quality of urbanization in the process of urbanization [36]. In this paper, evaluation indicators were primarily created in seven dimensions: economic development, people's lives, environmental protection, infrastructure, public services, urban vitality, and relationships between urban and rural areas [37–39] (Table 1). The regional ecosystem condition is an important basic condition for the smooth promotion of urbanization. The LXUA is an essential strategic support for maintaining China's ecological security and is situated in the crucial zone of transition from the first to the second terrain in China. It contains significant ecological security barriers such as the Source Region of Three Rivers, Qilian Mountains, and Gannan Plateau. Water provision services refer to the interception, infiltration, and storage of precipitation by ecosystems through their unique structures and water interactions, and the effective regulation of water circulation through evapotranspiration. The water provision services of the LXUA play an important role in mitigating surface runoff, supplementing groundwater, mitigating seasonal fluctuations in river water flow, and ensuring water source quality. Meanwhile, vegetation can effectively reduce the impact of precipitation on soil. Plant roots are intertwined with the soil, which can effectively fix the soil. The soil conservation services are crucial to reduce soil erosion, maintain soil fertility, prevent and control desertification, and reduce the occurrence of geological disasters such as landslides and debris flows. Carbon fixation services refer to the conversion of atmospheric carbon dioxide into organic carbon through photosynthesis, which are fixed in the plants or soil. Carbon fixation services can reduce the concentration of greenhouse gases such as carbon dioxide in the atmosphere, which plays an important role in maintaining the carbon oxygen balance and slowing down global warming. Integrated ecosystem services are the overall manifestation of various service functions and an important indicator reflecting the quality and condition of regional ecosystems. These four types of ecosystem services have an important impact on the ecological security of the Upper Yellow River and the country as a whole. Therefore, the study was based on the InVEST model, and the four aspects of water provision, soil conservation, carbon fixation, and integrated ecosystem services were selected for assessment. The analysis framework is shown in Figure 2.

Table 1. Indicators of urbanization quality.

Functional Layer	Indicator Layer	Attribute	Weight
Economic development level	Per capita GDP (RMB)	+	0.0422
	Per capita retail sales of consumer goods (RMB)	+	0.0368
	Per capita investment in fixed assets (RMB)	+	0.0454
	The proportion of tertiary industry (%)	+	0.0396
People's living standards	Employment rate of urban population (%)	+	0.0374
	Urban per capita disposable income (RMB)	+	0.0346
	The average wage of workers (RMB)	+	0.0227
	Urban per capita water consumption (ton)	−	0.0431
Environmental protection	Greening coverage of built-up area (%)	+	0.0424
	Comprehensive utilization rate of industrial solid waste (%)	+	0.0351
	Urban domestic sewage treatment rate (%)	+	0.0318
	Harmless disposal rate of domestic waste (%)	+	0.0347
Infrastructure	Residential area per capita (m^2)	+	0.0406
	Road area per capita (m^2)	+	0.0428
	Broadband penetration (%)	+	0.0251
	Gas penetration (%)	+	0.0242
	Park area per capita (m^3)	+	0.0352
Public service	Number of teachers per 10,000 people	+	0.0295
	Number of hospital beds per 10,000 people	+	0.0371
	Pension insurance coverage (%)	+	0.0355
	Number of books in a library	+	0.0439
Urban vitality	Population per square kilometer	+	0.0504
	Number of public buses per 10,000 people	+	0.0379
	Gross Domestic Product (RMB 10,000)	+	0.0299
	Number of enterprises with considerable scale	+	0.0466
Urban–rural relations	Average income ratio between urban and rural residents	−	0.0359
	Average consumption expenditure ratio between urban and rural residents	−	0.0396

The process of urbanization is one of the most important manifestations of the development and evolution of human society. The ecosystem is a natural background and supporting system for the subsistence and multiplying of human beings. Urbanization is closely related to ecosystems, and both are important components of the regional man–land system. The impact of urbanization on the ecosystem is bidirectional, with both negative stress and positive promotion effects. The coercive effect of urbanization on the ecosystem refers to the phenomenon of environmental pollution, imbalance between resource supply and demand, reduction of biodiversity, and degradation of ecosystem functions when the amount and speed of waste discharged by cities into the hinterland environment through production and living reach or exceed the speed of ecological environment decomposition and digestion. The promoting effect of urbanization on the ecosystem refers to the investment of more resources such as technology, funds, and manpower into urban economic construction activities within the range of ecological environment capacity and carrying capacity. Through policy intervention and the promotion of clean production technology, the economic development mode is transformed, resource utilization efficiency is improved, pollution emissions are reduced, citizens' lifestyles are transformed, the low-carbon and green development of cities is achieved, and the quality of the living environment is improved. The ecosystem also has a positive and negative impact on urbanization. On the one hand, resources such as water, soil, energy, and minerals mainly constrain urban scale, affect urban layout, limit urban industrial structure, and influence the speed of urbanization development, thereby exerting constraints on various aspects of urbanization; on the other hand, ecosystem services such as water provision, soil conservation, and carbon fixation are fundamental for supporting and guaranteeing conditions for urban development and residents' lives.

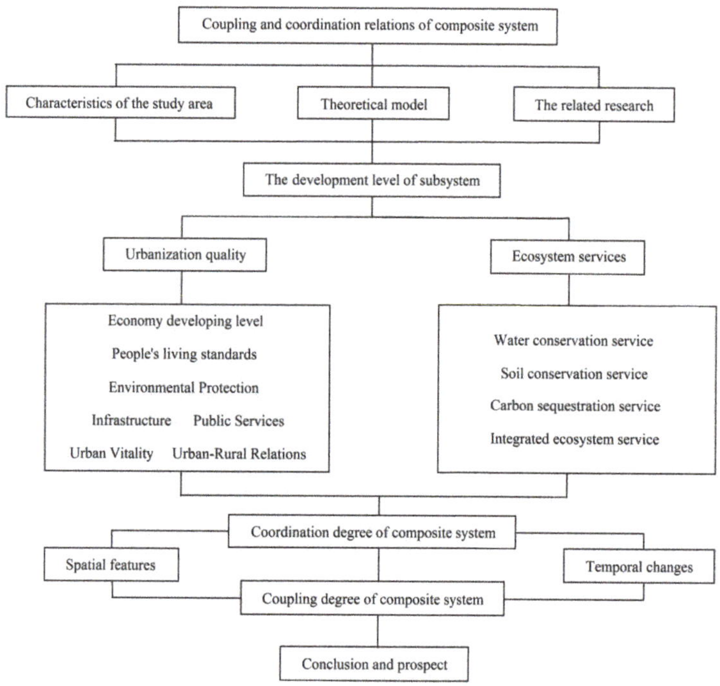

Figure 2. Analysis framework.

2.3. Data Sources

This study created an ecosystem classification map of the LXUA using the 100 m × 100 m land-use remote-sensing monitoring data in 2000 and 2020 provided by the Resource and Environment Science and Data Center (https://www.resdc.cn accessed on 22 August 2022) (Figure 3).

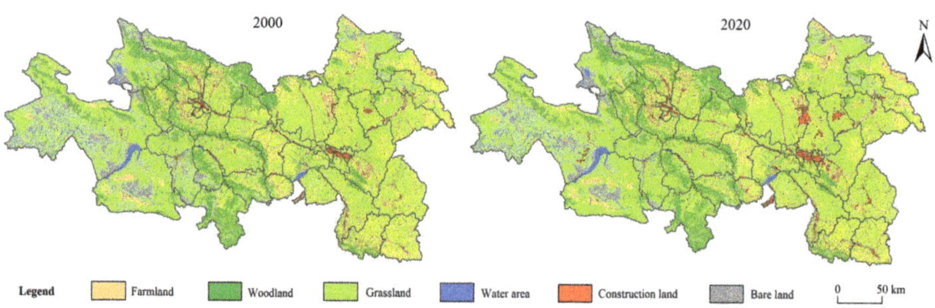

Figure 3. Land-use types of LXUA in 2000 and 2020.

The socio-economic data used in this study were mainly obtained from "China Statistical Yearbook", "Gansu Development Yearbook", "Qinghai Statistical Yearbook", "Gansu Urban Yearbook", statistical yearbooks, and statistical bulletins of various cities. Basic geographic data, including roads, rivers, administrative boundaries of counties, were sourced from the National Basic Information Center. Meteorological data, including temperature, precipitation, water pressure, etc., were obtained from the China Metrological Data Service Centre (http://data.cma.cn accessed on 28 August 2022), and the local 5-year average meteorological data were taken for the calculation considering the interannual fluctuation of the data. the soil data were sourced from National Earth System Science Data Center (http://www.geodata.cn accessed on 12 September 2022).

2.4. Methods

2.4.1. Exponential Efficacy Function Model

In previous studies, the linear efficacy function model was usually used to standardize the data. The linear efficacy function model defines the change of indicators as "uniform change", which is a relatively simplified form of processing. However, in the normal course of economic and social development, if an indicator increases continuously and reaches a certain number and scale, the actual utility provided by it will typically decrease over time, much like the well-known law of diminishing marginal utility. It will then be more challenging to maintain the indicator's growth or progress. The derivative of exponential efficacy function model is a lower convex function about independent variable. In practical applications, the exponential efficacy function model is chosen to better fit the development process and trend of the data. The formula is as follows:

$$d = Ae^{(x-x^s)/(x^h-x^s)B} \tag{1}$$

where d is the efficacy score, x^s is the theoretical minimum value, x^h is the theoretical maximum value. The parameters A and B can be determined by the critical points. When the index value and the theoretical minimum are the same, according to the linear efficacy function method, set d = 60, then A = 60. When the index value and the theoretical maximum are the same, set d = 100, then B = −ln0.6 [40,41].

$$d = 60e^{-(x-x^s)/(x^h-x^s)\ln 0.6} \tag{2}$$

2.4.2. The Entropy Method

The entropy method is an objective weighting method that determines the weight according to the dispersion degree of the indicator, which can deeply reflect the utility value of the indicator information and avoid the interference of human factors in the evaluation process. The greater the dispersion of the indicator value, the smaller its entropy value, the greater the amount of information provided by the indicator, and the greater the weight. The specific calculation process is as follows:

$$Y_{ij} = x_{ij}/\sum_{i=1}^{m} x_{ij} \tag{3}$$

$$k = \frac{1}{\ln m} \tag{4}$$

$$e_j = -k\sum_{i=1}^{m}(Y_{ij}\ln Y_{ij}) \tag{5}$$

$$D_j = 1 - e_j \tag{6}$$

$$W_j = D_j/\sum_{j=1}^{n} D_j \tag{7}$$

$$U_a = \sum(D_j \times W_j)(a = 1,2,3) \tag{8}$$

In the formula, m and n represent the number of samples and indicators, respectively, Y_{ij} is the i sample of the j indicators accounted for the proportion of the indicator of the total sample value. e_j, D_j and W_j are the entropy value, variability coefficient and weight value of the j indicator, and U_a is the subsystem orderliness. The specific weights of each index are shown in Table 1.

2.4.3. InVEST Model

The InVEST model is based on the distributed algorithm of "3S" technology. The spatial representation, dynamic assessment and quantitative evaluation of ecosystem service functions can be carried out quickly and accurately.

a. Water provision

Water provision is equal to the difference between precipitation and evapotranspiration, which is obtained from the water yield module of the InVEST model. The model is based on the Budyko water and heat coupling equilibrium assumption, taking into account factors such as terrain, climate, soil layer thickness and permeability [42]. The calculation formula is:

$$Z_{ij} = (1 - AETi_j/P_j) \times P_j \tag{9}$$

where Z_{ij} denotes the annual water yield of land-use type j in grid i (mm). $AETi_j$ represents the annual actual evapotranspiration of land-use type j in grid i (mm). P_i represents the average annual precipitation of grid i (mm).

b. Soil conservation

The modified general soil loss equation can be used to determine soil retention, which is equal to the difference between the amount of possible soil erosion and the amount of potential soil loss. The following is the calculation formula:

$$SD = R \times K \times LS \times (1 - C \times P) \tag{10}$$

where SD denotes the amount of soil conservation ($t \cdot hm^{-2} \cdot a^{-1}$). R is rainfall erosivity ($MJ \cdot mm \cdot hm^{-2} \cdot h^{-1} \cdot a^{-1}$). K is soil erodibility ($t \cdot hm^2 \cdot h \cdot hm^{-2} \cdot MJ^{-1} \cdot mm^{-1}$). LS is the gradient and slope length factor calculated by DEM. C is vegetation coverage and management factor. P is the engineering measure factor.

c. Carbon fixation

Carbon storage represents the carbon fixation capacity of terrestrial ecosystems. The calculation formula is:

$$C = C_{above} + C_{below} + C_{soil} + C_{dead} \tag{11}$$

where C is the underground carbon storage ($t \cdot hm^{-2} \cdot a^{-1}$). C_{above} is the aboveground carbon storage ($t \cdot hm^{-2} \cdot a^{-1}$). C_{below} is the underground carbon storage ($t \cdot hm^{-2} \cdot a^{-1}$). C_{soil} is the density of soil organic matter ($t \cdot hm^{-2} \cdot a^{-1}$). C_{dead} is carbon storage of litter ($t \cdot hm^{-2} \cdot a^{-1}$).

d. Integrated ecosystem services

We constructed the comprehensive index of regional ecosystem services by using the dispersion coefficient method, and the geometric average method was used for grid cells [43]. The calculation formula is:

$$ES_i = \frac{\sigma_{ik}}{x_{ik}} = \frac{1}{x_{ik}}\sqrt{\frac{1}{N}\sum_{k=1}^{N}(x_{ik} - \overline{x_{ik}})^2} \tag{12}$$

where ES_i is the ecosystem services of the i-th grid cell. σ_{ik} is the amount of the k-th ecosystem services on the i-th grid cell. x_{ik} is the normalized value of ecosystem services of the k-th category on the i-th grid unit in the region. $\overline{x_{ik}}$ is the average of normalized values of the k-th ecosystem services on the i-th grid cell. N is the main ecosystem service category.

2.4.4. Coupling Coordination Degree Model

Originally a physical concept, coupling refers to the phenomenon of two or more systems affecting each other through various interactions. The coupling degree can describe influence between systems or elements. The coupling action determines the state and structure of the system when it reaches the critical region, which determines the trend of the system from disorder to order. The impact of urbanization quality and ecosystem services interaction is defined as the coupling degree, which reflects the order of the system and the interaction strength between subsystems. The calculation formula is:

$$C = \left[\frac{\prod_{i=1}^{n} U_i}{\left(\frac{1}{n} \sum_{i=1}^{n} U_i \right)} \right]^{\frac{1}{n}} \tag{13}$$

where n is the number of subsystems; U_i is the value of each subsystem, C is the coupling degree, and the value of C is between 0 and 1. According to the value range of the coupling degree, it can be divided into four stages. When the CD is below 0.3, the system is in the low coupling stage; when the CD is between 0.3 and 0.5, the system is in the antagonistic stage; when the CD is between 0.5 and 0.8, the system is in the running-in stage; and when the CD is between 0.8 and 1.0, the system is in the high coupling stage. The measurement functions of urbanization quality and ecosystem services at stage t are $f\ (t, x)$ and $g\ (t, y)$, where x and y are the evaluation indexes of the two systems, respectively. The formula is as follows:

$$C = \left\{ [f(t,x) \times g(t,x)] / \left[\frac{f(t,x) + g(t,x)}{2} \right]^2 \right\}^{\frac{1}{2}} \tag{14}$$

$$D = \sqrt{C \times T} = \sqrt{C \times [af(t,x) \times bg(t,x)]} \tag{15}$$

The coupling degree can indicate the strength of interaction between urbanization quality and ecosystem services, but it is not possible to judge whether the coupling status is benign or not. When the development level of both systems is low, a high coupling degree can still be obtained. The degree of coordination adds a development coefficient to the degree of coupling, which is a composite reflection of the order of internal structure and the external scale of the system, where D is the coordination degree; C is the coupling degree; a and b are the weights. In this study, it is considered that both urbanization quality and ecosystem services are crucial to the evolutionary development of the composite system, and it was more beneficial to compare the actual conditions of different county subsystems horizontally by assigning the same pending coefficients to each system on the basis of the control variables [44], so a and b were each assigned weight of 0.5.

3. Results

3.1. Status of Subsystem Development

3.1.1. Urbanization Quality Subsystem

The enhancement of the LXUA's urbanization quality is crucial for advancing the coordinated development of Northwest China and serving as a crucial assurance for the construction of a multiethnic demonstration area of shared prosperity. According to the seventh national census, the resident population of the LXUA was 12.47 million, accounting for 40% of the total population of Gansu and Qinghai, among which the urbanization rate of the resident population in Lanzhou and Xining was 83.1% and 78.63%, respectively. In terms of distribution pattern, the urbanization quality of the LXUA in 2000 was in the barbell-shaped "double core" pattern, with areas higher than 0.4 mainly concentrated in Lanzhou and Xining urban areas, in the eastern and western cores of the urban agglomeration (Figure 4). At the same time, the urbanization quality of Baiyin and Linxia, the municipal and prefectural government locations, was also higher than 0.4. Compared with 2000, the number of medium- and high-value areas of urbanization quality in LXUA increased in

2020, including Haiyan of the Haibei urban area and Ping'an of the Haidong urban area, which had increased to more than 0.4; and the main urban areas of Lanzhou and Xining, and Baiyin increased to more than 0.6. It is worth noting that there were no counties with an urbanization quality higher than 0.6 in 2020 in Dingxi, Hainan or Huangnan urban areas within the city cluster.

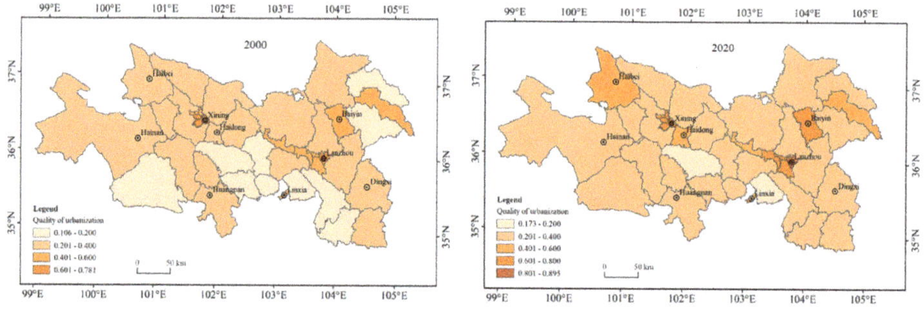

Figure 4. Urbanization quality of the LXUA in 2000 and 2020.

On the whole, the urbanization quality of the LXUA has fewer areas with medium and high values, and the development of sub-central cities in the urban agglomeration is relatively insufficient. At the same time, the average urbanization quality of all counties in Lanzhou in 2020 was only 0.536, while that of Xining was 0.539. As the leading and core of the urban agglomeration, its development potential has not been fully released, and the leading and driving effects of both on regional urbanization development need to be further strengthened. The urbanization quality of 25 counties in the urban agglomeration in 2020 was between 0.2 and 0.4, accounting for 64% of the total number of counties; among them, 10 counties had scores less than 0.3. The counties in the LXUA have a single level of economic development, insufficient development of secondary centers, and obvious characteristics of an extensive distribution of low-level counties. At present, Lanzhou and Xining are still in the stage of polarized development: the "siphon effect" is obvious, a large number of production factors are gathered in the core area, the development vitality of the secondary core areas around the provincial capital is insufficient, and the economic radiation and driving capacity of the central cities need to be strengthened.

In terms of temporal changes, the average urbanization quality of the counties in the LXUA was 0.328 in 2000, rising to 0.404 in 2020, with an overall increase of 23.17% and an average annual increase of 1.16%. There were big differences in the improvement of each county, among which Gaolan, Yuzhong, Jingyuan, Weiyuan, Dongxiang, Jishishan and Ledu's urbanization quality increased by more than 40%; and that of Yongdeng, Anding, Longxi, Lintao, Yongjing, Huangyuan and Huangzhong increased by more than 30%, mostly in the eastern Gansu section of the urban cluster (Figure 5). The junction of Yongdeng and Gaolan, Lanzhou New, was established in 2010. With the gathering of population and industry, infrastructure construction and production, and living services enhancement, its urbanization quality has achieved obvious improvement. Lintao, Yuzhong and Yongjing, close to Lanzhou, are radiated and driven by the core of the provincial capital, and the urbanization quality is also improved at a faster rate. The number of counties with an increase in urbanization quality of 20% to 30% was the largest, accounting for one-third of the total number of counties, mainly distributed in the urban agglomeration Qinghai area. The counties with less than a 20% increase were concentrated in the main urban areas of Lanzhou and Xining, as well as Baiyin, Pingchuan and Linxia. The urbanization quality of such counties in 2000 was higher than other surrounding counties. In the case of the same absolute value of growth, the larger the previous base, the lower the increase will be; at the same time, the development of regional urbanization also conforms to the law of marginal diminution, and the higher its development degree, the more difficult it will be to improve it.

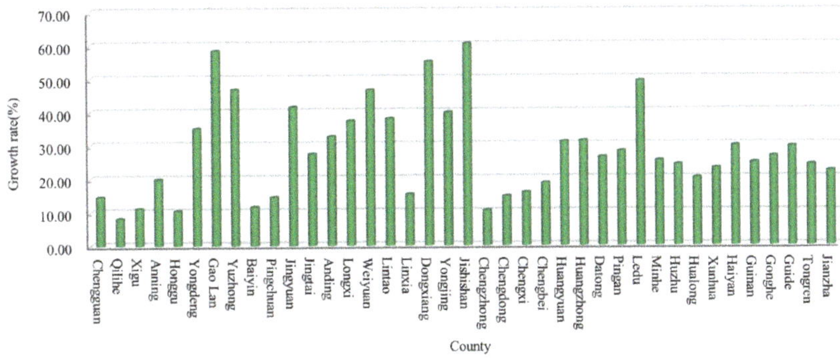

Figure 5. Growth rate of urbanization quality in the LXUA from 2000 to 2020.

3.1.2. Ecosystem Service Subsystem

Water provision services refer to the ability of an ecosystem to intercept or store water resources from rainfall. From 2000 to 2020, the water provision services of the LXUA improved significantly. The area of the high-value area in the southwestern Sanjiangyuan region had increased most significantly. The Laji Mountains and the Lianhua Mountains in the southeast evolved from the median area to the higher-value area. At the same time, the proportion of low-value areas for water provision services had increased in Yongdeng, Gaolan and Yuzhong counties in the northeast of the urban agglomeration (Figure 6); in 2020, water provision services increased from the northeast to the southwest, and high-altitude mountains became the main area for improving water provision services.

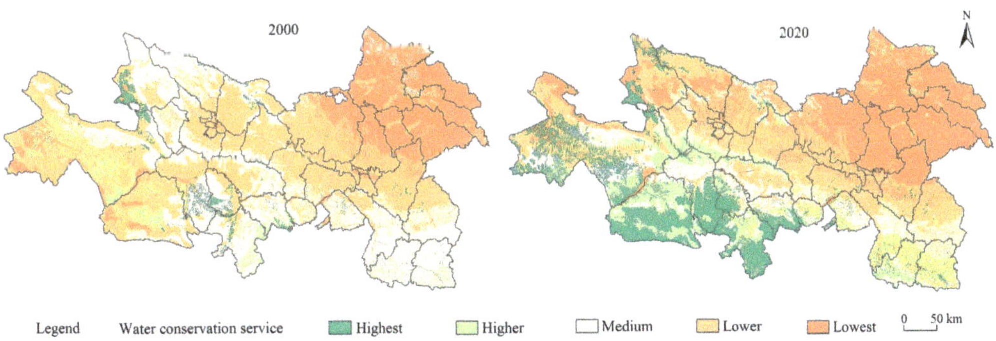

Figure 6. Water provision service of LXUA in 2000 and 2020.

Soil conservation services refer to the ability of the ecosystem to hold the soil in a given time, which is an important guarantee for regulating water and soil loss, preventing soil degradation, and reducing the risk of geological disasters. From 2000 to 2020, the overall change of soil conservation services was not significant, and the areas to be promoted were mainly concentrated in the west of urban agglomeration (Figure 7). The median area along the central Laji Mountains and the southern Xiqing Mountains increased, while the area of the low-value area decreased. In 2020, soil conservation services mainly served in low- and middle-value areas, with a single hierarchical structure and a lack of high-value areas, showing a differentiation pattern that was slightly higher in the middle and lower in the east and west.

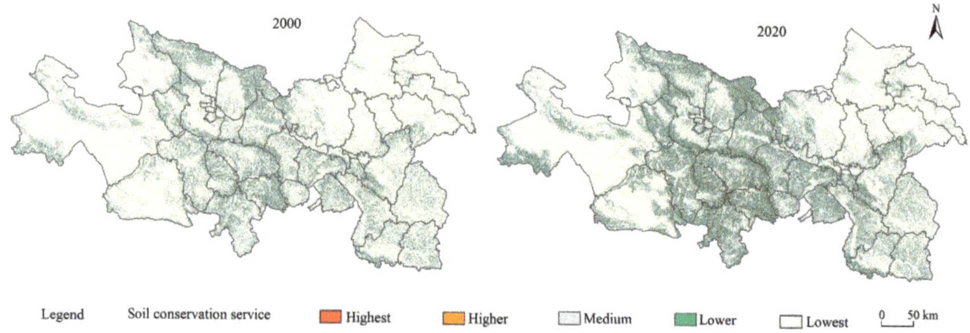

Figure 7. Soil conservation service of LXUA in 2000 and 2020.

Carbon fixation services refer to the capacity of terrestrial ecosystem to store carbon. There are two common ways of carbon fixation in nature: one is the photosynthesis of green plants and the other is the chemosynthesis of microorganisms such as nitrifying bacteria. The former, as the main mode of carbon fixation, is closely related to regional vegetation coverage and land-use type. The high-value areas of carbon fixation service in LXUA were mainly located in Daban Mountains in the north, Laji Mountains in the middle, the southern Xiqingshan residual vein and around Xinglong Mountains in the east (Figure 8). The low-value areas were mainly concentrated in the desert, water area and urban built-up areas of municipalities in the southwest. In terms of spatial distribution, the carbon fixation services in 2000 were mainly in the median area, with a relatively single hierarchical structure. In 2020, with the accelerated transformation of land-use types, the spatial distribution of carbon fixation services became more complex. The high and the low values were staggered, and the trend of "fragmentation" and "fragmentation" was obviously intensified.

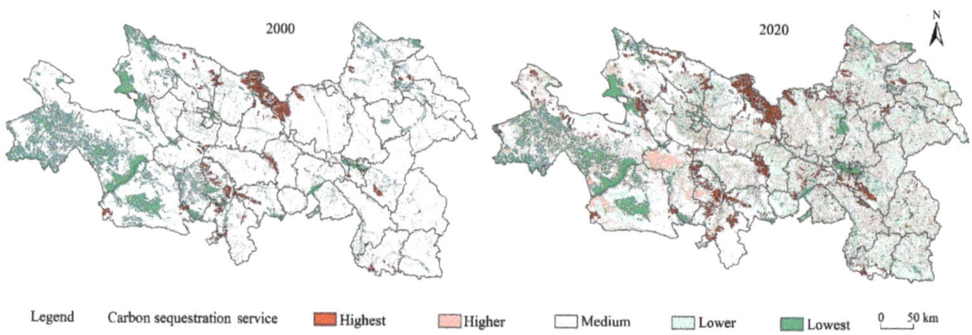

Figure 8. Carbon fixation service of LXUA in 2000 and 2020.

Overall, the integrated ecosystem services of the LXUA showed an overall stepped distribution pattern from northeast to southwest (Figure 9). The highest-value areas were mainly located in Hainan and Huangnan in the southwest. The terrain was dominated by high mountain landforms, with towering terrain and continuous mountain systems. The highest-value areas in the eastern urban agglomeration were small, scattered in the Maxian Mountains at the boundary between Yuzhong and Lintao, and Lianhua Mountains in the south of Weiyuan urban area. Through comparative analysis, the high-mountain area with an altitude of more than 2500 m in the urban agglomeration coincided with the high-value area of ecosystem services, showing a strong correlation. The land-use types in mountainous areas were mainly forestland and grassland. Their vegetation coverage was high, their root system was developed, and their ability to conserve water and soil was strong. Compared with 2000, the area of high-value ecosystem services in the southwest of urban agglomeration increased significantly in 2020. As an important part of the ecological barrier area of the Qinghai Tibet Plateau, Hainan and Huangnan are also the birthplace of the Yellow River and an important supply area of freshwater resources in China. Driven by the National Sanjiangyuan Ecological Protection and Construction Phases I and II, the forest coverage and wetland area in the region has increased significantly in the past two decades. The service level of regional integrated ecosystem had been significantly improved.

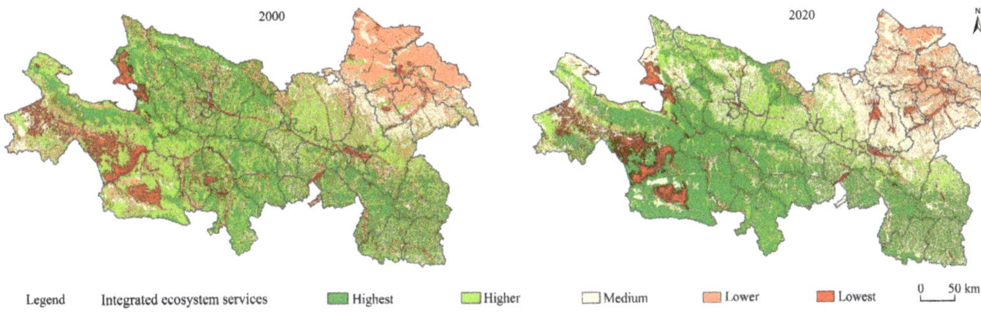

Figure 9. Integrated ecosystem services of LXUA in 2000 and 2020.

In 2020, the higher-value areas of ecosystem services were distributed along the Huangshui River, showing a belt pattern from northwest to southeast. Topographically, it mainly included Huangshui Valley, Lanzhou Basin, Zhuanglang Valley and Yuanchuan Valley. The land usage in this area is primarily made up of arable land and forest land, with the arable land being primarily dispersed along the rivers. These areas of the region are relatively low in elevation and have plentiful inflow water supplies. Crops, herbs and shrubs in cultivated land and forest land can increase soil organic matter content. The climate is regulated through carbon fixation and oxygen production, so as to reduce surface water and soil loss and effectively conserve water. The higher-value areas are strongly affected by human activities. So, the ecological environment of the valley basin should be effectively protected to reduce ecological damage and environmental pollution.

The medium- and lower-value areas of ecosystem services were located in the northeast of the LXUA, including Baiyin and Yongdeng, Gaolan and Yuzhong of the Lanzhou urban area. Land use in this area was mainly farmland and grassland. This region is located at the northwest edge of the Longxi Loess Plateau and the transitional zone from the eastern extension of the Qilian Mountains to the Tengger Desert. In this region, the climate is relatively dry, the ground vegetation is sparse, and the ecosystem services are under great pressure. The lowest-value areas of ecosystem services were highly overlapped with the water area and unused and construction land. The western part of the urban agglomeration was scattered in the Mugetan Desert of Guinan, the Tala Beach of Gonghe, the sand island at the northeast of Qinghai Lake and the Longyang Lake at the boundary between Gonghe and Guinan. The lowest-value areas in the east were mainly located in the urban built-up areas of each city, including the main urban areas of Lanzhou, Lanzhou New Area, and Baiyin.

3.2. *Coupling and Coordination Relationship of Composite System*

3.2.1. Coupling Degree (CD) of Composite System

According to the principle of entropy increase, the system always moves from order to disorder, and the process of entropy increase is irreversible. In the composite ecosystem, composed of urbanization and ecosystems, economic, social, and ecological subsystems continue to generate biomass cycles, energy flows, and information exchanges, promoting the evolution of the complex ecosystem towards disorder. According to the synergy theory, the synergy between elements leads to overall system identity, structural stability, evolutionary ordering, and functional optimization, which are the main forces that slow down the entropy increase process. The CD is a comprehensive measure of system synergy. According to the calculation results and relevant research [45], the CD is between 0.6 and 0.8, indicating that the system is in the running-in stage; and if the CD is between 0.8 and 1.0, the system is in a high coupling stage.

The CD between urbanization quality and ecosystem services in the LXUA was 0.793 in 2000 and 0.842 in 2020, with an overall growth of 6.18% over the past two decades. In terms of changes in each county, 31 counties in the urban agglomeration showed varying degrees of growth in coupling, with 11 counties rising from the grinding stage to the highly coupled stage (Figure 10). The growth of the CD reflects the improvement of the internal order degree of the composite system, that the overall evolution trend of the subsystems is consistent, and that the interaction is close. The CD of the eight counties decreased, with a decrease of less than 5%, mainly located in the centers of various cities. After 2000, the quality of urbanization in such counties has steadily improved, but the population is relatively dense, the proportion of construction land is relatively high, the area of natural vegetation is small, and ecosystem services are in a downward trend. The development direction of subsystems is contrary, making the CD of composite systems show a downward trend.

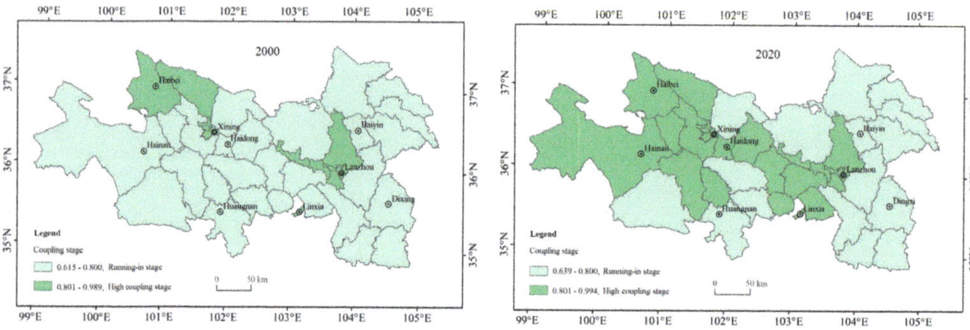

Figure 10. The coupling degree of composite ecosystem in LXUA in 2000 and 2020.

From the perspective of distribution pattern, in 2020, the CD of the Qinghai area in the west of the LXUA was higher than in the east. In 2020, there were 15 counties with a CD above 0.8, which were in a high coupling stage, mainly concentrated in Xining Haidong and Linxia urban areas. The CD of 24 counties was between 0.6 and 0.8, which indicates the running-in stage. The counties in the running-in stage can be divided into two categories in total. The first is Dingxi and Baiyin, which are affected by natural factors such as climate and terrain, and their ecosystem services are at a low level. Compared to the rapid development of urbanization, the ecosystem service capabilities centered on water conservation, soil conservation, carbon fixation and oxygen release have not been effectively improved, and even some regions have shown a downward trend. The low level of ecosystem services is the main reason affecting the CD of the system. The other is the counties in the southwest of the urban agglomeration, such as Guinan and Tongren. These areas are located in the Nature Preservation Zone of Sanjiangyuan, with abundant water resources, a wide variety of biological species, and a good ecological environment. In recent years, with the increase of national protection efforts and the promotion of relevant policies, the regional ecosystem service capacity has been significantly improved. However, due to the high altitude and relatively sparse population in this region, the development of urbanization is limited to some extent. The urbanization quality subsystem is significantly lagging behind the ecosystem service subsystem, and there is significant room for improvement in coupling.

3.2.2. Coupling Coordination Degree (CCD) of Composite System

The degree of coupling can better represent the overall order of the composite system and the strength of the interaction between the internal subsystems, but it is impossible to judge whether the coupling is benign; that is, when the quality of urbanization and the level of ecosystem services are low, a higher degree of coupling can still be obtained. The concept of ecological civilization emphasizes the interdependence and coexistence of production, life, and ecological development. If a subsystem in a composite system is in poor condition, it will affect the overall development of the system. Only when the two are at a high level and have a high degree of coupling can system coupling truly be achieved. The introduction of coordination as a comprehensive measure of coupling and development level can reflect the degree of order of the internal structure of the system and the accumulation of external scale. According to relevant research results, the coordination degree was divided into 10 intervals (Table 2), and the lowest scoring subsystem was defined as a lagging subsystem.

Table 2. Classification criteria for coordination degree.

Number	1	2	3	4	5
Coordination degree	0–0.09	0.10–0.19	0.20–0.29	0.30–0.39	0.40–0.49
Coordination level	Extreme disorder	Serious disorder	Moderate disorder	Mild disorder	On the verge of disorder
Number	**6**	**7**	**8**	**9**	**10**
Coordination degree	0.50–0.59	0.60–0.69	0.70–0.79	0.80–0.89	0.90–1.00
Coordination level	Barely coordination	Primary coordination	Intermediate coordination	Well coordination	High quality coordination

In 2000, the average coordination degree between urbanization quality and ecosystem service in 39 counties of the LXUA was 0.464. In 2020, it was 0.493, an overall increase of 6.25%. Among them, the ecosystem service index in 31 counties was lower than the urbanization quality index. The coordination degree of various counties ranged from 0.372 to 0.753, with a total of 11 counties higher than 0.5, including the intermediate coordination area Chengguan in Lanzhou urban area, the primary coordination area Qilihe and Anning, the central and western urban area of Xining, and 6 barely coordinated areas (Figure 11). Such counties belong to the central urban areas of various cities and are also the core and secondary core of the development of urban agglomerations. They have a relatively good industrial foundation and a densely distributed urban population. Under the strong impetus of regional urbanization, the overall coordination degree of the composite system is at a high level. At the same time, 7 lagging subsystems in 11 counties, including Chengguan, Anding, and Chengzhong, are ecosystem service subsystems, and there is significant room for improvement in regional ecosystem services.

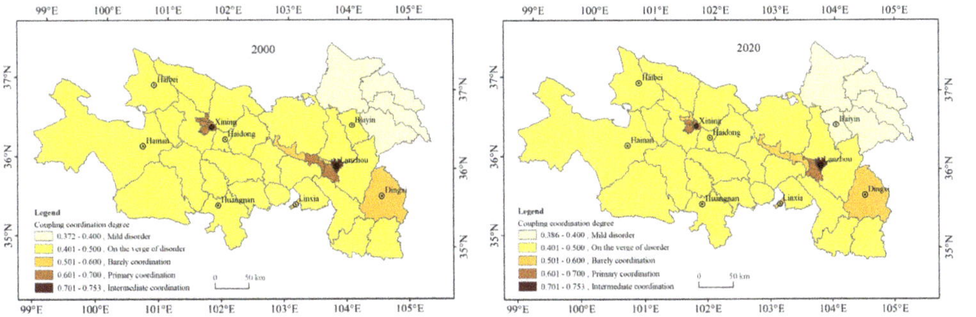

Figure 11. Coupling coordination degree of composite ecosystem in LXUA in 2000 and 2020.

In addition to Baiyin and the intermediate, primary, and barely coordinated areas, the remaining 24 counties were on the verge of imbalance. As the main type of urban agglomeration coordination level, its internal structure and external scale represent the overall state and development level of urban agglomeration. Except for Lintao, Weiyuan, and Datong, the urbanization quality index in the near-imbalance areas was significantly lower than the ecosystem service index. Based on the analysis of the status of subsystems and the degree of coupling, it was found that the key factor affecting the degree of coordination lies in the low comprehensive development score of the composite system, and more specifically, the low quality of urbanization. According to relevant indicators, the per capita GDP of the 24 counties on the brink of imbalance in 2020 was only about 60% of the national average. The proportion of high-tech industries is low, and the problems of industrial isomorphism and homogeneous competition in each county are more prominent. At the same time, the overall scale of each county is relatively small, mostly small- and medium-sized towns with less than 500,000 people, with many infrastructure weaknesses and low public service levels. Building a platform of distinctive and advantageous industries in counties, increasing their overall carrying capacity, and directing the orderly transfer of agricultural population are all essential to the high-quality development of counties in the current urban agglomeration.

The four counties belong to the mild-disorder area, mainly distributed in Baiyin urban area in the northeast of the urban agglomeration, including Baiyin, Pingchuan, Jingyuan and Jingtai, and the lagging subsystems are all ecosystem service subsystems. The CD of the four counties with mild disorder was at a relatively low level among the 39 counties, and the overall level of ecosystem services was low. Baiyin is located in the northwest of the LXUA, adjacent to the Tengger Desert in the north. The climate is relatively dry, with an annual rainfall of 180~450 mm and annual evaporation of 1500~1600 mm, and the forest coverage rate of the city is only 5.30%. Located in the transitional zone from the Loess Plateau to the inland denuded plateau, the ground is broken. The surface is mostly covered with loess and mainly composed of silty sand, with developed vertical joints, high porosity, and weak corrosion resistance. The fine soil covering layer on the surface is relatively thin and has a loose structure, which is prone to wind erosion and poplar blowing. At the same time, the backbone enterprises in Baiyin are still enterprises mainly engaged in raw materials and heavy chemicals, which were built during the "First Five-Year Plan" and "Third Line" periods, with high-energy-consumption industries such as nonferrous metals, chemical industry, thermal power, and building materials accounting for about 90%. The tasks of energy conservation and carbon reduction are relatively arduous. Strengthening environmental governance and protection and promoting the transformation and upgrading of industrial structure are currently the top priorities for the sustainable development of mild-disorder areas.

4. Discussion

By comparing previous studies, we believe that evaluating the scale of urbanization solely from population, economy, and land is not comprehensive in Northwest China. The assessment must consider aspects such as people's living standards, environmental protection, infrastructure, public service, urban vitality, and urban–rural relations. More than 60% of counties had an urbanization quality below 0.4 in 2020, indicating that the overall quality of urbanization in the LXUA is still low. It is necessary to change the development mode of simply pursuing the urbanization scale and focus on improving the quality and efficiency of urbanization. In the current research, statistical indicators are often used to evaluate the environmental condition, which can lead to higher evaluation results, reduce the spatial continuity, and enhance the mutability of the data. It is hard to reflect the real ecological status of the region using the evaluation results. Based on multivariate data and the InVEST model, we found that the overall spatial pattern of ecosystem services in the LXUA shows an increasing trend from northeast to southwest, while the level of ecosystem services is relatively low in city-and-town concentrated areas. The two systems in general show a "high coupling–low coordination" development status. This indicates that urbanization quality and ecosystem services are closely related, and the feedback effect is obvious in the northwest arid area. Many counties are still the antagonistic stage and the running-in stage between urbanization and ecosystems, and the overall coupling coordination degree is relatively low. It is necessary to reduce the negative impact of industry and life on ecosystems through policies, funds, technology, and other means.

The coupling mechanism between urbanization and ecosystems is very complex. This study starts from two aspects of urbanization quality and ecosystem services, and the evaluation and analysis of the composite system may still be incomplete. Subsequent research will continue to be supplemented and improved. The study area for this article is the LXUA. What are the different coupling mechanisms between this area and other regions? Whether the conclusions obtained in this study are applicable to other urban agglomerations, as well as other spatial scales at home and abroad, requires further research and verification in the future.

We found that there are significant differences in the development level of subsystems and the coupling coordination degree of composite systems in different regions. Each county should adopt measures suiting local conditions. The construction of public green space such as parkland, square land and urban protection greenbelt should be strengthened in the intermediate, primary, and barely coordinated areas. In addition, the regions also need to strengthen urban pollution control, promote waste recycling and low-carbon treatment, and improve urban resilience and ecological security guarantee capacity. Counties on the verge of disorder should rely on ecological security barriers such as the three-rivers source area and the Gannan Plateau, accelerate the construction of the ecological protection belt in the Upper Yellow River, and actively protect the Huangshui River, Datong River, Daban Mountains, Laji Mountains and other ecological corridors. Each county should actively improve its infrastructure and public services to enhance the quality of the county's residential environment. On the one hand, mild-disorder counties should strengthen their combat of desertification and water loss and governance of soil erosion, as well as reduce environmental risk and pollutant emissions. On the other hand, these counties should repair the abandoned mines, actively build the circular economy industrial system, and promote the green transformation of the economy and society.

5. Conclusions

This paper takes the LXUA in the Upper Yellow River as a case study area. We constructed a comprehensive evaluation index system for urbanization quality and ecosystem services. Through the exponential efficacy function model, the InVEST model and the coupling coordination model, we analyzed the respective development status of urbanization quality and ecosystem services in the LXUA and explored the coupling coordination relationship between them. This study is important to clarify the coevolution mechanism of the man–land areal system in the Upper Yellow River. The results of the study can pro-vide decision support and a theoretical basis for the ecological protection and high-quality development of the Yellow River Basin. The main conclusions are as follows:

(1) In 2020, the urbanization quality of the LXUA presented a barbell shaped "dual core" distribution pattern, with the insufficient development of secondary cores and prominent problems in widespread low-level counties. From 2000 to 2020, the overall urbanization quality of the LXUA increased by 23.25%, with each county showing a growth trend, but there were significant differences in the growth rate.

(2) The level of water provision, soil and water conservation, and carbon fixation ser-vices in urban agglomerations has shown an increasing trend since 2000, with different spatial distribution trends. Ecosystem services presented a stepwise distribution pattern that increases from northeast to southwest. High-value areas were mainly distributed in the Sanjiangyuan area in the southwest of the urban agglomeration, while low-value areas were mainly concentrated in Baiyin in the northeast of the urban agglomeration.

(3) The relationship between the urbanization quality and ecosystem services in the LXUA is strong and the interaction is significant. The CD in 2020 was 0.844, with a growth rate of 6.3% over the past twenty years. Fifteen counties were in the run-in stage, and 24 counties were in the highly coupled stage in the LXUA. The CD of most counties was increasing, and the reduced areas were mainly concentrated in the centers of various cities.
(4) For the counties within the LXUA, there is a lot of room to promote the coordination relationship between urbanization quality and ecosystem services. The CCD of the composite system in 2020 was 0.495. Among them, the number of counties on the verge of disorder was the largest and the area was the widest. The intermediate, primary, and barely coordinated areas were mainly distributed in the central urban areas of various cities, while the mild-disorder areas were distributed in Baiyin, northeast of the urban agglomeration. The analytical framework of the article is a guide to explore the evolution mechanism of the man-earth areal system in Northwest China. However, as the connotations of urbanization and ecosystem services continue to be enriched, an indicator system may still not be comprehensive. In our subsequent research, we will continue to supplement and improve them.

Author Contributions: Conceptualization, W.Z. and P.S.; data curation, W.Z.; formal analysis, W.Z. and P.S.; funding acquisition, P.S.; investigation, W.Z. and Y.Y.; methodology, W.Z. and Y.W.; project administration, W.Z. and P.S.; resources, P.S.; software, W.Z.; supervision, P.S.; validation, P.S.; visualization, W.Z.; writing—original draft, W.Z.; writing—review and editing, P.S. and Y.W. All authors have read and agreed to the published version of the manuscript.

Funding: This research was supported by the National Natural Science Foundation of China (Grant Number: 41771130).

Institutional Review Board Statement: Not applicable.

Informed Consent Statement: Not applicable.

Data Availability Statement: The data presented in this study are available on request from the first author.

Acknowledgments: We sincerely thank the reviewers for their helpful comments and suggestions about our manuscript.

Conflicts of Interest: The authors declare no conflict of interest.

Abbreviations

LUXA	Lanzhou-Xining urban agglomeration
CD	Coupling degree
CCD	Coupling coordination degree

References

1. Fang, C.; Gao, Q.; Zhang, X. Spatiotemporal characteristics of the expansion of an urban agglomeration and its effect on the eco-environment: Case study on the northern slope of the Tianshan Mountains. *Sci. China Earth Sci.* **2019**, *62*, 1461–1472. [CrossRef]
2. Li, Y.; Li, Y.; Zhou, Y. Investigation of a coupling model of coordination between urbanization and the environment. *J. Environ. Manag.* **2012**, *98*, 127–133. [CrossRef] [PubMed]
3. Howard, T.; Elisabeth, C. Modeling for all scales: An introduction to system simulation. *Acta. Ecol. Sin.* **2011**, *31*, 1925–1935.
4. Deosthali, V. Assessment of impact of urbanization on climate: An application of bio-climatic index. *Atmos. Environ.* **1999**, *33*, 4125–4133. [CrossRef]
5. Vester, F.; Von, H. Ecology and planning in metropolitan areas sensitivity model. *J. Ecol. Rural. Environ.* **2020**, *36*, 450–458.
6. Grimm, N.; Faeth, S. Global change and the ecology of cities. *Science* **2008**, *379*, 756–760. [CrossRef]
7. Yuan, Y.; Chen, D.; Wu, S. Urban sprawl decreases the value of ecosystem services and intensifies the supply scarcity of ecosystem services in China. *Sci. Total Environ.* **2019**, *697*, 134170. [CrossRef]
8. Fu, B.; Zhuang, X.; Jiang, G. Environmental problems and challenges in China. *Environ. Sci. Technol.* **2014**, *41*, 7597. [CrossRef]
9. Fang, C.; Liu, H.; Li, G. International progress and evaluation on interactive coupling effects between urbanization and the eco-environment. *J. Geogr. Sci.* **2016**, *26*, 1081–1116. [CrossRef]

10. Wang, J.; Zhou, W.; Pickett, A. A multiscale analysis of urbanization effects on ecosystem services supply in an urban megaregion. *Sci. Total Environ.* **2019**, *662*, 824–833. [CrossRef]
11. Berger, A.; Hodge, R. Natural change in the environment: A challenge to the pressure-state-response concept. *Soc. Indic. Res.* **1998**, *44*, 255–265. [CrossRef]
12. Grossman, G.; Krueger, A. Economic growth and the environment. *Q. J. Econ.* **1995**, *110*, 353–377. [CrossRef]
13. Rees, W.; Wackernagel, M. Urban ecological footprints: Why cities cannot be sustainable—And why they are a key to sustainability. *Environ. Impact Assess.* **1996**, *16*, 223–248. [CrossRef]
14. Clark, G.; Meric, G. Migration and Capital. *Ann. Am. Assoc. Geogr.* **1983**, *73*, 18–34. [CrossRef]
15. Remi, J.; Luc, C.; Marina, G. Demography, urbanization and development: Rural push, urban pull and ... urban push? *J. Urban Econ.* **2015**, *98*, 102–113.
16. Ehrl, P. Agglomeration economies with consistent productivity estimates. *Reg. Sci. Urban Econ.* **2013**, *43*, 751–763. [CrossRef]
17. Akihiro, O.; Mika, G. Regional determinants of energy intensity in Japan: The impact of population density. *Asia-Pac. J. Reg. Sci.* **2018**, *2*, 257–278.
18. Song, Y.; Xia, S.; Xue, D.; Luo, S.; Zhang, L.; Wang, D. Land Space Change Process and Its Eco-Environmental Effects in the Guanzhong Plain Urban Agglomeration of China. *Land* **2022**, *11*, 1547. [CrossRef]
19. Buyantuyev, A.; Wu, J.; Gries, C. Multiscale analysis of the urbanization pattern of the Phoenix Metropolitan landscape of USA: Time, space and thematic resolution. *Landscape Urban Plan.* **2010**, *94*, 206–217. [CrossRef]
20. Silva, J.; Wheeler, E. Ecosystems as infrastructure. *Perspectives Ecol. Conserv.* **2017**, *15*, 32–35. [CrossRef]
21. Carroli, L. Planning roles in infrastructure system transitions: A review of research bridging socio-technical transitions and planning. *Environ. Innov. Soc. Trans.* **2018**, *29*, 81–89. [CrossRef]
22. Costanza, R.; Arge, G. The value of the world' s ecosystem services and natural capital. *Nature* **1997**, *387*, 253–260. [CrossRef]
23. Li, Z.; Zhai, Y.; Zhang, T.; Zhou, X.; Cheng, Z.; Hong, J. How can ecosystem status be more comprehensively reflected? A case study of Jinan City, China. *Sci. Total Environ.* **2023**, *863*, 160970. [CrossRef] [PubMed]
24. Yan, Y.; Yao, L.; Lang, L. Revaluation of ecosystem services in inland river basins of China: Based on meta-regression analysis. *Acta Geogr. Sin.* **2019**, *74*, 1040–1057.
25. Bi, Y.; Zheng, L.; Wang, Y.; Li, J.; Yang, H.; Zhang, B. Coupling relationship between urbanization and water-related ecosystem services in China's Yangtze River Economic Belt and its socio-ecological driving forces: A county-level perspective. *Ecol. Indic.* **2023**, *146*, 109871. [CrossRef]
26. Liang, S.; Zhang, T. Analyzing integrated impacts of various policies on economic-environmental systems. *China Environ. Sci.* **2014**, *34*, 793–800.
27. Yin, D.; Yu, H.; Shi, Y.; Zhao, M.; Zhang, J. Matching supply and demand for ecosystem services in the Yellow River Basin, China: A perspective of the water-energy-food nexus. *J. Clean. Prod.* **2023**, *384*, 135469. [CrossRef]
28. Uehara, T.; Cordier, M.; Hamaide, B. Fully Dynamic Input-Output/System Dynamics Modeling for Ecological-Economic System Analysis. *Sustainability* **2018**, *10*, 1765. [CrossRef]
29. Zhai, C.; Wang, S. Spatio-temporal difference evolution and scenario simulation of multi-dimensional well-being in Poyang Lake area based on system dynamics. *Acta Ecol. Sin.* **2021**, *41*, 2954–2967.
30. Daily, G. Nature's services: Societal dependence on natural ecosystems. *Corp. Environ. Strategy* **1997**, *6*, 200–201.
31. Wang, Y.; Dai, E. Spatial-temporal changes in ecosystem services and the trade-off relationship in mountain regions: A case study of Hengduan Mountain region in Southwest China. *J. Clean. Prod.* **2020**, *264*, 121573. [CrossRef]
32. Chen, S.; Li, G.; Xu, Z.; Zhuo, Y.; Wu, C.; Ye, Y. Combined Impact of Socioeconomic Forces and Policy Implications: Spatial-Temporal Dynamics of the Ecosystem Services Value in Yangtze River Delta, China. *Sustainability* **2019**, *11*, 2622. [CrossRef]
33. Gao, J.; Bian, H.; Zhu, C.; Tang, S. The response of key ecosystem services to land use and climate change in Chongqing: Time, space, and altitude. *J. Geogr. Sci.* **2022**, *32*, 317–332. [CrossRef]
34. Wang, K.; Ouyang, X.; He, Q.; Zhu, X. Impact of Urban Land Expansion Efficiency on Ecosystem Services: A Case Study of the Three Major Urban Agglomerations along the Yangtze River Economic Belt. *Land* **2022**, *11*, 1591. [CrossRef]
35. Fang, C.; Liu, H.; Wang, S. The coupling curve between urbanization and the eco-environment China's urban agglomeration as a case study. *Ecol. Indic.* **2021**, *130*, 108107. [CrossRef]
36. Xu, H.; Jiao, M. City size, industrial structure and urbanization quality—A case study of the Yangtze River Delta urban agglomeration in China. *Land Use Policy* **2021**, *111*, 105735. [CrossRef]
37. Zhang, X.; Song, W.; Wang, J.; Wen, B.; Yang, D. Analysis on Decoupling between Urbanization Level and Urbanization Quality in China. *Sustainability* **2020**, *12*, 6835. [CrossRef]
38. Zhao, J.; Wang, M. A novel assessment of urbanization quality and its applications. *Phys. A* **2018**, *508*, 141–154. [CrossRef]
39. Wang, D.; Fang, C.; Gao, B. Measurement and spatio-temporal distribution of urbanization development quality of urban agglomeration in China. *Chinese Geogr. Sci.* **2011**, *21*, 695–707. [CrossRef]
40. Zhou, W.; He, J.; Hui, D.; Shen, W. Quantifying the short-term dynamics of soil organic carbon decomposition using a power function model. *Ecol. Process.* **2017**, *6*, 10–19. [CrossRef]
41. Jeyasothy, A.; Sundaram, S.; Sundararajan, N. SEFRON: A New Spiking Neuron Model with Time-Varying Synaptic Efficacy Function for Pattern Classification. *IEEE Neur. Net. Learn.* **2019**, *30*, 1231–1240. [CrossRef] [PubMed]

42. Yang, X.; Chen, R.; Ji, G.; Xu, J. Modelling water yield with the InVEST model in a data scarce region of northwest China. *Water Supply* **2020**, *20*, 1035–1045. [CrossRef]
43. Pan, X.; Shi, P.; Wu, N. Spatial-Temporal Interaction Relationship between Ecosystem Services and Urbanization of Urban Agglomerations in the Transitional Zone of Three Natural Regions. *Sustainability* **2021**, *12*, 10211. [CrossRef]
44. Tang, F.; Wang, L.; Guo, Y.; Fu, M.; Huang, N. Spatio-temporal variation and coupling coordination relationship between urbanisation and habitat quality in the Grand Canal, China. *Land Use Policy* **2022**, *117*, 106119. [CrossRef]
45. Huang, L.; Lin, S.; Chen, J. The spatial-temporal coupling analysis of China's regional innovation capability and energy utilization efficiency. *World Reg. Stud.* **2020**, *29*, 1161–1171.

land

Article

Assessing the Rural–Urban Transition of China during 1980–2020 from a Coordination Perspective

Shiwei Zhang [1,2], Jinghui Zhao [3,*], Yanfeng Jiang [4,*], Ali Cheshmehzangi [2] and Wenbin Zhou [5]

1 Institute of Geographic Sciences and Natural Resources Research, Chinese Academy of Sciences, Beijing 100101, China
2 Department of Architecture and Built Environment, University of Nottingham Ningbo China, Ningbo 315100, China
3 Hangzhou Yuanjie Space Technology Co., Ltd., Hangzhou 310030, China
4 School of Public Policy and Administration, Guangxi University, Nanning 530004, China
5 School of Public Affairs, Zhejiang University, Hangzhou 310058, China; wenbinzhou@zju.edu.cn
* Correspondence: zhaojinghui@yjkjkj.cn (J.Z.); yanfeng.jiang@gxu.edu.cn (Y.J.)

Abstract: Under the context of global rapid urbanization, exploring the dynamics of rural–urban transition in China can provide valuable experience for the Global South. In this study, we evaluate the rural–urban transition in China, from 1980 to 2020, based on socioeconomic data and a rural–urban transition coordination model by constructing a rural–urban development and integration index system. We identify the state and transition types, and we present optimization paths. The results show that, since the reform and opening-up, the rural–urban development index (URDI) in China has gradually expanded among regions while the rural–urban integrated index (URII) has experienced a trend of decline followed by an increase. Over the past 40 years, the spatial distribution characteristics of the $\Delta URDI$ have been "south high–north low", while the $\Delta URII$ has had a balanced spatial distribution. Over the first two decades of the past 40 years, the rural–urban transition in eastern coastal China was more coordinated, while regions with less coordination showed a two-tiered distribution pattern; over the last two decades, the coordination degree has increased. Over the past 40 years, the spatial distribution of high coordination presents "T-shaped" coastal and riverside characteristics. The transition types and coupling relationships of state regions are identified. Finally, optimization pathways are proposed for each type to further promote rural–urban integration.

Keywords: rural–urban transition; urban and rural integration; urban and rural relations; rural and urban disparities; China

Citation: Zhang, S.; Zhao, J.; Jiang, Y.; Cheshmehzangi, A.; Zhou, W. Assessing the Rural–Urban Transition of China during 1980–2020 from a Coordination Perspective. *Land* **2023**, *12*, 1175. https://doi.org/10.3390/land12061175

Academic Editors: Li Ma, Yingnan Zhang, Muye Gan and Zhengying Shan

Received: 18 April 2023
Revised: 24 May 2023
Accepted: 25 May 2023
Published: 1 June 2023

1. Introduction

Rural recession has become a global phenomenon with the rise in urbanization and industrialization [1,2]. In the Global South, developmental inequality is especially acute within cities and the countryside [3–5]. Under the influence of urban expansion, although the rural–urban spatial relationship has evolved with boundaries that are gradually blurred [6,7], a significant rural and urban divide still exists [8,9]. The United Nations has proposed the 2030 Agenda for sustainable development goals (SDGs), which promotes construction of sustainable cities, revitalization of the countryside, and enhancement of rural–urban linkages as keys to realizing sustainable rural–urban development [10,11].

China's rural–urban development has undergone rapid change since the 1980s [12]. This transition has influenced the transformation of rural–urban relations and human–land relations. China's social and economic development has rapidly changed with the urbanization rate and the share of non-agricultural economy has risen steadily. However, rural–urban imbalanced development has also grown into one of the most significant imbalances in China [13]. To address this imbalance and underdevelopment of rural areas, China's central governments have proposed several strategies for rural–urban integrated development over the past two decades [14]. As urbanization progresses, promoting land expansion to boost economic growth has been gradually running into a bottleneck as urban areas transition from incremental to stock development [15,16]. In the future, reducing rural–urban disparity and realizing rural–urban integrated development will become one of the important development paths in China [17].

In the early stage, theoretical exploration of rural–urban divide and transition has focused on the transformation of rural–urban economic relations, such as the dual-sector [18], Ranis–Fei [19], and Jorgensen [20] models. Moreover, in terms of rural–urban spatial structure, the polarization trickle-down effect [21], the center-periphery paradigm [22], and the Desakota model [23] have provided the theoretical foundation for rural–urban spatial relations. Under the context of rapid urbanization, scholars have studied rural–urban transition from various perspectives. The topics have included migration, industry, and landscape [24–28]. Furthermore, land use transition, human–land–industry systems, and rural–urban territorial systems have been applied to the analysis of rural–urban transition [29–32].

Recent studies have examined rural–urban transition from the view of rural–urban linkages [33–35], but little attention has been given to the dynamics of rural–urban diversity. In terms of spatial scale, these studies have focused on the economic zone, urban agglomerations, and country [36–39]. The time scale of these studies has been relatively small, which has lacked the depiction of comprehensive transition characteristics of rural–urban transitions. Existing research has focused on the overall characteristics of urban and rural areas but lacked discussion of the transition characteristics presented. Therefore, there remains to be a gap in understanding the rural–urban transition between rural–urban development and rural–urban integration, particularly from the perspective of exploring coupling relationships. To reveal rural–urban transition accurately, it is necessary to reflect the level of urban and rural development while emphasizing the gap between urban and rural individuals.

Rural–urban transition in China is of international significance and can provide valuable references for other developing countries. Since the reform and opening-up in 1978, China's social and economic development has been highly compressed in time and complex in content [37]. It has only taken 44 years to cross the threshold from low-income to high-income countries. Rural–urban transformation has been enormous, with the level of urbanization soaring from 17.92% to 63.89% [39,40]. Therefore, the issues that have arisen in China's rural–urban transition are instructive for other countries, especially Asian countries in the Global South.

Given the size of China's territory and the regional imbalance, it is necessary to examine the rural–urban transition at the national level. To fill the gap in rural–urban transition research with small temporal scales, in this study, we conduct a national-scale study of the rural–urban transition from 1980 to 2020. In this study, we also construct an evaluation index system, i.e., the urban–rural development index (URDI) and urban–rural integration index (URII), which not only measures the absolute level of rural–urban systems but also reveals rural–urban relative differences. The objectives of this study are: (1) to explore the characteristics and spatial patterns of the URDI and URII values in China, during 1980–2020; (2) to measure the intensity of rural–urban transition and its coordination degree in this period and to determine whether this relationship has any regional and temporal differences; (3) to identify the state and transition types during this period; and (4) to discuss implications and optimization pathways for each transition type.

2. Methodology

2.1. Conceptual Framework

The essence of rural–urban development is to enhance the living and production standards of urban and rural residents within a region. This is the foundation for maintaining the functioning of a rural–urban territorial system. Rural–urban integration is the difference between the levels of urban and rural development within a region. It is the key factor that determines the coordinated development of the rural–urban territorial system. Rural–urban development and rural–urban integration are interdependent. Improving the level of rural–urban integration can unlock the potential for economic growth, which positively affects rural–urban development [14].

The levels of rural–urban development and integration are representations of the rural–urban state at a specific point in time (Figure 1). They jointly determine the type of rural–urban state at that point in time, although they differ from each other to some extent. The transition from one state to another is examined by measuring the coupled coordination of rural–urban state changes within one period. The transition intensity is an indicator that quantifies the degree of rural–urban state changes at a specific time stage [13,41]. The type of state at the beginning and the end of the transition stage affects the intensity of the transition at that stage. In general, the URDI is low and the URII is relatively high in State 1. During transition stage B, which is the period of rapid urbanization, the URDI rises rapidly and the URII falls. In State 2, the URDI reaches an elevated level, but the URII is low. Upon entering transition stage B, the rural–urban disparity narrows, the URII rebounds to a high level, and the URDI is maintained at an elevated level. A high-quality rural–urban transition is achieved when both the URDI and the URII improve and increase in concert. A high-intensity rural–urban development transition accompanied by a low-intensity rural–urban integration transition or a high-intensity rural–urban integration transition combined with a low-intensity rural–urban development transition cannot form a high-quality rural–urban transition.

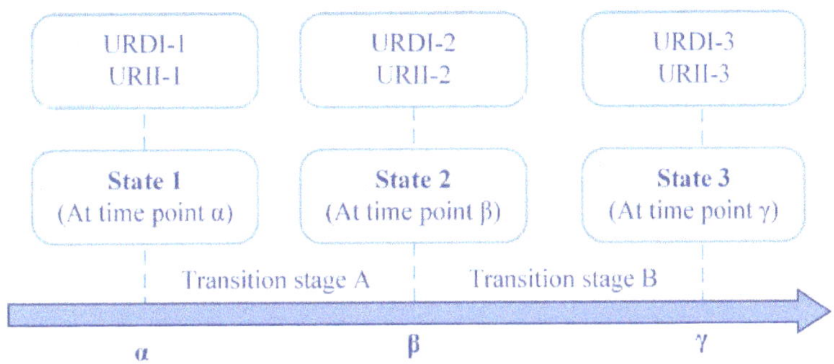

Figure 1. A conceptual model of rural–urban transition.

2.2. Index System Construction

Some studies have explored the coordination and relationships between rural and urban regional systems by constructing indicator systems of urban or rural systems [31,36,42], as well as the individual characteristics of regional development and integration [31,36]. However, the components of the indicator system are rather mixed. It is challenging to present the specific traits of regional rural–urban development and integration. The relevant indicators are chosen following the principles of completeness, representativeness, and accessibility of data. The indicator layers of the URDI and URII systems are established from four dimensions: production, income, consumption, and livelihood (Table 1). Urban and rural residents' incomes and consumption levels among them play significant roles in their wealth accumulation. To some extent, this can reflect the local levels of urban and rural development [43]. In addition to income and consumption levels, other significant factors in integrated rural–urban development include rural–urban disparity in living standards as well as coordinated growth of rural–urban production sectors [44]. The data of each indicator layer are processed by the extreme difference standardization method, which eliminates the influence of the data's dimensionality and allows for comparison of data from various indicators and years.

Table 1. The index system of rural–urban development and rural–urban integration.

Index	Indicator Layer	Definition	Weight
URDI	Income	Per capita disposable income	0.364
	Production	Per capita GDP	0.256
	Consumption	Per capita household expenditure	0.175
	Livelihood	Per capita food expenditure/per capita household expenditure	0.205
URII	Rural–urban income disparity	Rural per capita disposable income/Urban per capita disposable income	0.346
	Rural–urban production disparity	Rural per capita GDP/urban per capita GDP	0.292
	Rural–urban consumption disparity	Rural per capita household expenditure/urban per household expenditure	0.219
	Rural–urban livelihood disparity	Urban Engel's coefficient/rural Engel's coefficient	0.142

2.2.1. Weight Measurement

In this study, we employ the CRITICAL method to empower indicators. The CRITICAL method is an objective empowerment method that calculates the weight of indicators [45]. It is more objective than a traditional hierarchical analysis and entropy weight methods. The preliminary data are processed dimensionless to facilitate the comparison and weighting of indicators in different units or orders of magnitude.

2.2.2. Index Calculation

The URDI represents the absolute level of rural–urban development, while the URII represents the relative gap between rural and urban development. The formula for calculating these indices is as follows:

$$Y_i = \sum_{j=1}^{n} \omega_j X_{ij} \quad (1)$$

$$Z_i = \sum\nolimits_{j=1}^{n} \omega_j X_{ij} \cdot 100\% \quad (2)$$

where Y_i and Z_i represent the URDI and the URII for the i region, respectively; X_{ij} is the dimensionless data for the indicator; ω_j is the weight for the j indicator.

2.2.3. Measurement of ΔURDI and ΔURII

Rural–urban transition refers to the process of changing the state of rural–urban development or integration relative to a certain stage, and the intensity of the transition is used to determine its magnitude. The intensity of the rural–urban transition is judged by comparing the development status at the end of the period (T2) and the beginning of the period (T1) with the formula below:

$$\Delta URDI = Y_t \cdot \frac{1}{Y_{t-1}} \quad (3)$$

$$\Delta URII = Z_t - Z_{t-1} \quad (4)$$

where $\Delta URDI_t$ and $\Delta URII$ denote the intensity of rural–urban development transition and the intensity of rural–urban integration transition from stage $t-1$ to t, respectively; Y_t and Y_{t-1} are the URDI from stage $t-1$ to t, respectively; Z_t and Z_{t-1} are URII at time points t and $t-1$, respectively.

2.2.4. Measurement of Rural–Urban Transition Coordination Degree

Rural–urban integration aims to drive high-quality rural–urban development as a whole through rural development. Therefore, in this study, we construct a rural–urban development-integration coordination degree model to quantitatively measure the coordination state of rural–urban transition [46], which is calculated as:

$$C(YT_t{\cdot}ZT_t) = \left\{ \left(\frac{YT_t + ZT_t}{2} \right) \cdot \left[(YT_t{\cdot}ZT_t) \cdot \frac{1}{\left(\frac{YT_t+ZT_t}{2} \right)^2} \right]^{\frac{1}{2}} \right\} \tag{5}$$

where $C(YT_t{\cdot}ZT_t)$ denotes the coordination degree of rural–urban transition from stage $t-1$ to t, YT_t is $\Delta URDI$ from stage $t-1$ to t, and ZT_t is $\Delta URII$ from stage $t-1$ to t. A high degree of coordination means coordinated rural–urban transition.

2.2.5. Data Sources

For this study, to explore the rural–urban transition characteristics since the reform and opening-up, based on the availability of data, we selected 1980–2020 as the research period to largely correspond to the period of reform and opening-up. The relevant data were mainly obtained from the China Statistical Yearbook, China Population, and Employment Statistical Yearbook, China Rural Statistical Yearbook, China Compilation of Population Statistics 1949–1985, regional statistical yearbooks, etc. To reduce the impact of price changes during the inter-annual period on timing data, 1980 was taken as the base period. The Consumer Price Index (CPI) was used to correct the data of subsequent related years for urban and rural residents' incomes, consumption expenditures, and GDP. The statistics for some regions were lacking in 1980, and anomalies in the indicators were modified by data interpolation based on adjacent areas and years.

3. Results and Analysis

3.1. Characteristics of the URDI and the URII

At the national level, the $\Delta URDI$ has generally shown an upward trend since 1980. The first fluctuation occurred from 1992 to 1996, and an upward trend resumed after 2001. Since globalization and economic liberalization, the URII reached its highest value in 1984, and then showed a wave of decline until it reached its lowest value in 2003 [47]. The URII changed little between 2003 and 2009, with an upward trend after 2009 (Figure 2).

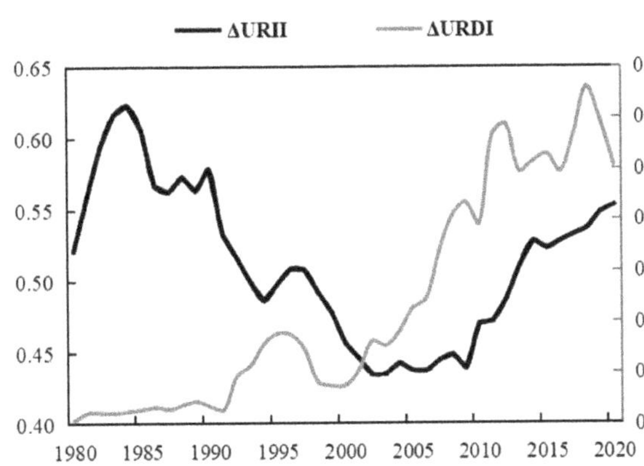

Figure 2. Changes in the URDI and URII of China during 1980–2020.

At the regional level, the URDI has grown significantly in each region over the past forty years. The western region has experienced a larger growth rate than the northeastern region (Figure 3). It is noteworthy that the URDI in the northeastern region was surpassed by the central and western regions from 2010 to 2020, and the gap between the western region and the northeastern region was slightly decreased by 2020. In terms of the URII, from 1980 to 1990, the URII of the western region showed a slight decline (Figure 4). During 1990–2000, the URII of all regions showed a significant decline, among which the northeastern region dropped the most, i.e., from 0.72 to 0.41, but it was still the highest. From 2000 to 2010, the URII in the eastern region declined slightly, while the URII in other regions increased slightly. The URII increased significantly across all regions between 2010 and 2020, with the central and western regions outpacing other regions.

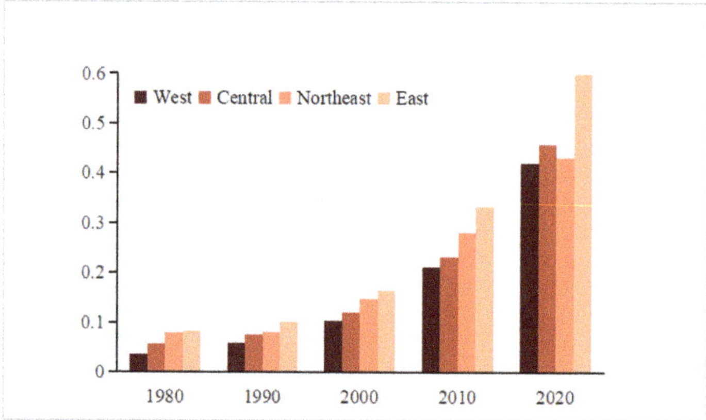

Figure 3. The URDI values of different regions during 1980–2020.

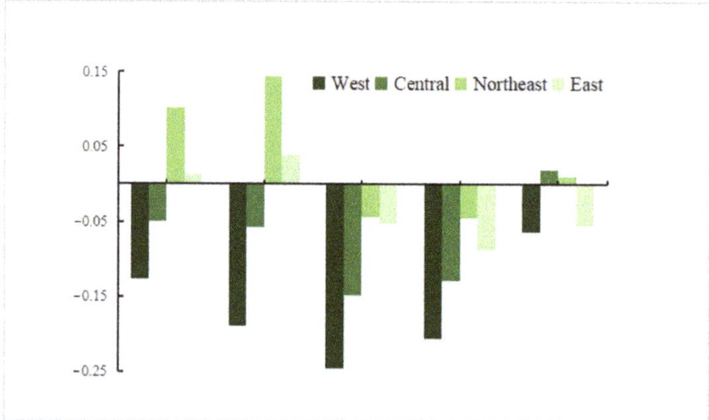

Figure 4. The URII values of different regions during 1980–2020.

The provincial URDI is arranged in reverse order using the maximum URDI value for each year as the benchmark (Figure 5). In general, the inter-provincial differences have gradually widened over the past four decades. First, the gap between Shanghai, Beijing and other regions has been gradually increasing, and the ratio between their URDI and the third place has risen from 1.09 (in 1980) to 1.18 (in 2000), and further expanded to 1.36 (in 2020). Second, with respect to the URDI, the gap between the central regions, western regions, and the northeast has been completely smoothed over the 40 years. The URDI values of some provincial regions in the central and western regions exceed those of the northeast region.

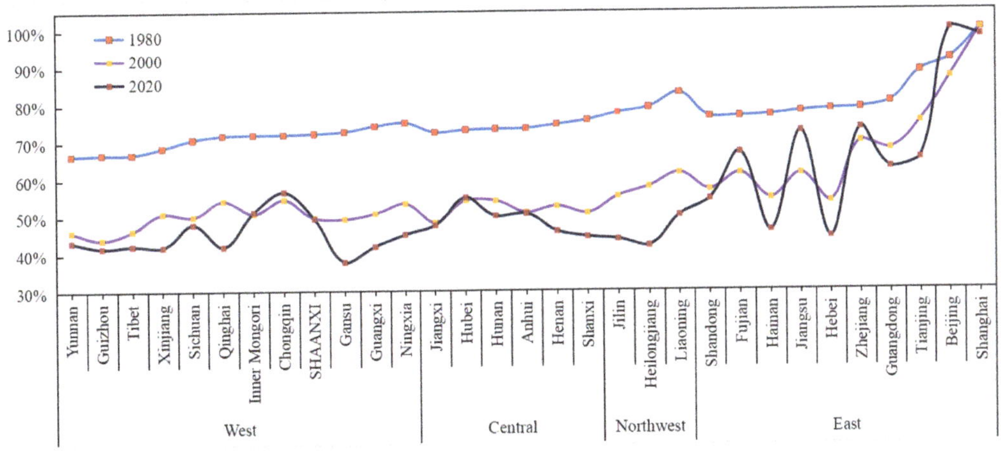

Figure 5. The comparison of provincial URDI values during 1980–2020.

3.2. Spatial Pattern of the URDI and the URII

In terms of the spatial distribution of the URDI, high-value areas exhibit a pattern of migration from north to south and from east to west over the course of 40 years. The low-value areas vary slightly and are all concentrated in the western region (Figure 6). The URDI gradually evolves from a coastal distribution in 1980 to a "T-shaped" distribution pattern along the coast in 2020, while the center of gravity is transferred from the Bohai Rim region to the southeast coastal area. In terms of the URII, it shows a spatial distribution of high values in the east and low values in the west, in 1980 and in 2020, with high-value areas in 1980 concentrated in the northeast and south. Furthermore, high-value areas of URII shift to coastal areas in east China in 2000 (Figure 7). While in 2020, the URII shows a scattered distribution with high- and low-value areas distributed across all regions, among which high-value areas show a continuous distribution in the central region.

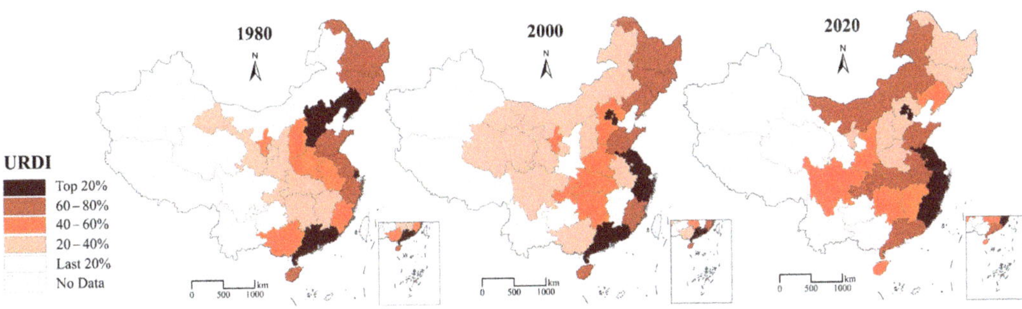

Figure 6. The spatial pattern of the URDI in China, during 1980–2020.

3.3. Coupling Relationships between the URDI and the URII

The URDI and the URII have both undergone significant changes over the last four decades. Scatter charts of the URDI and URII were constructed for 1980, 2000, and 2020 (Figure 8). Taking the average values of the URDI and the URII as the boundary to divide the high- and low-value regions, it can be seen that the scatter distribution varies little between 1980 and 2000, with the scatter primarily concentrating in the first and third quadrants in 1980–2000, and the scatter concentrating in the fourth quadrant in 2020. Between 2000 and

2020, the points in the high URDI—high URII decrease and in the low URDI—low URII increase. The positive correlation between URDI and the URII is gradually decoupled.

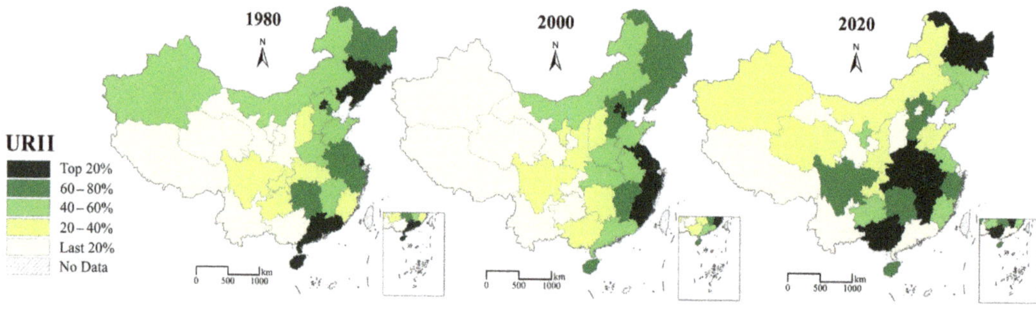

Figure 7. The spatial pattern of the URDI in China, during 1980–2020.

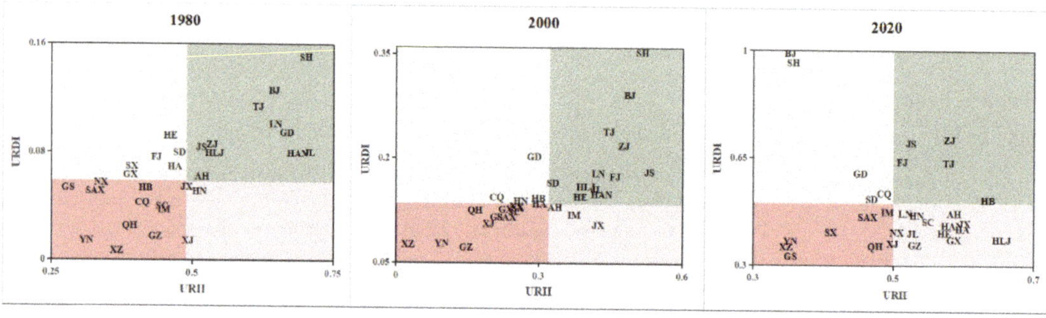

Figure 8. The scatter diagram of rural–urban development intensity during 1980–2020.

3.4. Spatial Pattern of Rural–Urban Transition

From 1980–2000, the $\Delta URDI$ showed a gradient difference from northwest to southeast (Figure 9). The intensity of the $\Delta URDI$ in the three southeastern coastal regions is significantly greater than that in the neighboring regions; the $\Delta URDI$ in the central and western regions is relatively low. From 2000 to 2020, the $\Delta URDI$ shows a balanced spatial pattern, with high-value areas are mainly concentrated in central and western China. Some of the low-value areas in the previous stage (Guizhou, Tibet, Sichuan, Shaanxi, Inner Mongolia, etc.) are transformed into high-value areas during this stage, while some high-value areas during the previous stage (Shanghai, Guangdong, Zhejiang, and Tianjin) are transformed into low-value areas during this stage. The rate of growth in the eastern region, especially in the northeast, has slowed down significantly.

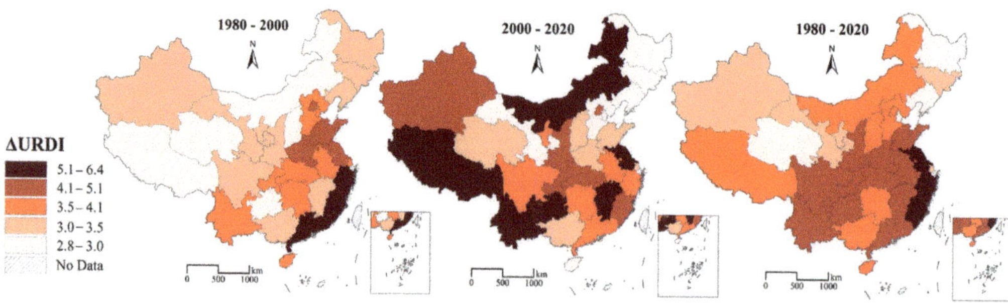

Figure 9. The spatial pattern of rural–urban development intensity in China, during 1980–2020.

The distribution pattern of the $\Delta URII$ differs from that of the $\Delta URDI$. From 1980 to 2000, the $\Delta URDI$ shows high spatial distribution characteristics of high in the east and low in the west (Figure 10). Specifically, Jiangsu and Fujian have $\Delta URII$ values of 0.015 and 0.023, respectively, while other regions have negative $\Delta URII$ values. The spatial distribution characteristics of the $\Delta URII$ in 2000–2020 are opposite to those of the previous two decades, showing the distribution characteristics of high in the west and low in the east. Among them, Shanghai (−0.162) and Beijing(−0.136) have lower $\Delta URII$ values and they are the only two regions with negative $\Delta URII$ values. Guizhou has the highest $\Delta URII$ value (0.384) and Guangxi follows (0.356). The areas with negative $\Delta URII$ values are in the eastern and northeastern regions: Shanghai (−0.348) and Beijing (−0.292). Guangdong, Jilin, Liaoning, Tianjin, and Hainan also have relatively low $\Delta URII$ values. Hubei (0.218) and Guangxi (0.194) have higher $\Delta URII$ values than other regions.

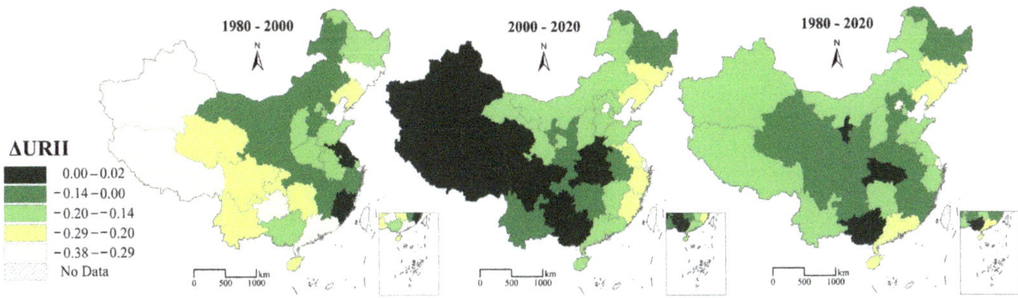

Figure 10. The spatial pattern of rural–urban integration transition intensity in China, during 1980–2020.

3.5. Coordination Degree Analysis of Rural–Urban Transition

Based on the above results, we explore the rural–urban coordination degree of rural–urban transition. The coordination degree is classified into five types based on the transition characteristics: severe incoordination (0.0–0.4), general incoordination (0.4–0.5), mild incoordination (0.4–0.5), general coordination, and high coordination. The results show that, from 1980 to 2020, the regions with high coordination are mainly concentrated in eastern coastal China, with Shanghai (0.829), Beijing (0.799), Zhejiang (0.7954), Fujian (0.7738), and Jiangsu (0.736) ranking in the top five. The regions with lower coordination show a two-tier distribution pattern: Guangdong (0.235) and Hebei (0.294), which have a high level of URDI in 1980, are at the bottom; Tibet (0.361) and Guizhou (0.356), which have a low level of URDI, also have a lower level of coordination. From 2000 to 2020, the national coordination degree has improved as compared with the previous two decades, having a higher coordination degree of more than one-third of provincial regions, and an even distribution in all regions except the northeastern region. There are only six regions with lower coordination degrees, among which the coordination degrees of Gansu (0.275) and Shanghai (0.297) are lower than 0.3. Throughout the four decades, the high-value areas of rural–urban development/integration and transition coordination show a coastal and riverine "T-shaped" pattern, while Gansu, Shanghai, Jilin, and Hebei have lower coordination degrees (Figure 11).

Based on the URDI and URII in 2020, the state type is divided by the mean value. According to the types of rural–urban transition coordination in 2020, the transition types of various states are identified (Figure 12). In 31 provincial regions, five provinces are high-quality transition types (Zhejiang, Jiangsu, Tianjin, Fujian, and Hubei), and their transition types are all coordinated. This indicates that the URDI and the URII values are rising with a high level of coordination. The lagging integration transition type also includes five regions. The higher the level of rural–urban development, the less coordinated the rural–urban transition is, indicating that the rural–urban divide in developed regions is widening. Thirteen regions are lagging development transition types; nine provinces and regions are double-lagging transition types. The lagging integration transition type regions show higher URDI values with lower coordination of rural–urban transition, while the double-lagging transition type regions show that the higher the URDI value, the higher the coordination of rural–urban transition.

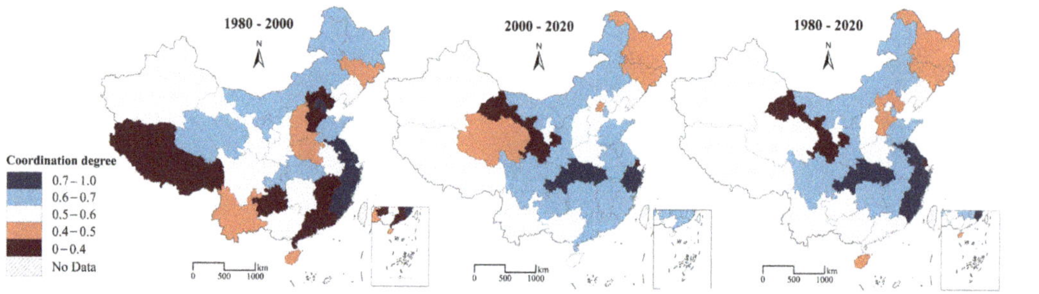

Figure 11. Spatial pattern of rural–urban transition coordination degree in China, during 1980–2020.

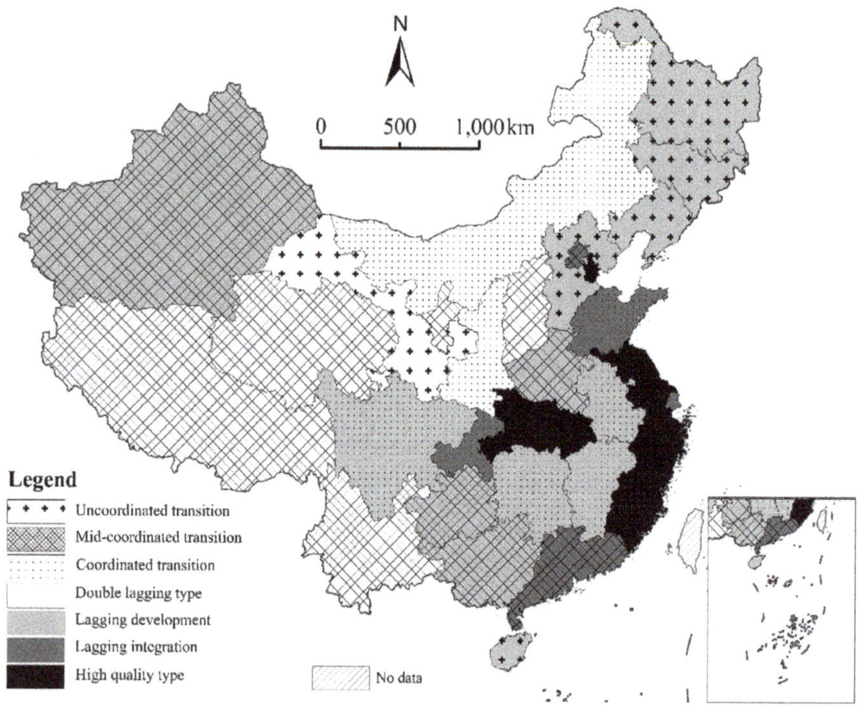

Figure 12. The spatial distribution of rural–urban state and transition types, in China, since reform and opening-up.

4. Discussion

4.1. The Optimal Path of Rural–Urban Transition

The lagging development regions are further classified into three subtypes: lagging development with transition coordination, lagging development with general incoordination, and lagging development with uncoordinated transition. The rural–urban lagging/transition coordination type includes Jiangxi, Sichuan, Anhui, and Hunan provinces, where the rural–urban transition is more coordinated and the rural–urban gap is smaller. The optimization strategy for these regions is to enhance the growth pole effect of urban areas, especially metropolises. The lagging development/transition incoordination type includes four provinces, namely Heilongjiang, Hebei, Jilin, and Hainan. These provinces had high-level URDI in the early stage of reform and opening-up, but the growth rate declined in the later stage. In these regions, the infrastructure is concentrated in urban areas. The optimal path forward is to focus on the redevelopment and improvement of small towns and cities, revitalization of old industrial bases, and releasing the development potential of urban areas.

In lagging integration regions, Shanghai is in an uncoordinated transition, while Beijing and Guangdong are generally incoordination, and Chongqing and Shandong belong to the coordinated transition. To optimize these regions, recommendations include improving the growth pole effect of small towns for the countryside, encouraging capital flow to rural areas, accelerating the development of rural tourism such as leisure farms, and promoting the integrated development of rural industries. Furthermore, broadening the channels of farmers' property incomes, improving the income structure of rural residents, creating a unified construction land market, and promoting the market of collective business construction land are all urgently needed. Local governments can promote the equalization of basic public services and build a comprehensive public service system.

Most double-lagging regions are concentrated in the west. Due to a western development strategy, coordinated rural and urban transition in the western region is prioritized. For these regions, infrastructure construction in remote areas and hilly regions should be prioritized to enhance transportation of the countryside, in particular, connecting roads from central towns to various villages. Moreover, an innovative industrial system should be established and advantageous industries such as new energy, tourism services, and ICT industry should be developed.

In general, the essence of coordinated rural–urban transition is to narrow the livelihood level of rural–urban residents [48]. In terms of income, there is a need to expand the channels of rural residents' property incomes and give peasants more sufficient property rights and interests [49]. In terms of production, there is a need to liberate the productivity of rural areas and actively explore the development path of rural industrialization and local urbanization. In terms of policies, funds, talents, and technologies should be encouraged to flow and backflow to the countryside [50], and local advantages should be capitalized on to develop high-value-added industries.

4.2. Policy Implications

The urban-biased and coastal opening development strategies have caused discoordination of rural–urban transition [47]. During the process of urbanization, rural–urban transition should be gradually promoted to further improve the current situation of an uncoordinated state, thereby achieving high-quality development. Policies should be targeted at the support of rural development. It is necessary to stimulate the initiative of endogenous growth of rural areas, especially regions of lagging development, to achieve high-quality coordinated development of areas.

Specifically, state-led policies have led to an increase in rural–urban disparities [51,52]. In the last 20 years, land-based finance has made the most substantial contribution to urban development in China [53]. Cities have achieved rapid development and expansion; however, it has also exacerbated the rural–urban divide [54]. Furthermore, land-based finance has resulted in pockets of accumulation of wealth for the wealthy, with fewer services for the poor [55]. The land-based finance model should be strengthened to help rural areas instead of exploiting the rural poor. For highly urbanized regions, policies can increase the intensity of levying land and real estate taxes. Moreover, capturing land value in cities can supplement funds for rural revitalization [56]. Governments need to provide affordable housing for low-income households to minimize the impact of increased land and housing prices owing to land-based finance.

In terms of rural development, increasing the value of rural land can be realized by rent, circulation, and mortgage of rural homesteads [57]. It can also be realized through the construction of rental housing for rural collective-owned land, reform of real estate tax, and control of the bubble of urban housing price [53,58,59]. Comprehensive land consolidation can further promote rural multi-functional development [60], and it can coordinate rural regional demand and income and improve the quality of rural living.

4.3. Limitations and Future Research

Although this study focuses on the rural–urban transition in China, it has general value for the governance of rural–urban development in the Global South. Developing countries are globally confronted with widespread problems due to rapid urbanization. At the same time, there are similarities in the development status of different countries and regions. Therefore, exploring large-scale spatial classification and zoning methods not only provides development paths for promoting sustainable development in China, but also provides valuable lessons for other regions in the Global South.

The Chinese example has implications for low and medium development levels and unbalanced rural–urban development areas in the Global South. However, it is not unique but rather a microcosm of many regions in developing countries. Notably, social and cultural backgrounds differ between China and other countries. Therefore, it is essential to introduce China's experience according to local conditions. Additionally, exploring reciprocal feedback between rural–urban transition and land use transition, especially concerning construction land, can control the development of rural–urban transition paths more specifically. This study focuses on macro-level spatial and temporal changes at a larger scale and is insufficient for specific impact mechanisms, a fact that deserves more academic attention in future studies.

5. Conclusions

This study evaluates the coordination level of rural–urban transition in each region of China, from 1980 to 2020, by constructing indicator systems for rural–urban development and integration. The conclusions are as follows.

(1) In general, China's URDI has increased rapidly since the 1980s, with regional disparity expanding substantially. The URII has experienced a trend of first decreasing and then increasing, and the difference between regions is gradually decreasing. The positive correlation between URDI and URII has gradually decoupled from 1980 to 2020.
(2) From 1980–2000, the intensity of the rural–urban transition in China has tremendous regional and temporal differences. The $\Delta URDI$ has shown the spatial characteristics of the "south high–north low", while the $\Delta URII$ is relatively evenly distributed. The high-value areas of coordination degree show the spatial pattern of coastal and riverine "T-shape".
(3) Ten types of regions are identified based on rural–urban state and transition. The decline of coordination is obvious in high-quality regions. In lagging integration regions, the higher the URDI, the transition is more uncoordinated; in double-lagging regions, high URDI means more coordination in the rural–urban transition.

The implementation of urban-biased and unbalanced regional development strategies can lead to an uncoordinated rural–urban transition. In the context of rapid urbanization, it is essential to control rural–urban transition to improve rural–urban disparity and to achieve high-quality development. Policies should be targeted at supporting rural development, and the initiative of rural endogenous growth is an effective way to realize a positive rural–urban transition.

Author Contributions: S.Z. contributed to the idea, data curation, methodology, formal analysis, and the original draft of the manuscript; J.Z. contributed to project administration, review and editing, and funding acquisition; Y.J. contributed to the conceptualization, original draft of the manuscript, review and editing; A.C. contributed to review and editing; W.Z. contributed to review and editing. All authors have read and agreed to the published version of the manuscript.

Funding: This research was supported by the Start-up research funding from Zhejiang University (126000-540501) and Research Project-"Measurement and optimal regulation of the integrated urban-rural development level in the Yangtze River Delta region" funded by Institute of Spatial Planning, Yangtze Delta Region Institute of Tsinghua University, Jiaxing.

Institutional Review Board Statement: Not applicable.

Informed Consent Statement: Not applicable.

Data Availability Statement: Not applicable.

Conflicts of Interest: The authors declare no conflict of interest with this work.

References

1. Kay, C. Development Strategies and Rural Development: Exploring Synergies, Eradicating Poverty. *J. Peasant Stud.* **2009**, *36*, 103–137. [CrossRef]
2. De Haan, L.J. The Question of Development and Environment in Geography in the Era of Globalisation. *GeoJournal* **2000**, *50*, 359–367. [CrossRef]
3. Montgomery, M.R. The Urban Transformation of the Developing World. *Science* **2008**, *319*, 761–764. [CrossRef]
4. Sahn, D.E.; Stifel, D.C. Urban-Rural Inequality in Living Standards in Africa. *J. Afr. Econ.* **2004**, *13*, 564–597. [CrossRef]
5. Sehrawat, M.; Giri, A.K. Panel Data Analysis of Financial Development, Economic Growth and Rural-Urban Income Inequality Evidence from SAARC Countries. *Int. J. Soc. Econ.* **2016**, *43*, 998–1015. [CrossRef]
6. Scott, M. Resilience: A Conceptual Lens for Rural Studies? *Geogr. Compass* **2013**, *7*, 597–610. [CrossRef]
7. Peou, C. Negotiating Rural-Urban Transformation and Life Course Fluidity: Rural Young People and Urban Sojourn in Contemporary Cambodia. *J. Rural Stud.* **2016**, *44*, 177–186. [CrossRef]
8. Delgado-Viñas, C.; Gómez-Moreno, M.-L. The Interaction between Urban and Rural Areas: An Updated Paradigmatic, Methodological and Bibliographic Review. *Land* **2022**, *11*, 1298. [CrossRef]
9. Spoor, M. Multidimensional Social Exclusion and the "Rural-Urban Divide" in Eastern Europe and Central Asia. *Sociol. Rural.* **2013**, *53*, 139–157. [CrossRef]
10. Somanje, A.N.; Mohan, G.; Lopes, J.; Mensah, A.; Gordon, C.; Zhou, X.; Moinuddin, M.; Saito, O.; Takeuchi, K. Challenges and Potential Solutions for Sustainable Urban-Rural Linkages in a Ghanaian Context. *Sustainability* **2020**, *12*, 507. [CrossRef]
11. UN. *Transforming Our World: The 2030 Agenda for Sustainable Development*; United Nations: New York, NY, USA, 2015; Volume 16301.
12. Yu, A.T.W.; Wu, Y.; Zheng, B.; Zhang, X.; Shen, L. Identifying Risk Factors of Urban-Rural Conflict in Urbanization: A Case of China. *Habitat Int.* **2014**, *44*, 177–185. [CrossRef]
13. Chen, K.; Long, H.; Liao, L.; Tu, S.; Li, T. Land Use Transitions and Urban-Rural Integrated Development: Theoretical Framework and China's Evidence. *Land Use Policy* **2020**, *92*, 104465. [CrossRef]
14. Zhu, J.; Zhu, M.; Xiao, Y. Urbanization for Rural Development: Spatial Paradigm Shifts toward Inclusive Urban-Rural Integrated Development in China. *J. Rural Stud.* **2019**, *71*, 94–103. [CrossRef]
15. Zhu, H.; Gao, X.; Lin, Y.; He, Y. Land-Development-Right Pricing Based on Spatial Characteristics in Urban Local Function Regeneration. *Land* **2022**, *11*, 2113. [CrossRef]
16. Khmara, Y.; Kronenberg, J. On the Road to Urban Degrowth Economics? Learning from the Experience of C40 Cities, Doughnut Cities, Transition Towns, and Shrinking Cities. *Cities* **2023**, *136*, 104259. [CrossRef]
17. Lu, Q.; Yao, S. From Urban-Rural Division to Urban-Rural Integration: A Systematic Cost Explanation and Chengdu's Experience. *China World Econ.* **2018**, *26*, 86–105. [CrossRef]
18. Lewis, W.A. Economic Development with Unlimited.d Supplies of Labour. *Manch. Sch.* **1954**, *22*, 139–191. [CrossRef]
19. Ranis, G.; Fei, J.C.H. A Theory of Economic Development. *Am. Econ. Rev.* **1961**, *51*, 533–565.
20. Jorgenson, D.W. The Development of Dual Economic. *Econ. J.* **1961**, *71*, 309–334. [CrossRef]
21. Myrdal, G.; Sitohang, P. *Economic Theory and Under-Developed Regions*; Duckworth: London, UK, 1957.
22. Friedmann, J. The Future of Periurban Research. *Cities* **2016**, *53*, 163–165. [CrossRef]
23. Mcgee, T. *The Emergence of Desakota Regions in Asia: Expanding a Hypothesis*; University of Hawaii Press: Honolulu, HI, USA, 1991.
24. Mauro, G. Rural–Urban Transition of Hanoi (Vietnam): Using Landsat Imagery to Map Its Recent Peri-Urbanization. *ISPRS Int. J. Geo-Inf.* **2020**, *9*, 669. [CrossRef]
25. Jedwab, R.; Christiaensen, L.; Gindelsky, M. Demography, Urbanization and Development: Rural Push, Urban Pull and ... Urban Push? *J. Urban. Econ.* **2017**, *98*, 6–16. [CrossRef]
26. Lagüera Díaz, S. Urban Rural Interaction: Processes and Changes in the Marina Oriental of Cantabria (Spain). *Land* **2023**, *12*, 166. [CrossRef]
27. Lerch, M. The Role of Migration in the Urban Transition: A Demonstration From Albania. *Demography* **2014**, *51*, 1527–1550. [CrossRef]
28. Zhao, P. Urban-Rural Transition in China's Metropolises: New Trends in Peri-Urbanisation in Beijing. *Int. Dev. Plan. Rev.* **2012**, *34*, 269–294. [CrossRef]

29. Li, T.; Long, H.; Liu, Y.; Tu, S. Multi-Scale Analysis of Rural Housing Land Transition under China's Rapid Urbanization: The Case of Bohai Rim. *Habitat Int.* **2015**, *48*, 227–238. [CrossRef]
30. Liu, Y.; Long, H.; Chen, Y.; Wang, J.; Li, Y.; Li, Y.; Yang, Y.; Zhou, Y. Progress of Research on Urban-Rural Transformation and Rural Development in China in the Past Decade and Future Prospects. *J. Geogr. Sci.* **2016**, *26*, 1117–1132. [CrossRef]
31. Yang, Y.; Liu, Y.; Li, Y.; Li, J. Measure of Urban-Rural Transformation in Beijing-Tianjin-Hebei Region in the New Millennium: Population-Land-Industry Perspective. *Land Use Policy* **2018**, *79*, 595–608. [CrossRef]
32. Handayani, W. Rural-Urban Transition in Central Java: Population and Economic Structural Changes Based on Cluster Analysis. *Land* **2013**, *2*, 419–436. [CrossRef]
33. Bosworth, G.; Venhorst, V. Economic Linkages between Urban and Rural Regions–What's in It for the Rural? *Reg. Stud.* **2018**, *52*, 1075–1085. [CrossRef]
34. Douglass, M. A Regional Network Strategy for Reciprocal Rural-Urban Linkages: An Agenda for Policy Research with Reference to Indonesia. *Int. Dev. Plan. Rev.* **1998**, *20*, 1–30. [CrossRef]
35. Bencardino, M.; Nesticò, A. Demographic Changes and Real Estate Values. A Quantitative Model for Analyzing the Urban-Rural Linkages. *Sustainability* **2017**, *9*, 536. [CrossRef]
36. Hu, Z.; Wang, Y.; Liu, Y.; Long, H.; Peng, J. Spatio-Temporal Patterns of Urban-Rural Development and Transformation in East of the "Hu Huanyong Line", China. *ISPRS Int. J. Geo-Inf.* **2016**, *5*, 24. [CrossRef]
37. Yang, Y.; Bao, W.; Wang, Y.; Liu, Y. Measurement of Urban-Rural Integration Level and Its Spatial Differentiation in China in the New Century. *Habitat Int.* **2021**, *117*, 102420. [CrossRef]
38. Niu, X.; Liao, F.; Liu, Z.; Wu, G. Spatial–Temporal Characteristics and Driving Mechanisms of Land–Use Transition from the Perspective of Urban–Rural Transformation Development: A Case Study of the Yangtze River Delta. *Land* **2022**, *11*, 631. [CrossRef]
39. Wang, Z.; Xu, X.; Wang, H.; Meng, S. Does Land Reserve System Improve Quality of Urbanization? Evidence from China. *Habitat Int.* **2020**, *106*, 102291. [CrossRef]
40. Chen, W.; Cheshmehzangi, A.; Mangi, E.; Heath, T. Implementations of China's New-Type Urbanisation: A Comparative Analysis between Targets and Practices of Key Elements' Policies. *Sustainability* **2022**, *14*, 6341. [CrossRef]
41. Long, H. *Land Use Transitions and Rural Restructuring in China*; Springer Nature: Singapore, 2020; ISBN 978-981-15-4923-6.
42. Ma, L.; Chen, M.; Fang, F.; Che, X. Research on the Spatiotemporal Variation of Rural-Urban Transformation and Its Driving Mechanisms in Underdeveloped Regions: Gansu Province in Western China as an Example. *Sustain. Cities Soc.* **2019**, *50*, 101675. [CrossRef]
43. Yan, L.; Hong, K.; Chen, K.; Li, H.; Liao, L. Benefit Distribution of Collectively-Owned Operating Construction Land Entering the Market in Rural China: A Multiple Principal–Agent Theory-Based Analysis. *Habitat Int.* **2021**, *109*, 102328. [CrossRef]
44. Wang, Y.; Liu, Y.; Li, Y.; Li, T. The Spatio-Temporal Patterns of Urban-Rural Development Transformation in China since 1990. *Habitat Int.* **2016**, *53*, 178–187. [CrossRef]
45. Diakoulaki, D.; Mavrotas, G.; Papayannakis, L. Determining Objective Weights in Multiple Criteria Problems: The Critic Method. *Comput. Oper. Res.* **1995**, *22*, 763–770. [CrossRef]
46. Zhang, Y.; Long, H.; Ma, L.; Ge, D.; Tu, S.; Qu, Y. Farmland Function Evolution in the Huang-Huai-Hai Plain: Processes, Patterns and Mechanisms. *J. Geogr. Sci.* **2018**, *28*, 759–777. [CrossRef]
47. Fujita, M.; Hu, D. Regional Disparity in China 1985–1994: The Effects of Globalization and Economic Liberalization. *Ann. Reg. Sci.* **2001**, *35*, 3–37. [CrossRef]
48. Choithani, C.; van Duijne, R.J.; Nijman, J. Changing Livelihoods at India's Rural–Urban Transition. *World Dev.* **2021**, *146*, 105617. [CrossRef]
49. Ma, Z.; Ran, R.; Xu, D. The Effect of Peasants Differentiation on Peasants' Willingness and Behavior Transformation of Land Transfer: Evidence from Sichuan Province, China. *Land* **2023**, *12*, 338. [CrossRef]
50. Dingde Xu Zhuolin Yong, X.D.L.Z.C.Q. Rural-Urban Migration and Its Effect on Land Transfer in Rural China. *Land* **2020**, *9*, 81. [CrossRef]
51. Pugh, C. "Urban Bias", the Political Economy of Development and Urban Policies for Developing Countries. *Urban Stud.* **1996**, *33*, 1045–1060. [CrossRef]
52. Kurnia, A.A.; Rustiadi, E.; Fauzi, A.; Pravitasari, A.E.; Ženka, J. Probing Regional Disparities and Their Characteristics in a Suburb of a Global South Megacity: The Case of Bekasi Regency, Jakarta Metropolitan Region. *ISPRS Int. J. Geo-Inf.* **2023**, *12*, 32. [CrossRef]
53. Liu, Z. Land-Based Finance and Property Tax in China. *Area Dev. Policy* **2019**, *4*, 368–382. [CrossRef]
54. Lin, G.C.S.; Zhang, A.Y. Emerging Spaces of Neoliberal Urbanism in China: Land Commodification, Municipal Finance and Local Economic Growth in Prefecture-Level Cities. *Urban. Stud.* **2015**, *52*, 2774–2798. [CrossRef]
55. Berrisford, S.; Cirolia, L.R.; Palmer, I. Land-Based Financing in Sub-Saharan African Cities. *Env. Urban.* **2018**, *30*, 35–52. [CrossRef]
56. Sinha, M. Harnessing Land Value Capture: Perspectives from India's Urban Rail Corridors. *Land Use Policy* **2021**, *108*, 105526. [CrossRef]
57. He, Q.; Zhou, J.; Tan, S.; Song, Y.; Zhang, L.; Mou, Y.; Wu, J. What Is the Developmental Level of Outlying Expansion Patches? A Study of 275 Chinese Cities Using Geographical Big Data. *Cities* **2020**, *105*, 102395. [CrossRef]
58. Zhang, S.; Wang, L.; Lu, F. Exploring Housing Rent by Mixed Geographically Weighted Regression: A Case Study in Nanjing. *ISPRS Int. J. Geo-Inf.* **2019**, *8*, 431. [CrossRef]

59. Hawley, Z.; Miranda, J.J.; Sawyer, W.C. Land Values, Property Rights, and Home Ownership: Implications for Property Taxation in Peru. *Reg. Sci. Urban. Econ.* **2018**, *69*, 38–47. [CrossRef]
60. Jiang, Y.; Long, H.; Ives, C.D.; Deng, W.; Chen, K.; Zhang, Y. Modes and Practices of Rural Vitalisation Promoted by Land Consolidation in a Rapidly Urbanising China: A Perspective of Multifunctionality. *Habitat Int.* **2022**, *121*, 102514. [CrossRef]

land

MDPI

Article

Sustainable Revitalization and Green Development Practices in China's Northwest Arid Areas: A Case Study of Yanchi County, Ningxia

Yongsheng Wang [1], Lulu Qu [2,*], Jieyong Wang [1], Qian Liu [2] and Zongfeng Chen [1]

1 Institute of Geographic Sciences and Natural Resources Research, Chinese Academy of Sciences, Beijing 100101, China
2 School of Public Policy and Administration, Chongqing University, Chongqing 400044, China
* Correspondence: qululu91@cqu.edu.cn

Abstract: Consolidating and expanding the achievements of poverty alleviation, and effectively connecting it with rural revitalization, are part of an important path to achieving sustainable poverty alleviation and common prosperity in China, especially in its northwest arid areas. In this paper, the human–earth system was employed to analyze the elemental composition, structural organization, and functional state of China's northwest arid areas. The results revealed the following: (1) poverty in northwest arid areas stems from the lack of a coupling and coordinating mechanism among humans, the economy, resources, and environmental elements; this is not conducive to transforming ecological advantages into regional development. (2) In the antipoverty stage, China's northwest arid areas innovate human–earth coupling and a coordinating mechanism through a series of targeted measures. (3) We found that three paths, namely "promoting the integration of featured advantageous industries and tourism culture, innovating the realization path according to local conditions, and paying attention to the subjectivity of farmers" broaden the means of sustainable livelihood, consolidate the achievements of poverty alleviation, and achieve rural revitalization. (4) In particular, it is necessary to practice the concept of green development and pursue ecological industrialization by establishing a policy system of green land-people-industry-right, thus building an endogenous growth mechanism of sustainable poverty alleviation and green development in China's northwest arid areas. The results provide theoretical support and model reference for the effective connection between consolidating and expanding the key achievements of poverty alleviation and rural revitalization in China's northwest arid areas.

Keywords: sustainable revitalization; green development practices; rural human–earth system science; arid areas man–land relation; Yanchi County; China

Citation: Wang, Y.; Qu, L.; Wang, J.; Liu, Q.; Chen, Z. Sustainable Revitalization and Green Development Practices in China's Northwest Arid Areas: A Case Study of Yanchi County, Ningxia. *Land* **2022**, *11*, 1902. https://doi.org/10.3390/land11111902

Academic Editors: Li Ma, Yingnan Zhang, Muye Gan and Zhengying Shan

Received: 13 September 2022
Accepted: 24 October 2022
Published: 26 October 2022

1. Introduction

Cities and villages interact and support each other [1,2], but rapid economic liberalization has led to socio-economic polarization between them. Furthermore, rural equal development has become a serious global issue [3,4]. Compared with a "city", a "village" is a region that has a vast rural land area outside the built-up area of a city, and a multi-level settlement space of different scales [5]. A village is a regional system that comprises different elements that, together with a city, constitutes a regional complex [6,7]. Rural development refers to the process of a continuous evolution of the economy, society, and settlements of rural areas, and is a current topic of interest in rural geography [8]. China's urban rural dual system and urban-biased development strategy have gradually widened the gap between urban and rural areas, especially with respect to the problem of rural decline. The shortage of rural elements, structural imbalances, and weak natural, economic, and technological degradation that are caused by rural poverty and other "rural diseases", are becoming more serious [9]. The data on "file establishment and registration" in 2013 revealed that illness, disability, education, and disasters are the main factors that cause poverty [10]. The differences in factors, such as water, soil, air, health, and people in rural areas, have led to human–earth relationships, industrial structure, and three living spaces, which make China's rural poverty multidimensional, contiguous and significant [11].

After the reform and opening up, China's agricultural and rural developments have made great achievements; moreover, farmers' lives have improved, from having insufficient food and clothing, to becoming well-off overall [12]. In the 1990s, agriculture, rural areas, and farmers (three rural) issues" were put forward for the first time [13]. In 2005, the Fifth Plenary Session of the 16th Central Committee of the Party pointed out that it is necessary to insist on solving the "three rural" issues as a top priority in the whole party's work. Since 2004, the No. 1 Central Documents have continuously focused on the issue of agriculture, rural areas, and farmers, for 18 years. The development of "three rural" in China was closely related to poverty. Poverty alleviation and development have achieved remarkable results, in promoting the development of "agriculture, rural areas and farmers".

China's rural areas have undergone tremendous changes. For instance, China's poverty alleviation and development have gone through five stages: rural economic system reform, regional development-style poverty alleviation, comprehensive poverty alleviation development, whole village promotion, participatory poverty alleviation, and targeted poverty alleviation [14]. Moreover, since the 18th National Congress of the Communist Party of China, according to the current standards, 98.99 million rural poor people in China have been lifted out of poverty; 832 poor counties emerged from poverty, 128,000 poor villages are no longer poor. Overall regional poverty has been resolved, and the United Nations 2030 poverty reduction goals that are related to the agenda for sustainable development, have been achieved [15]. The per capita net income of the poor increased from CNY 2982 in 2015, to CNY 10,740 in 2020, and absolute poverty has been eliminated.

After 2020, there will be relative poverty problems that are characterized by regional and urban–rural income differences; unequal access to public social services and multidimensional poverty [16]; rural backwardness problems, such as weak infrastructure, hollow villages, old and weak rural subjects, and weak governance capabilities [11], and the instability of poverty alleviation and return to poverty as a result of disasters, diseases, and epidemics. Therefore, the new poverty pattern requires the formulation of new assistance strategies to improve the efficiency of the use of assistance resources, enhance the self-development ability of poverty-stricken populations and to facilitate the endogenous development momentum of impoverished counties [17]. In 2018, China entered the intersection and integration period of poverty alleviation and rural revitalization [18]. After the victory of poverty alleviation, the "three rural" work center has turned to comprehensive rural revitalization. It is necessary to improve the dynamic monitoring and assistance mechanism, to prevent poverty from returning, and to do a good job of consolidating and expanding the results of poverty alleviation, in order to effectively connect with rural revitalization. The No. 1 Central Document of 2021 pointed out that the establishment of a five-year transition period will gradually lead to a smooth transition, from concentrated resources to supporting poverty alleviation, and comprehensively promote rural revitalization [19].

China has a vast territory, and there are differences between natural environmental factors such as topography, hydrology, climate, soil, and vegetation, and socio-economic factors, such as population and the economy [20], resulting in uneven levels of rural development [21]. It is necessary to adapt measures to local conditions in poverty alleviation areas, and carry out orderly efforts to consolidate and expand the achievements of poverty alleviation, in order to effectively connect with rural revitalization. Relevant research on the theory of convergence, regional models, and implementation paths needs to be promoted urgently. Therefore, this study was based on the analysis and linkage framework of the difference between poverty alleviation and rural revitalization, as well as on the investigation and mechanism refinement of the typical regional model of Yanchi County, in order to provide theoretical support and a model reference for the consolidation and expansion of poverty alleviation achievements in the northwest arid area, and to guide the effective connection of rural revitalization. The main contribution of this study is that it explores the causes of poverty alleviation and rural revitalization in China's northwest arid areas, investigates the mechanism of sustainable poverty alleviation in empirical case analyses, and subsequently puts forward a green development path and mode that are conducive to realizing structural optimization and functional improvement in arid areas.

2. Theoretical Analysis

2.1. A Theoretical Model for Connecting Poverty Alleviation and Rural Revitalization

Targeted poverty alleviation is a war of annihilation, and the working time node is 2015–2020, to focus on impoverished areas, counties, villages, and households. From the implementation of the rural revitalization strategy that was proposed at the 19th National Congress in 2017, to the comprehensive revitalization of all villages in 2050, targeted poverty alleviation aims at eliminating absolute poverty; solving the basic problems of poor farmers, such as income, food and clothing, housing, medical treatment and education; the first centennial goal of China's socialist construction; and building a well-off society in an all-round way. Rural revitalization aims at realizing agricultural and rural modernization, and solving relative poverty. It requires industrial prosperity, ecological livability, rural civilization, effective governance, and prosperity. Rural revitalization is inconsistent with the second centenary goal of China's socialist construction, that is, to help all people prosper together and build a prosperous, democratic, civilized, harmonious, and beautiful modern socialist country. Targeted poverty alleviation considers farmers as the basic unit, counties as the starting point and the central government as the overall planner. Meanwhile, rural revitalization takes villages as the basic unit, towns as the starting point, and national leadership as the overall planner [2,22]. Targeted poverty alleviation is a bottom-line constraint task. The state has carried out a large-scale third-party evaluation of effectiveness assessment, and rural revitalization is a task that is assessed for expected results (Table 1).

Table 1. The distinction between targeted poverty alleviation and rural revitalization.

Contents	Poverty Alleviation	Rural Revitalization
Strategy	Battle of annihilation	Protracted battle
Object	Poor areas, counties, villages, and households	All villages
Unit	Farmers are the basic unit, the county is the starting point, and the central government is responsible for overall planning	The village is the basic unit, the township is the starting point, and the state is coordinated
Aspect	Income, food and clothing, medical care, education, and housing security	Production, life, ecology, organization, culture, etc.
Time	Years 2015–2020	Years 2017–2050
Baseline	Eliminate absolute poverty	The goal is to realize agricultural and rural modernization, and solve relative poverty
Target	Build a well-off society in an all-round way	Realize the common prosperity of all people, and build a prosperous, strong, democratic, civilized country
Examine	Bottom line constraint assessment	Performance expectation assessment

2.2. The Mechanisms of Rural Revitalization from the Perspective of People-Land-Industry-Wealth-Right

Poverty alleviation is an important basis for rural revitalization. Rural revitalization can consolidate and enhance the effectiveness of poverty alleviation [23]. The challenges of linking poverty alleviation and rural revitalization are the linking units, contents, policies, and sequences of work. From the perspective of the work unit, rural revitalization needs to shift from farmers to villages. Furthermore, from the perspective of work content, rural revitalization involves the revitalization of ecology, culture, and organization, especially the guidance and drive of capable people, new business entities, and leading enterprises. From the perspective of the working team, the retention and upgrading of the first secretary of poverty alleviation, the team stationed in the village and the person in charge of assistance are related to the construction of rural revitalization cadres. From the perspective of work policies, the continuation and adjustment of various poverty alleviation policies affect the speed and quality of rural revitalization. From the perspective of work sequences, the preferred areas and priorities of rural revitalization [24] should be clarified, in order to reasonably formulate industrial, village, and regional revitalization plans.

The country has put forward a strategy for rural revitalization, and guided farmers to use land for construction, agriculture, and ecology, in order to develop rural industries through governments at all levels, social organizations, and rural talents. Through cooperative organizations, leading enterprises, and the integration of the tertiary industries, farmers' income would increase, the collective economy of the village would be strengthened, and an economic foundation will be provided for the prosperity of rural industries; these would lead to a prosperous life, and ecological livability. In addition, through innovative talent mechanisms and land use policies, the right prerequisites for rural civilization and governance can effectively be provided (Figure 1).

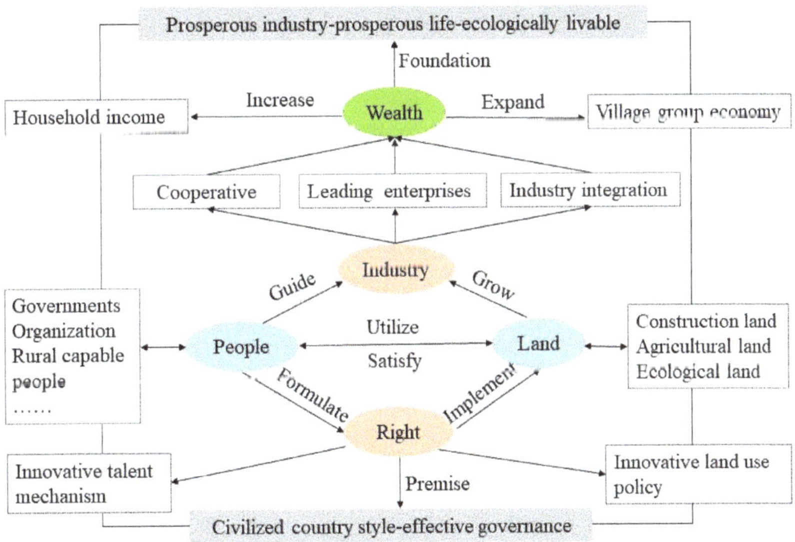

Figure 1. Mechanisms of rural revitalization in the perspective of people-land-industry-wealth-right.

2.3. The Internal Logic of the Effective Connection between Poverty Alleviation and Rural Revitalization

In order to achieve an effective connection between the consolidation and expansion of the results of poverty alleviation and rural revitalization, the internal logical relationship between the two should be understood. In practice, T-P-G (strategic target, core power and mechanism guarantee) interact and influence each other to form a synergistic whole of power support, guaranteed foundation, and integration and optimization (Figure 2). ① The dynamic support of T-P takes farmers as the main body, and activates farmers' endogenous power through production empowerment. ② The guarantee basis of T-G; the "five batches" and "six precision" of poverty alleviation; the "trinity" poverty alleviation pattern; the comprehensive "four inspections and four subsidies"; the monitoring and assistance to prevent the return of poverty; and a series of effective working mechanisms, are the institutional basis and guarantee for effective connection with rural revitalization to meet the needs of the new development stage of rural revitalization. ③ The integration and optimization of P-G, the effective connection between poverty alleviation and rural revitalization, and the core driving force, depend on farmers. It is also inseparable from the protection of government organizations and policy mechanisms, that is, taking farmers as the main body, enabling rural industrial development, building a stable and long-term mechanism, with overall management promoting the driving force of rural endogenous development. The advantages and combinations of the interaction and influence of different regional factors are different, forming different patterns [25].

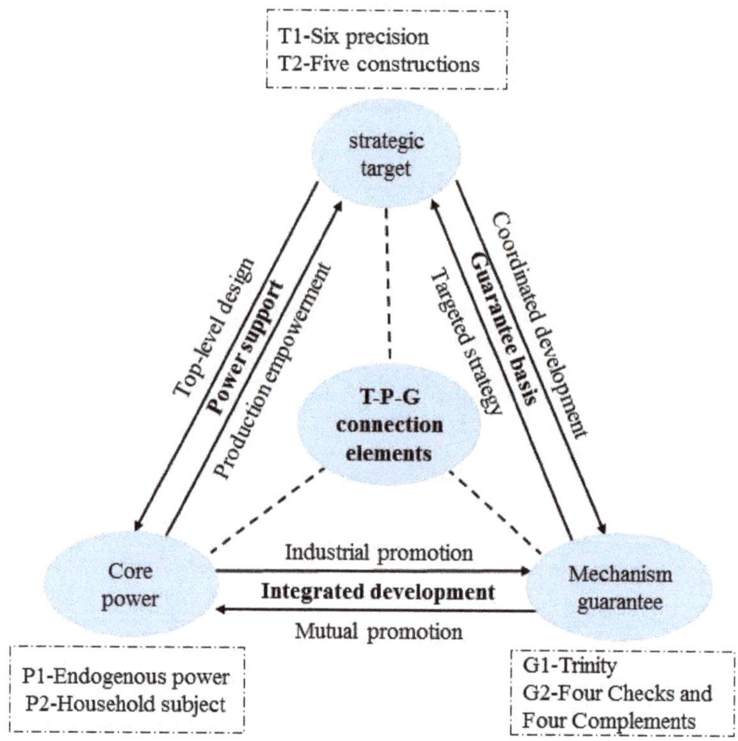

Figure 2. The internal logic of effective connection between poverty alleviation and rural revitalization.

3. Materials and Methods

3.1. Data Sources and Processing

This paper adopted multi-source data collection methods, such as desk research and government document files, in order to ensure the validity of the case study. Relevant data and information were obtained by conducting semi-structured interviews with some government departments: Rural Revitalization Bureau, Agriculture and Rural Bureau, Natural Resources Bureau, Housing and Urban–Rural Development Bureau, and the Development and Reform Bureau, in addition to members of townships, village committee directors, farmers, etc. (Table 2). From September 2020 to October 2021, the research team conducted three field investigations in the sampled districts and counties.

Table 2. Data sources and description.

Data Name	Data Source	Data Type
Land use change survey map	Yanchi Land and Resources Bureau	Vector
Administrative boundary data	National Geomatics Center of China (http://www.webmap.cn/) in 2019 (accessed on 3 September 2019)	Vector
Rural revitalization policies	Yanchi rural Revitalization Bureau	Text
Planning, water-saving irrigation	Township cadres and farmer interviews, and field research	Text
Socio-economic data	Yanchi statistical yearbook in 2021	Table

3.2. *Research Area*

Yanchi County is located in the eastern part of Ningxia Hui Autonomous Region (106°30′~107°47′ E, 37°04′~38°10′ N), with a total land area of 8522.2 km^2. It is located at the junction of Shaanxi, Gansu, Ningxia, and Inner Mongolia (Figure 3), and belongs to a typical farming-pastoral zone [26]. Yanchi County has great resource and environmental constraints. The north is located in the south edge of Maowusu sandy land, and the south is the Loess hilly and gully region, with a fragile ecological environment. The average annual rainfall in Yanchi county is 311 mm, while the annual evaporation in southwest China is 966–1030 mm, and 1030–1052 mm in northeast China. The lack of regional water resources, the sparsely populated land, and the fragile ecology are the main problems for rural development in the county [27].

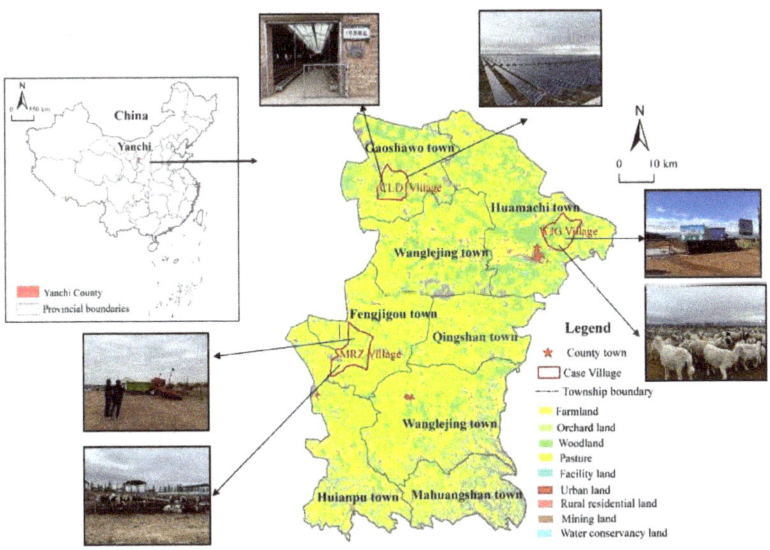

Figure 3. Location of Yanchi County and its land use.

3.3. *Yanchi County's Poverty Alleviation and Rural Revitalization Process*

Yanchi County is an old revolutionary base area, and a national poverty-stricken county. Before poverty alleviation, the problem of rural hollowing out was serious. The rural resident population accounted for only 47% of the total agricultural population. The development momentum of Tan sheep, day lily, and other leading industries was insufficient. The new business entities were scattered, weak and small, and the driving effect was not obvious. Since the implementation of the basic strategy of targeted poverty alleviation in 2013, the whole county has been involved in poverty alleviation work. In 2015, it was awarded the advanced collective of the national poverty alleviation system.

Furthermore, the experience of financial poverty alleviation, that is, the "Yanchi model" and "poverty alleviation insurance", were popularized throughout the country. In 2018, it passed the national special assessment and inspection on the withdrawal of poor counties, and became the first county in Ningxia to emerge from poverty. In 2019, it was selected as one of the first "demonstration counties for implementing the Rural Revitalization Strategy" in the autonomous region, and established the China Academy of targeted poverty alleviation and rural revitalization in northwest China, in order to provide intellectual support for the training of professionals in the fields of targeted poverty alleviation and rural revitalization, in agricultural and rural special research, and in local government decision-making. In 2021, Yanchi County was added to the poverty alleviation and rural revitalization regional case summary project of the National Rural Revitalization bureau, to summarize and consolidate the achievements of poverty alleviation and promote rural revitalization, and to provide case demonstrations for cities, counties, and villages in similar poverty alleviation areas (Figure 4).

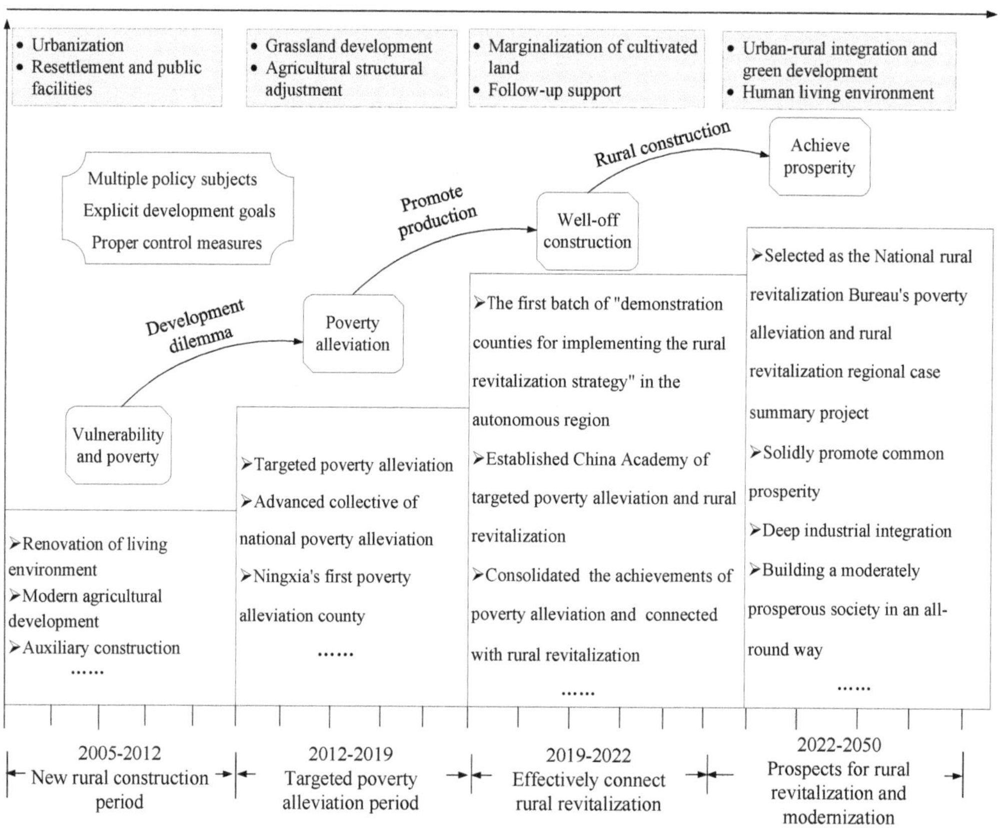

Figure 4. Poverty alleviation and rural revitalization in Yanchi County.

4. Results

4.1. Integrated Development Model Driven by the Photovoltaic Industry

Changliudun village is located 11 km southwest of Gaoshawo town in the northwest of Yanchi County. It governs three natural villages, with a total population of 265 households and 736 people. Changliudun village has explored the development model of "branch + enterprise + farmer households", in order to allow more villagers to find jobs and increase their income. It has also actively connected with photovoltaic enterprises, organized idle labor to work in the photovoltaic park nearby, which has been exporting more than 200 laborers to the photovoltaic park every year (with 12 fixed employees), and increased the per capita income by about CNY 4000. Relying on the photovoltaic industry, the village has an area of 953 hm^2, with a transfer fee of more than CNY 50 million, benefiting 250 people in 89 households, with a per capita income of about CNY 200,000. A centralized photovoltaic poverty alleviation village-level power station has been built, with a power generation scale of 3 MW, and an average annual power generation of 4.2 million kWh; it increased the village's average collective annual income by CNY 220,000 for 20 years, to effectively transform it from "blood transfusion" poverty alleviation to "hematopoietic" poverty alleviation.

In addition, efforts have been made to explore the integration of agriculture, culture, and tourism, as well as raise sheep under photovoltaic panels, in order to develop the collective economy and broaden the channels for the villagers to become wealthier. Firstly, the village explored the development of a tertiary industry dominated by photovoltaic tourism, and developed photovoltaic popular science tourism and rural experience tourism; it also built tourism supporting facilities, such as a cultural stage, folk culture corridor, and tourist experience mill, in order to guide farmers to develop farmhouse entertainment and create high-quality home stays. The next step is to develop photovoltaic science popularization tourism and rural experience tourism; integrate tourism, science popularization, experience, and entertainment into cultural tourism elements; and comprehensively develop "food, housing, transportation, travel, shopping and entertainment". Secondly, the village gives full play to the guiding role of breeding cooperatives, forage processing plants, and leaders in becoming wealthy; it improved the model of "power generation on the board and sheep raising under the board". The village continuously expanded the scale of high-quality forage and small grains, and promoted the formation of Tan sheep forage. The ultimate goal is to promote large-scale and specialized breeding, in order to form an "integrated production and marketing" business model, and to expand the employment space of farmers (Figure 5).

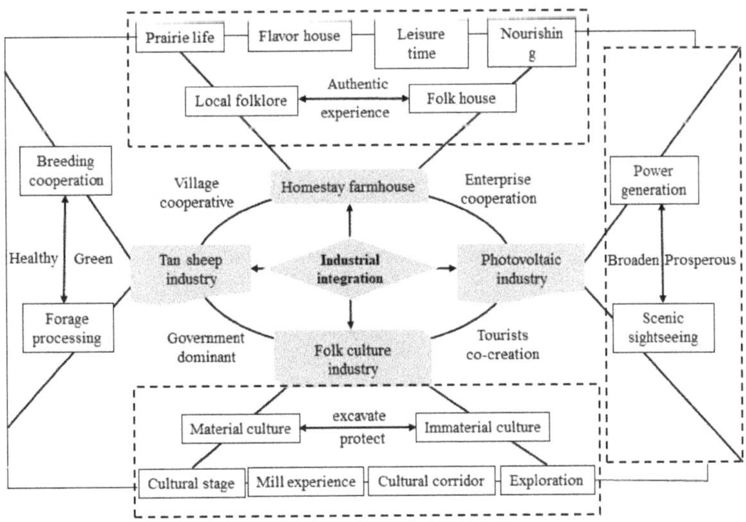

Figure 5. The photovoltaic industry drives integrated development mode in Changliudun village.

4.2. *Efficient Utilization of Water Resources and Integrated Development Model of Grass and Livestock*

The total amount of water resources in Yanchi County is 73 million m^3. The per capita water resources are less than one-fifth of the national average, and less than one-third of Ningxia's average. The development of agriculture and animal husbandry is restricted by water resources. The Tan sheep industry is a key to poverty alleviation and rural revitalization in Yanchi County. With the continuous increase in Tan sheep, it is facing problems with high-quality forage and standard production [28]. Ma'erzhuang village in Fengjigou town is located in the transition zone between Maowusu sandy land and Loess Plateau. It is dry and rainy, and the traditional planting industry depends on the elements. Tan sheep breeding is the main source of income for farmers. The industry is vulnerable to feed cost increases and fluctuations in mutton market prices. Farmers are faced with the dual problem of unstable income and poor natural environment (Figure 6).

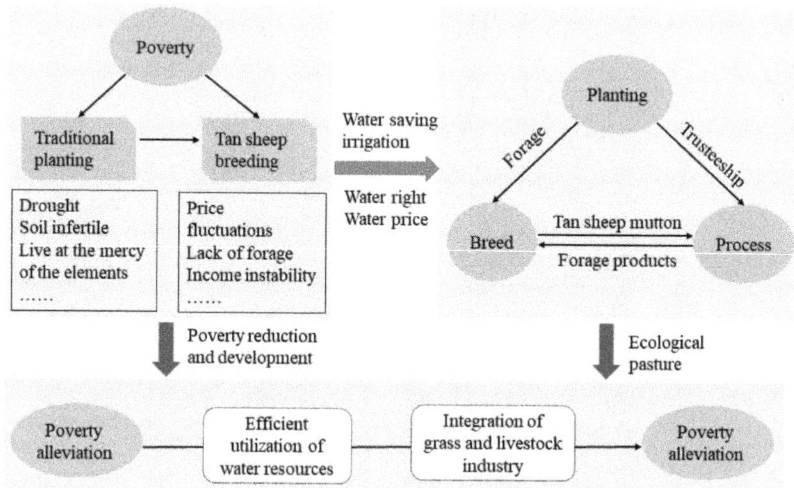

Figure 6. Efficient utilization of water resources and integrated development mode in Ma'erzhuang.

Since 2017, Ma'erzhuang village has transformed and improved its field facilities, agronomic production, and agricultural management through a comprehensive agricultural development project. Integrating all kinds of agriculture and water-related funds of CNY 27.72 million, the irrigation system of the 10,700 Mu irrigation area in the village has been transformed and upgraded, water pipelines and drip irrigation belts have been laid, water- and fertilizer-integrated irrigation systems as well as an automatic irrigation control center have been built. Through the full coverage of drip irrigation in the village, water, fertilizer, labor, and machinery costs have been reduced, and the planting area and output of corn have increased. In addition, Ma'erzhuang established a high-efficiency water-saving irrigation professional cooperative; established irrigation, water fee collection, unit operation, and other systems; and provided socialized services to farmers through unified cultivation, sowing, procurement, irrigation, fertilization, loss prevention, and the harvesting of corn, as well as water right distribution and water price innovation and reform. This has gradually formed a forage processing plant, a corn silage trusteeship planting base, and a Tan sheep ecological pasture planting, which promote the integrated development of the grass and livestock industry, and increase farmers' income.

In 2020, the disposable income of farmers in the whole village reached CNY 14,900, an increase of CNY 4850 over that in 2017; the collective income of the village reached CNY 1 million. The efficient utilization of water resources and the integrated development mode of grass and livestock in Ma'erzhuang village not only effectively solved the problems of forage shortage and the short industrial chain faced by the development of the Tan sheep industry, but it also effectively avoided the ecological risk faced by the development of the grass and livestock industry in arid areas, and effectively ensured the connection between poverty alleviation and rural revitalization.

4.3. Grid Management and Organizational Revitalization Mode

Grassroots governance is the foundation of national governance, and is key to meeting the vital needs of villagers. Based on actual characteristics of large areas and scattered farm households, Yanchi County implemented a four-level grid management mechanism of "county, township, village, and group", in order to comprehensively improve the level of grassroots governance. It also implemented effective monitoring and precise assistance to prevent poverty from returning, by providing a strong support mechanism guarantee for the effective connection between poverty alleviation and rural revitalization. In order to prevent a return to poverty, and to fully implement the rural revitalization strategy in Wanjigou village in Huamachi town, on the basis of original poverty alleviation teams, the village created the "party construction + four-level grid" management mode by implementing such measures as the first secretary, the village working team, and helping cadres and grassroots teams, such as village cadres, villagers' representatives, and group leaders (Figure 7). These measures effectively connected the teams. Among them, the leader of Baocun village acts as the commander to coordinate local governance of the village. The village Party Secretary and the first secretary stationed in the village serve as deputy commanders, and assist the commander to do a good job of overseeing the grid. The first secretary, village cadre, and full-time network member stationed in the village, serve as the grid head; meanwhile, village cadres, team members stationed in the village, members of the village Supervision Committee, and villagers' representatives, serve as grid members who comprehensively check the basic information of residents in the grid.

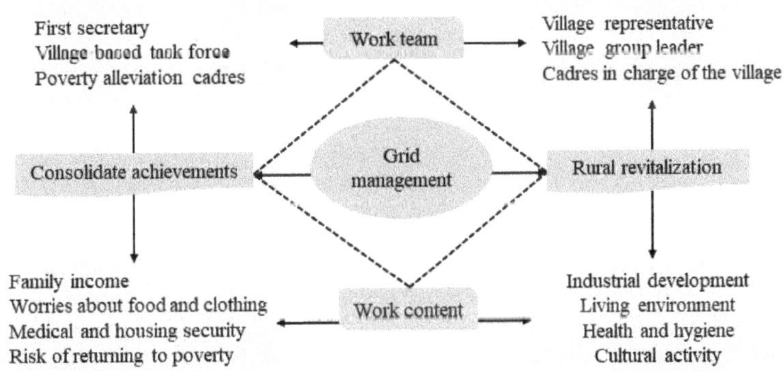

Figure 7. Grid management and organizational revitalization in Wanjigou village.

According to the principle of "moderate population scale, convenient service management and effective resource allocation", as well as "the relationship between kinship, geopolitics and industry", rural households in Anhui Province are divided into 17 basic grids based on the registered residence of natural villages. The work of grid members not only includes the monitoring of farmers' family incomes and poverty return risks, but also the development of poverty alleviation industries related to rural revitalization, the improvement of human settlements and sanitation, comprehensive social management, folkway construction and migration, rural cultural activities, talent skill training, and realizing the effective connections of work contents (Figure 7). The main advantages of the grid management system are as follows: (1) realizing dynamic monitoring and effectively preventing a return to poverty, which provides a strong guarantee for consolidating the results of poverty alleviation; (2) reducing the burden of grassroots cadres, and focusing on rural party construction and industrial development; (3) reducing the costs of farmers' affairs, efficiently serving the various demands of farmers, and providing strong support for effective rural governance; (4) establishing a training mechanism for village-level grassroots reserve cadres, which provide a team guarantee for the revitalization of rural organizations.

5. Summary and Implications

5.1. General Law of Rural Revitalization in Arid Areas

During the transition period, the core work of poverty alleviation is to consolidate and expand the effective connection between achieving poverty alleviation and rural revitalization; the two have inherent consistency and logical continuity [29]. Whether in terms of development planning, working mechanisms and policy guarantees, or in terms of major measures, assistance mechanisms and industrial advantages [30], the effective connection is still inseparable from the continuity that is driven by external markets and rural elites [31], and from the continuity of the convergence of poverty alleviation achievements, regional strategies, and policy objects. According to the reality of poverty alleviation in rural areas, it is necessary to embed administrative forces, strengthen the driving of villagers and market linkage, better identify the track, and make good use of the situation, in a timely manner.

Furthermore, continuous innovations are made in the mechanisms of element integration, power activation, and goal guidance. For example, in the element integration mechanism, with green as the background, the realization path of green development is innovated and integrated with modern elements [7], in order to develop modern organic green low-carbon industries. It is necessary to build a reserve fund for "green rights", to build a deep processing platform, improve the quality of agricultural products, increase employment opportunities, and improve farmers' incomes [32]. It is also necessary to cultivate green-collar workers, decision makers, practitioners, and managers, who are engaged in the modern agriculture and green industry, to lay the organizational foundation for rural revitalization.

Protecting the living environment in arid areas, promoting the sustainable use of natural resources, and achieving regional sustainable development, are important policies for consolidating poverty alleviation achievements and promoting rural revitalization in the new era. It is necessary to adhere to the path of comprehensive and green development, develop conservation-oriented agriculture, circular agriculture and ecological agriculture, according to the requirements of high-quality construction, strengthening ecological environment protection, combining the consolidation of poverty alleviation achievements with environmental governance and ecological protection, and avoiding the vicious circle of "poverty-ecological environment destruction-natural disasters-poverty" [33]. Green rights and interests must be protected, and organization and responsibility are required for the acquisition. As a bridge that connects farmers and the agricultural product market, and ensures the sale of agricultural products, green rights and interests will be important to strengthen the green industry through ecological compensation, and make ecological industrialization a fashionable concept. The three mechanisms are modern factor upgrading, development power activation, and strategic goal guidance [34]. Based on original industrial development, giving full play to the organization, finances, and technology, and continuously promoting industrial transformation and upgrading, are also fundamental guarantees for consolidating and expanding the achievements of poverty alleviation and rural revitalization [35]. A follow-up study will refine the reasons that led to this development, based on natural and human resources, trade, etc., which is a unique and feasible perspective.

5.2. Policy and Practical Implications of Rural Green Development for Arid Areas

China is currently in an important period of consolidating and expanding the achievements of poverty alleviation and rural revitalization. In order to better enhance rural revitalization and development, it needs to promote the integration of characteristic and advantageous industries, in addition to tourism culture [36,37]. It is necessary to develop the tourism economy of characteristic industries, expand the development of characteristic industrial chains, build characteristic industrial belts, activate industrial resources, create rural cultural tourism products with high participation and strong experience, and promote the integration of the advantages of industrial resource and cultural tourism [38,39]. China has a vast territory, with diverse natural geography, obvious differences in resource endowment and social economy, and different levels of rural development [40]. We should innovate development ideas, interact with each other, from top to bottom, and make reasonable plans, develop leading industries, and strive to achieve large-scale integration and standardization on the basis of the meteorological, hydrological, topographic, and other resources available; this should be combined with the reality of industrial development and regional policies. Moreover, it is necessary to pay attention to the subjectivity of farmers. Poverty alleviation is focused on assisting a poverty-stricken population under the current standard, whereas rural revitalization pays more attention to issues, such as poverty-stricken populations no longer returning to poverty, and sustaining the livelihood of poverty-stricken farmers [41,42]. Therefore, it is necessary to adhere to the dominant position of farmers, and focus on empowering farmers, improving their self-development capabilities, giving them the right to participate in rural construction, and assisting them to fully obtain a fair distribution of benefits.

An effective connection between poverty alleviation and rural revitalization is inseparable from the agricultural industrial policy. The internal logical framework of the connection between poverty alleviation and rural revitalization that was proposed in this study still needs the identification framework of the main body of agricultural industrial policy, in order to realize the matching between different agricultural and industrial policies, and the main body of production, operation, and service [43]. In addition, it is necessary to rely on the market mechanism to "find the track" and "make the best use of the situation", in order to promote the integration of regional resources and the improvement in production efficiency. The photovoltaic industry of Yanchi County exerts the scale effect of "photovoltaic + housing renovation", "photovoltaic + tourism", and "photovoltaic + employment, which not only promotes the income of the masses, but also provides a sustainable development path for the collective development of the village [44,45]. In the future, it is necessary to fully mobilize social forces, to strengthen the leading role of social capital such as enterprises, actively practice and explore new models of cooperative development, give full play to multi-party forces and professional and market-oriented platforms, provide technical and talent training support for local development, stimulate new drivers of local development, and to achieve common prosperity.

6. Conclusions

Based on the internal logical framework of the effective connection between poverty alleviation and rural revitalization, taking Yanchi County as a case, the internal connection process of poverty alleviation and rural revitalization was analyzed. The following conclusions are drawn: ① Yanchi County's photovoltaic industry-driven integrated development model, efficient water resource utilization, grass livestock-integrated development model, grid management and organizational revitalization model, and other cases, support the theoretical and logical analysis of the effective connection between poverty alleviation and rural revitalization, and provide a typical sample for poverty alleviation villages to consolidate the achievements of poverty alleviation and rural revitalization. ② During the effective connection period, attention should be paid to the coordinated promotion of the "three paths", namely promoting the integration of characteristic and advantageous industries and tourism culture, innovating the realization path according to local conditions, and focusing on the subjectivity of farmers to broaden the sustainable livelihood path and realize rural development. ③ It is necessary to practice the concept of green development, and to take the path of ecological industrialization by establishing a policy system of green land-people-industry-right, thus building an endogenous growth mechanism of sustainable poverty alleviation and green development in China's northwest arid areas.

Author Contributions: Conceptualization, Y.W.; formal analysis, L.Q. and J.W.; writing—original draft, Y.W. and L.Q.; writing—review and editing, Q.L. and Z.C. All authors have read and agreed to the published version of the manuscript.

Funding: This research was funded by the National Natural Science Foundation of China (Grant No. 42101202, 42101203, and 42201283), National Natural Science Foundation of China and National Science Foundation of the United States Sustainable Regional System Cooperation Research Project, (No. T221101034), the China postdoctoral Science Foundation Program (No. 2022M710015), the Chongqing Social Science Planning Project (No. 2022ZTYB02), and Fundamental Research Funds for the Central Universities (No. 2022CDJSKJC29).

Institutional Review Board Statement: Not applicable.

Informed Consent Statement: Not applicable.

Data Availability Statement: Not applicable.

Acknowledgments: The insightful and constructive comments and suggestions from anonymous reviewers were greatly appreciated.

Conflicts of Interest: The authors declare no conflict of interest.

References

1. Liu, Y.; Li, Y. Revitalize the world's countryside. *Nature* **2017**, *548*, 275. [CrossRef] [PubMed]
2. Liu, Y. Research on the urban-rural integration and rural revitalization in the new era in China. *Acta Geogr. Sin.* **2018**, *73*, 637–650. (In Chinese)
3. Rizzo, A.; Khan, S. Johor Bahru's response to transnational and national influences in the emerging Straits Mega-City Region. *Habitat Int.* **2013**, *40*, 154–162. [CrossRef]
4. Munya, P.W. A Bayesian Abduction Model for Sensemaking. Ph.D. Thesis, North Carolina Agricultural and Technical State University, Greensboro, NC, USA, 2014.
5. Cloke, P.; Marsden, T.; Mooney, P.H. *Handbook of Rural Studies*; Sage: Thousand Oaks, CA, USA, 2006; pp. 18–28.
6. Loures, L.; Vaz, E. Exploring expert perception towards brownfield redevelopment benefits according to their typology. *Habitat Int.* **2018**, *72*, 66–76. [CrossRef]
7. Liu, Y. Modern human-earth relationship and human-earth system science. *Sci. Geogr. Sin.* **2020**, *40*, 1221–1234. (In Chinese)
8. Long, H.; Zou, J.; Liu, Y. Differentiation of rural development driven by industrialization and urbanization in eastern coastal China. *Habitat Int.* **2009**, *33*, 454–462. [CrossRef]
9. Liu, Y.S.; Zang, Y.Z.; Yang, Y.Y. China's rural revitalization and development: Theory, technology and management. *J. Geogr. Sci.* **2020**, *30*, 1923–1942. [CrossRef]
10. Guo, Y.; Wang, J. Identifying the determinants of nongrain farming in China and its implications for agricultural development. *Land* **2021**, *10*, 902. [CrossRef]
11. Wang, Y.; Wen, Q.; Liu, Y. Achieving effective connection between rural revitalization and targeted poverty alleviation in poverty-stricken regions. *Sci. Geogr. Sin.* **2020**, *40*, 1840–1847. (In Chinese)
12. Chen, X. On the proposal of Rural Revitalization Strategy from the perspective of 40 years of rural reform. *Adm. Reform* **2018**, *4*, 4–10.
13. Wen, T. "Three rural issues": Reflection at the end of the century. *Reading* **1999**, *12*, 3–11.
14. Liu, Y.; Zhou, Y.; Liu, J. Regional differentiation characteristics of rural poverty and targeted- poverty alleviation strategy in China. *Bull. Chin. Acad. Sci.* **2016**, *31*, 269–278. (In Chinese)
15. Wu, X.; Fu, B.; Wang, S.; Song, S.; Li, Y.; Xu, Z.; Wei, Y.; Liu, J. Decoupling of SDGs followed by re-coupling as sustainable development progresses. *Nat. Sustain.* **2022**, *5*, 452–459. [CrossRef]
16. Li, X. There Will Be an Era without "Poverty" after 2020. Available online: http://www.sohu.com/a/128816336_481285 (accessed on 24 October 2021).
17. Zhou, Y.; Guo, Y.; Liu, Y. Comprehensive measurement of county poverty in China and the goal of poverty reduction after 2020. *Acta Geogr. Sin.* **2018**, *73*, 1478–1493.
18. Wang, S.; Feng, Z. The organic connection between poverty alleviation and Rural Revitalization: Logical relationship, connotation and key content. *J. Nanjing Agric. Univ.* **2019**, *19*, 8–14.
19. Huang, W.; Zhu, W. Multidimensional evaluation of the consolidation and expansion of targeted poverty alleviation: An empirical analysis of poverty alleviation practice in B city of G province in China. *Manag. World* **2021**, *37*, 111–128.
20. Zhou, Y.; Huang, H.; Liu, Y. The spatial distribution of Chinese villages and its influencing factors. *Acta Geogr. Sin.* **2020**, *75*, 2206–2223.
21. Yang, Y.; Bao, W.; Wang, Y.; Liu, Y. Measurement of urban-rural integration level and its spatial differentiation in China in the new century. *Habitat Int.* **2021**, *117*, 102420. [CrossRef]
22. Zhong, Y.; Chen, S. The difference and internal relationship between poverty alleviation and Rural Revitalization. *Agric. Econ.* **2020**, *3*, 94–96.
23. Wang, Y.; Li, Y. Promotion of degraded land consolidation to rural poverty alleviation in the agro-pastoral transition zone of northern China. *Land Use Policy* **2019**, *88*, 104114. [CrossRef]

24. Liu, Y. The basic theory and methodology of rural revitalization planning in China. *Acta Geogr. Sin.* **2020**, *75*, 1120–1133. (In Chinese)
25. Zhang, Y.; Long, H.; Ma, L.; Tu, S.; Chen, K. Research progress of urban- rural relations and its implications for rural revitalization. *Geogr. Res.* **2019**, *38*, 578–594.
26. Wang, Q.; Wang, Y.; Du, G. Geographic exploration of spatial differentiation characteristics and driving mechanism of cultivated land transfer in arid area based on man land relationship. *J. Agric. Resour. Environ.* **2012**, *38*, 241–248.
27. Qu, L.; Wang, Y.; Liu, Y.; Ma, Q. Evaluation of water and land resources system bearing capacity and path optimization for rural revitalization. *J. Nat. Resour.* **2021**, *36*, 300–314. (In Chinese) [CrossRef]
28. Hu, Y. Present situation and Countermeasures of Tan sheep industry in Yanchi County. *Mod. Agric. Sci. Technol.* **2017**, *12*, 261–263.
29. Guo, Y.; Zhou, Y.; Liu, Y. Targeted poverty alleviation and rural revitalization in poverty-stricken areas: Internal logic and mechanism. *Geogr. Res.* **2019**, *38*, 2819–2832. (In Chinese)
30. Wang, Y.; Xie, Y.; Wang, M. Research on rural e-commerce relationship network and local construction under the background of targeted Poverty Alleviation—A case study of Wuhua County, Guangdong Province. *World Reg. Stud.* **2017**, *26*, 119–130.
31. Huang, J. The Historical Shift of the Focus of the Work of "Three Rural Issues": One of No. 1 Central Document Commentary in 2021. *Economic Daily*, 24 February 2021.
32. Chen, Z.; Liu, X.; Lu, Z. The Expansion Mechanism of Rural Residential Land and Implications for Sustainable Regional Development: Evidence from the Baota District in China's Loess Plateau. *Land* **2021**, *10*, 172. [CrossRef]
33. Liu, Y.; Wang, Y. Rural land engineering and poverty alleviation: Lessons from typical regions in China. *J. Geogr. Sci.* **2019**, *29*, 643–657. [CrossRef]
34. He, R. Urban-rural integration and rural revitalization: Theory, mechanism and implementation. *Geogr. Res.* **2018**, *37*, 2127–2140.
35. Liu, Y.; Long, H.; Li, Y. Human geography research based on the new thinking of global rural-urban relationship. *Acta Geogr. Sin.* **2021**, *76*, 2869–2884. (In Chinese)
36. Yu, H.; Zhou, Z.; Wu, B. The gap between urban and rural areas, the evolution of agricultural productivity and agricultural subsidies—An analysis from the perspective of new structural economics. *China Rural. Econ.* **2019**, *10*, 40–59.
37. Alexander Wandl, D.I.; Nadin, V.; Zonneveld, W.; Rooij, R. Beyond urban–rural classifications: Characterising and mapping territories-in-between across Europe. *Landsc. Urban Plan.* **2014**, *130*, 50–63. [CrossRef]
38. Caschili, S.; Montis, A.D.; Trogu, D. Accessibility and rurality indicators for regional development. *Comput. Environ. Urban Syst.* **2015**, *49*, 98 114. [CrossRef]
39. Shi, L.; Wang, Y.; Liu, Y.; Li, Y. The poverty evolution of typical countries along the Belt and Road and implications from China's poverty reduction experiences. *J. Geogr. Sci.* **2022**, *32*, 458–476. [CrossRef]
40. Long, H.; Liu, Y.; Li, X.; Chen, Y. Building new countryside in China: A geographical perspective. *Land Use Policy* **2010**, *27*, 457–470. [CrossRef]
41. Li, Y.; Liu, Y.; Long, H.; Cui, W. Community-based rural residential land consolidation and allocation can help to revitalize hollowed villages in traditional agricultural areas of China: Evidence from Dancheng County, Henan Province. *Land Use Policy* **2014**, *39*, 188–198. [CrossRef]
42. Qu, L.; Li, Y.; Feng, W. Spatial-temporal differentiation of ecologically-sustainable land across selected settlements in China: An urban-rural perspective. *Ecol. Indic.* **2020**, *112*, 105783. [CrossRef]
43. Wen, Q.; Zheng, D. Identification and revitalization of rural poverty-stricken areas in northwest China. *Geogr. Res.* **2019**, *38*, 509–521.
44. Yang, R.; Xu, Q.; Zhou, J. Analysis of spatial transformation mechanism of traditional villages in Fengjian Village based on actor network theory. *Sci. Geogr. Sin.* **2018**, *38*, 1817–1827.
45. Yu, L.; Li, X. Local practice of industrial poverty alleviation and pro-poor mechanism. *Agric. Econ. Manag.* **2020**, *4*, 5–12.

land

MDPI

Article

Function Evolution of Oasis Cultivated Land and Its Trade-Off and Synergy Relationship in Xinjiang, China

Tianyi Cai [1,2], Xinhuan Zhang [1,*], Fuqiang Xia [1] and Danni Lu [1,2]

[1] State Key Laboratory of Desert and Oasis Ecology, Xinjiang Institute of Ecology and Geography, Chinese Academy of Sciences, Urumqi 830011, China
[2] University of Chinese Academy of Sciences, Beijing 100049, China
* Correspondence: zhangxh@ms.xjb.ac.cn; Tel.: +86-991-7827314

Abstract: Cultivated land (CL) performs irreplaceable manifold functions in social stability, economic development, and ecological security, which is also essential for the accomplishment of the United Nations Sustainable Development Goals 2030 Agenda. China is the world's most populous country, and it has important reference significance for the realization of the multi-function synergistic management of CL in China by revealing the evolution characteristics of cultivated land functions (CLFs) and the interaction between CLFs. However, the research to date has tended to focus on the eastern coastal areas and the central traditional agricultural areas of China. This study focuses specifically on Xinjiang, the main area of the arid region of northwest China. The connotations of social, economic, and ecological functions of oasis cultivated land (OCL) in Xinjiang were first discussed from a system theory perspective. Then, an evaluation index system of CLFs was constructed. On this basis, the evolution characteristics of CLFs and the interaction between CLFs in Xinjiang from 1990 to 2018 were quantitatively evaluated. Findings suggest that: (1) the economic function of the OCL in Xinjiang is strengthening, while the ecological function is degrading and the social function remains stable. Overall, the evolution of CLFs in Xinjiang was first dominated by ecological and social functions and then became economic-function-oriented; (2) the synergistic relationship between CLFs is weakening and the trade-off relationship is increasing over time. The trade-off effect between the economic function and other functions of OCL is strengthened gradually due to the OCL-use activities dominated by the economic function. This study not only enriches the regional content of CL multi-function research but can also provide reference for decision-making for the sustainable utilization and multi-function synergistic management of OCL in Xinjiang, China.

Keywords: oasis; cultivated land function; trade-off and synergy; multi-function management; Xinjiang

Citation: Cai, T.; Zhang, X.; Xia, F.; Lu, D. Function Evolution of Oasis Cultivated Land and Its Trade-Off and Synergy Relationship in Xinjiang, China. *Land* **2022**, *11*, 1399. https://doi.org/10.3390/land11091399

Academic Editor: Jianjun Zhang

Received: 5 August 2022
Accepted: 24 August 2022
Published: 26 August 2022

1. Introduction

A roadmap for "the future we want" in terms of human well-being and environmental sustainability has been constructed under the United Nations Sustainable Development Goals 2030 Agenda [1]. Satisfying the rising food consumption demand of humans on the premise of neither compromising multiple ecosystem services nor exceeding the safe limits of planetary boundaries will be a global challenge in sustainable land management [2–5]. As a natural resource essential for human survival and development, cultivated land (CL) has irreplaceable functions in safeguarding food security, maintaining ecosystem balance, supporting the livelihoods of farmers and herdsmen, and promoting economic development in rural areas [6]. Furthermore, these functions are correlated and interrelated [7]. The investigation of the sustainable management of CL from a multifunctional perspective has become a new paradigm in existing research on land science worldwide [8–11].

China is the world's most populous and the largest developing country. In addition to concerning the food security of the 1.4 billion Chinese people, China's sustainable management of CL is also immediately correlated with global food security and the United Nations Sustainable Development Goals 2030 Agenda. In that case, the way of achieving sustainable utilization and multifunctional synergistic management of CL in China during agricultural development is a crucial problem to be solved. To this end, various cultivated land functions (CLFs) should be first evaluated in a scientific manner. Then, the law of temporal and spatial evolution of CLFs and the interaction between functions and formation mechanisms need to be revealed. These are the scientific basis for promoting the realization of the multi-functional synergistic management of CL in China.

Considerable literature has grown up around the widely recognized multi-function property of CL. In terms of the functional connotation of CL, scholars hold the view that CL is a joint production system that produces both commercial and non-commercial outputs. Its underlying function is to provide products and services in conformity with the demands of human survival and development, which is normally associated with society, economy, and environment [12–16]. In terms of the classification of CLFs, there are currently two main classification approaches. One is to categorize CLFs in provisioning, regulating, supporting, and cultural functions with reference to the classification of ecosystem services [12,15–18]. The other is either to divide CLFs into social, economic, ecological, and cultural functions by inserting regional elements such as society, economy, ecology, and culture into the CL utilization system from the aspect of system theory [13,19–24] or to split CLFs into producing, living, and ecological functions [25,26]. In terms of the evaluation method of CLFs, the comprehensive functional index evaluation method is the most used quantitative method at present. Given its advantages of the application at multiple scales, the ability to describe multiple function types, and facilitating comparison between functions, it has been widely used in existing studies [12,13,16,19,20,22,25,26]. In addition, CLFs have also been evaluated using the value evaluation method [27], the material quality evaluation method [28], the energy analysis method [29], and the policy back-stepping method [30].

Affected by the changes in the products and services required for human survival and development in different periods, CLFs have shown distinct characteristics in different socio-economic stages [25,31], which is reflected in the shift from an emphasis on productivism-oriented functional needs to the more diversified functional requirements such as economy, society, culture, and ecology [9,10,32]. The manifestation of CLFs also varies in different regions. CLFs in Western developed countries, for instance, have been transformed markedly, with more emphasis on the functional requirements of CL in ecological services, landscape aesthetics, cultural entertainment, and the promotion of rural sustainable development [6,32]. By comparison, CL in China, a populous developing country, plays a momentous part in safeguarding food security and supporting farmers' livelihoods [30]. Further, significant spatial differences can also be found in the CLFs in different regions of China [23,26]. To be specific, the economic function of CL in developed urban agglomerations and their hinterland areas is declining [19,31], while the recreational functions of their CL are on the rise [24,33]. In the meantime, farmers in major agricultural production regions cause particular stress on the functions of elderly care and employment of CL [34]. Conversely, the ecological function of CL is more significant in key ecological function areas to the west of the "Hu Line" [26]. In terms of the interaction between CLFs, the research methods of ecosystem service trade-off and synergy have been adopted for measurement, such as the correlation coefficient method [21,23], bivariate spatial autocorrelation method [22,28], gray T-relational model [35], and coupling coordination degree [19], etc. However, in specific research processes, the above quantitative methods need to be selectively used in combination with their respective research scales and data support. The correlation between functions becomes more complex because of the diversity of CLFs and the preferences of human demands and choices. The synergy and trade-off relationships can be observed extensively between CLFs, which are presented in significant spatio-temporal heterogeneity [19,21–23,36,37]. Hence, it is of great necessity to research multiple functions of CL based on local conditions.

To sum up, previous studies have suffered from shortcomings, albeit while producing fruitful results. Regarding research content, a unified classification and evaluation system has not yet been formed in academia, since the multi-functional connotation, classification, and evaluation of CL is under continuous discussion for improvement, which should be further investigated and perfected. Regarding research areas, coastal regions in eastern China and traditional agricultural regions in central China with high levels of urbanization and modernization have been studied in most cases, with few studies concentrated on economically backward and ecologically vulnerable areas in western China. Regarding research perspectives, many scholars focus their research on the evaluation of CL multi-functionality and the revelation of the spatial pattern characteristics of a single time section, lacking intensive research on the long-term multifunctional evolution characteristics of CL and the interaction between various functions. In fact, this should have been an important basis for measuring whether the utilization of CL in an area is reasonable and sustainable and also important decision-making reference for propelling the synergistic management of multiple CLFs.

Located in the northwest inland of China, Xinjiang is the largest provincial-level administrative region and the main area of the arid region in northwest China [38]. Oases are the substantial basis for human survival and development in the arid region [39]. As the main body of the modern oasis of Xinjiang, CL with distinct regional features performs irreplaceable multi-fold functions in social stability, economic development, and ecological security. However, Xinjiang has become a hot spot in China's increment of CL over the past three decades [40,41]. The highly intensive CL activities lead to severe environmental problems while bringing about remarkable social–economic benefits [42]. Apparently, oasis cultivated land (OCL) utilization in Xinjiang is confronted with challenges in food, ecological, and social security, which should be transformed urgently towards sustainable utilization and multi-function synergistic management.

This study presents a case study of Xinjiang, the main area in the arid region of northwest China. Multi-function evolution characteristics of CL over a long period and the interaction between the functions are extensively examined in this paper. This study seeks to enrich the regional content of CL multi-function research and sets out to provide a decision-making reference for the sustainable utilization of CL and the multi-function synergistic management of CL in Xinjiang. The specific objectives of this study are: (i) investigating the functional connotation and classification of OCL in Xinjiang and constructing a multi-function evaluation index system for OCL, (ii) quantitatively evaluating and presenting the temporal and spatial evolution laws of the functions of OCL in Xinjiang from four temporal sections (1990, 2000, 2010, and 2018), (iii) investigating the trade-off and synergy relationship between functions of OCL in Xinjiang and their formation mechanisms, and (iv) making policy suggestions for the multi-function synergistic management of OCL in Xinjiang in accordance with the conclusions.

2. Materials and Methods

2.1. Study Area

Xinjiang, short for the Xinjiang Uygur Autonomous Region, 73°40′–96°23′ E and 34°25′–49°10′ N, is located in the middle of Eurasia and in the northwest of the People's Republic of China, which has a typical temperate continental arid climate. It has three mountain ranges, namely the Altai Mountains, Tianshan Mountains, and Kunlun Mountains, as well as two inland basins—the Junggar Basin and the Tarim Basin. They form a topography and landform of "three mountain ranges surrounding two basins" and an ecological landscape pattern of "mountain–oasis–desert", as shown in Figure 1; among them, CL is the main body of the artificial oasis of Xinjiang [38].

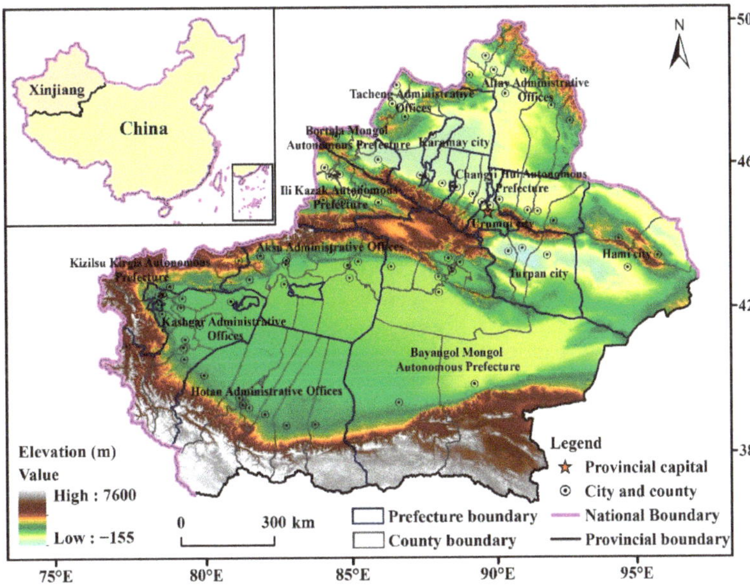

Figure 1. Geographical location and administrative division of Xinjiang.

As of 2018, CL in Xinjiang reached an area of 5.24 million hectares, accounting for roughly 3% of the total land area of Xinjiang. The total population of the region has reached 24.87 million, with the share of the urban population being 50.91% [43]. Nearly half of Xinjiang's population lives in rural areas due to lagging urbanization, with almost 80% of the total rural employees engaged in agriculture [43]. As the main spatial carrier of agricultural development and the production system most closely associated with oasis ecological security and rural social stability, OCL in Xinjiang bears typical regional characteristics and important research significance.

2.2. Methods

2.2.1. Connotation, Classification, and Evaluation Index of the OCL Function in Xinjiang

The arid region is a geographical whole consisting of three landscape systems (mountains, oases, and deserts) that are interrelated and mutually restricted in the ecological process [38,39]. To be specific, the mountain system provides resources such as water and soil to support the formation and development of oasis areas [38]. The oasis system, as an area with relatively higher productivity and a center of human production and life in the arid region, is a hub of gathering and exchanging materials and energy between mountain and desert systems. The oasis system includes four subsystems of population, society, economy, and the ecological environment [44]. The desert system has a vast area and harsh environment and is ecologically vulnerable [38].

The CL system is a complex system coupled with "nature–society–economy" [13,14,19,25]. The function of CL refers to the ability of CL to provide products and services that meet the needs of human survival and development during a certain period of development, utilization, and protection and under the comprehensive action of its constituent elements [25]. OCL is a key role in connecting the population, economy, society, and ecological subsystems of the oasis area [44]. When the population, society, economy, and ecological environment are embedded in the process of OCL utilization, the social, economic, and ecological functions of OCL are correspondingly formed (Figure 2). The use of OCL is essentially to meet the needs of human beings for the above three functions of CL. The rationality of the human's CL-use activities in the oasis area is also indirectly reflected by the changes of the three functions of CL and the interaction between them.

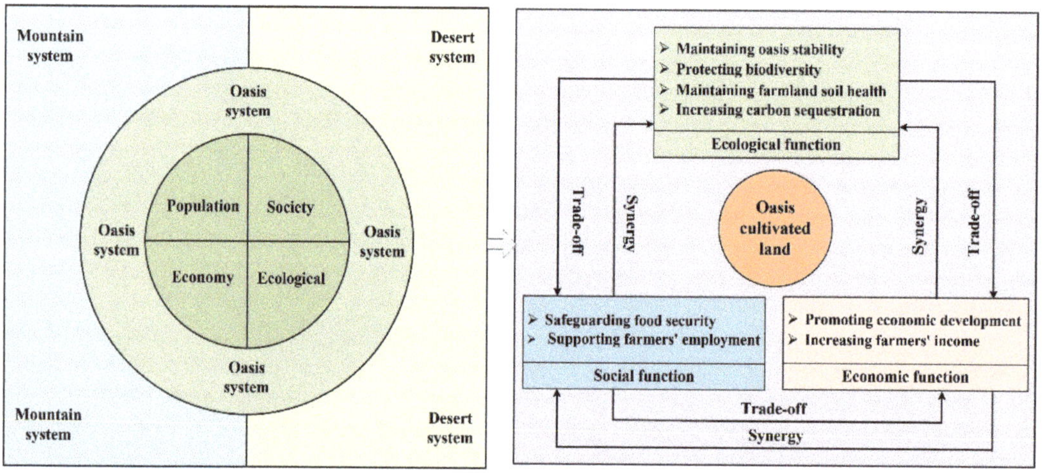

Figure 2. Functional connotation and classification of OCL in Xinjiang.

Specifically, the social function of cultivated land (SFC) in Xinjiang is reflected in the important role of OCL in safeguarding food security and supporting farmers' employment in the process of social and economic development of the oasis region, which is directly associated with the stability of Xinjiang. The Xinjiang oasis region is separate from the main grain-producing areas of China. Moving grain over long distances increases transportation costs and risks. Meanwhile, there are also uncertainties in grain production due to climatic aridity and vulnerable ecology. Therefore, OCL plays an important role in ensuring food self-sufficiency in oasis areas. The contracted farmland is the basic means of production and the workplace for farmers since the household contract responsibility system was implemented by the Chinese government in rural areas in the 1980s [45]. In particular, the employment of farmers is extremely reliant on OCL and agriculture due to the lagging development of secondary and tertiary industries in the oasis area of Xinjiang. Therefore, the OCL is prominent in promoting the active workforce in rural areas. The function of OCL in grain production and supply was described by two indicators, including the grain yield per unit area and the grain self-sufficiency rate. The function of OCL in employment security was described by two indicators, including the agricultural dependency of employment and the population-carrying capacity of OCL. The specific calculation methods of the above indicators are shown in Table 1.

The economic function of cultivated land (ENFC) in Xinjiang is reflected in the important role of OCL in promoting regional economic development and increasing farmers' income by producing agricultural products and participating in market transactions to obtain economic benefits. The oasis is an area with relatively higher productivity in the arid region. Furthermore, agriculture is the initial economic form of the oasis, the foundation of national economic development, and the safeguard of industrial development in oasis areas [44]. On the one hand, agricultural producers increase the output and economic benefits of agricultural products through fully utilizing agricultural production conditions such as water, soil, light, and heat in the oasis. On the other hand, the economic output of OCL is also propelled by expanding the sown area of cash crops based on the continuous adjustment of the agricultural planting structure. The function of OCL in propelling regional economic development was described by two indicators, including the contribution of agriculture to the national economy and the management benefits of OCL. Furthermore, the function of OCL in increasing farmers' incomes was described by two indicators, including the per capita net income of rural households and the planting proportion of non-food crops. The specific calculation methods of the above indicators are shown in Table 1.

Table 1. Evaluation index system of oasis cultivated land function in Xinjiang.

Function Layer	Index Layer	Calculation Method	Unit
Social function	Grain yield per unit area	Total grain output/sown area of crops	ton/hm^2
	Grain self-sufficiency ratio	Total grain output/(total regional population × 400 kg) × 100%	ton/hm^2
	Agricultural dependency of employment	Number of agricultural employees/rural employees × 100%	%
	Population-carrying capacity of OCL	(Rural population − animal husbandry population)/CL area	person/hm^2
Economic function	Contribution of agriculture to the national economy	Gross output of the planting industry/GDP × 100%	%
	Management benefit of OCL	Gross output of the planting industry/CL area	10^4 yuan/hm^2
	Farmer's income level	Per capita net income of rural households	yuan
	Planting proportion of non-food crops	Sown area of non-grain crops/total sown area of crops × 100%	%
Ecological function	Ecological service value of CL conversion area [a]	Calculation based on ArcGis software and Xie's method	10^9 yuan
	Farmland biodiversity [b]	$-\sum pi \cdot \ln(pi)$, pi is the proportion of the sown area of various crops	-
	Agrochemical load of farmland [c]	(Chemical fertilizer + pesticide + plastic film) /CL area	10^{15} sej/hm^2
	Net carbon sink level of farmland [d]	(CL carbon sink − CL carbon release)/CL area	ton C/hm^2

Notes: The attributes of the evaluation indicators in the above table are all positive indicators, except the agrochemical load of farmland. [a]: The coefficients of ecological service value in the study period are calculated as 16,599, 59,085, 24,521, 95,289, 115,082, and 2921 for cultivated land, forests, grasslands, rivers/lakes, wetlands, and unused lands, respectively, in Xinjiang, using Xie's calculation method [46], in combination with the food production and price levels in Xinjiang. The unit is yuan/hm^2. [b]: The biodiversity of cultivated land is calculated using Song's calculation method [30], with twelve crop planting types including wheat, rice, corn, beans, potatoes, cotton, oil seeds, sugar beets, vegetables, melons for fruit, and alfalfa [43]. [c]: Usages of chemical fertilizers, plastic films, and pesticides were converted into solar energy values with a unified dimension before the sum calculation with reference to the method of Chang [47], and their conversion rates of solar energy values are 4.7×10^9 sej/g, 6.38×10^7 sej/g, and 2.69×10^9 sej/g, respectively. [d]: The carbon sink of cultivated land in Xinjiang is calculated by using Tian's calculation method [48]. Furthermore, the carbon emission of cultivated land involves the carbon emission caused by the utilization of cultivated land, with the carbon emission coefficient listed as fertilizer 1.285 kg C/kg [49], pesticide 4.9341 kg C/kg, plastic film 5.18 kg C/kg [48], agricultural diesel 0.5927 kg C/kg, irrigation 20.476 kg C/hm^2, and tillage 3.126 kg C/kg hm^2 [50].

The ecological function of cultivated land (ELFC) in Xinjiang is reflected in the important role of OCL in maintaining oasis stability, protecting biodiversity, maintaining farmland soil health, and increasing carbon sequestration. Oasisization and desertification are two elementary geographical processes with mutual feedback [39]. As the expansion of CL is a main form of oasisization, large quantities of natural ecological land will inevitably be sacrificed due to the excessive expansion of CL. As a result, it will result in breaking the dynamic balance between oasisization and desertification, threatening the stability of the oasis. Since diversified farmland crop is an ecological attribute of the OCL resource system, planting various crops is essential for protecting the biological diversity in the OCL [30]. Soil health is the basis for guaranteeing the function of OCL. Specifically, the increased soil nutrients, the enhanced soil fertility, and the control of weeds and pests can benefit from the rational utilization of agricultural chemicals. However, the irrational and excessive application of agricultural chemicals might bring forth negative feedback, such as damaging soil structure and exacerbating soil environmental load [11]. Oasis crops store carbon dioxide absorbed from the atmosphere via photosynthesis in an organic form, greatly contributing to carbon sequestration. Moreover, humans can also increase their carbon sink outputs through the management measures of OCL. Furthermore, there are also carbon emissions in the utilization of OCL in the agriculture production process [48]. Hence, the ELFC in Xinjiang was described by four indicators, including the ecological service value of the OCL conversion area, farmland biodiversity, farmland agrochemical load, and farmland net carbon sink level. The specific calculation methods of the above indicators are shown in Table 1.

2.2.2. Measurements for CLFs

To begin with, raw data were processed using MinMaxScaler to remove the influence of the index dimension. It is calculated as follows:

$$\text{Positive index}: \quad x_{ij} = \frac{X_{ij} - \min\{X_j\}}{\max\{X_j\} - \min\{X_j\}} \tag{1}$$

$$\text{Negative index}: \quad x_{ij} = \frac{\max\{X_j\} - X_{ij}}{\max\{X_j\} - \min\{X_j\}} \tag{2}$$

When the index is positive, the greater the better; when the index is negative, the smaller the better. County-level data of the same index at different temporal sections were processed for standardization to safeguard the comparability of data in different temporal sections.

Based on the function index, three functions of OCL were then calculated as shown below:

$$F(SFC) = \sum_{j=1}^{4} W(SFC_j) \times f(SFC_j) \tag{3}$$

$$F(ENFC) = \sum_{j=1}^{4} W(ENFC_j) \times f(ENFC_j) \tag{4}$$

$$F(ELFC) = \sum_{j=1}^{4} W(ELFC_j) \times f(ELFC_j) \tag{5}$$

where $F(SFC)$, $F(ENFC)$, and $F(ELFC)$ are the social, economic, and ecological function indices of CL, respectively; $W(SFC_j)$, $W(ENFC_j)$, $W(ELFC_j)$ and $f(SFC_j)$, $f(ENFC_j)$, $f(ELFC_j)$ are the weight of the j-th characterization index and the standardized indices for the evaluations of social, economic, and ecological function evaluation of OCL, respectively. Moreover, all function indices ranged from 0 to 1, and the larger the index, the stronger the corresponding function. Characteristic indices corresponding to three CLFs were summarized using the equal-weight method. In other words, the weight value of each index was 0.25.

2.2.3. Analysis Method for the Evolution of CLFs

The spatio-temporal characteristics of the function evolution of OCL at the county scale of Xinjiang were described using the function grading method and the longitudinal comparison coefficient method. All CLFs indices in four-time sections (1990, 2000, 2010, 2018) of 84 counties in Xinjiang were first converted into a sequence of 336 units for each function through accumulation during function grading, and each CLF index was then divided into a high-value area, mid-value area, and low-value area using the natural breaking point grading method based on ArcGIS10.7 software.

The longitudinal comparison coefficient of the function is calculated as follows:

$$LCC_{ij,t} = F_{ij,t2}/F_{ij,t1} \tag{6}$$

where $LCC_{ij,t}$ is the longitudinal comparison coefficient of the j-th function of OCL at the period t in the county (city) i in the study area; $F_{ij,t1}$ and $F_{ij,t2}$ are the j-th CLF indices in the county i of the study area in the base and end periods, respectively. If $LCC_{ij,t}$ is greater than 1, it indicates that the CLF is enhanced in the study period; if not, it indicates a gradual decline.

2.2.4. Analysis Method of Interaction between CLFs

As only four time sections (incl. 1990, 2000, 2010 and 2018) were involved in the calculation of the CLFs of varying counties in the study, the demands for sample size in the typical mathematical statistics methods cannot be satisfied. In that case, the trade-off and synergy among the functions of OCL were analyzed using the gray T-relational model improved by Huang and Wang [51]. With no strict requirement for sample size, the model adopted describes the strength, size, and order of the relations between factors using the order of gray correlation, which is superior in terms of uniqueness, symmetry, stability, and standardization. It can also be used for reflecting the positive and negative correlations of the sequence [51]. The model can be calculated as follows:

For the temporal section, $[a, b], b > a \geq 0$, assuming

$$\Delta t_k = t_k - t_{k-1}, [a, b] = \bigcup_{k=2}^{n} \Delta t_k, \Delta t_k \cap \Delta t_{k-1} = \phi, k = 2, 3, \cdots, n \tag{7}$$

The two original time series set on $[a, b]$ are:

$$X_1 = \{x_1(t_1), x_1(t_2), \cdots, x_1(t_n)\}; X_2 = \{x_2(t_1), x_2(t_2), \cdots, x_2(t_n)\} \tag{8}$$

Of which, $x_i(t_k), i = 1, 2; k = 2, 3, \cdots, n$ is not equal to 0, then

$$y_i(t_k) = \frac{x_i(t_k) - x_i(t_{k-1})}{x_i(t_{k-1})}, i = 1, 2; k = 2, 3, \cdots, n \tag{9}$$

It represents the relative increment of the sequence X_i from the time point t_{k-1} to the time point t_k, that is, the relative initialization is performed on the original sequence.

$$D_i = \frac{\sum_{k=2}^{n} |y_i(t_k)|}{n-1}, i = 1, 2 \tag{10}$$

It represents the mean of the absolute value of the relative increment of the sequence in various periods of time.

$$z_i(t_k) = \frac{y_i(t_k)}{D_i}, i = 1, 2; k = 2, 3, \cdots, n \tag{11}$$

It represents the equalization of the relative increment of the sequence X_i from the time point t_{k-1} to the time point t_k.

The improved gray T-relational coefficient $\xi(t_k)$ of the sequences X_1 and X_2 from the time point t_{k-1} to the time point t_k is presented below:

i When $z_1(t_k)$ and $z_2(t_k)$ are not zero simultaneously,

$$\frac{sign(z_1(t_k) \cdot z_2(t_k))}{\frac{1}{2} + \frac{1}{2}||z_1(t_k)| - |z_2(t_k)|| + \frac{1}{2}\left[1 - \frac{\min(|z_1(t_k)|, |z_2(t_k)|)}{\max(|z_1(t_k)|, |z_2(t_k)|)}\right] + \frac{1}{2}\left[\frac{\max(|y_1(t_k)|, |y_2(t_k)|)}{\min(|y_1(t_k)|, |y_2(t_k)|)}\right]} \tag{12}$$

ii When $z_1(t_k)$ and $z_2(t_k)$ are not zero simultaneously, or when $y_1(t_k)$ and $y_2(t_k)$ are zero simultaneously, $\xi(t_k) = 1$; when $y_1(t_k)$ and $y_2(t_k)$ are not zero simultaneously, then $\xi(t_k) = 0$.

Where $sign(z_1(t_k) \cdot z_2(t_k))$ is the sign function, showing the positive and negative correlation. To be specific, when $sign(z_1(t_k) \cdot z_2(t_k)) > 0$, it indicates a positive correlation; when $sign(z_1(t_k) \cdot z_2(t_k)) < 0$, it shows a negative correlation.

The improved gray T-relational coefficient r of the sequences X_1 and X_2 from the time point t_{k-1} to the time point t_k is presented below:

$$r = \frac{1}{b-a} \sum_{k=2}^{n} \Delta t_k \cdot \xi(t_k) \tag{13}$$

When $t_k = kk = 1, 2, \cdots, n$, it can be simplified as:

$$r = \frac{1}{n-1} \sum_{k=2}^{n} \xi(t_k) \tag{14}$$

When $-1 \leq r < 0$, the sequences X_1 and X_2 are negatively correlated. That is to say, there is a trade-off relationship between both, and the closer the value approaching to -1, the stronger the trade-off. When $0 < r \leq 1$, the sequences X_1 and X_2 are positively correlated. In other words, there is a synergistic relationship between both. The closer the value approaches 1, the stronger the synergy is. When $r = 0$, the sequences X_1 and X_2 are independent of each other.

2.3. Data Sources and Processing

Data of CL area in 1990 and 2000 were acquired from the Statistical Yearbook of Xinjiang Uygur Autonomous Region and those in 2010 and 2018 were from the Ledger Data of Land Change Survey of Xinjiang Uygur Autonomous Region. Other relevant socio-economic statistical data were obtained from the Xinjiang Statistical Yearbook and the Statistical Yearbook of Xinjiang Prefectures of the corresponding years. The missing data of some indices were replaced by the mean of adjacent years or the data of adjacent years. Outputs of the national economy and the planting industry were converted into comparable prices, with 1990 as the base period according to the price index of the corresponding year. Furthermore, the per capita net income of rural households was converted into the comparable price, with 1990 as the base period according to the consumer price index of rural residents of the corresponding year. The land-use remote sensing monitoring data of Xinjiang were obtained from the Resource and Environment Science Data Center of the Chinese Academy of Sciences (http://www.resdc.cn (accessed on 10 June 2021)) with a spatial resolution of 30 m. Administrative boundary vector data were from the basic geographic databases of China at a 1:1 million scale released by the China National Catalog Service for Geographic Information (http://www.webmap.cn (accessed on 5 January 2021)).

3. Results

3.1. Spatial and Temporal Evolution Characteristics of CLFs

3.1.1. General Characteristics

Evolution trends of the SFC, ENFC, and ELFC of Xinjiang varied significantly in the study period (Figure 3). The ENFC was increasing, with its function index increasing from 0.23 in 1990 to 0.44 in 2018, showing a relative increment of 91.30%. The ELFC was degrading gradually, with its function index declining from 0.51 in 1990 to 0.43 in 2018, showing a relative decrement of 15.69%. The SFC was stable as a whole, showing an increasing trend and then a decreasing trend. Its function index first increased from 0.40 in 1990 to 0.47 in 2010 and then decreased to 0.42 in 2018, showing a relative increment of 5%. To sum up, the evolution of CLFs in Xinjiang was first dominated by ecological and social functions and then became economic-function-oriented.

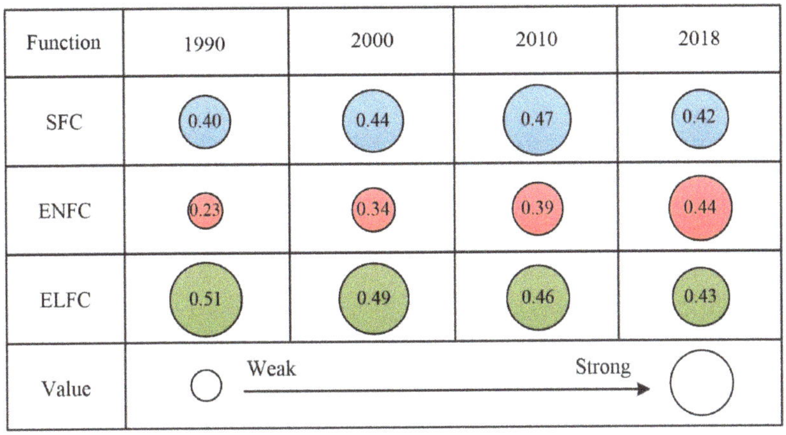

Figure 3. Overall evolution trends of oasis cultivated land functions of Xinjiang from 1990 to 2018.

3.1.2. Spatial Differences

1. Social function;

The spatio-temporal evolution characteristics of the SFC at the county level in Xinjiang are reported in Figure 4a–d. The numbers of counties included in the high-value, mid-value, and low-value areas of the SFC were 25, 38, and 21, respectively, in 1990. Specifically, the high-value area was mainly distributed in the Hotan Administrative Offices (HTAO) and Kashgar Administrative Offices (KSAO) in southern Xinjiang, in which most counties are national-level poverty-stricken areas, with OCL as the main source of livelihood. The low-value area was mainly distributed in the Altay Administrative Offices (ATAO) in northern Xinjiang. This is a traditional grassland pastoral area in Xinjiang, and its grassland resources performs a more significant social function than OCL. Furthermore, the mid-value area was mainly distributed in the main agricultural production regions along the southern and northern slopes of the Tianshan Mountains. By 2018, 31, 36, and 17 counties were contained in the high-value, middle-value, and low-value areas, respectively, for the SFC. Furthermore, their corresponding spatial distribution also changed. The high-value area changed into a scattered distribution from a centralized distribution. The high-value area in southern Xinjiang was mainly distributed in HTAO, western KSAO, and the northern Aksu Administrative Offices (ASAO), where the OCL was the main source of farmers' employment and income increase due to the dense population and low urbanization. Furthermore, the high-value area in northern Xinjiang was mainly distributed in Tacheng Administrative Offices (TCAO), the Ili River Valley, and the Changji Hui Autonomous Prefecture (CHAP), where the social function of OCL was significant due to its rapid growth in grain production and supply capacity. Besides, the low-value area was scattered and located in animal husbandry counties along the border, and the mid-value area was distributed throughout the southern and northern regions of Xinjiang in clusters.

The longitudinal comparison coefficient of the SFC ranged from 0.51 to 6.06 from 1990 to 2018, as shown in Figure 4e. The SFC was strengthening in 48 counties and decreased in the remaining 36 counties in 2018, compared with that in 1990. Most of the 48 counties with strengthened social functions were distributed in northern Xinjiang, while the 36 counties with decreased social functions were mostly in southern Xinjiang. The SFC at the county level in Xinjiang presents the regional characteristics of "increasing in the north and decreasing in the south", with an unchanged pattern of "weak in the north and strong in the south".

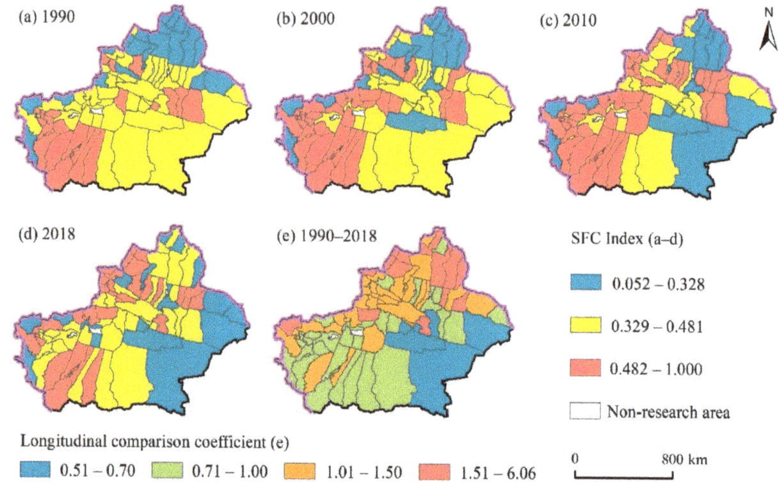

Figure 4. Temporal and spatial evolution of the SFC at the county level in Xinjiang from 1990 to 2018: (**a**) SFC in 1990; (**b**) SFC in 2000; (**c**) SFC in 2010; (**d**) SFC in 2018; and (**e**) Changes in SFC in 2018 compared to 1990.

2. Economic function;

The spatio-temporal evolution characteristics of the ENFC at the county level in Xinjiang are reported in Figure 5a–d. There were 0, 19, and 65 counties included in the high-value, mid-value, and low-value areas in 1990, respectively, presenting a low ENFC as a whole. The mid-value area was distributed in Turpan, KSAO, and western HTAO. Other counties fell into the low-value area. The ENFC at the county level has been enhanced significantly by 2018, with 29, 43, and 12 counties included in the high-value, mid-value, and low-value areas for the ENFC, respectively. A part of the high-value area was distributed in the striped area in the Manas River Basin and the Tarim River, which are mostly agricultural production areas with satisfactory agricultural production suitability and high-level productivity of OCL. Another part is distributed in Bayangol Mongol Autonomous Prefecture (BYMAP), Turpan city, and Hami city, which presents a more prominent ENFC through shifting the planting structure of OCL from traditional food crops to high-yield economic crops, such as cotton and special forest fruits. The low-value area was dominated by part of the animal husbandry counties, which are rich in grassland resources, with a high proportion of animal husbandry in the agricultural industrial structure. The mid-value area was mainly distributed in HTAO, eastern CHAP, and ATAO.

The longitudinal comparison coefficient of the ENFC ranged from 1.01 to 5.51 from 1990 to 2018, as shown in Figure 5e. In 41 counties of Xinjiang, the ENFC has improved faster than Xinjiang's average level, mainly distributed in ATAO, CHAP, Turpan city, Hami city, and BYMAP. To sum up, the ENFC at the county level in Xinjiang sees a generally upward trend, which is "faster in the northeast and slower in the southwest", whereas the ENFC shows a distribution pattern that is "stronger in the southeast and weaker in the northwest".

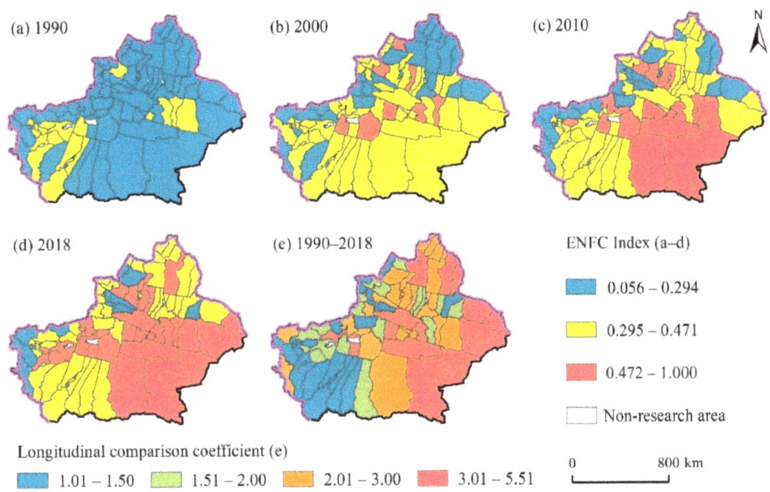

Figure 5. Temporal and spatial evolution of the ENFC at the county level in Xinjiang from 1990 to 2018: (**a**) ENFC in 1990; (**b**) ENFC in 2000; (**c**) ENFC in 2010; (**d**) ENFC in 2018; and (**e**) Changes in ENFC in 2018 compared to 1990.

3. Ecological function.

The spatio-temporal evolution characteristics of the ELFC at the county level in Xinjiang are reported in Figure 6a–d. There were 29, 37, and 18 counties included in the high-value, mid-value, and low-value areas, respectively, for the ELFC in 1990. The high-value area was distributed along the northern slope of the Tianshan mountains, ASAO, and eastern KSAO; the mid-value area was distributed in various prefectures in southern Xinjiang; the low-value area was distributed in northern and southern Xinjiang. As of 2018, there have been 29, 43, and 12 counties included in the high-value, mid-value, and low-value areas for the ELFC, respectively. The high-value area was distributed in eastern CHAP and the Ili River Valley, which has good biodiversity of farmland and a high level of net carbon sink of OCL. A part of the low-value area was located in BYMAP, KSAO, and ASAO, which have large populations and a high utilization rate of agricultural water and soil resources with a prominent imbalance between population and land. The other part of the low-value areas was distributed in Turpan city and Hami city, where economic crops such as cotton, special fruits, and vegetables have been substituted for food crops, with higher usage of agrochemicals for economic crops than that of food crops [52], causing more significant damages to ELFC due to the high environmental load of OCL in these areas.

The longitudinal comparison coefficient of the ELFC ranged from 0.32 to 1.68, as shown in Figure 6e. The ELFC was declining in 50 counties distributed in the western end of the northern slope of the Tianshan Mountains, BYMAP, ASAO, KSAO, Turpan city, and Hami city, and increasing in the remaining 34 counties. To sum up, the ELFC at the county level in Xinjiang is generally dominated by a downward trend, which is "faster in the southeast and slower in the northwest", whereas the ELFC shows a distribution pattern that is "stronger in the northwest and weaker in the southeast".

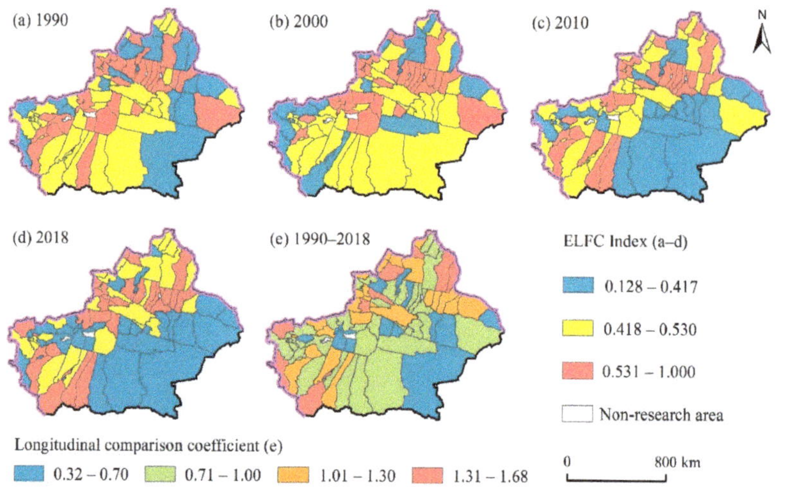

Figure 6. Temporal and spatial evolution of the ELFC at the county level in Xinjiang from 1990 to 2018: (**a**) ELFC in 1990; (**b**) ELFC in 2000; (**c**) ELFC in 2010; (**d**) ELFC in 2018; and (**e**) Changes in ELFC in 2018 compared to 1990.

3.2. *Trade-Off and Synergy Relationship between CLFs*

3.2.1. Temporal Scale

Gray T-relational coefficients between two out of the three CLFs in Xinjiang are shown in Table 2. The findings show that the relationship between CLFs in Xinjiang is mainly synergistic, but the synergistic relationship between CLFs is weakening and the trade-off relationship is increasing over time. Also, there are differences in the interaction relationship between CLFs.

Table 2. Gray T-relational coefficients between CLFs in Xinjiang.

Function Group	Gray T-Relational Coefficients			
	1990	2000	2010	2018
SFC—ENFC	0.1812	0.1039	0.0685	−0.0491
SFC—ELFC	0.0859	0.0530	0.1025	0.1160
ENFC—ELFC	0.1814	0.0716	−0.0779	−0.1021

The SFC and ENFC were synergistic in 1990 (r = 0.1812), which gradually weakened in 2000 and 2010. As of 2018, the relationship between both has changed from synergy to trade-off (r = −0.0491). The SFC and ELFC were synergistic, while the synergistic intensity first increased and then decreased throughout the research period. As a whole, the synergy between the two is strengthened as the gray T-relational coefficient (r) developed into 0.1160 in 2018 from 0.0859 in 1990. The synergy between ENFC and ELFC in 1990 (r = 0.1814) weakened in 2000 and changed into a trade-off in 2010 (r = −0.0779) that further strengthened in 2018 (r = −0.1021).

In the early stage of the research period, OCL in Xinjiang was adopted to satisfy the regional food self-sufficiency, with a low proportion of economic crops. The agricultural production method was backward, and the increase in OCL production relied on intensive labor input and the low level of soil and water development. There were fewer inputs of production-increasing factors such as pesticides, chemical fertilizers, and plastic film; human activities exerted less influence on the ecosystem of OCL. Affected by the above factors, the three functions of OCL in Xinjiang showed a relatively low-level synergy at this stage.

In the later stage of the research period, some rural labor forces were replaced by labor-saving agricultural machines, along with the improved urbanization level and the progress in agricultural mechanization, weakening the SFC's ability to absorb farmer employment. With the continuous improvement of grain production and self-sufficiency, local governments and farmers have expanded the proportion of cash crops in the planting structure under the inducement of economic benefits and widely used agrochemicals to increase farmland production. Besides, a large amount of ecological land was reclaimed to expand the scale of OCL and increase the benefits. In consequence, the ELFC showed a decline with the continuous strengthening of ENFC under the disturbance of high-intensity human agricultural activities. The SFC and ENFC have also gradually changed from a synergistic relationship to a trade-off relationship.

3.2.2. Spatial Scale

Trade-offs and synergies relationship between the three CLFs at the county level in Xinjiang and their spatial differences are reported in Figure 7.

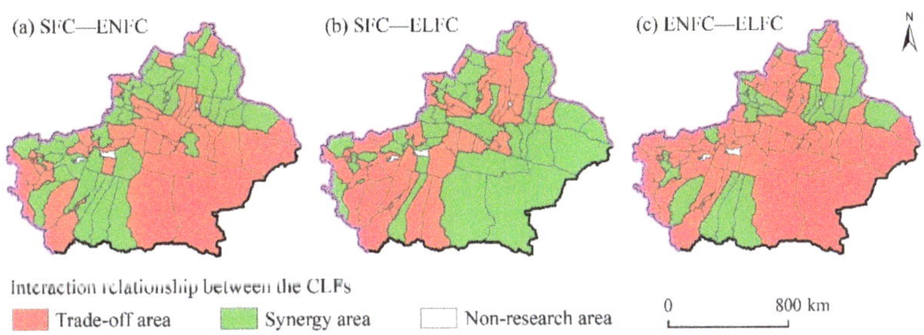

Figure 7. Pattern of interaction between CLFs at the county level in Xinjiang: (**a**) The interaction between SFC and ENFC; (**b**) The interaction between SFC and ELFC; and (**c**) The interaction between ENFC and ELFC.

The SFC and ENFC at the county level in Xinjiang were dominated by synergy (incl. 45 counties) and concurred with trade-offs (incl. 39 counties), as shown in Figure 7a. Those counties with a synergistic relationship between the two are distributed in TCAO, Ili River Valley area, ATAO, eastern CHAP, HTAO, Kizilsu Kyrgyz Autonomous Prefecture, and eastern KSAO in Xinjiang. Those counties with a trade-off relationship are distributed in BYMAP, Turpan city, Hami city, western CHAP, and western KSAO.

The SFC and ELFC at the county level in Xinjiang showed a pattern of equivalence between synergies (incl. 42 counties) and trade-offs (incl. 42 counties), as shown in Figure 7b. Those counties pertaining to the synergy are distributed in the TCAO, Bortala Mongol Autonomous Prefecture (BTMAP), Turpan city, Hami city, eastern CHAP, and southern BYMAP. Those counties belonging to the trade-off are distributed in HTAO, KSAO, northern Ili Valley, and ATAO.

The ENFC and ELFC at the county level in Xinjiang showed a pattern dominated by trade-offs (incl. 53 counties) and concurred with synergies (incl. 31 counties), as shown in Figure 7c. Those counties belonging to synergy are distributed in Ili Valley, the eastern section of the northern slope of Tianshan mountains, and eastern HTAO. Those counties belonging to trade-off are distributed in the Tacheng Basin, the western end of the northern slope of the Tianshan Mountains, Turpan city, Hami city, BYMAP, ASAO, KSAO, and Kizilsu Kyrgyz Autonomous Prefecture.

4. Discussion

4.1. Similarities and Differences between Xinjiang and Other Regions in the Functional Evolution of CL

Previous studies have shown that the overall evolution of CLFs has obviously different characteristics in different socio-economic stages [9,25,31,32]. This conclusion has also been ascertained in this case study of Xinjiang. The SFC and ELFC of Xinjiang were relatively significant between 1990 and 2010; the ENFC has been strengthened since 2010, more significantly than the SFC and ELFC. This also reflects that the CLFs in Xinjiang has shifted from serving food and survival needs in the past to serving the new needs of increasing farmers' income and promoting rural economic development currently. The difference is that the CLFs in Western developed countries has been significantly transformed from productivism to post-productivism [6,32], and the CLFs in China has also transformed since 2006 and are moving towards the synergy of multiple functions [30]. However, the CLFs in Xinjiang are characterized by polarized evolution (with the strengthening economic functions and weakening ecological and social functions). The utilization of CL focusing on the pursuit of economic functions inevitably has a trade-off effect on other cultivated land functions, thereby affecting the sustainable use of OCL.

From the perspective of changes in various CLFs, previous studies have shown that the SFC are constantly weakening with the progress of urbanization and agricultural modernization, which has been ascertained in studies conducted nationwide [30], in the eastern coastal areas [31,35,36], and in the central traditional agricultural areas of China [19,28]. However, the SFC in Xinjiang was stable as a whole in the research period. It is different from the findings in other regions of China. There are two causes for this special phenomenon. On the one hand, Xinjiang is far away from China's main grain-producing areas, so the SFC for Xinjiang's grain self-sufficiency is relatively stable. On the other hand, the urbanization in Xinjiang (50.91% as of 2018) is low with the backward development of secondary and tertiary industries. Oasis farmland remains a stabilizer for carrying the employment of the majority of farmers when the abilities to absorb employment in urban and non-agricultural industries are insufficient.

The ENFC in Xinjiang has been enhanced. The trend is similar to previous findings in other regions of China [19,28]. However, there is a difference that the growth of economic function is more prominent in OCL. That is because the oasis area shows favorable collaboration between water, soil, light, and heat resources with flat and contiguously distributed land, as well as high productivity of CL [44]. More than that, the proportions of economic crops such as cotton, oilseeds, melons, and fruits are rising in the agricultural planting structure of Xinjiang on the basis of food self-sufficiency, contributing to the manifestation of the ENFC.

The degradation of the ELFC is commonly accompanied by the strengthening of the ENFC in different areas of China. Such a phenomenon is more prominent in Xinjiang, which can be boiled down to two reasons. First, Xinjiang has been a hot spot in China's increment of CL over in the past 30 years [41]. The significant expansion of CL in Xinjiang has occupied a large amount of natural ecological land and scarce water resources, which has caused a severe negative impact on the ecological stability of the oasis area [39,42]. Second, the increased output in the CL of Xinjiang has a strong dependence on chemical fertilizers and pesticides. The mean application intensity of chemical fertilizers (358.23 kg) and pesticides (4.23 kg) per hectare of CL in Xinjiang is much higher than that of the fertilizers (125.53 kg) and pesticides (3.68 kg) used worldwide in the same period [37]. Meanwhile, Xinjiang is also a main cotton production area and the province with the largest consumption of plastic film in China. Its residual plastic film content of farmland is 4–5 times higher than the mean level in China [53], which leads to a high agrochemical load on CL and damages the ecological environment.

OCL in Xinjiang is essential for ensuring food security, supporting farmers' employment and social stability, promoting economic development in rural areas, and maintaining ecological security. The rationality of regional CL-use activities can be reflected indirectly by the interaction between various functions of OCL. According to the research findings, the synergy among the three functions of Xinjiang's CL is weakening over time, whereas the trade-off is strengthening. This indicates that the interaction between CLFs changes with human preferences for CLFs and changes in CL-use patterns and intensity, which has also been verified in some research in other regions [19,21–23,36].

4.2. Policy Recommendations of Multi-Function Synergistic Management of OCL in Xinjiang

The trade-off effect between the economic function and other functions of CL is strengthened gradually due to the CL-use activities dominated by the economic function in Xinjiang. This is in fact a revelation to us, that the unreasonable utilization of OCL in Xinjiang should be addressed by managing the trade-offs between CLFs and guiding the synergistic development of social, economic, and ecological functions of OCL in the process of future utilization. Four countermeasures and suggestions are proposed in response to the above research contents.

First, the SFC in the food production and self-sufficiency of Xinjiang should be stabilized. It is the geographical location of Xinjiang that determines the high cost and risk of long-distance grain transportation. Also, there are also uncertain risks in food production under global climate change and the vulnerability of regional ecological backgrounds. Therefore, it is imperative to depend on the OCL to deal with the problem of food supply of Xinjiang. On the one hand, the permanent basic CL protection system should be strictly implemented, with strict control over the occupation of high-quality CL resources during urbanization. On the other hand, the planting structure of CL should be optimized and adjusted in combination with the dynamics of population and grain demand in different counties and cities, preventing the excessive "non-grain" orientation of CL. Moreover, the construction of high-standard farmland should be also promoted in the oasis area to strengthen the irrigation security and agricultural mechanization of grain production, contributing to increasing the grain production capacity of OCL.

Second, the SFC in ensuring farmers' employment of Xinjiang should be restrained. The level of urbanization in Xinjiang is relatively low, and the development of the non-agricultural economy is relatively backward. This has resulted in many rural laborers relying on OCL resources for their livelihoods, exacerbating artificial pressure on OCL and resulting in the dysfunction of OCL. Therefore, it is necessary to step efforts in channeling rural surplus labor to non-agricultural employment and reduce the pressure on oasis farmland. To be specific, Xinjiang's population urbanization rate is far lower than the national level in the same period, which has a large room for improvement. Hence, new urbanization and the development of non-agricultural industries should be propelled for attracting surplus rural laborers to work in cities and industrial parks. In addition, villages and towns developing labor-intensive industries based on their actual conditions should be supported based on the policy of various provinces aiding Xinjiang, so as to guide rural laborers to participate in non-agricultural employment nearby.

Third, the ELFC in maintaining oasis stability and the soil health of Xinjiang should be restored. The past agricultural development model characterized by large-scale expansion in Xinjiang led to severe coercive effects on ecological land and scarce water resources in the arid region and threatened the stability of the oasis. Also, non-point source pollution of farmland and jeopardies to soil health were caused due to the over-reliance on agrochemicals for increasing the production of CL yields. In response to these, on the one hand, the strictest possible water resource management system should be implemented to ensure an appropriate scale of OCL through setting CL with water. On the other hand, the layout of OCL should be optimized based on agricultural suitability, with Xinjiang's land space planning taken into consideration. Meanwhile, the encroachment of ecological space by CL should be strictly controlled. In addition, agricultural producers should rely on the progress of agricultural science and technology to improve the green and low-carbon utilization of CL, thereby reducing the unreasonable application of chemical fertilizers and pesticides and improving the recycling and resource utilization of waste agricultural film.

Fourth, the ENFC in promoting rural economic development and increasing farmers' income in Xinjiang should be optimized. The fast growth of the ENFC in Xinjiang is achieved by the expansion of the agricultural scale and the adjustment of planting structure (the increasing proportion of economic crops planted). However, the growth potential is limited due to the single source of the economic benefits of OCL. Therefore, the integrated development of agriculture and the secondary and tertiary industries should be guided to extend the means of increasing farmers' income from low-level CL planting to other links, such as processing and sales. By doing so, the income-increasing chain can be expanded by extending the industrial chain. Besides, ecological, green, and organic agriculture should be developed on the basis of advantageous resources, such as light, heat, and ecology in Xinjiang. Meanwhile, regional green agricultural brands should also be created to improve the market competitiveness of agricultural products, contributing to the conversion of ecological advantages to economic advantages.

4.3. *Advantages, Limitations, and Prospects*

This study selects Xinjiang, the main area of the arid region of northwest China, with unique regional features, as the study area. The connotations of social, economic, and ecological functions of OCL were first discussed based on the system theory. Then, an evaluation index system of CLFs was constructed. On this basis, the evolution characteristics of and the interaction between oasis CLFs in Xinjiang from 1990 to 2018 were investigated. The analytical framework proposed is applicable to related research on oasis CLFs in other arid regions apart from providing insights into the regional perspective of CL multi-function research.

Though the research has obtained certain results, there are also some deficiencies. The evaluation index system established in this paper might be deficient in some aspects due to the limited understanding of oasis farmland functions and the difficulty in data acquisition. It remains to be improved by considering the connotation and scientific evaluation methods of social, economic, and ecological functions of OCL together with multidisciplinary knowledge. What's more, this paper only considers the change of CLFs at the county level in Xinjiang. In the future, more in-depth studies at the level of typical villages and towns in oasis should be considered, which will provide a corresponding decision-making reference for the differentiated policy formulation of the multi-function synergistic management of CL at different scales.

5. Conclusions

This paper constructs a multi-function evaluation index system for OCL on the basis of discussing the function connotation and classification of OCL in Xinjiang. On this basis, the evolution characteristics of and the interaction between CLFs in Xinjiang from 1990 to 2018 are quantitatively evaluated. The main conclusions include:

(1) Evolution Features of CLFs: There were obvious differences in the evolution trends of the three functions of OCL in Xinjiang from 1990 to 2018. Among them, the economic function continued to increase, the ecological function was gradually degraded, and the social function was relatively stable. In general, the evolution of CLFs in Xinjiang was first dominated by ecological and social functions and then became economic-function-oriented. At the county level of Xinjiang, the SFC presents the regional characteristics of "increasing in the north and decreasing in the south"; the growth rate of the ENFC shows the regional characteristics of "growing fast in the northeast and slow in the southwest"; the deceleration of the ELFC presents the regional characteristics of "growing fast in the southeast and slow in the northwest".

(2) Interaction between CLFs: The relationship between SFC and ENFC in Xinjiang has evolved into a trade-off from synergy. The evolution characteristics of the relationship between ENFC and ELFC are the same as above. Although the relationship between SFC and ELFC are synergistic, it shows a different synergistic intensity tendency that first increased and then decreased. Spatially, the SFC and ENFC at the county level

in Xinjiang are dominated by synergy. The ENFC and ELFC at the county level in Xinjiang shows a pattern dominated by trade-offs. The SFC and ELFC at the county level in Xinjiang show a pattern of equivalence between synergies and trade-offs.

Author Contributions: Conceptualization, T.C. and X.Z.; methodology, T.C.; software, D.L.; formal analysis, T.C.; data curation, F.X.; writing—original draft preparation, T.C. and X.Z.; writing—review and editing, T.C., X.Z., F.X. and D.L.; visualization, D.L.; supervision, X.Z.; funding acquisition, X.Z. All authors have read and agreed to the published version of the manuscript.

Funding: This research was funded by the Third Xinjiang Scientific Expedition Program, grant number 2021xjkk0900.

Institutional Review Board Statement: Not applicable.

Informed Consent Statement: Not applicable.

Data Availability Statement: All data used for the study appear in the data source section of the submitted article.

Acknowledgments: The authors acknowledge all colleagues and friends who voluntarily reviewed the translation of the survey and study manuscript.

Conflicts of Interest: The authors declare no conflict of interest.

Abbreviations

Item	Descriptions
CL	Cultivated land
CLFs	Cultivated land functions
OCL	Oasis cultivated land
SFC	The social function of cultivated land
ENFC	The economic function of cultivated land
ELFC	The ecological function of cultivated land

References

1. UN. *Transforming Our World: The 2030 Agenda for Sustainable Development*; United Nations: New York, NY, USA, 2015.
2. Gao, L.; Bryan, B.A. Finding Pathways to National-Scale Land-Sector Sustainability. *Nature* **2017**, *544*, 217–222. [CrossRef] [PubMed]
3. Heck, V.; Hoff, H.; Wirsenius, S.; Meyer, C.; Kreft, H. Land Use Options for Staying Within the Planetary Boundaries—Synergies and Trade-offs Between Global and Local Sustainability Goals. *Glob. Environ. Chang.* **2018**, *49*, 73–84. [CrossRef]
4. Smith, P. Managing the Global Land Resource. *Proc. R. Soc. B Biol. Sci.* **2018**, *285*, 20172798. [CrossRef] [PubMed]
5. Steffen, W.; Richardson, K.; Rockstrom, J.; Cornell, S.E.; Fetzer, I.; Bennett, E.M.; Biggs, R.; Carpenter, S.R.; de Vries, W.; de Wit, C.A.; et al. Planetary Boundaries: Guiding Human Development on a Changing Planet. *Science* **2015**, *347*, 1259855. [CrossRef]
6. OECD. *Multifunctionality: Towards an Analytical Framework*; Organization for Economic Cooperation and Development: Paris, France, 2001.
7. Long, H.; Ma, L.; Zhang, Y.; Qu, L. Multifunctional Rural Development in China: Pattern, Process and Driving Mechanism. *Habitat Int.* **2022**, *121*, 102530. [CrossRef]
8. Bouma, J. How to Realize Multifunctional Land Use as a Contribution to Sustainable Development. *Front. Environ. Sci.* **2021**, *9*, 620285. [CrossRef]
9. Renting, H.; Rossing, W.A.H.; Groot, J.C.J.; Van der Ploeg, J.D.; Laurent, C.; Perraud, D.; Stobbelaar, D.J.; Van Ittersum, M.K. Exploring Multifunctional Agriculture. A Review of Conceptual Approaches and Prospects for an Integrative Transitional Framework. *J. Environ. Manag.* **2009**, *90*, S112–S123. [CrossRef]
10. Hodbod, J.; Barreteau, O.; Allen, C.; Magda, D. Managing Adaptively for Multifunctionality in Agricultural Systems. *J. Environ. Manag.* **2016**, *183*, 379–388. [CrossRef]
11. Li, Y.; Zhang, J.; Jia, J.; Fan, F.; Zhang, F.; Zhang, J. Research Progresses on Farmland Soil Ecosystem Multifunctionality. *Acta Pedol. Sin.* **2022**, *59*, 1177–1189. (In Chinese)
12. Andersen, P.S.; Vejre, H.; Dalgaard, T.; Brandt, J. An Indicator-Based Method for Quantifying Farm Multifunctionality. *Ecol. Indic.* **2013**, *25*, 166–179. [CrossRef]
13. Jiang, G.; Wang, M.; Qu, Y.; Zhou, D.; Ma, W. Towards Cultivated Land Multifunction Assessment in China: Applying the "Influencing Factors-Functions-Products-Demands" Integrated Framework. *Land Use Policy* **2020**, *99*, 104982. [CrossRef]
14. Song, X.; Ouyang, Z. Connotation of Multifunctional Cultivated Land and Its Implications for Cultivated Land Protection. *Prog. Geogr.* **2012**, *31*, 859–868. (In Chinese)

15. de Groot, R. Function-Analysis and Valuation as A Tool to Assess Land Use Conflicts in Planning for Sustainable, Multi-functional Landscapes. *Landsc. Urban Plan* **2006**, *75*, 175–186. [CrossRef]
16. Fagioli, F.F.; Rocchi, L.; Paolotti, L.; Słowiński, R.; Boggia, A. From the Farm to the Agri-Food System: A Multiple Criteria Framework to Evaluate Extended Multi-functional Value. *Ecol. Indic.* **2017**, *79*, 91–102. [CrossRef]
17. Coyle, C.; Creamer, R.E.; Schulte, R.P.O.; O'Sullivan, L.; Jordan, P. A Functional Land Management Conceptual Framework Under Soil Drainage and Land Use Scenarios. *Environ. Sci. Policy* **2016**, *56*, 39–48. [CrossRef]
18. Firbank, L.; Bradbury, R.B.; McCracken, D.I.; Stoate, C. Delivering Multiple Ecosystem Services from Enclosed Farmland in the UK. *Agric. Ecosyst. Environ.* **2013**, *166*, 65–75. [CrossRef]
19. Zhang, Y.; Long, H.; Ge, D.; Tu, S.; Qu, Y. Spatio-Temporal Characteristics and Dynamic Mechanism of Farmland Functions Evolution in the Huang-Huai-Hai Plain. *Acta Geogr. Sin.* **2018**, *73*, 518–534. (In Chinese)
20. Zhang, S.; Hu, W.; Huang, L.; Du, H. Exploring the Effectiveness of Multifunctional Cultivated Land Protection Linking Supply to Demand in Value Engineering Theory: Evidence from Wuhan Metropolitan Area. *Sustainability* **2019**, *11*, 6229. [CrossRef]
21. Guo, B.; Jin, X.; Yang, X.; Guan, X.; Lin, Y.; Zhou, Y. Determining the Effects of Land Consolidation on the Multifunctionlity of the Cropland Production System in China Using a SPA-Fuzzy Assessment Model. *Eur. J. Agron.* **2015**, *63*, 12–26. [CrossRef]
22. Qian, F.; Chi, Y.; Lal, R. Spatiotemporal Characteristics Analysis of Multifunctional Cultivated Land: A Case Study in Shenyang, Northeast China. *Land Degrad. Dev.* **2020**, *31*, 1812–1822. [CrossRef]
23. Tao, J.; Fu, M.; Sun, J.; Zheng, X.; Zhang, J.; Zhang, D. Multifunctional Assessment and Zoning of Crop Production System Based on Set Pair Analysis-A comparative Study of 31 Provincial Regions in Mainland China. *Commun. Nonlinear Sci.* **2014**, *19*, 1400–1416. [CrossRef]
24. Su, Y.; Su, C.; Xie, Y.; Li, T.; Li, Y.; Sun, Y. Controlling Non-Grain Production Based on Cultivated Land Multifunction Assessment. *Int. J. Environ. Res. Public Health* **2022**, *19*, 1027. [CrossRef] [PubMed]
25. Zou, L.; Li, Y.; Liu, Y.; Wang, J. Theory Building and Empirical Research of Production-Living-Ecological Function of Cultivated Land Based on the Elements. *Geogr. Res.* **2021**, *40*, 839–855. (In Chinese)
26. Xiong, C.; Zhang, Y.; Wang, Y.; Luan, Q.; Liu, X. Multi-Function Evaluation and Zoning Control of Cultivated Land in China. *China Land Sci.* **2021**, *35*, 104–114. (In Chinese)
27. Cao, S.; Zhang, J.; Liu, Y.; Yu, Z.; Liu, X. Net Value of Farmland Ecosystem Services in China. *Land Degrad. Dev.* **2018**, *29*, 2291–2298. [CrossRef]
28. Zhu, Q.; Hu, W.; Zhao, Z. Dynamic Analysis on Spatial-Temporal Pattern of Trade-Offs and Synergies of Multifunctional Cultivated Land—Evidence from Hubei Province. *Econ. Geogr.* **2018**, *38*, 143–153. (In Chinese)
29. Fonseca, A.M.P.; Marques, C.A.F.; Pinto-Correia, T.; Guiomar, N.; Campbell, D.E. Emergy Evaluation for Decision-Making in Complex Multifunctional Farming Systems. *Agric. Syst.* **2019**, *171*, 1–12. [CrossRef]
30. Song, X.; Huang, Y.; Wu, Z.; Ouyang, Z. Does Cultivated Land Function Transition Occur in China? *J. Geogr. Sci.* **2015**, *25*, 817–835. [CrossRef]
31. Song, X.; Li, X. Theoretical Explanation and Case Study of Regional Cultivated Land Use Function Transition. *Acta Geogr. Sin.* **2019**, *74*, 992–1010. (In Chinese)
32. Wilson, G.A. From 'Weak' to 'Strong' Multifunctionality: Conceptualising Farm-Level Multifunctional Transitional Pathways. *J. Rural Stud.* **2008**, *24*, 367–383. [CrossRef]
33. He, S.; Su, Y.; Shahtahmassebi, A.R.; Huang, L.; Zhou, M.; Gan, M.; Deng, J.; Zhao, G.; Wang, K. Assessing and Mapping Cultural Ecosystem Services Supply, Demand and Flow of Farmlands in the Hangzhou Metropolitan Area, China. *Sci. Total Environ.* **2019**, *692*, 756–768. [CrossRef] [PubMed]
34. Li, H.; Cai, Y.; Wang, Y. Spatial Heterogeneity and Individual Differences of Arable Land Use Functions in Peasant Household: A Case Study of Representative Area in Hubei Province. *J. Nat. Resour.* **2016**, *31*, 228–240. (In Chinese)
35. Yang, X.; Tan, M. Changes and Relationships of Arable Land Functions in Beijing in Recent Years. *J. Nat. Resour.* **2014**, *29*, 733–743. (In Chinese)
36. Zhu, C.; Li, W.; Du, Y.; Xu, H.; Wang, K. Spatial-Temporal Change, Trade-Off and Synergy Relationships of Cropland Multifunctional Value in Zhejiang Province, China. *Trans. Chin. Soc. Agric. Eng.* **2020**, *36*, 263–272. (In Chinese)
37. Zhan, X.; Shao, C.; He, R.; Shi, R. Evolution and Efficiency Assessment of Pesticide and Fertiliser Inputs to Cultivated Land in China. *Int. J. Environ. Res. Public Health* **2021**, *18*, 3771. [CrossRef]
38. Chen, X. *Physical Geography of Arid Areas in China*; Science Press: Beijing, China, 2010.
39. Xue, J.; Gui, D.; Lei, J.; Sun, H.; Zeng, F.; Mao, D.; Jin, Q.; Liu, Y. Oasification: An Unable Evasive Process in Fighting Against Desertification for the Sustainable Development of Arid and Semiarid Regions of China. *Catena* **2019**, *179*, 197–209. [CrossRef]
40. Liu, J.; Kuang, W.; Zhang, Z.; Xu, X.; Qin, Y.; Ning, J.; Zhou, W.; Zhang, S.; Li, R.; Yan, C.; et al. Spatiotemporal Characteristics, Patterns, and Causes of Land-Use Changes in China Since the Late 1980s. *J. Geogr. Sci.* **2014**, *24*, 195–210. [CrossRef]
41. Kuang, W.; Liu, J.; Tian, H.; Shi, H.; Dong, J.; Song, C.; Li, X.; Du, G.; Hou, Y.; Lu, D.; et al. Cropland Redistribution to Marginal Lands Undermines Environmental Sustainability. *Natl. Sci. Rev.* **2021**, *9*, nwab091. [CrossRef]
42. Cai, T.; Zhang, X.; Xia, F.; Zhang, Z.; Yin, J.; Wu, S. The Process-Mode-Driving Force of Cropland Expansion in Arid Regions of China Based on the Land Use Remote Sensing Monitoring Data. *Remote Sens.* **2021**, *13*, 2949. [CrossRef]
43. Statistics Bureau of Xinjiang Uygur Autonomous Region. *Xinjiang Statistical Yearbook*; Chinese Statistics Press: Beijing, China, 2019.
44. Hang, D. *Artificial Oasis in Xinjiang*; China Environmental Science Press: Beijing, China, 2001.

45. Bi, G.; Yang, Q.; Zhang, J.; Cheng, X. China's Rural Land Institutional Reform in the Past 40 Years Since the Reform and Opening up and Its Future Directions. *China Land Sci.* **2018**, *32*, 1–7. (In Chinese)
46. Xie, G.; Zhen, L.; Lu, C.; Xiao, Y.; Li, W. Applying Value Transfer Method for Eco-Service Valuation in China. *J. Resour. Ecol.* **2010**, *1*, 51–59. [CrossRef]
47. Chang, T. Study on The Mechanism and Efficiency of Emergy Metabolism of Oasis System in Xinjiang. Ph.D. Thesis, Xinjiang Institute of Ecology and Geography, Chinese Academy of Sciences, Urumqi, China, 2018.
48. Tian, Y. The Development of China's Low-Carbon Agriculture: Production Efficiency, Spatial Differences and Influencing Factors. Ph.D. Thesis, Huazhong Agricultural University, Wuhan, China, 2015.
49. Zhang, C.; Wang, L.; Hua, C.; Jin, S.; Liu, P. Potentialities of Fertilizer Reduction for Grain Produce and Effects on Carbon Emissions. *Resour. Sci.* **2016**, *38*, 790–797. (In Chinese)
50. Li, B.; Zhang, J.; Li, H. Research on Spatial-Temporal Characteristics and Affecting Factors Decomposition of Agricultural Carbon Emission in China. *China Popul. Resour. Environ.* **2011**, *21*, 80–86. (In Chinese)
51. Huang, H.; Wang, H. Improvements on the Gray T-related Dergee. *Math. Pract. Theory* **2016**, *46*, 131–139. (In Chinese)
52. Lai, B.; Tan, M.; Chai, Z.; Chen, B.; Li, Q.; Dong, J.; Wang, F.; Tian, C. Investigation and Evaluation of the Chemical Fertilizer Application Situation of Farmland in Xinjiang. *Arid Zone Res.* **2014**, *31*, 1024–1030. (In Chinese)
53. Xu, Y.; Fang, S.; Ma, X.; Zhu, Q. Prevention and Control Strategy for the Pollution of Agricultural Plastic Film. *Chin. Eng. Sci.* **2018**, *20*, 96–102. (In Chinese) [CrossRef]

land

MDPI

Article

China's Urban and Rural Development Significantly Affects the Pattern of Human Appropriation of Net Primary Production

Tian Zhang [1], Jian Peng [2] and Xiaoshu Cao [1,*]

[1] Northwest Land and Resource Research Center, Shaanxi Normal University, Xi'an 710119, China
[2] College of Urban and Environmental Sciences, Peking University, Beijing 100871, China
* Correspondence: caoxsh@snnu.edu.cn

Abstract: Increasing human activities have greatly influenced the ecosystem and the use of ecological resources, and the unbalanced urban–rural development in China (urban and rural areas being two major bases of human activities) has always been accompanied by heterogeneous ecological effects. Human appropriation of net primary production (HANPP) is an integrated indicator quantifying the human domination of productivity and harvest in the biosphere. Identifying the unbalanced constraints of urban and rural development on HANPP has become necessary for improving human–land relationships. This study analyzed the spatial distribution and regional differentiations of the HANPP in China in 2015 and investigated how HANPP and its components responded to unbalanced regional urban–rural development. The results show that the total amount of HANPP was 2.68 PgC and gradually decreased from the southeast to the northwest of China in 2015, representing 60.33% of the NPP_{pot}. In addition, $HANPP_{luc}$, harvest through cropland, livestock grazing, and forestry contributed 60.70%, 29.86%, 8.53%, and 0.91%, respectively, to the total HANPP, with $HANPP_{luc}$ playing the dominant role in 21 provinces. There was a significant differentiation ($p < 0.05$) in the spatial distribution of HANPP (gC/m^2), $HANPP_{harv}$ (gC/m^2), and $HANPP_{luc}$ (gC/m^2), especially between the Huanyong Hu Line and the western–eastern part of China, fundamentally resulting from uneven regional development. In addition, biomass production–consumption decoupling existed in most regions in China, 17 provinces were identified as consumption type, and a universal positive correlation ($p < 0.05$) was identified between the production–consumption ratio of occupied biomass and $HANPP_{harv}$ (%HANPP). Different drive mechanisms were found between urban–rural development and HANPP, and each HANPP index was more likely to be affected by urban economy (UE), rural population (RP), and rural agricultural technology (RA) in China. The higher regional average nighttime light intensity, the proportion of the built-up area, and the urban road area corresponded with a large $HANPP_{luc}$ value. Conversely, HANPP would decrease as the proportion of urban green spaces increased. Furthermore, HANPP (%NPP_{pot}) and HANPP (gC/m^2) mostly depended on the rural development index, while $HANPP_{luc}$ and $HANPP_{harv}$ were mainly controlled by urban and rural development, respectively. Our findings help understand, first, how unbalanced regional development influences human-induced biomass occupation, the comprehensive urban ecological construction, and rural ecological restoration and, second, that the overall planning of urban–rural integration development must be strengthened to face greater ecological pressures in the future.

Keywords: human appropriation of net primary production (HANPP); urban and rural development; land use change; human activities; China

Citation: Zhang, T.; Peng, J.; Cao, X. China's Urban and Rural Development Significantly Affects the Pattern of Human Appropriation of Net Primary Production. *Land* **2023**, *12*, 1062. https://doi.org/10.3390/land12051062

Academic Editors: Li Ma, Yingnan Zhang, Muye Gan and Zhengying Shan

Received: 12 April 2023
Revised: 11 May 2023
Accepted: 11 May 2023
Published: 12 May 2023

1. Introduction

The increase in human activity (urbanization, deforestation, cultivation, hunting, etc.) has had a tremendous impact on ecosystems and ecological resources. Excessive interference has led to a series of contradictions in human–land relationships and has become the core of geography research [1]. Intensified human–land conflict and the excessive transformation and occupation of ecological resources have posed severe challenges to sustainable global development. In view of this situation, there is an urgent need to study ecological pressure and the factors that lead to it. However, quantifying the magnitude of the substantial amount of human interference and its consequences for the ecosystem is a challenge [2]. Human appropriation of net primary production (HANPP) is a credible aggregated indicator that quantifies the magnitude of human-activity-related alterations in the ecological energy flows from the perspective of the carbon cycle [3]. HANPP is always represented as the ratio of the ecological sources used by human activities to the potential ones, and can measure regional ecological pressure and sustainability. Net primary productivity (NPP) refers to the amount of photosynthetic active radiation fixed by green vegetation as the carbon storage per unit area [4], which determines the ecological resource capacity that humans can occupy [5].

Urban and rural areas are considered as two major bases of human activities, and there is a big difference between their construction and economic levels [6–8]. As the most typical way by which the ecosystem is altered by humans, urbanization can not only directly consume multiple ecological resources through land conversions, but also drive the exchange of the regional population, economy, and materials [9,10]. Meanwhile, as suppliers of materials, labor, and various services to cities [11], expansive rural areas bear the pressure of ecological resource consumption resulting from urban population growth. With the continuous concentration of population in cities, there is a local and telecoupling phenomenon between the consumption and production of ecological resources in urban and rural areas [12]. The unbalanced development between urban and rural areas directly affects the supply–demand relationship of ecological resources and regional functions, especially the food security and ecological security in the areas where ecological resources are supplied are under increasing pressure. Whether human activities in some areas have exceeded the threshold of the ecosystems and what are the influencing factors are the topical questions that need to be answered urgently. Therefore, exploring the differentiation of HANPP under the influence of urban and rural development is helpful to identify the areas of human activity overload, and to clarify the contributions of urban and rural areas to the sustainable development of an ecosystem. In recent years, with the implementation of strategies for rural revitalization and urban–rural integration, China, as a large developing country, has entered a critical period of transformation in urban–rural relations. However, the long-standing "dual system" has resulted in the division of urban and rural areas and unbalanced development [13] accompanied by heterogeneous ecological effects of human activities. Therefore, determining the internal driving mechanisms of the unbalanced urban–rural development on HANPP is helpful to judge the constraints affecting the ecological sustainable development, which is a crucial topic to alleviate the ecological contradictions between urban and rural areas and to improve human–land relations globally.

HANPP is an effective method of assessing human–land relationships and research on HANPP has received increasing attention internationally; however, related research in China is still in its infancy. Relevant studies mostly focus on the improvement of the simulation model [14], comparison with other similar indicators [15], and the evaluation and spatial–temporal dynamic analysis of HANPP on global [16–18], national [5,19–21], or regional scales [2,22–24]. Embodied HANPP, an indicator closely related to the import and export of ecological resources, is an emerging research field [25]. Different results can be found based on various global-scale models and periods, i.e., HANPP was estimated to be 20.32% in 1995 [26], 23.8% in 2000 [16], and 13–25% from 1910 to 2005 [18]. Existing national-scale studies mostly focus on European countries and always show a relatively high level of HANPP (over 50%), with a tendency to decline over a long time series [27–29]. However, case studies in Asian countries show that the ecological pressure has significantly increased [5,30], representing the imbalanced effect of human activities on HANPP in densely inhabited regions. Moreover, as a biophysical indicator for measuring regional sustainable development, the relationship between HANPP and ecological footprint [31,32] as well as biodiversity [33,34], has always been focused on. Studies on the driving mechanism of HANPP are mostly limited to the single factor of social economy, among which the decoupling of HANPP and population or GDP has attracted more attention [35]. Population density has been found to have a significant influence on the spatial distribution of HANPP [36]. However, less attention has been paid to the comprehensive studies of how HANPP responds to multidimensional indicators for urban and rural development, such as land conversion, population growth, economy development, technology progress, and ecological construction, and studies on the variation in HANPP during urban–rural relationship transformation are also particularly lacking in recent years.

The past decades have been characterized by rapid economic development and intensive human activities in China, and the duality of rural–urban development is a common problem in most regions, inducing an obvious spatial heterogeneity of ecological resources consumed by human activities. If the relationship between urban–rural development and HANPP was quantified, it would fill research gaps and help answer the essential question of how to reduce the ecological pressure of human activities and promote harmonious human–land relationships through coordinated urban–rural development, and thus be referable for further studies in other regions. Therefore, with the ultimate goal of identifying the influence mechanism of urban–rural development on HANPP and investigating the factors driving its heterogeneity, three detailed objectives were defined in this study: (1) to evaluate HANPP in China in 2015 and its spatial imbalance characteristics; (2) to depict the relationships between HANPP and China's urban–rural development and investigate HANPP responses to the degree of regional urban–rural coordination; and (3) to identify the dominant driving factors of different dimensions of urban and rural construction on HANPP and its components.

2. Materials and Methods

2.1. Data Sources

In this paper, three categories of datasets were chosen from 31 regions in China (22 provinces, 5 autonomous regions, and 4 municipalities). This excluded unobtainable statistical datasets from Hong Kong, Macau, and Taiwan. The dataset included remote sensing images, statistical yearbook data, and vector datasets. To estimate the terrestrial potential NPP, yearly land cover data MCD12Q1 with the resolution of 500 m were obtained from NASA Earthdata (https://lpdaac.usgs.gov/products/mcd12q1v061/ (accessed on 12 May 2022)) and MOD17A3HGF data were used to characterize the actual NPP (https://lpdaac.usgs.gov/products/mod17a3hgfv061/ (accessed on 23 May 2022)). The Monthly Normalized Difference Vegetation Index (NDVI) with a resolution of 1 km was obtained from the MOD13A3 dataset (https://lpdaac.usgs.gov/products/mod13a3v061/ (accessed on 6 June 2022)) and was mainly used to spatialize the harvested NPP of each province by virtue of its linear relationship with regional crop and livestock production. Among the statistical data, the annual average temperature and the annual total precipitation data of 830 meteorological stations in China used in this study were collected from the China Meteorological Data Service Centre (http://data.cma.cn/ (accessed on 15 June 2022)) and used to drive the Thornthwaite Memorial model to evaluate the potential NPP in China. For the missing data of several individual stations, this study selected the threshold of 80% and used Lagrange's theorem to interpolate and fill in the gaps [37]. If the number of missing months exceeded 20%, the data of the station in that year would be taken as a null value. The dataset of 31 provinces in China on biomass harvested or destroyed during harvest for crop yield, wood harvest, and livestock, available in the China statistical yearbooks in the National Bureau of Statistics (http://www.stats.gov.cn/tjsj/ (accessed on 17 June 2022)), was quantified. Indicators representing urban and rural population growth, economic development, land expansion, ecological construction, and agricultural technology, which comprehensively reflect the level of urban–rural development in China, were also obtained from these annual statistical yearbooks.

2.2. Estimation Model of HANPP

On the basis of different research backgrounds and spatial scales, several definitions of HANPP have been used before [16,26,38,39], which caused a slight obstacle when the HANPP estimation results were being compared. In this study, we chose the definition proposed by Haberl et al. (2007) [16], which has been widely recognized in recent years. According to the definition of HANPP [16], two major parts were considered when human activities appropriated the ecosystem NPP, the harvested biomass and the HANPP caused by land use and land conversion. Specifically, agricultural harvest, forestry product, and livestock grazing are defined as principal combinations of the harvested part ($HANPP_{harv}$) [40]. The NPP of the potential natural vegetation without human disturbance is called NPP_{pot} and the one of the currently prevailing vegetation is called NPP_{act}, and the difference between them ($HANPP_{luc}$) was defined as the change in biomass caused by human-induced land use. Some amount of biomass still remains in the ecosystem after harvest (i.e., straw and stump), which is called NPP_{eco}. Therefore, HANPP could be calculated using Formula (1) and transformed into more detailed forms using Formula (2). Figure S1 presents a conceptual model of HANPP.

$$HANPP = NPP_{pot} - NPP_{eco} = HANPP_{luc} + HANPP_{harv} \tag{1}$$

$$HANPP_{luc} = NPP_{pot} - NPP_{act} \tag{2}$$

Accurate estimation of NPP_{pot} is the key to quantifying $HANPP_{luc}$. In terms of the selection of estimation models, many studies have used the Lund–Potsdam–Jena dynamic global vegetation model (LPJ-DGVM) [16,41]. The Integrated biosphere simulator (IBIS) and LPJ-GUESS (general ecosystem simulator) models can also simulate the vegetation dynamics under different climate scenarios [42]. However, these models are mainly applied on the global scale, and the resolution is generally $0.5° \times 0.5°$, which is not suitable for the accurate characterization of small-scale HANPP when used alone. In addition, some studies have established the transformation relationship between NPP_{act} and NPP_{pot} in stable-vegetation areas based on climatic elements to achieve an extended estimation of non-vegetated areas. Among them, O'Neill et al. (2007) divided all pixels of land use data into ones disturbed and undisturbed by human activities and assumed that the NPP_{pot} of undisturbed pixels is equal to NPP_{act}, whereas the NPP_{pot} of all disturbed pixels could be replaced by the average NPP_{act} of the surrounding undisturbed pixels [2]. Models such as Miami, Thornthwaite Memorial, and Chikugo have been widely used for estimating NPP_{pot} by establishing the relationship between meteorological indicators and potential vegetation [5,23,43]. To integrate the advantages of each model, improve the accuracy of NPP_{pot} estimation, and represent the optimal NPP_{pot} in their research, Erb et al. (2016) used the average value of the Miami model, the LPJ-DGVM model, and a vegetation-approach model [44]. Therefore, in this study, NPP_{pot} was derived by combining the Thornthwaite Memorial model, the LPJ model, and the vegetation-approach model, among which the result of the LPJ model came from the potential NPP simulation with a resolution of $0.5° \times 0.5°$ under the S2 scenario provided by the TRENDY program. In addition, according to the research of Erb et al. (2016) on global vegetation biomass [44], the conversion rules of NPP_{act} and NPP_{pot} of various land use types were extracted (Table S3) so as to drive the vegetation-approach model. The spatial patterns of NPP_{pot} simulated by the three models are highly similar, and the results of each model and the average value of NPP_{pot} in China are shown in Figure S2.

$HANPP_{harv}$ is the biomass harvested or destroyed during harvest (harvest of crops and crop residues, harvest through livestock grazing, harvest through forestry, etc.), which is derived from the statistics yearbooks from the Chinese National Bureau of Statistics database, and $HANPP_{harv}$ in this study was converted into carbon (gC) by assuming an average carbon content of dry matter (DM) biomass of 50% [18]. $HANPP_{harv}$ on cropland consists of crop harvest and the used or unused residues [45]. The annual yields of 15 kinds of crops widely cultivated in China were collected, and their harvested fresh weights were converted into dry matter equivalents with the moisture content (%). The moisture content of the specific crop was obtained from the International Network of Food Data Systems provided by FAO (http://www.fao.org/infoods/infoods/tables-and-databases/ (accessed on 20 March 2022)) and the literature [3,5]. Furthermore, the crop residues were calculated from crop yield data with the harvest factors and recovery rates (the former is the ratio of crop residual to primary crop harvest, and the latter is the ratio of used crop residues to available residues), which were obtained from the literature [18]. The grazed biomass was calculated as the difference between feed demand and supply [45]. In this study, 5 grass-feeding livestock species (asses, horses, mules, cows, and sheep) were transformed into standard sheep units and the feed demand was calculated by the forage consumption (1.8 kg of dry matter coarse forage per sheep unit per day). In addition, the feed supply was derived by the conversion coefficients from croplands to straw from 6 crops species (i.e., rice, wheat, maize, millet, sorghum, and soybeans) and the 25% straw utilization factor [5]. The forestry harvest includes the removal of timber and felling losses. The total timber and bamboo production was collected from the Chinese National Bureau of Statistics and the National Forestry Statistical Yearbook; 49.5% of wood density (0.495 t/m^3) was used to convert fresh weight to dry matter [46], and the felling losses were estimated using recovery rates defined as the ratio of wood removal to felling [18]. In addition, considering the method of Haberl et al. (2007) [16], $HANPP_{harv}$ on built-up land was assumed to be 50% of NPP_{act}. The detailed calculation methods of $HANPP_{harv}$ are provided in Supplementary File S1.

2.3. *Quantification of Urban and Rural Development*

In this study, the focus was how HANPP responds to regional development in China, so multi-indicators were selected to quantify the unbalanced urban–rural growth. Urban areas expand spatially when people crowd into these areas as the economy improves. According to this, 10 urbanization indicators from three perspectives (economy, population, and spatial expansion) were chosen to detect their relationship with HANPP. The corresponding rural development could also be measured using those 10 indicators but from the perspectives of economy, population, and agricultural technology. The indicators from 31 Chinese provinces, presented in Table 1, which were collected from China's statistical yearbooks of 2015, provide a comprehensive reflection of China's urban–rural development. The level of coordinated urban–rural development in China was also assessed using these indicators. After standardizing with the data range, the weight of each indicator in Table 1 was calculated using the entropy method [47]. Then, the urban and rural development indexes (expressed as $f(u)$ and $f(r)$, respectively) in 31 provinces in China could be quantified. A larger value of $f(u)$ or $f(r)$ means a higher level of rural–urban development, with which the comprehensive urban–rural development index (T) was also calculated. In addition, referring to the coupled coordination model in physics, the coupling degree ($C \in [0, 1]$) expressed by $f(u), f(r)$, and T was used to analyze whether the whole urban–rural system is ordered. Since the C value only reflects the interaction within each system, which cannot indicate the coordinated order [48], the coordinated development degree ($D \in [0, 1]$) in each province was evaluated by the coordination degree model, which was divided into 9 levels in this study [49] (Table 2). The detailed formulas of all the indicators are presented in Supplementary File S2.

Table 1. Quantification indicators of urban–rural development in China.

Objects	Perspective	Indexes	Symbols *	Weights
Urban	Economy	Added value of the secondary and tertiary industries (10^8 CNY)	UE1	0.065
		Proportion of added value of the secondary and tertiary industries (%)	UE2	0.019
		Added value of transportation, warehousing, and postal services (10^8 CNY)	UE3	0.051
		Per capita consumption expenditure of urban residents (CNY)	UE4	0.084
	Population	Urban population (10^4 people)	UP1	0.045
		Proportion of urban population (%)	UP2	0.020
	Space	Nighttime light intensity (nW/cm^2/sr)	US1	0.194
		Proportion of built-up area (%)	US2	0.147
		Proportion of urban green spaces (%)	US3	0.243
		Proportion of urban road area (%)	US4	0.134
Rural	Economy	Added value of the primary industry (10^8 CNY)	RE1	0.103
		Proportion of added value of the primary industry (%)	RE2	0.058
		Per capita consumption expenditure of rural residents (CNY)	RE3	0.073
		Engel coefficient of rural residents (%)	RE4	0.134
	Population	Rural population (10^4 people)	RP1	0.118
		Proportion of rural population (%)	RP2	0.037
	Agricultural technology	Total power of agricultural machinery (10^4 kW)	RA1	0.129
		Application amount of agricultural fertilizer (10^4 tons)	RA2	0.112
		Effective irrigated area (10^3 hectare)	RA3	0.125
		Total sown area of crops (10^3 hectare)	RA4	0.111

* UE, UP, and US are the abbreviations for urban economy, urban population, and urban space, respectively, and RE, RP, and RA are the abbreviations for rural economy, rural population, and rural agricultural technology, respectively.

Table 2. Classification criteria of coordinated urban–rural development (*D*).

D Value	Implications	*D* Value	Implications	*D* Value	Implications
$0 < D \leq 0.2$	Severe imbalance	$0.4 < D \leq 0.5$	Near imbalance	$0.7 < D \leq 0.8$	Intermediate coordination
$0.2 < D \leq 0.3$	Moderate imbalance	$0.5 < D \leq 0.6$	Barely coordination	$0.8 < D \leq 0.9$	Well coordination
$0.3 < D \leq 0.4$	Inchoate imbalance	$0.6 < D \leq 0.7$	Primary coordination	$0.9 < D \leq 1$	Super coordination

2.4. Statistical Methods

Spatial analysis in ArcMap 10.3 and multi-statistical methods were applied to the evaluation of spatial heterogeneity and driving factors of HANPP in China. In this study, the independent-samples *t* test was used to explore the differentiation of HANPP and its components and investigate whether HANPP differs significantly among different zones in China (i.e., western, eastern, northern, and southern area; coastal land; and inland area). Furthermore, stepwise regression analysis helped verify what kind of constraints the urban–rural development in China exerted on HANPP and then we extracted the factors dominating the differentiation of HANPP based on the standard partial regression coefficient in the model.

3. Results

3.1. Spatial Distribution of HANPP in China

The total amount of human appropriation of biomass was 2.68 PgC, representing 60.33% of the NPP_{pot} in China in 2015. Figure 1a shows that HANPP (gC/m^2) gradually decreased from the southeast to the northwest of China, showing an obvious difference on both sides of the Hu Huanyong Line (the mutation line of China's population density). Hunan, Hubei, Henan, Jiangxi, Guangxi, Guangdong, and Jilin Provinces had high HANPP values (greater than 1000 gC/m^2). The negative value of HANPP (gC/m^2) came from the contribution of $HANPP_{luc}$, mainly concentrated in central Xinjiang, western Yunnan, and southern Tibet, reflecting that the actual carbon sequestration of vegetation was higher than the potential sequestration under the same climatic conditions. Among these, the negative HANPP values in Xinjiang Province mainly appeared in the cultivated land. In addition, two high-value strips were identified in the east–west direction of HANPP (gC) in each province (Figure 1b), which consisted of Inner Mongolia (0.224 PgC), Heilongjiang Province (0.207 PgC), Sichuan Province (0.170 PgC), Tibet (0.146 PgC), and Henan Province (0.143 PgC).

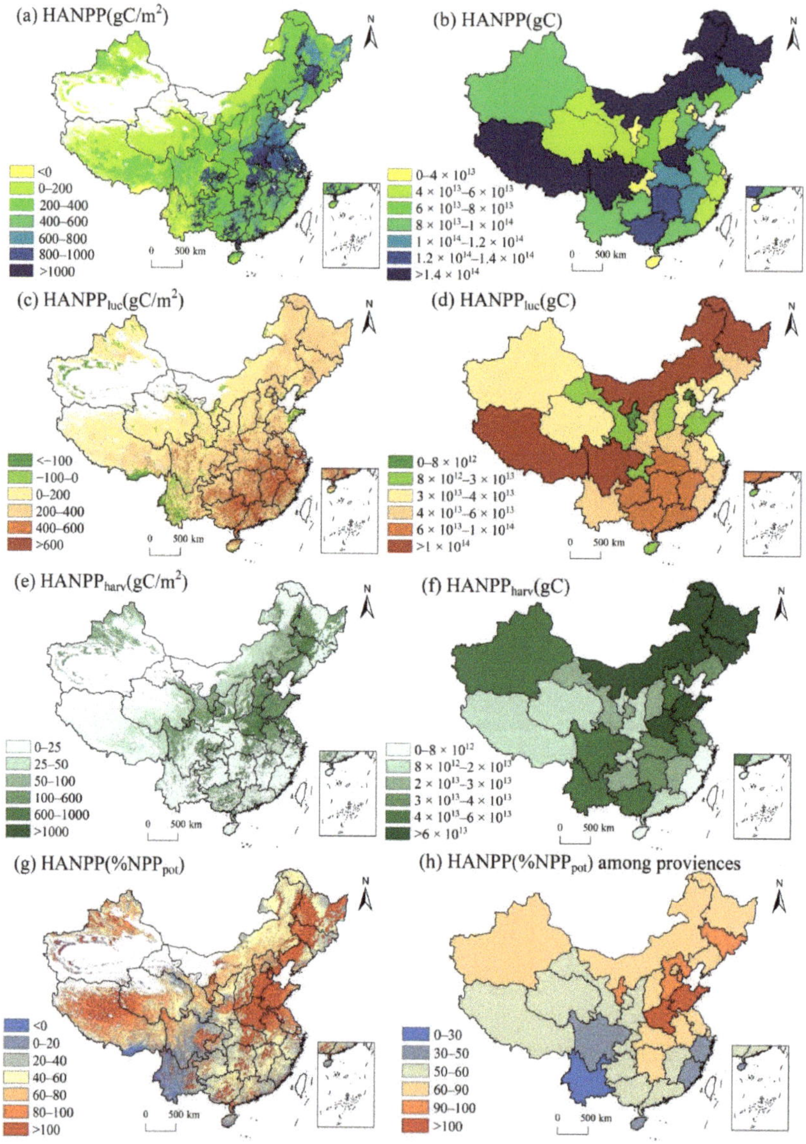

Figure 1. HANPP, $HANPP_{luc}$, $HANPP_{harv}$, and HANPP (% NPP_{pot}) in China (**a**,**c**,**e**,**g**) and among provinces (**b**,**d**,**f**,**h**) in 2015.

Figure 1c shows obvious positive and negative differences in $HANPP_{luc}$ (gC/m^2) in China in 2015, indicating that land use change played a bidirectional role on ecosystem NPP. An $HANPP_{luc}$ (gC/m^2) value of less than 0 means that the production capacity of vegetation after human intervention could be higher than the potential capacity under the same climatic conditions. The negative values of $HANPP_{luc}$ (gC/m^2) in Xinjiang, Ningxia, and Gansu Provinces were mainly the result of cultivated land, indicating that in regions with a relatively insufficient natural background, cultivation based on artificial management could break through the climate limit of biomass accumulation to a certain extent, which has been confirmed by an existing study [18]. On the contrary, positive values of $HANPP_{luc}$ (gC/m^2) were dominant across the whole country, indicating that human activity has destroyed the original green spaces to a certain extent, and $HANPP_{luc}$ (gC/m^2) values in big cities (i.e., Beijing, Tianjin, Shanghai, Shenzhen, and Guangzhou Provinces) were significantly higher than those in surrounding areas. Total $HANPP_{luc}$ (gC) values of over 0.1 PgC are mainly concentrated in Inner Mongolia (0.148 PgC), Tibet (0.130 PgC), Heilongjiang Province (0.118 PgC), and Sichuan Province (0.112 PgC) (Figure 1d). Although China has continuously strengthened the protection of grasslands and vigorously promoted the policy of "returning grazing land to grassland" in recent years, most grassland was still overloaded and overgrazed [50]. Inner Mongolia and Tibet are the key pastoral areas of China, and the high values of $HANPP_{luc}$ (gC) in Inner Mongolia and Tibet were mainly related to grassland degradation caused by grazing.

The North China Plain and northeast provinces in China had relatively high $HANPP_{harv}$ (gC/m^2) values (Figure 1e), with an average value of more than 600 gC/m^2, and the biomass harvest in these regions was mainly provided by cultivated land. Among them, Shandong and Henan Provinces had the most extensive distribution of cultivated land with stable production. In addition, there was significant spatial heterogeneity of $HANPP_{harv}$ (gC/m^2) among the provinces in southern China, where the harvest of crops and the amount of crop residues were much higher than those in the surrounding woodlands. The limited ecological land in northern Xinjiang Province showed relatively high $HANPP_{harv}$ (gC/m^2), mainly resulting from grassland and cultivated land. Figure 1f shows that the high-value areas of $HANPP_{harv}$ (gC) in each province in 2015 expanded from the North China Plain to the entire northeast of China. Among them, Henan (0.097 PgC), Heilongjiang (0.089 PgC), and Shandong (0.077 PgC) were the top three provinces of $HANPP_{harv}$ (gC), while Beijing, Tianjin, and Shanghai Province were at the bottom since they did not undertake the production function and their area is relatively small.

HANPP ($\%NPP_{pot}$) can reflect the ecological pressure caused by the use of resources resulting from human activities, and the low-value areas of HANPP ($\%NPP_{pot}$) were mainly distributed in the southeast coastal areas and southwest China (Figure 1g). Specifically, the mountainous area accounted for more than 80% of the total area in Yunnan Province, which limited the destruction of natural vegetation and land reclamation by humans. In addition, the $HANPP_{harv}$ in southeastern coastal provinces mainly resulted from the low-level harvest of forestry, which also led to a relatively low HANPP ($\%NPP_{pot}$). It is worth noting that Tibet is a major pasturing area in China, but the NPP_{act} in Tibet was only 137.38 gC/m^2, while the harvest through livestock grazing was already close to the NPP_{pot}, so the HANPP ($\%NPP_{pot}$) in most areas of Tibet was generally high in 2015. Estimating the HANPP ($\%NPP_{pot}$) of each province helps identify the regions with oversaturated ecological occupation in China, which provides scientific support for formulating urban development strategies and ecological construction measures in corresponding areas. It can be seen from Figure 1h that HANPP was high ($\%NPP_{pot}$) in the grain-producing provinces in the North China Plain, while HANPP was low in the southwestern provinces with complex terrain. With the major grain-producing provinces benefiting from improved agricultural technology and efficient manual management, the high value of HANPP ($\%NPP_{pot}$) in these provinces was actually a sign of a significant increase in land production efficiency.

3.2. Contributions of HANPP Components

The total amounts of $HANPP_{luc}$ and $HANPP_{harv}$ were found to be 1.63 PgC and 1.05 PgC, respectively, in China in 2015. Specifically, the harvest through cropland, livestock grazing, and forestry contributed 29.86%, 8.53%, and 0.91% to the total HANPP, respectively, while the land-use-induced biomass appropriation contributed 60.70%. However, there were great differences in the contributions of HANPP components among provinces (Figure 2). $HANPP_{luc}$ played the dominant role among 21 provinces (over 50% of HANPP), especially in the cities with rapid urban expansion (i.e., Shanghai and Beijing, with 74.82% and 74.78% of $HANPP_{luc}$ (%HANPP), respectively) and some coastal cities (i.e., Fujian, Zhejiang, and Guangdong, with 86.07%, 85.99%, and 79.07% of $HANPP_{luc}$ (%HANPP), respectively). On the contrary, the harvest on cropland ($HANPP_{harv}$ (crop)) contributed most to HANPP in Shandong (62.33%), Henan (57.30%), Jilin (51.66%), Jiangsu (50.58%), and Hebei (50.46%), the traditional grain-producing provinces, and livestock grazing in the traditional pastoral regions (i.e., Gansu, Xinjiang, Ningxia, and Qinghai) resulted in 20% of HANPP. It is worth noting that different urban functions and development planning resulted in various ecological occupation structures, and expansive impervious surface replaced almost all ecological land in Shanghai, the economic center in China, where $HANPP_{luc}$ contributed the highest to NPP_{pot} in China (50.55%). Compared with the high HANPP (%NPP_{pot}) due to harvested biomass through cropland, livestock grazing, and forestry, the substantial compression of ecological land due to excessive urban development needs more attention.

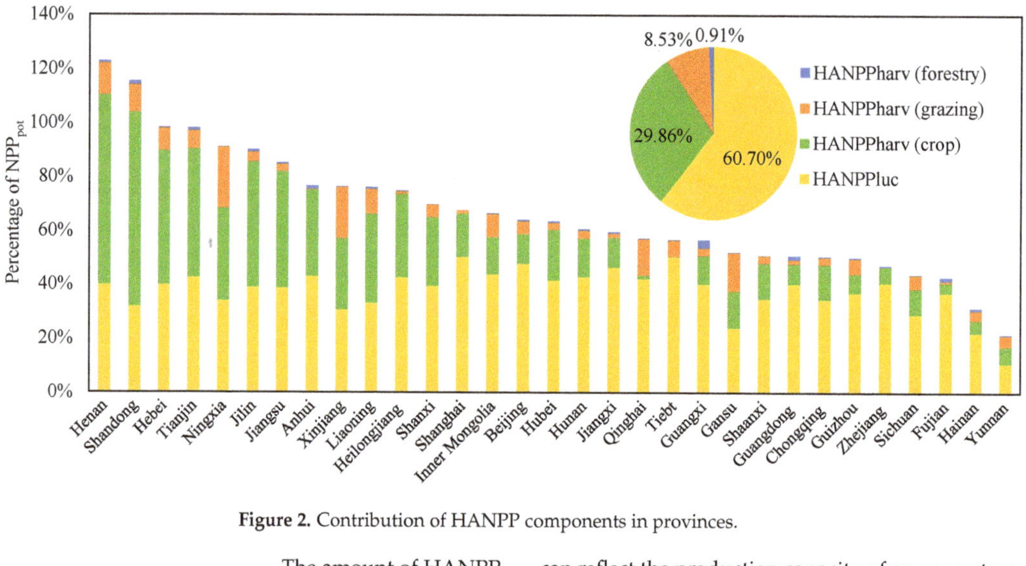

Figure 2. Contribution of HANPP components in provinces.

The amount of $HANPP_{harv}$ can reflect the production capacity of an ecosystem, and the regional population directly determines the consumption demand of the local biomass. Without considering the local import and export, we built a production–consumption ratio, which was calculated as the ratio of $HANPP_{harv}$ to the multiplication of national biomass consumption per capita and the regional population. The production–consumption ratio was divided into three types: an area was self-sufficient if the ratio was close to one, a ratio over one indicated adequate biomass production, and a ratio under one indicated inadequate biomass production. The latter two values lead to ecological resources being circulated among regions. On the basis of $HANPP_{harv}$ in 2015, a significant imbalance could be found between the supply of and demand for biomass in China (Figure 3a). In all, 14 provinces in the north and southwest China were defined as the production type, and ratios over the 95th percentile were found in the sparsely populated Tibet (6.36) and Inner

Mongolia (4.07). Conversely, the remaining 17 provinces in the middle and southeastern coastal areas in China needed external biomass supplies. Specifically, the production–consumption ratios of rapidly growing cities such as Shanghai (0.06) and Beijing (0.09) were under the fifth percentiles. Furthermore, it should be clarified that a telecoupling of ecological resources always occurs in remote spaces, and the gradient differences of production–consumption ratios in Figure 3a represent interregional resource flows; more attention should be paid to this in future studies.

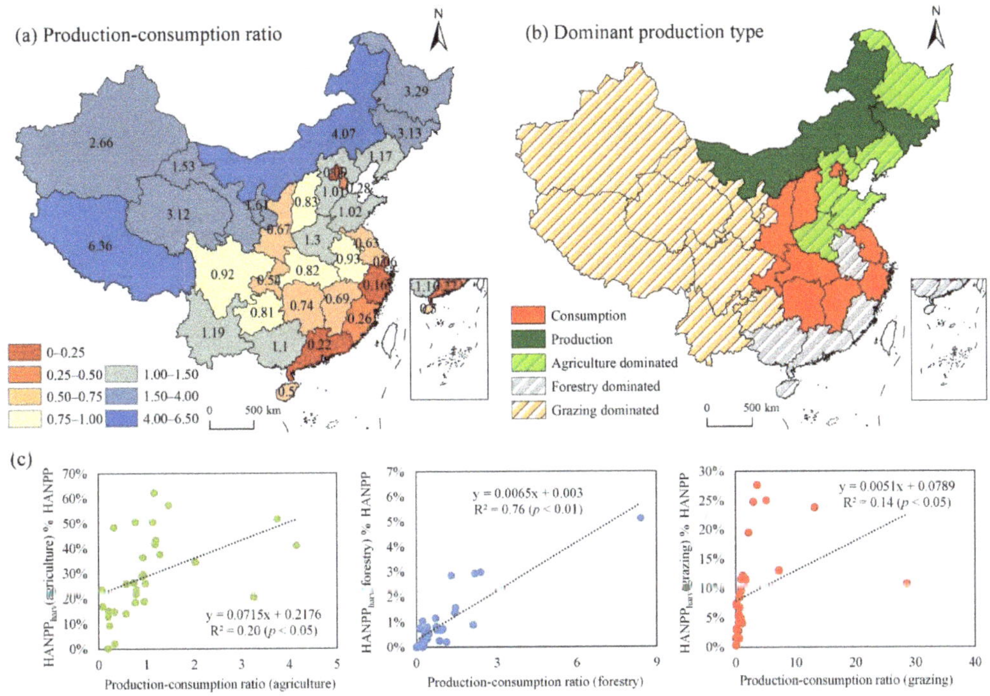

Figure 3. The production–consumption ratio (**a**), the dominant production types among provinces in China (**b**), and the relationships between production–consumption ratios and HANPP$_{harv}$ (%HANPP) (**c**).

Five dominant production types were identified among 31 provinces with production–consumption ratios of HANPP$_{harv}$ through agriculture, forestry, and grazing (Figure 3b), and a significant spatial aggregation could be found among all these types. A great number of ecological resources were needed among consumption regions with ratios under 1, which included the metropolis with high population densities (i.e., Shanghai, Beijing, and Tianjin) and several provinces in the middle of China. On the contrary, only two production regions (i.e., Inner Mongolia and Jilin) were found in the northeast of China where the production–consumption ratios of the components of HANPP$_{harv}$ were over 1. Furthermore, huge gaps existed among the three ratios in the remaining provinces and the highest ratio was selected as the dominant production type. Specifically, five agriculture-dominant provinces, situated mostly in the circum-Bohai Sea region, and five provinces in south China with better climate conditions, provided more biomass through forestry. HANPP$_{harv}$ through grazing was found as the dominant production type among eight traditional husbandry provinces in the west of China. Figure 3c shows that a universal positive correlation could be found between the production–consumption ratio and the HANPP$_{harv}$ (% of HANPP) through agriculture, forestry, and grazing in 31 provinces in China. The aggregation characteristics of sample points declared whether the HANPP$_{harv}$ was consumed locally, showing that the interregional flows and the exchange of agricultural and livestock products were more complicated than the forestry products.

3.3. Regional Differentiations in HANPP

An obvious spatial differentiation of HANPP was found in China. However, the focus should be to identify whether there is a significant difference among regionalization zones, and nine HANPP indexes were chosen for further discussion (Table 3). Here, the independent-samples T test was used to detect the differentiation of nine HANPP indexes across the four zones (Zone 1 to Zone 4; Figure S3). The result showed that HANPP (gC/m^2), $HANPP_{harv}$ (gC/m^2), and $HANPP_{luc}$ (gC/m^2) were significantly different on both sides of the Huanyong Hu line (Zone 1) but their total amount and contribution to NPP_{pot} did not pass the test. The same regional differentiation of HANPP indexes could be found between the western and eastern parts of China (Zone 2) as Zone 1, and the ecological pressure caused by human occupation (HANPP (%NPP_{pot})) was also significantly different in Zone 2. $HANPP_{harv}$ and $HANPP_{luc}$ have different sensitivities to meteorological elements, so the corresponding indexes showed various characteristics between the north and south parts of China (Zone 3), which was divided on the basis of regional temperature and precipitation. There is a difference between economic development and population agglomeration in China's coastal and inland regions, so the $HANPP_{harv}$ indexes related to land productivity did not show significant differentiations in Zone 4. Furthermore, from the perspective of the nine HANPP indexes, it is worth noting that HANPP (gC/m^2) showed significant differences among all the regionalization zones (Table 3). The complex characteristics of HANPP indexes in China indicated that they were not only restricted by population density and terrestrial natural conditions but that a complex mechanism of regional comprehensive development (i.e., economy, urbanization, and agrotechnique) also had an impact on them.

Table 3. Regional differentiations of HANPP indexes.

Indexes		Regional Differentiation *							
		Zone 1		Zone 2		Zone 3		Zone 4	
		t Value	*p* Value	t Value	*p* Value	t Value	*p* Value	t Value	*p* Value
HANPP	(gC)	−0.433	0.668	−0.780	0.441	0.293	0.772	−2.323	0.027 *
	(%NPP_{pot})	−0.009	0.993	2.141	0.041 *	3.672	0.001 ***	0.474	0.639
	(gC/m^2)	5.933	0.001 ***	4.924	0.001 ***	−2.341	0.026 *	2.170	0.039 *
$HANPP_{harv}$	(gC)	0.191	0.850	0.258	0.798	1.667	0.109	−1.649	0.110
	(%NPP_{pot})	−0.035	0.973	1.450	0.158	3.721	0.001 ***	0.308	0.760
	(gC/m^2)	2.647	0.013 *	3.730	0.001 ***	0.786	0.441	0.942	0.354
$HANPP_{luc}$	(gC)	−0.531	0.616	−1.335	0.192	−0.654	0.518	−2.167	0.039 *
	(%NPP_{pot})	0.066	0.984	1.939	0.062	0.525	0.604	0.498	0.623
	(gC/m^2)	8.477	0.00 1 ***	3.314	0.002 **	−5.817	0.001 ***	1.881	0.070

* Zone 1 refers to two regions between the Huanyong Hu Line; Zones 2 and 3 refer to the western/eastern and northern/southern parts of China, respectively; and Zone 4 refers to the coastal land and the inland area, which are also shown in Figure S3. *: $p < 0.05$, **: $p < 0.01$, and ***: $p < 0.001$.

4. Discussion

4.1. Credibility of HANPP

In existing studies, HANPP (%NPP_{pot}) has attracted more attention than HANPP (gC/m^2) or HANPP (gC) due to its ability to represent regional ecological sustainability and the ecological pressure caused by human occupation. However, the lack of a uniform standard in the selection of estimation models of NPP_{pot} and NPP_{act} hinders the comparison of HANPP results among different regions. For instance, in related studies, a variety of models have been used to estimate NPP_{pot}, e.g., LPJ-DGVM [16,41], the Guangsheng Zhou model [5], the new calibrated Miami model [45], and multiple-iterations modification based on the disturbed pixels in NPP_{act} [2]. Rather than comparing the similarities of HANPP

(%NPP_{pot}) between regions, it is better to focus on whether its spatial distribution and long-term variation characteristics are similar.

In China, HANPP (%NPP_{pot}) was found to be 60.33% in 2015, which is higher than it was in 2010 (57.80%), estimated by Chen et al. (2015) [5], as well as the global HANPP (%NPP_{pot}) values in 2000 (23.80%), 2005 (25.00%), and 2010 (20.70%) [16–18]. Huang et al. (2020) found that HANPP (%NPP_{pot}) was 72.4% in the Yangtze River Delta in 2015 [51], and we found a slightly lower value (68.92%) in the same region in this study. From the perspective of the spatial distribution characteristics of HANPP, the grid-scaled HANPP (%NPP_{pot}) in this study is highly similar to the results of Haberl et al. (2007) and Kastner et al. (2021) [16,17]. In general, the simulation of HANPP among provinces and cities in China is still in its infancy and the deviations are mainly due to different methods and datasets used during NPP_{pot} and $HANPP_{harv}$ estimation within limited studies.

4.2. Relationships between HANPP and China's Urban and Rural Development

Significant spatial differentiation of HANPP was found under the impact of unbalanced human activities and the ecological environment in China, especially the duality of rural–urban development. In this study, 20 indictors reflecting regional economy, population, urban expansion, and rural agricultural technology were chosen, as shown in Table 1, and the correlation coefficients and nine HANPP indexes (Table 3) of 31 provinces were summarized (Figure 4). From the perspective of urban economy, significant positive correlations were identified between the added value of the secondary and tertiary industries (UE1) and HANPP (gC/m^2), $HANPP_{harv}$ (gC/m^2), and $HANPP_{luc}$ (gC/m^2). The added value of transportation, warehousing, and postal services (UE3) had a significant influence on most HANPP indexes (Figure 4a), which guarantees the circulation of regional ecological resources. However, UE2 and UE4 showed a negative correlation with the total amount of HANPP, $HANPP_{harv}$, and $HANPP_{luc}$, which is because a larger proportion of UE1 and a higher consumption expenditure of urban residents always appears in highly urbanized areas, which usually have a relatively weak biomass production (low $HANPP_{harv}$); the $HANPP_{luc}$ would also not increase significantly when urban construction is saturated, so the HANPP indexes would be suppressed instead. High urban population (UP1) resulted in a great amount of biomass occupation per unit area, whereas the negative correlations between urban population proportion (UP2) and HANPP (gC), $HANPP_{harv}$ (gC), and $HANPP_{luc}$ (gC) reflected the polarization characteristics of the urban population agglomeration and ecological resource production. Among the urban spatial expansion indicators, the higher regional average nighttime light intensity (US1), the proportion of the built-up area (US2), and the proportion of the urban road area (US4) always corresponded with large $HANPP_{luc}$ (% NPP_{pot}) and $HANPP_{luc}$ (gC/m^2) values, which also directly led to the significant positive correlations ($p < 0.01$) and relatively high correlation coefficients with HANPP (% NPP_{pot}) and HANPP (gC/m^2). Conversely, urban ecological construction gained valuable space to counter the degradation of natural vegetation caused by urban expansion, and most of the HANPP indexes, especially the $HANPP_{luc}$ indexes, decreased as the proportion of urban green spaces increased (US3).

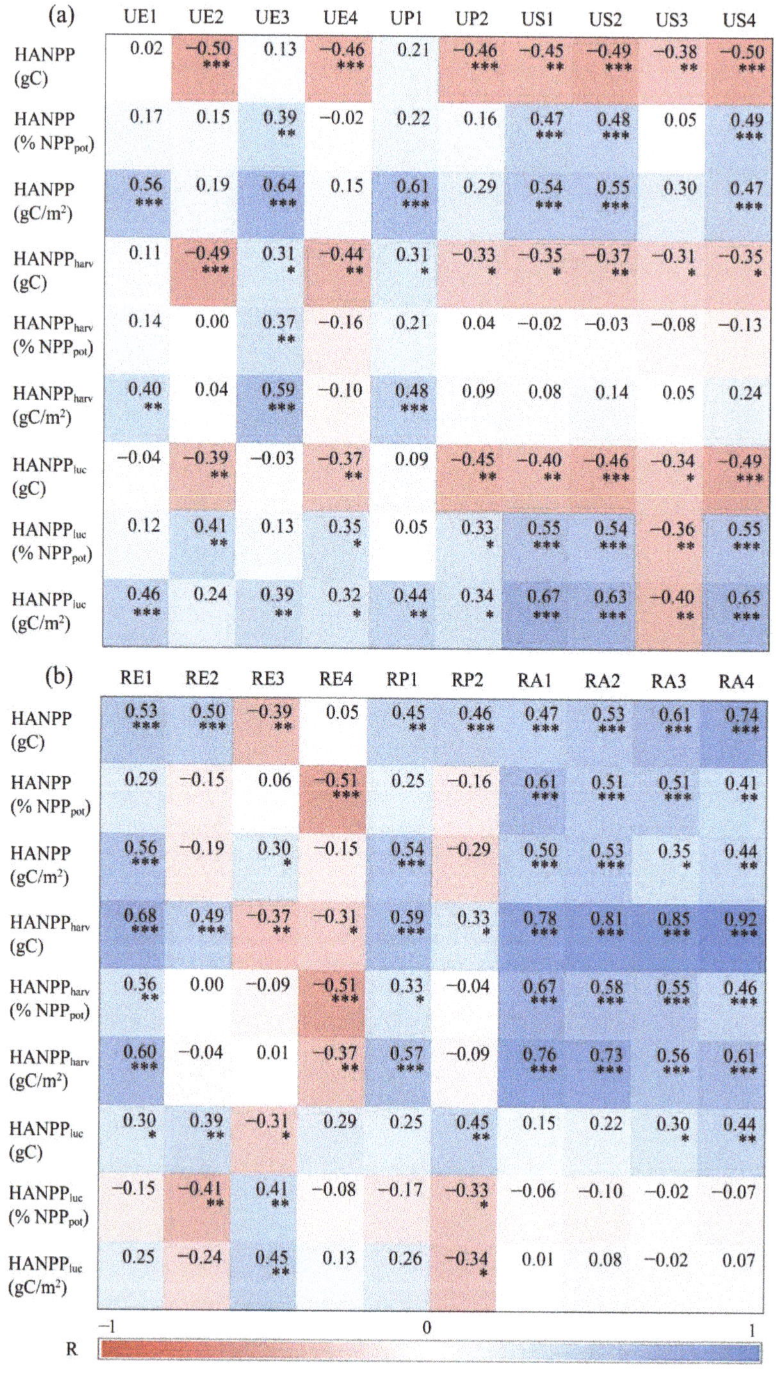

Figure 4. Relationship between HANPP indexes and urban (**a**) and rural (**b**) development indicators. (UE1–4, UP1–2, and US1–4 and RE1–4, RP1–2, and RA1–4; refer to Table 1; *: $p < 0.1$, **: $p < 0.05$, and ***: $p < 0.01$).

HANPP indexes responded to rural development in different ways (Figure 4b). Stronger positive correlations existed between the added value of the primary industry (RE1) and all $HANPP_{harv}$ indexes than UE1, and the total amount of HANPP, $HANPP_{harv}$, and $HANPP_{luc}$ in each province was more sensitive to the proportion of the added value of the primary industry (RE2) than other HANPP indexes. Higher consumption expenditure of rural residents (RE3) appeared not only in highly urbanized areas but also in the agriculture, forestry, and animal husbandry developed areas, which showed relatively lower correlations with the total amount of HANPP, $HANPP_{harv}$, and $HANPP_{luc}$ than UE4. The higher the Engel coefficient of rural residents (RE4), the lower the HANPP (% NPP_{pot}) and all $HANPP_{harv}$ indexes. This shows that the main article consumed by residents was food, the regional economy was relatively lagging in terms of development, and the influence of human activities on ecology was relatively weak. Rural population (RP1) had significant and higher positive correlations with all the $HANPP_{harv}$ indexes than UP1 but no statistical correlations with $HANPP_{luc}$ indexes. Moreover, the biomass occupied through land use per unit area and its percentage in NPP_{pot} would be reduced if the rural population was the main component (RP2) in a region. All the rural agricultural technology indicators showed similar impacts on the nine HANPP indexes, especially in terms of the significant positive effects ($p < 0.01$) on and high correlation coefficients among $HANPP_{harv}$ (gC), $HANPP_{harv}$ (% NPP_{pot}), and $HANPP_{harv}$ (gC/m^2). Adequate effective irrigated area (RA3) and total sown area of crops (RA4) are necessary for adequate food supply, the total power of the agricultural machinery (RA1) reflects the regional agricultural machinery ownership, and the harvest biomass of grain and forestry would benefit from these elements as well as the amount of agricultural fertilizer (RA2) applied.

An interaction effect always exists between regional economy improvement and biomass occupation, and industrial structure would alter the HANPP proration. Overdevelopment of the secondary and tertiary industries would restrain harvest biomass, but adequate ecological supply depends on the primary industry, so balanced industrial development should be the focus among regions. Harvested biomass ($HANPP_{harv}$) is mostly determined by the rural population rather than the urban population, while the urban or rural population proportion has hardly any influence on it, indicating that urban people mainly play the role of the consumer of ecological resources. Urban sprawl, a result of land surface modification by humans, usually leads to the irrecoverable loss of NPP. Cultivated land offers additional products along with the improvement of agricultural technology, although it replaces part of the natural vegetation. Moreover, worldwide, the increasing populations and economies are inevitably accompanied by rapid urbanization, and the corresponding requirements of ecological resources must be sustainable. Thus, through policies and agriculture techniques, it is necessary to improve agriculture production and the utilization efficiency of NPP per area, as well as properly control urban expansion.

4.3. HANPP Responses to Coordinated Regional Urban–Rural Development

As two major systems indispensable to regional development, urban areas and rural areas play an important role in promoting sustainable use of ecological resources. Therefore, coordinated urban–rural development and the impact on the HANPP in 31 provinces in China was quantified. The quantification of the urban development index ($f(u)$) indicated that highly urbanized areas (Figure 5a) were mostly concentrated in the coastal provinces but high values of the rural development index ($f(r)$) mostly appeared in southern China (Figure 5b). Specifically, Shanghai ($f(u) = 0.90$), Beijing ($f(u) = 0.43$), Henan ($f(r) = 0.73$), and Shandong ($f(r) = 0.68$) showed the peak values over the 95th percentiles of $f(u)$ and $f(r)$. In contrast, the valley values under the fifth percentiles were in Qinghai ($f(u) = 0.04$), Tibet ($f(u) = 0.02$), Tianjin ($f(r) = 0.09$), and Beijing ($f(r) = 0.08$). Furthermore, the degree of coordinated urban–rural development (D) among provinces was evaluated by the comprehensive urban–rural development index (T) and the coupling degree (C) (Table S4) and showed an obvious spatial differentiation between coastal and inland areas (Figure 5c). In this study, the D value was divided into nine levels (Table 2) to map coordinated urban–rural development types in China (Figure 5d). Only a few coastal provinces had achieved primary coordination (Jiangsu and Guangdong) and barely coordination (Shandong and Zhejiang). Inversely, most regions showed a universal imbalance and urban–rural development gap. Notably, there was severe imbalance in Heilongjiang Province, Tibet, Gansu Province, and Xinjiang Province. In short, the ideal urban–rural coordination did not exist in China in 2015.

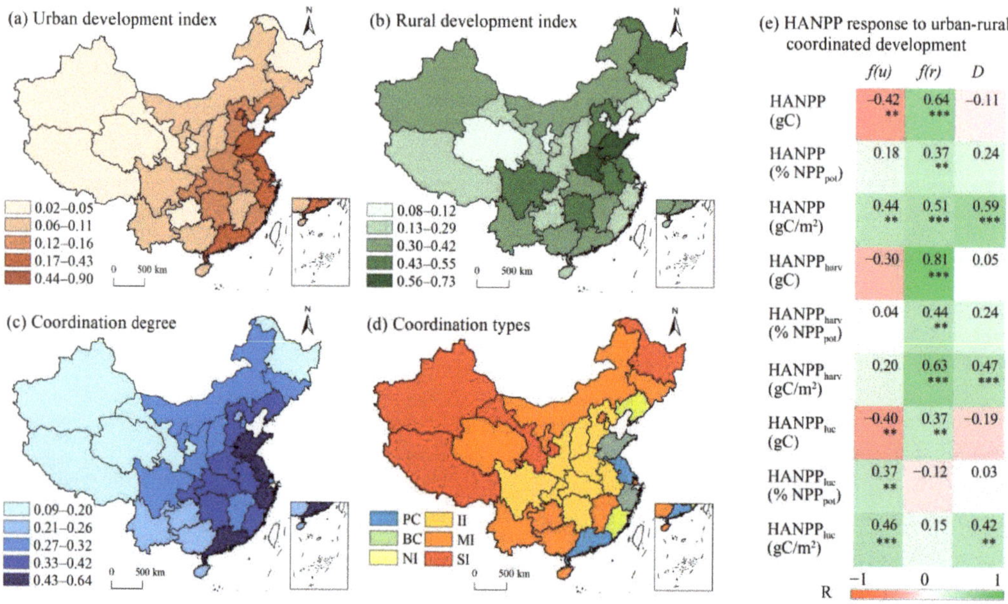

Figure 5. Urban and rural development index (**a**,**b**) and coordinated urban–rural development degree (**c**) and types (**d**) in China. HANPP responses to coordinated regional urban–rural development (**e**). (PC, primary coordination; BC, barely coordination; NI, near imbalance; II, inchoate imbalance; MI, moderate imbalance; SI, severe imbalance; **: $p < 0.05$, and ***: $p < 0.01$).

The relationship between HANPP indexes and coordinated urban–rural development (Figure 5e) helped create verifications and supplements for Figure 4. It is worth noting that $f(u)$ and $f(r)$ showed opposite driving effects on HANPP (gC), $HANPP_{harv}$ (gC), and $HANPP_{luc}$ (gC) and the regional ecological pressure characterized by HANPP (%NPP_{pot}) and HANPP (gC/m^2) mostly depended on $f(r)$. The highly urbanized regions with high $f(u)$ usually appeared in provinces with limited territorial areas (i.e., Shanghai, Beijing, and Tianjin), in which the total amount of occupied biomass was at a relatively low level. On the contrary, provinces with high levels of $f(r)$ had relatively large territorial areas in China and developed production capacity. As a component of HANPP, $HANPP_{luc}$ was mainly controlled by the urban development index ($f(u)$). Conversely, $f(r)$ made significant positive impacts on $HANPP_{harv}$. All the occupied biomass indexes per unit area showed the most sensitive response to coordinated urban–rural development degree (D) and good urban–rural coordination obviously improved HANPP (gC/m^2) and $HANPP_{harv}$ (gC/m^2) but relatively weaker urban–rural coordination influenced $HANPP_{luc}$ (gC/m^2) variation. From the perspective of six coordination types among provinces in China, the sample points of $HANPP_{harv}$ (gC/m^2) tended to be discrete, along with a higher degree of coordination (Figure 6), and the samples with moderate imbalance (MI) urban–rural coordination showed the highest degree of dispersion of HANPP (gC/m^2) and $HANPP_{luc}$ (gC/m^2).

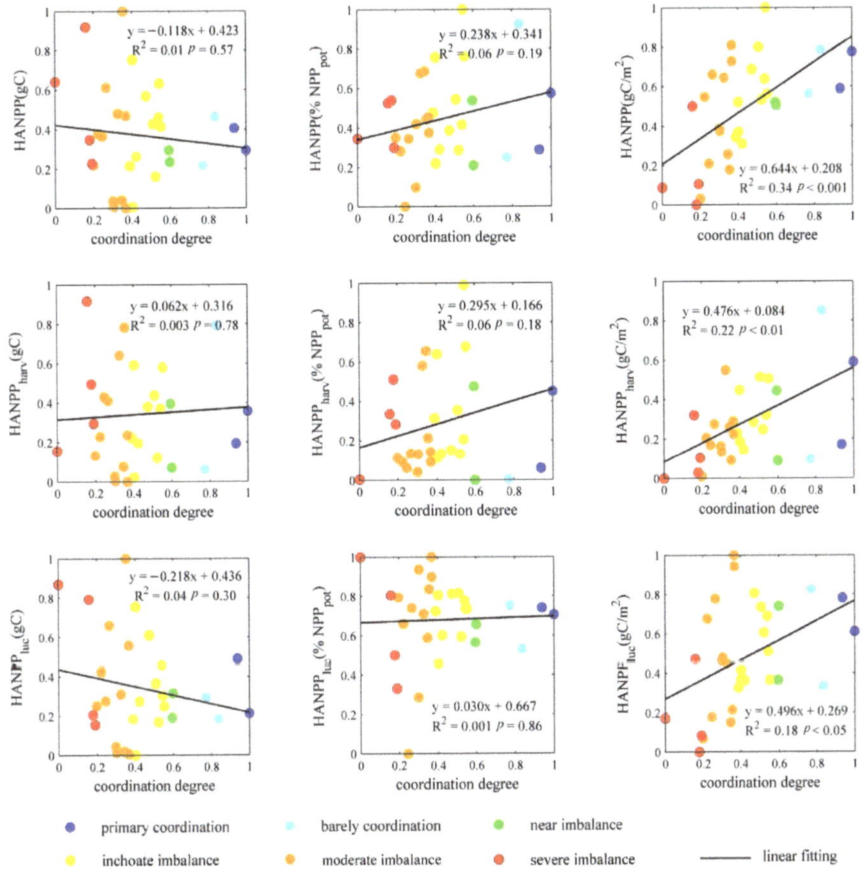

Figure 6. Correlations between HANPP indexes and the degree of coordinated urban–rural development.

4.4. Dominant Driving Factors of Each HANPP Index

Bidirectional effects were found among urban–rural development indictors of biomass appropriation, and the dominant driving factors of each HANPP index were detected by stepwise regression to eliminate the indirect correlation. Table 4 shows the fitting models and correlation coefficients from three perspectives (urban, rural, and urban–rural development), in which stronger decisive effects were displayed by the driving factor with larger standardized coefficients, and a higher R referred to a better fitting model. Among all the driving factors in the 27 fitting equations in Table 4, the rural development indicators (34 times) appeared more frequently than the urban development indicators (21 times). At the same time, in different dimensions of urban–rural development, each HANPP index was more likely to be affected by urban economy (UE) than rural economy (RE), more likely to be affected by rural population (RP) than urban population (UP), and more likely to be affected by rural agricultural technology (RA) than urban spatial expansion (US). From the analog effect of each equation, driven by 20 urban and rural development indicators, the analog effect of $HANPP_{harv}$ indexes was the best, followed by that of HANPP indexes, and $HANPP_{luc}$ indexes were the weakest; the selected urban and rural development indicators could explain their variation characteristics only to a certain extent. From the perspectives of urban, rural, and comprehensive urban–rural development, the analog effects of all the nine HANPP indexes were the best when only considering rural development indicators, followed by the regression models of comprehensive urban–rural development indicators, and the ability to explain HANPP indexes was relatively limited when only considering urban development indicators. From the perspective of the explanatory ability of each independent variable, the impact of urban economy (UE) on occupied biomass was much stronger than that of urban population (UP) and urban spatial expansion (US). On the contrary, rural population (RP) and rural agricultural technology (RA) had more significant explanatory effects on each HANPP index than rural economy (RE). Among all the independent variables, UP2, US1, US2, US3, and RA3 did not appear in any of the fitting models.

Table 4. Dominant driving factors of each HANPP index.

Dependent Variables		Independent Variables	Model with Standardized Coefficients	R
HANPP	(gC)	Urban	= −0.599UE2 + 0.366UP1	0.613
		Rural	= 1.111RA4 + 0.348RE4 − 0.365RP1	0.823
		Urban–rural	= 0.650RA4 − 0.309US4	0.797
	($\%NPP_{pot}$)	Urban	= 1.223UE3 − 0.924UE1	0.567
		Rural	= 1.007RA1 − 1.044RP1 − 0.384RE2 + 0.599RA2	0.849
		Urban–rural	= 1.007RA1 − 1.045RP1 + 0.384UE2 + 0.599RA2	0.849
	(gC/m^2)	Urban	= 0.642UE3	0.642
		Rural	= 0.628RE1 − 0.401RP2	0.682
		Urban–rural	= 0.642UE3	0.642
$HANPP_{harv}$	(gC)	Urban	= −0.672UE2 + 0.543UE3	0.706
		Rural	= 0.897RA4 − 0.371RP1 + 0.345RA2	0.940
		Urban–rural	= 0.897RA4 − 0.371RP1 + 0.345RA2	0.940
	($\%NPP_{pot}$)	Urban	= 1.245UE3 − 0.978UE1	0.566
		Rural	= 1.189RA1 − 0.641RP1	0.763
		Urban–rural	= 1.189RA1 − 0.641RP1	0.763
	(gC/m^2)	Urban	= 1.210UE3 − 0.688UE1	0.665
		Rural	= 0.846RA1 − 0.315RP2	0.822
		Urban–rural	= 0.340RA1 + 0.771US4 − 0.480UE4 + 0.425RA2	0.897
$HANPP_{luc}$	(gC)	Urban	= −0.486US4	0.486
		Rural	= 0.356RP2 + 0.912RA4 − 0.698RA1	0.678
		Urban–rural	= −0.486US4	0.486
	($\%NPP_{pot}$)	Urban	= 0.412UE2	0.412
		Rural	= −0.412RE2	0.412
		Urban–rural	= −0.412RE2	0.412
	(gC/m^2)	Urban	= 0.458UE1	0.458
		Rural	= 0.513RE3 + 0.357RP1	0.568
		Urban–rural	= 0.458UE1	0.458

Comprehensively considering the dominant driving factors and standardized coefficients of each HANPP index, it was found that UE3 had the strongest explanation for the differentiation characteristics of HANPP (gC/m^2) and that the restriction of RP and RA indicators on HANPP ($\%NPP_{pot}$) was much stronger than that of urban development indicators. The total amount of HANPP (gC) was affected by the opposite effects of RA4 and US4, and RA4 showed a stronger control of it. In addition, RP and RA indicators showed absolute driving advantages in $HANPP_{harv}$ ($\%NPP_{pot}$) and $HANPP_{harv}$ (gC) and RA indicators had a stronger positive driving effect on them. From the perspective of comprehensive urban–rural development, urban development indicators had no significant driving effect on the above two indexes. The influence of each independent variable on $HANPP_{harv}$ (gC/m^2) was relatively complicated, in consideration of RA, UE, and US indicators, and a better-fitting model could obtain $HANPP_{harv}$ (gC/m^2); the standardized coefficient showed that the urban development indicators played a more important role. Table 4 shows that the fitting equations between the independent variables and $HANPP_{luc}$ were relatively simple. US4, RE2, and UE1 were the dominant driving factors for $HANPP_{luc}$

(gC), $HANPP_{luc}$ (%NPP_{pot}), and $HANPP_{luc}$ (gC/m^2), respectively, and none of the population factors were correlated. It should be noted that the ability of selected independent variables to interpret all $HANPP_{luc}$ indexes was significantly weaker than their ability to interpret HANPP and $HANPP_{harv}$, indicating that there are driving factors besides those selected in this study. Further analysis of the positive or negative driving differences of each independent variable showed that UE3, RA1, RA2, and RA4 had a significant positive influence in each fitting model, which could be verified and compared with the results in Figure 4. In summary, HANPP (%NPP_{pot}), $HANPP_{harv}$ (%NPP_{pot}), and $HANPP_{luc}$ (%NPP_{pot}) were more dependent on rural development, while urban development was the decisive factor for all HANPP indexes per unit area, and the total amount of occupied biomass of each province was more affected by rural development than urban.

5. Conclusions

In this study, we estimated the HANPP in China in 2015 based on multi-source datasets and quantified the contribution characteristics of HANPP components as well as the biomass production–consumption types in 31 provinces. The regional differentiations of HANPP indexes among typical regionalization zones were also identified. In addition, an evaluation framework of urban–rural development was built to investigate how HANPP indexes respond to the regional unbalanced development and we identified the dominant driving factors of HANPP indexes from the perspectives of economy, population, urban expansion, and agricultural technology. The results showed that the total amount of HANPP was 2.68 PgC and gradually decreased from the southeast to the northwest of China, which represented 60.33% of the NPP_{pot}. The grain-producing provinces had high HANPP (%NPP_{pot}) values, while the southwestern provinces with complex terrain had low HANPP (%NPP_{pot}) values. HANPP (gC/m^2), $HANPP_{harv}$ (gC/m^2), and $HANPP_{luc}$ (gC/m^2) were significantly different ($p < 0.05$) between the Huanyong Hu Line and the western–eastern part of China, while $HANPP_{harv}$ did not show significant differentiations in the coastal land and inland areas.

From the perspective of HANPP components, in China, the total amount and the average value of $HANPP_{luc}$ and $HANPP_{harv}$ were found to be 1.63 PgC and 216.97 gC/m^2 and 1.05 PgC and 126.53 gC/m^2, respectively, and the harvest through cropland, livestock grazing, and forestry contributed 29.86%, 8.53%, and 0.91% to the total HANPP, respectively. Land use change had a bidirectional influence on $HANPP_{luc}$ (gC/m^2); the negative values were mostly concentrated in a small number of cultivated land patches in northwest China, which indicated that the biomass production under intensive artificial management could be higher than the potential production. Moreover, the North China Plain and northeast provinces had relatively high $HANPP_{harv}$ (gC/m^2) values, mainly provided by cultivated land. Different urban functions resulted in various ecological occupation structures. $HANPP_{luc}$ played the dominant role among 21 provinces (over 50% of HANPP), especially in the cities with rapid urban expansion and some coastal ones. With the production–consumption ratio of biomass based on HANPP, 17 provinces in the middle and southeastern coastal areas in China were defined as the consumption type. Specifically, the ratios in Shanghai (0.06) and Beijing (0.09) were under the fifth percentiles, and a great number of ecological resources were needed from surrounding regions. In addition, a universal positive correlation ($p < 0.05$) was found between the production–consumption ratio and the $HANPP_{harv}$ (% of HANPP) through agriculture, forestry, and grazing in 31 provinces in China.

In different dimensions of urban–rural development, each HANPP index was more likely to be affected by urban economy (UE) than rural economy (RE); the impact of rural population (RP) was much stronger than that of urban population (UP); and rural agricultural technology (RA) had more significant explanatory effects on each HANPP index than urban spatial expansion (US). Specifically, in the dimension of economic development, the added value of transportation, warehousing, and postal services (UE3) had a significant influence on most HANPP indexes. From the demographic perspective, rural population (RP1) had significant and higher positive correlations with all the $HANPP_{harv}$ indexes than UP1 but no statistical correlation with $HANPP_{luc}$. Meanwhile, agricultural technology indicators showed significant positive effects ($p < 0.01$) on and high correlations with $HANPP_{harv}$. The higher regional average nighttime light intensity (US1), the proportion of the built-up area (US2), and the proportion of urban road area (US4) always corresponded with large $HANPP_{luc}$ values. Conversely, most HANPP indexes decreased as the proportion of urban green spaces increased (US3). Furthermore, the regional ecological pressure characterized by HANPP (%NPP_{pot}) and HANPP (gC/m^2) mostly depended on the rural development index ($f(r)$) and $HANPP_{luc}$ and $HANPP_{harv}$ were mainly controlled by $f(u)$ and $f(r)$, respectively. A higher degree of urban–rural coordination (D) obviously improved HANPP (gC/m^2) and $HANPP_{harv}$ (gC/m^2) but had a relatively weaker effect on the variation in $HANPP_{luc}$ (gC/m^2).

Supplementary Materials: The following supporting information can be downloaded at https://www.mdpi.com/article/10.3390/land12051062/s1: Figure S1: Conceptual model of human appropriation of net primary production (HANPP); Figure S2: Spatial distribution of NPP_{pot} in China; Figure S3: Typical regionalization of China to detect the HANPP differentiation; Table S1: Moisture contents (%), harvest factors and recovery rates of typical crops; Table S2: Conversion coefficients from crop harvest to straw; Table S3: Regulation for assessing NPP_{pot} by vegetation-approach model; Table S4: Urban–rural coordinated development indexes; Supplementary File S1: Estimation of $HANPP_{harv}$; Supplementary File S2: Evaluation of urban–rural coordinated development.

Author Contributions: Conceptualization, T.Z. and J.P.; methodology, T.Z.; software, T.Z.; validation, T.Z. and J.P.; formal analysis, T.Z.; resources, J.P.; data curation, X.C.; writing—original draft preparation, T.Z.; writing—review and editing, J.P.; visualization, T.Z.; supervision, X.C.; funding acquisition, T.Z. and X.C. All authors have read and agreed to the published version of the manuscript.

Funding: This research was funded by the National Natural Science Foundation of China (Grant No. 42001097) and the National Natural Science Foundation of China (Grant No. 41831284).

Data Availability Statement: The data presented in this study are available on request from the corresponding author.

Acknowledgments: We gratefully acknowledge Helmut Haberl from the University of Natural Resources and Life Sciences for his advice on the research methods and the HANPP estimation model used in this manuscript.

Conflicts of Interest: The authors declare no conflict of interest.

References

1. Fan, Y.P.; Fang, C.L. Eco-city and man-land relationship. *Acta Ecol. Sin.* **2022**, *42*, 4313–4323.
2. O'Neill, D.W.; Tyedmers, P.H.; Beazley, K.F. Human appropriation of net primary production (HANPP) in Nova Scotia, Canada. *Reg. Environ. Chang.* **2007**, *7*, 1–14. [CrossRef]
3. Schwarzlmueller, E. Human appropriation of aboveground net primary production in Spain, 1955–2003: An empirical analysis of the industrialization of land use. *Ecol. Econ.* **2009**, *69*, 282–291. [CrossRef]
4. Piao, S.L.; Fang, J.Y.; Zhou, L.M.; Zhu, B.; Tan, K.; Tao, S. Changes in vegetation net primary productivity from 1982 to 1999 in China. *Glob. Biogeochem. Cycles* **2005**, *19*, 1605–1622. [CrossRef]
5. Chen, A.F.; Li, R.Y.; Wang, H.L.; He, B. Quantitative assessment of human appropriation of aboveground net primary production in China. *Ecol. Model.* **2015**, *312*, 54–60. [CrossRef]
6. Epstein, T.S.; Jezeph, D. Development-There is another way: A rural-urban partnership development paradigm. *World Dev.* **2001**, *29*, 1443–1454. [CrossRef]
7. Jiang, C.J.; Li, J.T.; Liu, J.L. Does urbanization affect the gap between urban and rural areas? Evidence from China. *Socio-Econ. Plan. Sci.* **2022**, *82*, 101271. [CrossRef]
8. Gao, Y.Y.; Zheng, J.H.; Bu, M.L. Rural-urban income gap and agricultural growth in China: An empirical study on the provincial panel data, 1978–2010. *China Agric. Econ. Rev.* **2014**, *6*, 92–107. [CrossRef]
9. Yu, B.B. Ecological effects of new-type urbanization in China. *Renew. Sustain. Energy Rev.* **2021**, *135*, 110239. [CrossRef]
10. Du, X.J.; Huang, Z.H. Ecological and environmental effects of land use change in rapid urbanization: The case of Hangzhou, China. *Ecol. Indic.* **2017**, *81*, 243–251. [CrossRef]
11. Haase, D.; Toetzer, T. Urban-rural linkages-analysing, modelling, and understanding drivers, pressures, and impacts of land use changes along the rural-to-urban gradient. *Environ. Plan. B-Urban* **2012**, *39*, 194–197. [CrossRef]

12. Ren, Y.F.; Fang, C.L.; Li, G.D.; Sun, S.A.; Bao, C.; Liu, R.W. Progress in local and tele-coupling relationship between urbanization and eco-environment. *Acta Geogr. Sin.* **2020**, *75*, 589–606.
13. Chen, Y.; Luo, P.; Chang, T. Urbanization and the urban–rural income gap in China: A continuous wavelet coherency analysis. *Sustainability* **2020**, *12*, 8261. [CrossRef]
14. Pan, Y.; Yu, C.Q.; Zhang, X.Z.; Chen, B.X.; Wu, J.X.; Tu, Y.L.; Miao, Y.J.; Luo, L.M. A modified framework for the regional assessment of climate and human impacts on net primary productivity. *Ecol. Indic.* **2016**, *60*, 184–191. [CrossRef]
15. Teixido-Figueras, J.; Steinberger, J.K.; Krausmann, F.; Haberl, H.; Wiedmann, T.; Peters, G.P.; Duro, J.A.; Kastner, T. International inequality of environmental pressures: Decomposition and comparative analysis. *Ecol. Indic.* **2016**, *62*, 163–173. [CrossRef]
16. Haberl, H.; Erb, K.H.; Krausmann, F.; Gaube, V.; Bondeau, A.; Plutzar, C.; Gingrich, S.; Lucht, W.; Fischer-Kowalski, M. Quantifying and mapping the human appropriation of net primary production in earth' terrestrial ecosystems. *Proc. Natl. Acad. Sci. USA* **2007**, *104*, 12942–12945. [CrossRef] [PubMed]
17. Kastner, T.; Matej, S.; Forrest, M.; Gingrich, S.; Haberl, H.; Hickler, T.; Krausmann, F.; Lasslop, G.; Niedertscheider, M.; Plutzar, C.; et al. Land use intensification increasingly drives the spatiotemporal patterns of the global human appropriation of net primary production in the last century. *Glob. Chang. Biol.* **2021**, *28*, 307–322. [CrossRef]
18. Krausmann, F.; Erb, K.H.; Gingrich, S.; Haberl, H.; Bondeau, A.; Gaube, V.; Lauk, C.; Plutzar, C.; Searchinger, T.D. Global human appropriation of net primary production doubled in the 20th century. *Proc. Natl. Acad. Sci. USA* **2013**, *110*, 10324–10329. [CrossRef]
19. Niedertscheider, M.; Gingrich, S.; Erb, K.H. Changes in land use in South Africa between 1961 and 2006: An integrated socio-ecological analysis based on the human appropriation of net primary production framework. *Reg. Environ. Chang.* **2012**, *12*, 715–727. [CrossRef]
20. Plutzar, C.; Kroisleitner, C.; Haberl, H.; Fetzel, T.; Bulgheroni, C.; Beringer, T.; Hostert, P.; Kastner, T.; Kuemmerle, T.; Lauk, C.; et al. Changes in the spatial patterns of human appropriation of net primary production (HANPP) in Europe 1990–2006. *Reg. Environ. Chang.* **2016**, *16*, 1225–1238. [CrossRef]
21. Mahbub, R.B.; Ahmed, N.; Rahman, S.; Hossain, M.M.; Sujauddin, M. Human appropriation of net primary production in Bangladesh, 1700–2100. *Land Use Policy* **2019**, *87*, 104067. [CrossRef]
22. Huang, X.; Yang, Y.; Chen, C.; Zhao, H.; Yao, B.; Ma, Z.; Ma, L.; Zhou, H. Quantifying and mapping human appropriation of net primary productivity in Qinghai grasslands in China. *Agriculture* **2022**, *12*, 483. [CrossRef]
23. Andersen, C.B.; Donovan, R.K.; Quinn, J.E. Human appropriation of net primary production (HANPP) in an agriculturally-dominated watershed, southeastern USA. *Land* **2015**, *4*, 513–540. [CrossRef]
24. Zhang, Y.J.; Pan, Y.; Zhang, X.Z.; Wu, J.X.; Yu, C.Q.; Li, M.; Wu, J.S. Patterns and dynamics of the human appropriation of net primary production and its components in Tibet. *J. Environ. Manag.* **2018**, *210*, 280–289. [CrossRef] [PubMed]
25. Roux, N.; Kastner, T.; Erb, K.H.; Haberl, H. Does agricultural trade reduce pressure on land ecosystems? Decomposing drivers of the embodied human appropriation of net primary production. *Ecol. Econ.* **2021**, *181*, 106915. [CrossRef]
26. Imhoff, M.L.; Bounoua, L.; Ricketts, T.; Loucks, C.; Harriss, R.; Lawrence, W.T. Global patterns in human consumption of net primary production. *Nature* **2004**, *429*, 870–873. [CrossRef]
27. Niedertscheider, M.; Kuemmerle, T.; Müller, D.; Erb, K.H. Exploring the effects of drastic institutional and socio-economic changes on land system dynamics in Germany between 1883 and 2007. *Glob. Environ. Chang.* **2014**, *28*, 98–108. [CrossRef]
28. Krausmann, F. Land use and industrial modernization: An empirical analysis of human influence on the functioning of ecosystems in Austria 1830–1995. *Land Use Policy* **2001**, *18*, 17–26. [CrossRef]
29. Gingrich, S.; Niedertscheider, M.; Kastner, T.; Haberl, H.; Cosor, G.; Krausmann, F.; Kuemmerle, T.; Müller, D.; Reith-Musel, A.; Jepsen, M.R.; et al. Exploring long-term trends in land use change and aboveground human appropriation of net primary production in nine European countries. *Land Use Policy* **2015**, *47*, 426–438. [CrossRef]
30. Kastner, T. Trajectories in human domination of ecosystems: Human appropriation of net primary production in the Philippines during the 20th century. *Ecol. Econ.* **2009**, *69*, 260–269. [CrossRef]
31. Dorninger, C.; Wehrden, H.V.; Krausmann, F.; Bruckner, M.; Feng, K.S.; Hubacek, K.; Erb, K.H.; Abson, D.J. The effect of industrialization and globalization on domestic land-use: A global resource footprint perspective. *Glob. Environ. Chang.* **2021**, *69*, 102311. [CrossRef]
32. Haberl, H.; Steinberger, J.K.; Plutzar, C.; Erb, K.H.; Gaube, V.; Gingrich, S.; Krausmann, F. Natural and socioeconomic determinants of the embodied human appropriation of net primary production and its relation to other resource use indicators. *Ecol. Indic.* **2012**, *23*, 222–231. [CrossRef]
33. Marull, J.; Tello, E.; Bagaria, G.; Font, X.; Cattaneo, C.; Pino, J. Exploring the links between social metabolism and biodiversity distribution across landscape gradients: A regional-scale contribution to the land-sharing versus land-sparing debate. *Sci. Total Environ.* **2018**, *619–620*, 1272–1285. [CrossRef]
34. Haberl, H.; Plutzar, C.; Erb, K.H.; Gaube, V.; Pollheimer, M.; Schulz, N.B. Human appropriation of net primary production as determinant of avifauna diversity in Austria. *Agric. Ecosyst. Environ.* **2005**, *110*, 119–131. [CrossRef]
35. Zhou, C.; Elshkaki, A.; Graedel, T. Global human appropriation of net primary production and associated resource decoupling: 2010–2050. *Environ. Sci. Technol.* **2018**, *52*, 1208–1215. [CrossRef] [PubMed]
36. Krausmann, F.; Haberl, H.; Erb, K.H.; Wiesinger, M.; Gaube, V.; Gingrich, S. What determines geographical patterns of the global human appropriation of net primary production? *J. Land Use Sci.* **2009**, *4*, 15–33. [CrossRef]

37. Das, B.; Chakrabarty, D. Lagrange's interpolation formula: Representation of numerical data by a polynomial curve. *Int. J. Math. Trends Technol.* **2016**, *34*, 64–72. [CrossRef]
38. Wright, D.H. Human impacts on energy flow through natural ecosystems, and implications for species endangerment. *Ambio* **1990**, *19*, 189–194.
39. Rojstaczer, S.; Sterling, S.M.; Moore, N.J. Human appropriation of photosynthesis products. *Science* **2001**, *294*, 2549–2552. [CrossRef]
40. Haberl, H.; Erb, K.H.; Krausmann, F. Human appropriation of net primary production: Patterns, trends, and planetary boundaries. *Annu. Rev. Environ. Resour.* **2014**, *39*, 363–391. [CrossRef]
41. Fetzel, T.; Gradwohl, M.; Erb, K.H. Conversion, intensification, and abandonment: A human appropriation of net primary production approach to analyze historic land-use dynamics in New Zealand 1860–2005. *Ecol. Econ.* **2014**, *97*, 201–208. [CrossRef]
42. Yuan, Q.Z.; Wu, S.H.; Dai, E.F.; Zhao, D.S.; Ren, P.; Zhang, X.R. NPP vulnerability of China's potential vegetation to climate change in the past 50 years. *Acta Geogr. Sin.* **2016**, *71*, 797–806.
43. Du, J.S.; Yu, D.Y. Impacts of climate change and human activities on net primary productivity of grassland in agro-pastoral transitional zone in northern China. *J. Beijing Norm. Univ.* **2018**, *54*, 365–372.
44. Erb, K.H.; Fetzel, T.; Plutzar, C.; Kastner, T.; Lauk, C.; Mayer, A.; Niedertscheider, M.; Körner, C.; Haberl, H. Biomass turnover time in terrestrial ecosystems halved by land use. *Nat. Geosci.* **2016**, *9*, 674–678. [CrossRef]
45. Zhang, F.Y.; Pu, L.J.; Huang, Q. Quantitative assessment of the human appropriation of net primary production (HANPP) in the coastal areas of Jiangsu, China. *Sustainability* **2015**, *7*, 15857–15870. [CrossRef]
46. Schwarzlmüller, E. Human appropriation of net primary production (HANPP) in Spain, 1955–2003: A socio-ecological analysis. In *Social Ecology Working Paper 99*; Faculty for Interdisciplinary Studies, Klagenfurt University: Vienna, Austria, 2008.
47. Wang, F.X.; Mao, A.H.; Li, H.L.; Jia, M.L. Quality measurement and regional difference of urbanization in Shandong Province based on the entropy method. *Sci. Geogr. Sin.* **2013**, *33*, 1323–1329.
48. Wang, W. The evaluation of coordinated development level between urban and rural and its spatial-temporal pattern in Yangtze River economic zone. *Econ. Geogr.* **2017**, *37*, 60–66.
49. Wang, Y.; Sun, P.J.; Li, C.G.; Liu, H.; Zhou, J. Spatial-temporal evolution features of urban and rural harmonious in northeast China since 2003. *Econ. Geogr.* **2018**, *38*, 59–66.
50. Liu, Y. Grassland monitoring report in China in 2015. *Chin. Animat. Anime* **2016**, *6*, 18–35.
51. Huang, Q.; Zhang, F.; Zhang, Q.; Ou, H.; Jin, Y. Quantitative assessment of the impact of human activities on terrestrial net primary productivity in the Yangtze River Delta. *Sustainability* **2020**, *12*, 1697. [CrossRef]

land

MDPI

Article

Spatiotemporal Patterns in and Key Influences on Cultivated-Land Multi-Functionality in Northeast China's Black-Soil Region

Heyang Gong [1], Zhibo Zhao [2], Lei Chang [1], Guanghui Li [1], Ying Li [1] and Yuefen Li [1,*]

1 College of Earth Sciences, Jilin University, Changchun 130061, China; hygong19@mails.jlu.edu.cn (H.G.); changlei2217@mails.jlu.edu.cn (L.C.); ligh20@mails.jlu.edu.cn (G.L.); yingli20@mails.jlu.edu.cn (Y.L.)
2 Faculty of Built Environment, University of Malaya, Kuala Lumpur 50603, Malaysia; 17220374@siswa.um.edu.my
* Correspondence: yfli@jlu.edu.cn; Tel.: +86-137-5600-9975

Abstract: Cultivated-land multi-functionality has become an important way to achieve sustainable cultivated-land protection, and it has become a hot spot in the field of land-management policy. Taking the cultivated black soils in the grain-producing area of Jilin Province, Northeast China, as a case study, this paper assessed the multi-functions of cultivated land over the past 30 years by applying the improved TOPSIS model. Furthermore, the key limiting factors and influencing factors of the multi-functions of cultivated land were identified through the obstacle-degree model and the Geo-detector. The results show that the level of multi-functionality rose from 1990 to 2020, but an increase in both economic and social functions hindered improvements in the ecological function of cultivated land. There were obvious spatial differences in the functions of cultivated land in different counties, with ecological functions showing the highest degree of differentiation, followed by social and economic functions. The per capita agricultural output, the degree of agricultural mechanization, the average output from cultivated land, and the agricultural-labor productivity had the most restrictive effects on the functions of cultivated land, with barrier-degree values of 15.90, 13.90, 11.76, and 10.30, respectively. Coupling–coordination in the multi-functions and sub-functions of cultivated land showed an upward trend, from "low coupling coordination–antagonistic coupling coordination" to "high coupling coordination-optimal coupling coordination". The government should include the level of multi-functional utilization in future policies for the management and utilization of cultivated land and take measures to reduce the differences in the functions of cultivated land among regions. Quantifying the multi-functional value of cultivated land and subsidizing land cultivation should encourage farmers to protect the land and help to strengthen multi-functional planning and functional design, improve ecological utilization, and promote the sustainable use of cultivated land.

Keywords: multi-functionality of cultivated land; breadbasket; spatiotemporal variation; coupling–coordination degree; influencing factors

Citation: Gong, H.; Zhao, Z.; Chang, L.; Li, G.; Li, Y.; Li, Y. Spatiotemporal Patterns in and Key Influences on Cultivated-Land Multi-Functionality in Northeast China's Black-Soil Region. *Land* **2022**, *11*, 1101. https://doi.org/10.3390/land11071101

Academic Editors: Li Ma, Yingnan Zhang, Muye Gan and Zhengying Shan

Received: 8 June 2022
Accepted: 14 July 2022
Published: 19 July 2022

1. Introduction

As a scarce and non-renewable resource, cultivated land provides many essential products and services for human society [1–3]. With the development of more urbanized societies and economies, cultivated land is not just limited to the traditional function of supplying food products but also carries many other non-productive functions, such as an economic-return function, a social-security function, an ecological function, and a landscape function [4–6]. However, the different functions of cultivated land have not been paid enough attention to in the utilization and management of cultivated land, which makes the contradiction between the supply and demand of cultivated-land functions and eventually leads to the occurrence of unsustainable conditions such as cultivated-land degradation, non-grain cultivation, and abandonment [7–9]. The future expansion of cultivated-land production is likely to encounter a complex situation of competing demands and trade-offs [10]. Effective measures must be taken to balance the supply and demand of multi-functional cultivated land [11,12]. This requires in-depth knowledge of the level and changing characteristics of cultivated-land functionality and the factors influencing them to provide scientific support for the sustainable utilization and protection of cultivated land.

Multi-functional research originates from studies of agricultural multi-functionality [13], referring to the fact that, in addition to food production, agriculture also has a role in ecological services, landscape maintenance, employment security, and cultural heritage [14,15]. However, because of the differences in the types of crops grown and the responsibilities between cultivated land and agriculture, the multi-functionality of cultivated land expands the implications of economic, social, and ecological functions based on agricultural multi-functionality [16]. Particularly in the context of family-based agricultural production in China (a household-responsibility system), cultivated land has many participants who need to produce food to ensure food security, and the connotations of multi-functional cultivated land are rich and complex [17,18].

Quantitative evaluation is key to the study of multi-functional cultivated land and has been applied since the implementation of the Land Use and Land Cover Change (LUCC) program [19]. Currently, research is centered on two main aspects: evaluating a single function of cultivated land and a more comprehensive evaluation of the multi-functionality of cultivated land. The former includes the social value of cultivated land [20], and ecological [21] and monetary compensation [22]. The latter includes spatiotemporal analyses and understanding the driving factors behind the multiple functions of cultivated land [23–26]. However, the emphasis is often on the imbalance of a single function or a specific time point, and it is difficult to effectively trace temporal and spatial variations in the characteristics of cultivated land and its functions. As a result, our understanding of the multi-functionality of cultivated land is still poor. Long-term studies can be used to examine the rates of change over time and test the effectiveness of policies [27], yet there is a lack of long-term research on the multi-functionality of cultivated land. Equally, in terms of research application, the majority of studies focus on analyzing and evaluating the results, and rarely propose measures and policies to improve the function of cultivated land. Currently, the research outputs do not provide any guidance on the actual management of cultivated land, and the focus is usually on developed urbanized areas, where the conflict with cultivated land is more pronounced. Less attention has been paid to the multi-functionality of cultivated land in important grain-producing areas. To improve the shortcomings of existing research, a clear understanding of the historical change in cultivated-land functionality in major grain-producing regions is needed, and the obstacles and driving factors behind the changes in cultivated-land functions need to be identified, so that effective policies can be adopted for the future. The multi-functional utilization of cultivated land should, therefore, become the focus for the protection of cultivated land and the goal of sustainable utilization.

One of the world's major black-soil regions is found in northeast China. This fertile soil represents a key grain-producing area in China, and northeast China is an important exporter of commercial grains. The agricultural functional areas identified in "National Main Functional Area Plan" are also important in maintaining China's food security [28]. However, this region faces serious cropland degradation, including soil-nutrient loss [29], the thinning of the cultivated layer [30], and the loss of soil physicochemical properties. Simultaneously, unfavorable conditions for land cultivation, such as population outflow, low food prices, and reduced agricultural production efficiency, have begun to emerge [25], directly threatening the future sustainable use of cultivated land and the development of the region's economy and society. While these problems have existed across China for some time, they are particularly prominent in the main grain-producing areas. Cultivated land has various functions, but no one is investing in them. The government has not prioritized the issue of multi-functionality and currently lacks effective control measures, which is eroding farmers' rights and interests and reducing their enthusiasm for cultivated-land protection.

Counties represent the smallest unit with complete administrative power in China and are the basic unit for policy formulation and implementation. Carrying out research at the county level could yield direct and targeted suggestions for formulating practical and effective cultivated-land-use policies. Breadbaskets are the core unit for grain production in China's major grain-producing areas and are also important county-level units for cultivated-land management, making the black-soil region ideal for multi-functional research. Considering the research gaps highlighted above, this study used the breadbaskets in Jilin Province, in the hinterland of Northeast China, as a research area to evaluate the multi-functionality of cultivated land, analyze the obstacles to and driving factors behind the multi-functionality, and ultimately put forward suggestions for improving the multi-functionality of cultivated land. An improved Technique for Order Preference by Similarity to Ideal Solution (TOPSIS) model was used to evaluate the multi-functionality of cultivated land in the breadbaskets in 1990, 2000, 2010, and 2020. An obstacle-degree model was then used to determine the key obstacles limiting the functionality of the cultivated land. Geographic detectors were used to analyze the main factors changing cultivated-land functions. Finally, based on the analyses, effective measures to improve the multi-functional utilization of cultivated land in this area are proposed as a reference point for future developments.

2. Materials and Methods

2.1. Study Area

The study area was located in the geographic geometric center of Northeast Asia, spanning from 121°38′ to 131°19′ E and from 40°50′ to 46°19′ N, and is known as "Hometown of Black Soil", representing one of the world's three major black-soil belts. The soil in this region is fertile, with a high organic matter content (the average organic matter content being > 27 g/kg) and abundant cultivated-land resources.

The breadbaskets of China are the top-ranked counties based on the proportion of commercial grain output, overall grain production, and area sown for grain, accounting for 50%, 25%, and 25% by weight of all grain produced. In 2009, China's State Council promulgated "National Plan for Newly Increased Grain Production Capacity of 100 Billion catties (2009–2020)", and a total of 800 breadbaskets were identified as the core areas for grain production across the country. The breadbaskets chosen for this study were located in Jilin Province, northeast China (Figure 1); in total, 28 research units were represented and nine prefecture-level cities, including Changchun, Jilin, and Siping, for a total land area of 118,259.42 km^2. Together, the breadbaskets account for 78.25% of the cultivated land in Jilin Province and contribute 89.7% of the province's grain output. However, while this area has made significant contributions to national and regional food security, its economic and social development faces serious challenges. It is the most important grain-production base in China, but over the last 10 years the population of the study area fell by an astonishing 25.08%, twice the overall rate for Jilin Province. Over the last 10 years, the gross domestic product (GDP) decreased by 10.54%, and the per capita income decreased by 3%. In contrast, over the same time period China's GDP and per capita income considerably grew, by 146.53% and 133.05%, respectively. The problems highlighted in this region are prevalent in many major grain-producing regions in the country; this study, therefore, can provide a reference point for similar regions.

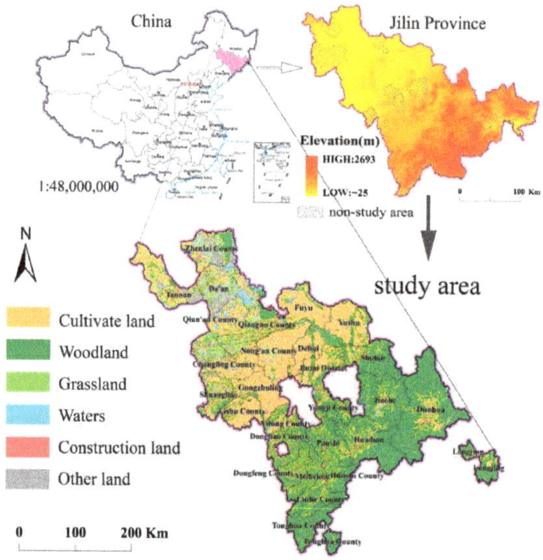

Figure 1. The location and main land types in the black-soil breadbaskets in Jilin Province, Northeast China (The map of China in the figure is produced under the supervision of the Ministry of Natural Resources of the People's Republic of China, drawing number: GS (2019) No. 1673).

2.2. Data Collection and Pre-Processing

The details of the data used for the study are presented in Table 1. The Gauss–Kruger projection and 2000 National Geodetic Coordinate System (CGCS2000) were used, and the scale was unified to counties.

Table 1. Descriptions of the data sources.

Data Type	Data Source	Time Series	Resolution
Land use/land cover	Resource and Environment Science Data Center, Chinese Academy of Sciences	1990, 2000, 2010, 2020	30 m × 30 m
River and road data	Extracted from land-use and -cover data from Data Center for Resources and Environmental Sciences, Chinese Academy of Sciences	1990, 2000, 2010, 2020	Same as land-use/land-cover data
Digital elevation model (DEM)	Geospatial Data Cloud (http://www.gscloud.cn/, accessed on 21 December 2021)	2009	30 m × 30 m
Slope	Calculated from DEM data	2009	30 m × 30 m
Meteorological	Meteorological Data Center, China Meteorological Administration	1990, 2000, 2010, 2020	Site
Socioeconomic	Jilin and counties (cities) statistical yearbooks	1990, 2000, 2010, 2020	County level
Agricultural	Jilin rural statistical yearbooks	1990, 2000, 2010, 2020	County level
Cultivated-land quality	Agricultural-land grading and projections	2009, 2019	1:100,000

2.3. Methods

2.3.1. Classification and Quantification of the Multi-Functional Value of Cultivated Land

The varied classification criteria used for cultivated-land functions can be grouped into three main categories: economic, social, and ecological (Figure 2).

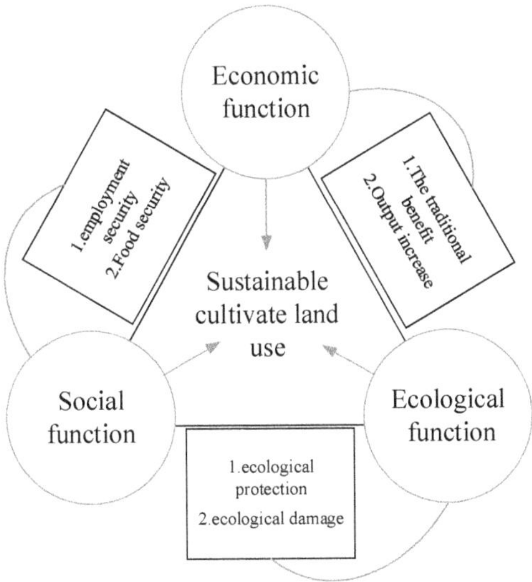

Figure 2. A classification framework for the functions of cultivated land.

However, because of the differential development of human societies and different research areas, the functions of cultivated land and the strength of each function vary significantly, and the selection of indicators for a particular project must adhere to the principles of correlation and availability [31]. Table 2 presents a summary of the indicators chosen for this study.

Table 2. Indices for assessing cultivated-land functions. The overall per capita grain demand was determined as 400 kg per person per year [32]; the safety standard for chemical-fertilizer application followed the international chemical fertilizer application safety standard of 225 kg per hectare [33]; the degree of fragmentation of cultivated land was represented by the ratio of the number of cultivated-land patches to defined area, which was calculated based on an ArcGIS platform; the ecological value of cultivated land was calculated by referring to the ecological service value coefficient table compiled by Xie et al. [34].

Function	Indicator	Calculation Method	Unit	Trend	Weight
Economic	Average grain output	Grain production/Cultivated-land area	kg/hm^2	Positive	0.0422
	Average output of cultivated land	Output value of primary industry/Cultivated-land area	Ten thousand CNY/hm^2	Positive	0.1034
	Percentage of cultivated-land value	Gross plantation output/Gross regional product	Dimensionless	Positive	0.0701
	Agricultural-labor productivity	Output value of primary industry/Employees of primary industry	CNY/person	Positive	0.1003
	Per capita agricultural output	Gross agricultural output/Total population	CNY/person	Positive	0.1208

Table 2. *Cont.*

Function	Indicator	Calculation Method	Unit	Trend	Weight
Social	Per capita cultivated land	Cultivated-land area/Total population	hm^2/person	Positive	0.0394
	Grain commodification index	Grain production/(Per capita grain demand × Population)	Dimensionless	Positive	0.078
	Per capita grain production	Total grain production/Total population	kg/person	Positive	0.078
	Degree of agricultural mechanization	Total power of agricultural machinery/Cultivated-land area	kW/hm^2	Negative	0.0976
	Labor-transfer index	Non-agricultural population/Total population	Dimensionless	Positive	0.0284
Ecological	Fertilizer-use-intensity index	The total amount of fertilizer applied on the ground/Safety standard for fertilizer use	Dimensionless	Negative	0.013
	Production-value energy consumption	Agricultural electricity consumption/Gross agricultural output value	kW/CNY 10,000	Negative	0.007
	Effective-irrigation index	Effective irrigated cultivated-land area/Cultivated-land area	Dimensionless	Positive	0.0689
	Fragmentation of cultivated land	Parameter calculation	Dimensionless	Negative	0.0306
	Proportional ecological value of cultivated land	Ecological service value of cultivated-land area/Ecological service value of total land area	Dimensionless	Positive	0.0737
	Land-reclamation coefficient	Cultivated-land area/Total land area	Dimensionless	Positive	0.0486

The grain output of an area of land represents the traditional grain yield of cultivated land. As well as this, the economic function should also consider the increase in output value generated by cultivating the land, based on the average rate of rural-labor output and the value of the per capita agricultural output. The social function includes employment security for the local farmers and food security. The function of food security can be further divided into two types: intra-regional and extra-regional guarantees, which are expressed as per capita grain output and the grain-commercialization index, respectively. The per capita cultivated-land area, the degree of agricultural mechanization, and the labor-transfer index represent the employment-security function of cultivated land.

Cultivated land is also a part of an ecosystem and has an ecological function. Cultivated land has a positive effect on the ecological needs of human beings and supports biodiversity but can also have a negative impact on the environment if used unsympathetically. Positive effects can be quantified by the proportional ecological value of cultivated land, the effective-irrigation index, and the coefficient of land reclamation. The proportional land ecological value is calculated from the ecological-service-value coefficient [34], which reflects the ecological contribution of cultivated-land systems in all ecosystems and is an important indicator reflecting the basic ecological attributes of cultivated land. Negative effects mainly include the overuse of pesticides, an increase in agricultural energy consumption, and the fragmentation of cultivated land as a result of overuse. These are quantified using the fertilizer-use-intensity index, energy consumption per CNY ten thousand of output value, and the degree of fragmentation of cultivated land, respectively. The specific calculations used for each index are shown in Table 2.

2.3.2. Calculating the Multi-Functional Utilization of Cultivated Land and Determining any Obstacles

An improved TOPSIS ("the distance method between superior and inferior solution") model was used to evaluate the functional value of cultivated land [35]. The TOPSIS model is a commonly used multi-objective decision-making method that considers the advantages and disadvantages of a scheme by determining the distance between the index, and the "positive ideal solution" and "negative ideal solution". If the scheme is close to the "positive ideal solution" and far from the "negative ideal solution", it is superior; the converse means it is inferior. This method not only overcomes the lack of objectivity of, for example, the AHP and Delphi methods, but also the information-loss problem in factor analyses and mutation analyses [36]. TOPSIS models are widely used for decision analyses, environmental assessments, and land evaluations. However, the traditional TOPSIS model does not consider the weights of the indicators, as the weight of each indicator is the same by default. This is inconsistent with real-life situations and widens the difference between the model results and empirical data. In this study, the weight determined with information entropy was used to modify the decision matrix, making the TOPSIS calculations more objective [35]. The specific steps applied were as detailed below.

Step 1: Build a decision matrix. First, the data were normalized, and indicator weights calculated. The range-standardization method was used to eliminate the differences between the dimensions and data levels for each indicator, and the entropy-weight method was used to determine the weight of each indicator (see Table 2 for the calculated weights). Both the range-standardization method and the entropy-weight method are objective and are widely used in statistics, geography, and elsewhere [26,36,37]. When applying the range-standardization method, data translation must be performed to eliminate the interference of extreme values and make the results as accurate as possible. The value of each item of standardized matrix P is then multiplied by its weight vector matrix W to obtain improved decision matrix G:

$$G = P \times W = \begin{bmatrix} p_{11} & p_{12} & \cdots & p_{1j} \\ p_{21} & p_{22} & \cdots & p_{2j} \\ \vdots & \vdots & \ddots & \vdots \\ p_{i1} & p_{i2} & \cdots & p_{ij} \end{bmatrix} \times \begin{bmatrix} w_1 \\ w_2 \\ \vdots \\ w_i \end{bmatrix} = \begin{bmatrix} p_{11} \times w_1 & p_{12} \times w_1 & \cdots & p_{1j} \times w_1 \\ p_{21} \times w_2 & p_{22} \times w_2 & \cdots & p_{2j} \times w_2 \\ \vdots & \vdots & \ddots & \vdots \\ p_{i1} \times w_i & p_{i2} \times w_i & \cdots & p_{ij} \times w_i \end{bmatrix} = \begin{bmatrix} g_{11} & g_{12} & \cdots & g_{1j} \\ g_{21} & g_{22} & \cdots & g_{2j} \\ \vdots & \vdots & \ddots & \vdots \\ g_{i1} & g_{i2} & \cdots & g_{ij} \end{bmatrix} \quad (1)$$

where p_{ij} refers to the standardized value of index i in research unit j; w_i refers to the weight of index i; and g_{ij} refers to the improved value of index i in research unit j.

Step 2: Calculate the ideal solution and the ideal value distance. The "positive ideal solution", V_i^+, and "negative ideal solution", V_i^-, of index i in the improved decision matrix were determined, and distances D_j^+ and D_j^- from research unit j to V_i^+ and V_i^- were measured. The closeness of the ideal solution to research unit j, Degree T_j, was calculated as:

$$\begin{aligned} V_i^+ &= \{maxg_{ij} | i = 1, 2, \cdots m\} = \{g_1^+, g_2^+, \cdots g_m^+\} \\ V_i^- &= \{ming_{ij} | i = 1, 2, \cdots m\} = \{g_1^-, g_2^-, \cdots g_m^-\} \end{aligned} \quad (2)$$

$$\begin{aligned} D_j^+ &= \sqrt{\sum_{i=1}^{m} (g_{ij} - g_i^+)^2} (i = 1, 2, \cdots m) \\ D_j^- &= \sqrt{\sum_{i=1}^{m} (g_{ij} - g_i^-)^2} (i = 1, 2, \cdots m) \end{aligned} \quad (3)$$

$$T_j = \frac{D^-}{D^- + D^+} (1 \le j \le n) \quad (4)$$

where T_j is the closeness of index j. With $0 \le T_j \le 1$, the larger the value for T_j is, the better the overall effect of the multi-functional evaluation of cultivated land in the region is; conversely, the smaller the value is, the worse the effect is. When T_j is closer to 1, the index is closer to the "positive ideal solution", indicating that the multi-functionality of the cultivated land is optimal, and the multi-functional use of the cultivated land has reached the expected goal. When T_j is closer to 0, the index is closer to the "negative ideal solution", indicating that the multi-functionality of the cultivated land is poor, and the full potential multi-functionality of the cultivated land has not been reached.

Step 3: Determine any obstacles. Based on the multi-functional evaluation, an obstacle-degree model was used to identify any obstacles affecting the multi-functionality of the cultivated land. This can be used as a baseline for the scientific and practical utilization of cultivated land and can improve the feasibility and effectiveness of cultivated-land protection policies and utilization. The obstacle degree was calculated as:

$$O_j = R_{ij} w_j / \left(\sum_{i=1}^{m} R_{ij} w_j \right), R_{ij} = 1 - b_j \quad (5)$$

where O_j is the obstacle degree of cultivated-land function i and index j.

2.3.3. Analysis of the Coupling and Coordination Relationships among Various Sub-Functions of Cultivated Land

A coupling–coordination-degree model was introduced to quantitatively analyze the interactions among various functions of cultivated land and the degree of coupling and coordination among them. Coupling is a physical concept that can describe the strength of an interaction between two or more systems or motions, but it cannot characterize the level of cooperation among systems. The coordination degree makes up for this deficiency by measuring the level of coordinated development among the systems. It is widely used in studies of system relationships among land, economy, and society [37]. The specific steps used were as detailed below.

Step 1: Measure the coupling degree. After standardizing the data, the coupling–coordination degree was measured as:

$$C_t = \sqrt[n]{\prod_{i=1}^{n} Y_m / \left(\sum_{i=1}^{n} Y_m\right)^n} \quad (6)$$

where C_t is the coupling degree in year t. With $0 \leq C_t \leq 1$, the closer the value is to 1, the stronger the interaction between systems is, while the converse is true. Y_m is the comprehensive score of system m, and n is the number of subsystems. When the relationship among three systems is measured, $n = 3$, and when the relationship between two subsystems is measured, $n = 2$.

Step 2: Measure the coordination index:

$$T = \sum_{i=1}^{n} \alpha_i Y_m, \sum_{i=1}^{n} \alpha_i = 1 \quad (7)$$

where T is the coordination index and α_i is the weight of subsystem i. When measuring the coupling–coordination degree of each subsystem, the entropy-weight method was used to calculate the weight of each index, and the weight was then calculated [36].

Step 3: Measure the coupling–coordination degree:

$$D = \sqrt{C \times T} \quad (8)$$

where D is the coupling–coordination degree. The higher the coupling–coordination score is, the better the coupling–coordination relationship between the two systems is.

2.3.4. Analysis of the Key Drivers of Multi-Functionality of Cultivated Land

A change in the functionality of cultivated land represents a part of a large, complex system, and the factors influencing cultivated-land evolution during different developmental stages and different regions vary. Development processes, urban construction, and policies are jointly affected, and each factor has a variable degree of influence. To examine the breadbasket regions in Jilin Province, 18 influencing factors were chosen, as shown in Table 3.

Based on the multi-functional evaluation of cultivated land, the key factors affecting change in multi-functionality were identified using Geodetector, a statistical method that detects spatial heterogeneity and reveals the drivers behind it [38]. The strengths of the driving factors were determined following the method by Wang and Xu [39].

Table 3. Definition of factors influencing the multi-functionality of cultivated land. The per capita construction-land standard in rural areas was 150 m^2/person, and the urban per capita construction-land standard was 120 m^2/person. The average weighting method was used to measure the overall pressure; the management and control levels of permanent basic farmland, prohibited construction areas, restricted construction areas, conditional construction areas, and permitted construction areas in the agricultural-policy zoning decreased in order, and were assigned values of 5, 4, 3, 2, and 1, respectively, in the calculations. Because of data limitations in 1990 and 2000, Jilin Province did not produced agricultural-land-grading data; the average grain yield was used to correct the cultivated-land utilization to obtain graded data for the corresponding years.

Factor	Indicator	Abbreviation	Calculation
Physical geography	Elevation	Elevation	Elevation values from DEM
	Slope	slope	The actual slope of the cultivated land
	Annual precipitation	annual precipitation	Regional averages from weather-station data
	Distance from major rivers	DFMR	Euclidean distance based on the distribution of the water system
	Cultivated-land quality	cultivated-land quality	Agricultural-land grading
	Distance from provincial capital	DFPC	Distance from the main city of Changchun
	Distance from central city	DFCC	Distance from the main city of the prefecture-level city
Economic development	Per capita GDP	GDP per capita	Total GDP/Total population
	Per capita agricultural output of farmers	PCAOVF	Output value of primary industry/Rural population
	Proportion of secondary and tertiary industries	PSTI	Output value of secondary and tertiary industries/Total GDP
	Fixed asset investment per land	FAIPL	Total fixed asset investment/Regional land area
Urban construction	Urbanization rate	urbanization rate	Urban population/Total population
	Population density	Population density	Total population/Regional land area
	Percentage of built-up area	PBA	Built-up area/Total land area
	Road-network density	road network density	Proportion of road-network length to total area
Policy	Agricultural policy division	APD	Permanent basic farmland, prohibited construction areas, restricted construction areas, conditional construction areas, permitted construction areas
	Construction-land-index pressure	CLIP	Rural per capita construction land/Rural construction land standard + Urban per capita construction land/Urban per capita land standard
	Proportion of financial support to agriculture	PFFSA	Proportion of expenditure on agriculture, forestry, and water affairs in public finance

3. Results

3.1. *Spatiotemporal Variation in and Main Obstacles to the Multi-Functionality of Cultivated Land*

3.1.1. Temporal Variation in the Multi-Functionality of Cultivated Land

The evaluation, degree of change, and coefficient of variation of each function of cultivated land during different periods in the breadbasket regions are shown in Table 4. The average multi-functional scores for cultivated land for the four research time nodes were 0.222 (1990), 0.227 (2000), 0.361 (2010), and 0.451 (2020), representing an increase of 102.92% during the 30 years from 1990 to 2020. Although the overall multi-functional level of cultivated land was not high, the increase was large with clearly differentiated phases, showing a trend from basically unchanged to rapid improvement to steady increase. The economic sub-function increased the most, from 0.148 in 1990 to 0.484 in 2020, an increase of 227.51%. The social function was second, with an increase of 152.69%, from 0.159 in 1990 to 0.401 in 2020. The ecological function was almost stagnant, with current levels comparable to those of 30 years ago, only rising from 0.441 in 1990 to 0.456 in 2020. The rapid improvement in the economic and social functions of cultivated land in the breadbaskets led to an increase in the level of multi-functionality, to which the ecological function did not contribute.

Table 4. Trends in multi-functional changes in cultivated land in the black-soil breadbaskets of Jilin Province, Northeast China, from 1990 to 2020.

Function	Score				Change (%)				CV			
	1990	2000	2010	2020	1990–2000	2000–2010	2010–2020	1990–2020	1990	2000	2010	2020
Economic	0.148	0.148	0.359	0.484	0.24%	142.51%	34.73%	227.51%	0.267	0.265	0.310	0.252
Social	0.159	0.165	0.289	0.401	3.75%	75.27%	38.96%	152.69%	0.338	0.309	0.343	0.332
Ecological	0.441	0.451	0.459	0.456	2.16%	1.76%	−0.52%	3.41%	0.368	0.393	0.430	0.450
Multi	0.222	0.227	0.361	0.451	1.97%	59.08%	24.97%	102.72%	0.300	0.212	0.290	0.283

3.1.2. Spatial Variation and Differentiation in the Multi-Functionality of Cultivated Land

There were obvious and significant spatial differences in the multi-functionality and sub-functions of cultivated land in the 28 breadbaskets. The coefficient of variation of the ecological functions was the largest, showing a continuous increase over the past 30 years, from 0.368 in 1990 to 0.450 in 2020 (Table 4), indicating that the ecological functions of cultivated land varied among different breadbaskets. the differences in time-progressed ecological functions were more pronounced than those in other functions.

The spatial differences in the social function of cultivated land were also obvious but to a lesser extent than the ecological function. The coefficient of variation dropped from 0.338 in 1990 to 0.332 in 2020, indicating little overall change, although there were some differences. The differences in the economic function of cultivated land were the smallest compared with the other functions, with the average coefficient of variation decreasing from 0.267 in 1990 to 0.252 in 2020, indicating that the economic attributes of cultivated land were well balanced among the breadbaskets.

Based on the current multi-functionality of cultivated land and the changes recorded over the last 30 years, although the multi-functionality increased, it showed an unbalanced development of the Matthew effect, i.e., higher levels of multi-functionality led to higher levels, while lower levels led to yet lower levels. The range of the multi-functionality and sub-functions of cultivated land among the breadbaskets significantly increased, showing an overall increase of 220%, with the range in economic and social functions increasing by more than 300%. Even the much weaker ecological function also expanded by 12%. Overall, the use of cultivated land over the last 30 years increased the spatial differences in multi-functional utilization to an exaggerated degree. This directly reflects the lack of the consideration of cultivated-land functions in cultivated-land protection (Figures 3 and 4). In addition, the changes in the ecological functions of cultivated land were polarized (Figure 3c) between the east and the west (Figure 4c). On average, the former (with a sum score of about 9.108) had a value about three times that of the latter (sum score of about 3.674) (Figure 3c).

3.1.3. The Main Obstacles to Multi-Functionality of Cultivated Land

The obstacle degree was calculated with the evaluation index for the four time nodes for each breadbasket to determine what factors were limiting the functions of cultivated land. Over the last 30 years, the economic function of cultivated land restricted the multi-functionality of cultivated land the most, followed by the social function and then the ecological function (Table 5). However, the dominant restrictive effect of the economic function weakened (from 13.46 in 1990 to 11.56 in 2020), while the restrictive effect of the ecological function increased (from 4.78 in 1990 to 6.56 in 2020, a rise of 37.24%). This suggests that how the ecological function is managed in the future could be the key to improving the functions of cultivated land. After sorting the index obstacle scores for the breadbaskets, the per capita agricultural output, the degree of agricultural mechanization, the average output of cultivated land, and the agricultural-labor productivity had the strongest restrictive effects on the functions of cultivated land. However, based on the changes in the index values, the per capita agricultural output showed a decreasing trend (from 4.29 in 1990 to 3.63 in 2020), and the degree of agricultural mechanization may soon become the biggest obstacle. Barriers to national contribution and effective irrigation are rising, highlighting the fact that attention must be paid to the multi-functional utilization of cultivated land in the future.

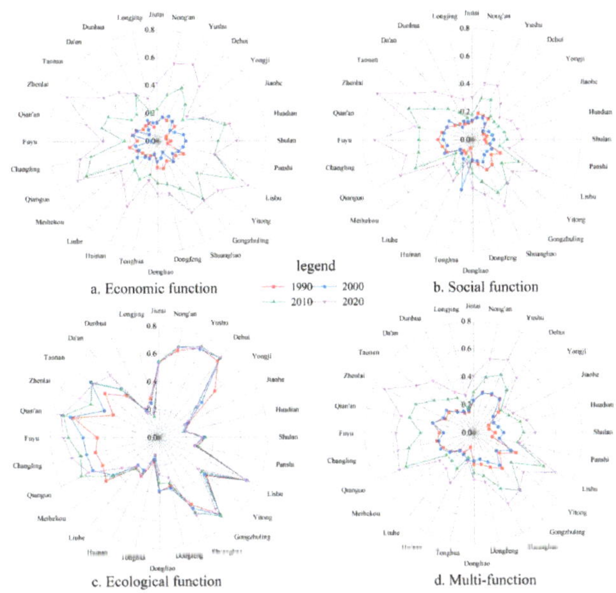

Figure 3. Multi-functional composition of cultivated land in 28 black-soil breadbaskets, Jilin Province, Northeast China, from 1990 to 2020. (**a**) Economic function. (**b**) Social function. (**c**) Ecological function. (**d**) Multi-function.

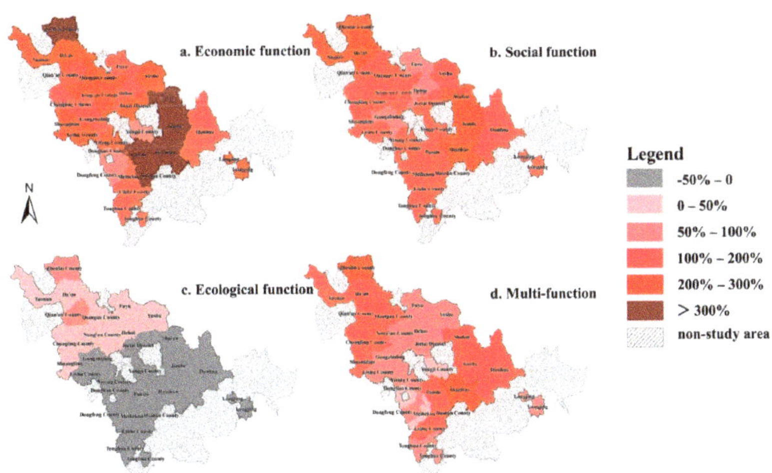

Figure 4. Changes in function of cultivated land in the black-soil breadbaskets of Jilin Province, Northeast China, from 1990 to 2020. (**a**) Economic function. (**b**) Social function. (**c**) Ecological function. (**d**) Multi-function.

Table 5. Calculated obstacle degrees for factors influencing cultivated-land functions in the black-soil breadbaskets of Jilin Province, Northeast China, from 1990 to 2020.

Year	Item	Function Order			Index Order				
		1	2	3	1	2	3	4	5
1990	Barrier indicator	Economic function	Social function	Ecological function	PCAO	AOV	ALP	DAM	FCI
	Handicap	13.46	9.76	4.78	4.29	3.64	3.55	3.42	2.36
2000	Barrier indicator	Economic function	Social function	Ecological function	PCAO	AOV	DAM	ALP	FCI
	Handicap	13.51	9.75	4.74	4.11	3.22	3.21	3.00	2.53
2010	Barrier indicator	Economic function	Social function	Ecological function	PCAO	DAM	PCLV	AOV	FCI
	Handicap	12.31	10.09	5.60	3.87	3.67	2.77	2.60	2.29
2020	Barrier indicator	Economic function	Social function	Ecological function	PCAO	DAM	PCLV	EII	AOV
	Handicap	11.56	9.88	6.56	3.63	3.60	3.02	2.50	2.30
Whole period	Barrier indicator	Economic function	Social function	Ecological function	PCAO	DAM	AOV	ALP	FCI
	Handicap	50.84	39.49	21.68	15.90	13.90	11.76	10.30	9.35

3.2. Coupling and Coordination Relationships among Various Sub-Functions of Cultivated Land

A coupling–coordination degree model was used to compare the sub-functions of cultivated land in the breadbaskets from 1990 to 2020 and was divided into 10 levels (Figure 5). The degree of coupling–coordination among various functions of cultivated land showed an overall upward trend, from "low coupling coordination–antagonistic coupling coordination" to "high coupling coordination–optimal coupling coordination". Nine breadbaskets in eastern Jilin with low levels of coupling–coordination all improved over time, reaching antagonistic coupling–coordination or even high coupling–coordination. Nineteen breadbaskets in central and western Jilin Province all attained a high degree and optimal level of coupling and coordination. Among them, Lishu County and Zhenlai County achieved good coupling–coordination, but no breadbasket reached high-quality coupling–coordination. From 1990 to 2000, overall, the coupling–coordination degree of the multi-functional value of cultivated land remained at a low level, and in fact, the coupling–coordination degree for Gongzhuling City, Yongji County, Qian'an County, Changling County, and Da'an City all declined. During the 20-year period between 2000 and 2020, the coupling and coordination degree of the multi-functional value of cultivated land did improve, although the increase was generally faster in the western region than the eastern region.

3.3. Key Factors Influencing the Multi-Functionality of Cultivated Land

Figure 6 and Table 6 rank the driving factors affecting the multi-functionality of cultivated land in the 28 breadbaskets at the four time nodes. In general, physical and geographical factors were the main factors influencing the multi-functional changes in cultivated land, while the influence of economic factors showed a sharp downward trend over the 30 years (the total q-value decreased from 1.55 in 1990 to 0.92 in 2020). However, urban construction and policy factors had very limited effects on the changes in multi-functionality. The factors that had a greater impact in 1990 were the quality of cultivated land and the proportion of secondary and tertiary industries. In 2000, the pressure of construction land, annual precipitation, and proportion of built-up area began to have a greater impact on the multi-functionality of cultivated land. From 2010, the quality of cultivated land and proportion of secondary and tertiary industries once again dominated the changes in multi-functionality. Annual precipitation and elevation in 2020 were particularly important for cultivated-land multi-functionality.

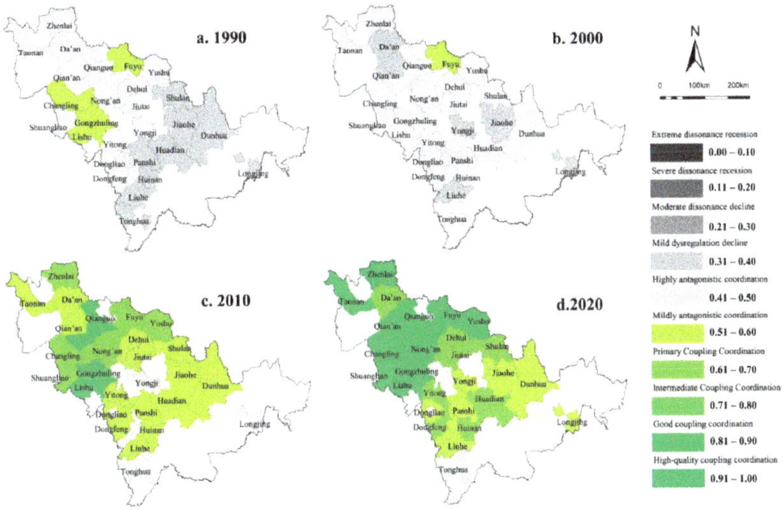

Figure 5. Coupling–coordination among sub-functions of cultivated land in the black-soil breadbaskets of Jilin Province, Northeast China, from 1990 to 2020. Data from Wang et al. [40] were used to divide the coupling–coordination degree into 10 types.

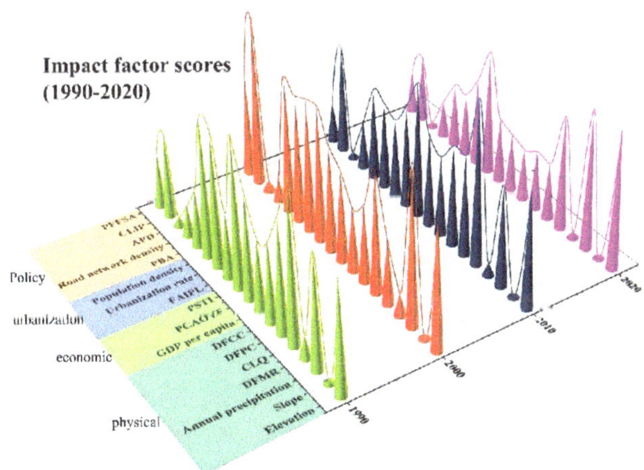

Figure 6. Impact scores for multi-functional factors affecting cultivated land in the black-soil breadbaskets of Jilin Province, Northeast China, from 1990 to 2020. The meanings of the abbreviations used in the figures are shown in Table 3.

Table 6. Indicator values for factors influencing the multi-functionality of cultivated land in the black-soil breadbaskets of Jilin Province, Northeast China, from 1990 to 2020.

Factor	Indicator	1990	2000	2010	2020	Average
Physical geography	Elevation	0.54	0.62	0.56	0.65	0.59
	Slope	0.01	0.01	0.01	0.00	0.01
	Annual precipitation	0.55	0.75	0.48	0.66	0.61
	Distance from major rivers	0.11	0.11	0.06	0.03	0.08
	Cultivated-land quality	0.76	0.31	0.78	0.54	0.59
	Distance from provincial capital	0.54	0.57	0.47	0.26	0.46
	Distance from central city	0.38	0.37	0.47	0.28	0.37
	Subtotal	2.88	2.73	2.81	2.42	2.71
Economic development	Per capita GDP	0.34	0.31	0.37	0.24	0.32
	Per capita agricultural output of farmers	0.50	0.43	0.43	0.27	0.41
	Proportion of secondary and tertiary industries	0.71	0.50	0.63	0.30	0.53
	Subtotal	1.55	1.23	1.44	0.81	1.26
Urban construction	Fixed asset investment per land	0.31	0.64	0.48	0.59	0.50
	Urbanization rate	0.64	0.63	0.37	0.41	0.51
	Population density	0.65	0.60	0.41	0.29	0.49
	Subtotal	1.60	1.86	1.26	1.29	1.50
Policy	Percentage of built-up area	0.27	0.65	0.43	0.30	0.41
	Road-network density	0.14	0.10	0.22	0.17	0.16
	Agricultural-policy division	0.04	0.05	0.03	0.02	0.03
	Construction-land-pressure index	0.40	0.79	0.51	0.31	0.50
	Proportion of financial support to agriculture	0.26	0.48	0.20	0.20	0.28
	Subtotal	1.11	2.07	1.39	1.00	1.39

4. Discussion

4.1. Policy Suggestions for Multi-Functional Management of Cultivated Land

For the mismatch between the supply and demand of multi-functional cultivated land, proper interference and regulation are needed [41]. Implementing smaller-scale land-use management or land-use planning is often more efficient in solving the problem and at a low cost [42]. This research study deepened our understanding of the function of cultivated land and can inform policies for the long-term protection of cultivated land. The results of this study are applicable not only to Northeast China but also to other major grain-producing areas that are under pressure to protect cultivated land.

Incorporating multi-functional utilization into policy considerations for cultivated-land management can help balance differences in cultivated-land functions among regions. Research on cultivated-land protection has a long history and has received special attention in recent years [43], especially in the context of prominent global land-use contradictions and serious threats to food security [44]. However, the protection of cultivated-land quantity, quality, productivity, etc., has perhaps been overemphasized [45], while the sustainability of cultivated-land use has received less attention, although it has started to come into focus [46,47], and the relatively hidden multi-functional attributes of cultivated land have rarely been considered. Differences in the use of cultivated land often lead to differences in management policies, which ultimately affect the future sustainability of land use. The multi-functional imbalance of cultivated land in the 28 breadbaskets studied here highlights the need to address this issue. A regional imbalance results in a lack of drive for cultivated-land protection. The direction and perception of cultivated-land protection need to change towards using the sustainable protection and utilization of cultivated land as a starting point for cultivated-land management, and to integrate multi-functionality into the formulation of cultivated-land-management policies and planning.

The multi-functional value of cultivated land needs to be demonstrated, and sufficient compensation offered, to encourage cultivated-land protection. The economic benefits of growing grain on cultivated land are relatively low, and the comparative benefits of growing grain are likely to continue to decrease with the development of economy and society. When the income of agricultural production is unbalanced, the government can improve and protect farmers through subsidy programs [48]. However, without reasonable compensation for multi-functional land use, there is no incentive for farmers to protect the land [49]. Obstacles to multi-functionality compound the problem. Among the four main obstacles to multi-functionality of cultivated land in the studied breadbaskets, the per capita agricultural output and the average output of cultivated land were directly related to the income level generated by cultivating land, while the other two barriers, agricultural-labor productivity and agricultural mechanization, were highly correlated with the level of economic income from cultivated land. These indicators can only improve if income levels are high enough. Currently, farmers are presented with a stark choice between protecting the land and remaining on lower incomes or giving up the land and moving to cities to achieve higher incomes. The government needs to take effective measures to recalibrate the value of cultivated-land resources, highlighting the multi-functional value of cultivated land, and give farmers proper compensation for land protection.

The ecological utilization of cultivated land in breadbaskets needs to be improved and the long-term sustainable utilization of cultivated land needs to be promoted. Ecological issues with cultivated land are closely related to the presence of people, compared with other natural resources, because of the long history of human settlement around areas of cultivation [50]. People benefit directly from the positive effects of cultivation, such as environmental improvement and biodiversity, but there are also negative effects, such as water pollution and straw-burning pollution [51]. The ecological function of breadbaskets has not been improved for many years, and this has increasingly restricted the multi-functionality of cultivated land. At a time when global ecological security is under threat, effective measures should be taken to curb negative trends, and the future ecological utilization of cultivated land is necessary. When considering the cultivated-land output, recycling and reducing the use of pesticides, fertilizers, herbicides, etc., should be promoted to save production costs and improve agro-ecological benefits.

4.2. Limitations and Future Research

Based on long-term research, we analyzed the unbalanced state of the multi-functionality of cultivated land and gave a feasible solution to balance the multi-functionality of cultivated land. However, with the continuous emergence of global ecological threats and the rapid urbanization of agricultural areas, the multi-functional utilization of cultivated land still requires further attention and more in-depth research, and our study still has two important limitations that need to be addressed [52].

The evaluation criteria used to assess the multi-functionality of cultivated land at different research scales are different, and policy formulation needs to be adjusted accordingly. The multi-functional mismatch of cultivated land is multi-scaled, and the analysis of different scales is an important method to determine the reasons for this mismatch between supply and demand [53]. Based on the scales being considered, relevant indicators for evaluation need to be selected [31]. As a county-level study, this research study addresses problems in county-level farmland protection policies, and can suggest countermeasures, but it is difficult to extrapolate the results to larger or smaller study areas. For example, understanding the imbalance of cultivated-land functions among regions and taking corresponding measures require the evaluation of the functions of cultivated land at a national level. Conversely, engineering and utilization measures to improve the function of cultivated land need to be considered at a plot scale. Evaluation forms the basis for policy-making, and the choice of indicators forms the basis for evaluation [54]. Therefore, in future research at different scales, different evaluation-index systems should be selected according to local conditions, to make the results more robust.

The functions of cultivated land represent a large, complex system, and it is difficult to describe all the relevant aspects of cultivated land with the existing macro data. The functions of cultivated land are intertwined with economic, social, and ecological systems, and macro-statistical data and geographic data can only reflect the overall functions of cultivated land to a certain extent. The different scales and standards of the statistical data used are also likely to cause discrepancies between evaluation results and empirical data. To eliminate such errors, data encompassing as many perspectives as possible need to be collated [55]. For example, regarding the ecological function of cultivated land, the level of farmland pollution may be affected by factors such as straw burning, agricultural non-point source pollution, and plastic pollution [56]. These indicators may require additional investigation and research at a micro level. Equally, extrapolating policy recommendations from cultivated-land utilization through research on multi-functionality and facilitating farmers in recognizing the multi-functionality of cultivated land are topics worthy of long-term research.

5. Conclusions

Multi-functionality is an objective attribute of cultivated land that influences the sustainability of cultivated-land utilization and the long-term protection of cultivated land. To date, the importance of black-soil cultivation and protection has not received sufficient attention because of an overemphasis on grain output. An effective way of addressing this situation is to understand the changes in the multi-functionality of cultivated land and identify the key driving and limiting factors behind any changes. We used an improved TOPSIS model to measure the temporal and spatial variation in the cultivated-land functions over 30 years, from 1990 to 2020, in 28 breadbaskets in Jilin Province, Northeast China. The key driving factors behind and major obstacles to cultivated-land functions were determined using an obstacle degree model and a geographic detector model. The coupling and coordination relationships among the various functions of cultivated land were also analyzed. Suggestions are made to help improve future land-management policies. This study provides a baseline for the multi-functional utilization of cultivated land and the long-term protection of cultivated land in China and even in the world's major grain-producing regions and expands the perspective of cultivated-land protection.

The multi-functionality of cultivated land in the breadbaskets increased significantly in the 30 years from 1990 to 2020 (by 102.92%). Whether a function is used to its full potential is an important limiting factor for cultivated-land utilization, and the ecological function of cultivated land is likely to be more important in the future. There was an obvious spatial differentiation in cultivated-land functions across the breadbaskets; this difference was the largest for the ecological function, followed by the social function, and was the smallest for the economic function. The coupling–coordination degree for each sub-function generally showed an upward trend, but there was again an obvious spatial differentiation. The multi-functionality of cultivated land in the breadbaskets presented an unbalanced growth trend. In addition, the sub-functions of cultivated land were not well coordinated, leading to an imbalance in promoting cultivated-land protection. Cultivated-land protection in this region over the last 30 years also exacerbated social inequality. The government must intervene with effective measures, especially in the processes of cultivated-land utilization and management, and the multi-functionality of cultivated land should be taken as a starting point for formulating policies. The multi-functional value of cultivated land needs to be evaluated, and farmers should be offered reasonable compensation to reduce the imbalance and differences in cultivated-land functions among regions and improve farmers' enthusiasm for cultivated-land protection. The ecological utilization of cultivated land in breadbaskets should be improved and sustainable utilization should be promoted.

Author Contributions: H.G., Conceptualization, Methodology, Software, Formal analysis, Visualization, Investigation, and Writing—original draft; L.C. and Y.L. (Ying Li), Methodology and Writing—review and editing; G.L., Formal analysis and Data curation; Z.Z., Visualization and Validation. Y.L. (Yuefen Li), Conceptualization, Supervision, Funding acquisition, and Writing—review and editing. All authors have read and agreed to the published version of the manuscript.

Funding: This work was supported by National Natural Science Foundation of China (grant No. 42177447) and Science and Technology Development Plan Project of Jilin Province (grant No. 20210203010SF).

Institutional Review Board Statement: Not applicable.

Informed Consent Statement: Not applicable.

Data Availability Statement: The data presented in this study are available in the article.

Conflicts of Interest: The authors declare no conflict of interest.

References

1. Yang, H.; Li, X.B. Cultivated land and food supply in China. *Land Use Policy* **2000**, *17*, 73–88. [CrossRef]
2. Foley, J.A.; DeFries, R.; Asner, G.P.; Barford, C.; Bonan, G.; Carpenter, S.R.; Chapin, F.S.; Coe, M.T.; Daily, G.C.; Gibbs, H.K.; et al. Global consequences of land use. *Science* **2005**, *309*, 570–574. [CrossRef] [PubMed]
3. Grafton, R.Q.; Daugbjerg, C.; Qureshi, M.E. Towards food security by 2050. *Food Secur.* **2015**, *7*, 179–183. [CrossRef]
4. Marshall, E.J.R.; Moonen, A.C. Field margins in northern Europe: Their functions and interactions with agriculture. *Agric. Ecosyst. Environ.* **2002**, *89*, 5–21. [CrossRef]
5. Granvik, M.; Lindberg, G.; Stigzelius, K.A.; Fahlbeck, E.; Surry, Y. Prospects of multifunctional agriculture as a facilitator of sustainable rural development: Swedish experience of Pillar 2 of the Common Agricultural Policy (CAP). *Nor. Geogr. Tidsskr.* **2012**, *66*, 155–166. [CrossRef]
6. Zhang, R.; Zheng, H.; Liu, Y. Evaluation on cultivated land ecological security based on the PSR model and matter element analysis. *Acta Ecol. Sin.* **2013**, *33*, 5090–5100. [CrossRef]
7. Lai, Z.H.; Chen, M.Q.; Liu, T.J. Changes in and prospects for cultivated land use since the reform and opening up in China. *Land Use Policy* **2020**, *97*, 104781. [CrossRef]
8. Deng, X.Z.; Huang, J.K.; Rozelle, S.; Zhang, J.P.; Li, Z.H. Impact of urbanization on cultivated land changes in China. *Land Use Policy* **2015**, *45*, 1–7. [CrossRef]
9. Su, Y.; Qian, K.; Lin, L.; Wang, K.; Guan, T.; Gan, M. Identifying the driving forces of non-grain production expansion in rural China and its implications for policies on cultivated land protection. *Land Use Policy* **2020**, *92*, 104435. [CrossRef]
10. Lambin, E.F.; Gibbs, H.K.; Ferreira, L.; Grau, R.; Mayaux, P.; Meyfroidt, P.; Morton, D.C.; Rudel, T.K.; Gasparri, I.; Munger, J. Estimating the world's potentially available cropland using a bottom-up approach. *Glob. Environ. Chang.* **2013**, *23*, 892–901. [CrossRef]
11. Maes, J.; Egoh, B.; Willemen, L.; Liquete, C.; Vihervaara, P.; Schaegner, J.P.; Grizzetti, B.; Drakou, E.G.; La Notte, A.; Zulian, G.; et al. Mapping ecosystem services for policy support and decision making in the European Union. *Ecosyst. Serv.* **2012**, *1*, 31–39. [CrossRef]
12. Saarikoski, H.; Primmer, E.; Saarela, S.-R.; Antunes, P.; Aszalos, R.; Baro, F.; Berry, P.; Garcia Blanko, G.; Gomez-Baggethun, E.; Carvalho, L.; et al. Institutional challenges in putting ecosystem service knowledge in practice. *Ecosyst. Serv.* **2018**, *29*, 579–598. [CrossRef]
13. Peters, M.K.; Hemp, A.; Appelhans, T.; Becker, J.N.; Behler, C.; Classen, A.; Detsch, F.; Ensslin, A.; Ferger, S.W.; Frederiksen, S.B.; et al. Climate-land-use interactions shape tropical mountain biodiversity and ecosystem functions. *Nature* **2019**, *568*, 88–92. [CrossRef]
14. Herath, G. Sustainable development of multifunctional landscapes. *Ecol. Econ.* **2004**, *50*, 319–321. [CrossRef]
15. Wiggering, H.; Dalchow, C.; Glemnitz, M.; Helming, K.; Muller, K.; Schultz, A.; Stachow, U.; Zander, P. Indicators for multifunctional land use—Linking socio-economic requirements with landscape potentials. *Ecol. Indic.* **2006**, *6*, 238–249. [CrossRef]
16. Costanza, R.; dArge, R.; deGroot, R., Farber, S.; Grasso, M.; Hannon, B.; Limburg, K.; Naeem, S.; ONeill, R.V.; Paruelo, J.; et al. The value of the world's ecosystem services and natural capital. *Nature* **1997**, *387*, 253–260. [CrossRef]
17. Song, X.; Li, X. Theoretical explanation and case study of regional cultivated land use function transition. *Acta Geogr. Sin.* **2019**, *74*, 992–1010.
18. Song, X.Q.; Ouyang, Z. Connotation of multifunctional cultivated land and its implications for cultivated land protection. *Prog. Geogr.* **2012**, *31*, 859–868.
19. Sohl, T.L.; Sleeter, B.M.; Zhu, Z.; Sayler, K.L.; Bennett, S.; Bouchard, M.; Reker, R.; Hawbaker, T.; Wein, A.; Liu, S.; et al. A land-use and land-cover modeling strategy to support a national assessment of carbon stocks and fluxes. *Appl. Geogr.* **2012**, *34*, 111–124. [CrossRef]
20. Chen, L.; Qu, F.T.; Shi, X.Y. The social value of cultivated land resources: A case in Liulin County of Shanxi Province. *Resour. Sci.* **2006**, *28*, 86–90.
21. Li, D.; Ren, Z.; Liu, X.; Lin, Z. Dynamic change of ecological service value of cultivated land in Shaanxi Province. *J. Arid. Land Resour. Environ.* **2013**, *7*, 40–45.
22. Zhu, P.X.; Bu, T.T.; Wu, Z.T. Study on compensation standard of land expropriation based on comprehensive value of cultivated land. *China Popul. Rec. Environ.* **2011**, *09*, 32–37.
23. Liu, H.F.; Bi, R.T.; Guo, Y.L.; Wang, J. Protection zoning of cultivated land based on form-structure-function multidimensional evaluation system. *Trans. Chin. Soc. Agric. Eng.* **2021**, *52*, 168–177.
24. Luo, C.; Cai, Y. The stage characteristics and spatial heterogeneity of cultivated land resource function evolution in agricultural producing areas of Hubei Province. *Econ. Geogr.* **2016**, *36*, 153–161.
25. Yuan, H. A Study on cultivated land quality protection and agricultural sustainable development in the main grain producing areas of northeast China. *Econo. Rev.* **2017**, *11*, 106–111.

26. Zhang, Y.; Long, H.; Ma, L.; Ge, D.; Tu, S.; Qu, Y. Farmland function evolution in the Huang-Huai-Hai Plain: Processes, patterns and mechanisms. *J. Geogr. Sci.* **2018**, *28*, 759–777. [CrossRef]
27. Scown, M.W.; Winkler, K.J.; Nicholas, K.A. Aligning research with policy and practice for sustainable agricultural land systems in Europe. *Proc. Natl. Acad. Sci. USA* **2019**, *116*, 4911–4916. [CrossRef]
28. Xia, H.; Zhang, W.S.; He, L.; Ma, M.Y.; Peng, H.; Li, L.; Ke, Q.; Hang, P.P.; Wang, X.T. Assessment on China's urbanization after the implementation of main functional areas planning. *J. Environ. Manag.* **2020**, *264*, 110381. [CrossRef]
29. Kang, R.; Ren, Y.; Wu, H.; Zhang, S. Changes in the nutrients and fertility of black soil over 26 years in Northeast China. *Sci. Agric. Sin.* **2016**, *49*, 2113–2125.
30. Liu, X.B.; Zhang, X.Y.; Wang, Y.X.; Sui, Y.Y.; Zhang, S.L.; Herbert, S.J.; Ding, G. Soil degradation: A problem threatening the sustainable development of agriculture in Northeast China. *Plant Soil Environ.* **2010**, *56*, 87–97. [CrossRef]
31. Jiang, G.H.; Wang, M.Z.; Qu, Y.B.; Zhou, D.Y.; Ma, W.Q. Towards cultivated land multifunction assessment in China: Applying the "influencing factors-functions-products-demands" integrated framework. *Land Use Policy* **2020**, *99*, 104982. [CrossRef]
32. Tang, H.; Li, Z. Study on per capita grain demand based on Chinese reasonable dietary pattern. *Sci. Agric. Sin.* **2012**, *45*, 2315–2327.
33. Li, T.; Zhang, F.; Hu, H. Authentication of the Kuznets Curve in agriculture non-point source pollution and its drivers analysis. *China Popul. Resour. Environ.* **2011**, *21*, 118–123.
34. Xie, G.; Zhen, L.; Lu, C.X.; Xiao, Y.; Chen, C. Expert knowledge based valuation method of ecosystem services in China. *J. Nat. Resour.* **2008**, *23*, 911–919.
35. Mir, M.A.; Ghazvinei, P.T.; Sulaiman, N.M.N.; Basri, N.E.A.; Saheri, S.; Mahmood, N.Z.; Jahan, A.; Begum, R.A.; Aghamohammadi, N. Application of TOPSIS and VIKOR improved versions in a multi criteria decision analysis to develop an optimized municipal solid waste management model. *J. Environ. Manag.* **2016**, *166*, 109–115.
36. Xin, G.; Yang, C.; Yang, Q.; Li, C.; Wei, C. Post-evaluation of well-facilitied capital farmland construction based on entropy weight method and improved TOPSIS model. *Trans. Chin. Soc. Agric. Eng.* **2017**, *33*, 238–249.
37. Zhang, H.; Feng, S.; Qu, F. Research on coupling coordination among cultivated land protection, construction land intensive use and urbanization. *J. Nat. Resour.* **2017**, *32*, 1002–1015.
38. Wang, J.F.; Zhang, T.L.; Fu, B.J. A measure of spatial stratified heterogeneity. *Ecol. Indic.* **2016**, *67*, 250–256. [CrossRef]
39. Wang, J.; Xu, C. Geodetector: Principle and prospective. *Acta Geogr. Sin.* **2017**, *72*, 116–134.
40. Wang, S.; Kong, W.; Ren, L.; Zhi, D.; Dai, B. Research on misuses and modification of coupling coordination degree model in China. *J. Nat. Resour.* **2021**, *36*, 793–810. [CrossRef]
41. Gonzalez-Garcia, A.; Palomo, I.; Gonzalez, J.A.; Lopez, C.A.; Montes, C. Quantifying spatial supply-demand mismatches in ecosystem services provides insights for land-use planning. *Land Use Policy* **2020**, *94*, 104493. [CrossRef]
42. Pierce, S.M.; Cowling, R.M.; Knight, A.T.; Lombard, A.T.; Rouget, M.; Wolf, T. Systematic conservation planning products for land-use planning: Interpretation for implementation. *Biol. Conserv.* **2005**, *125*, 441–458. [CrossRef]
43. Zhai, W.; Huang, X. Analysis on the effect of policies operation of cultivated land protection in China. *China Land Sci.* **2003**, *17*, 8–13.
44. Smith, P. Delivering food security without increasing pressure on land Pete Smith. *Glob. Food Secur.* **2013**, *2*, 18–23. [CrossRef]
45. Xu, X.L.; Wang, L.; Cai, H.Y.; Wang, L.Y.; Liu, L.; Wang, H.Z. The influences of spatiotemporal change of cultivated land on food crop production potential in China. *Food Secur.* **2017**, *9*, 485–495. [CrossRef]
46. Qi, X.; Li, J.; Yuan, W.; Wang, R.Y. Coordinating the food-energy-water nexus in grain production in the context of rural livelihood transitions and farmland resource constraints. *Resour. Conserv. Recycl.* **2021**, *164*, 105148. [CrossRef]
47. Wang, Q.; Lang, H.; Zhong, J.; Chen, M.; Zhang, B.; Zeng, S. Progress review on land science research in 2021 and prospects for 2022. *China Land Sci.* **2022**, *36*, 104–115.
48. Alizamir, S.; Iravani, F.; Mamani, F. An Analysis of Price vs. Revenue Protection: Government Subsidies in the Agriculture Industry. *Manage. Sci.* **2019**, *65*, 32–49. [CrossRef]
49. Cao, Z.; Hao, J.; Liang, L. The value accounting of cultivated land resources in Huang—Huai—Hai Region. *J. Arid Land Resour. Environ.* **2009**, *9*, 5–10.
50. Ellis, E.C. Land Use and Ecological Change: A 12,000-Year History. *Annu. Rev. Environ. Resour.* **2021**, *46*, 1–33. [CrossRef]
51. Wang, Q.; Jin, X.B.; Zhou, Y.K. Cultivated land ecological security and spatial aggregation pattern in Hebei Province. *Trans. Chin. Soc. Agric. Eng.* **2011**, *27*, 338–344.
52. Huang, J.; Yu, H.; Han, D.; Zhang, G.; Wei, Y.; Huang, J.; An, L.; Liu, X.; Ren, Y. Declines in global ecological security under climate change. *Ecol. Indic.* **2020**, *117*, 106651. [CrossRef]
53. Burger, J.R.; Allen, C.D.; Brown, J.H.; Burnside, W.R.; Davidson, A.D.; Fristoe, T.S.; Hamilton, M.J.; Mercado-Silva, N.; Nekola, J.C.; Okie, J.G.; et al. The Macroecology of Sustainability. *PLoS Biol.* **2012**, *10*, e1001345. [CrossRef] [PubMed]
54. Bongaarts, J. Summary for policymakers of the global assessment report on biodiversity and ecosystem services of the Intergovernmental Science-Policy Platform on Biodiversity and Ecosystem Services. *Popul. Dev. Rev.* **2019**, *45*, 680–681. [CrossRef]
55. Peng, J.; Liu, Z.C.; Liu, Y.X.; Hu, X.X.; Wang, A. Multifunctionality assessment of urban agriculture in Beijing City, China. *Sci. Total Environ.* **2015**, *537*, 343–351. [CrossRef] [PubMed]
56. Zhang, D.; Ng, E.L.; Hu, W.L.; Wang, H.Y.; Galaviz, P.; Yang, H.D.; Sun, W.T.; Li, C.X.; Ma, X.W.; Fu, B.; et al. Plastic pollution in croplands threatens long-term food security. *Glob. Chang. Biol.* **2020**, *26*, 3356–3367. [CrossRef]

land

MDPI

Article

Potential Land-Use Conflicts in the Urban Center of Chongqing Based on the "Production–Living–Ecological Space" Perspective

Tian Liang [1,2], Peng Du [1], Fei Yang [1,2], Yuanxia Su [1], Yinchen Luo [2,3], You Wu [4] and Chuanhao Wen [5,*]

1 College of Public Management, Chongqing Finance and Economics College, Chongqing 401320, China
2 Institute of Green Development, Chongqing Finance and Economics College, Chongqing 401320, China
3 College of Economics, Chongqing Finance and Economics College, Chongqing 401320, China
4 Chongqing United Equity Exchange Group Co., Ltd., Chongqing 401121, China
5 College of Economics, Yunnan University, Kunming 650091, China
* Correspondence: chhwen1972@ynu.edu.cn; Tel.: +86-185-0234-0601

Abstract: With the rapid population growth and accelerating urbanization process, people compete for the scarce land resources to pursue their incompatible interests. Thus, a series of land-use conflicts (LUCs) problems are caused. Scientifically identifying the intensity of LUCs is the basis for coordinating the man-land relations. We selected the urban center of Chongqing (UCC) as the study area and chose the landscape ecological risk assessment to estimate the level of LUCs by using the hot-spot analysis and neighborhood analysis to analyze the spatiotemporal evolution characteristics and potential risk of LUCs in the UCC over the past 20 years. The results show that the conversion between the living–production space (LPS) and other spaces was most frequent. The assessment model based on the theoretical framework of landscape ecological risk assessment could effectively measure LUCs. The average conflict level of UCC has increased from 0.62 to 0.69. The area of the out-of-control zone has increased, forming hot spots in the concentrated areas of social and economic activities. In contrast, the area of the controllable zone has decreased, forming cold spots in the high-altitude forest areas. The entire area faces the potential risk of the LUCs, but not seriously. The area of the high and extreme potential conflict zones has increased and is concentrated in the northern region of the study area. Targeted management strategies and policy recommendations for regional development should be adopted for different LUCs zones in UCC at international and national levels. Our research can be extended to other areas under rapid urbanization to assess and better manage their land resources for sustainable use, and further to promote the harmonious development of regional man-land relations.

Keywords: land-use conflicts; "production–living–ecological space"; landscape ecological risk; man-land relations

Citation: Liang, T.; Du, P.; Yang, F.; Su, Y.; Luo, Y.; Wu, Y.; Wen, C. Potential Land-Use Conflicts in the Urban Center of Chongqing Based on the "Production–Living–Ecological Space" Perspective. *Land* **2022**, *11*, 1415. https://doi.org/10.3390/land11091415

Academic Editors: Li Ma, Yingnan Zhang, Muye Gan and Zhengying Shan

Received: 26 July 2022
Accepted: 26 August 2022
Published: 28 August 2022

1. Introduction

The land is an indispensable and scarce resource to meet human production and life. Land-use conflicts (LUCs) describe the incoordination of land-use structure when meeting the diverse human demands under the deterioration of the natural environment, which is a sensitive indicator of human–environment interaction [1]. Its essence is the process of game and competition between the human system and natural system in time and space in the same area [2]. It is a manifestation of the uncoordinated development of the human–environment system, which will have many negative effects. LUCs are common in different regions of the world. Due to the rapid population growth and accelerating urbanization process, the demand for different land-use types is increasing [3]. The kinds of LUCs are also more and more diverse. Similar to the conflict between construction land expansion and essential farmland protection, there is a conflict between ecological land protection and production land expansion. The conflict between the increase in land demand and the degradation of land quality is becoming more and more prominent [4].

The drastic spatial change in land use is one of the most critical manifestations of LUCs. The disorderly spatial pattern of PLES in the same region will place tremendous pressure on limited land resources and cause waste of spatial resources [5]. The LUCs pose severe challenges to the sustainability of the land system and regional coordinated development, which raises excellent concerns about LUCs.

It is widely believed that the multifunctional of land-use, land resource scarcity, and diversity of human needs are the fundamental causes of LUCs [1,6–9]. It occurs when different land users pursue their incompatible interests, they will compete for the scarce land resources. With the competition of spatial resources by humans, a series of LUC problems are caused, such as the land spatial pressure increasing, the landscape ecological stability weakening, spatial interference strengthening, etc. [10]. These problems were widespread in the world, especially in the rapid urbanization area. Therefore, how to identify the spatial–temporal evolution characteristics and the potential risk of LUCs scientifically is the essential work and the core focus of LUCs research [11]. It was of great significant to alleviate the negative influence of conflict on land sustainability, promote the optimal allocation of regional land resources, and coordinate man-land relations [12].

"Production–living–ecological space" (PLES) is a theory proposed by the Chinese government in the strategy of ecological civilization construction, aiming at realizing sustainable utilization and focusing on the perspective of land multifunctional utilization [13]. The report of the 18th National Congress of the Communist Party of China (CPC) and the 14th 5-Year Plan of the CPC Central Committee on National Economic and Social Development both proposed the goal of developing the nation's PLES. According to the muti-functional attributes of the land, the urban center of Chongqing (UCC)'s land was divided into ecological–production space (EPS), production–ecological space (PES), living–production space (LPS), and ecological space (ES). The EPS is the space with ecological and production function, and the main function is ecological. The PES is the space that is used primarily for agricultural production function and has an ecological function at the same time. The LPS is the space that meets the needs of human life and entertainment, and contains the highest economic value. The ES is the space that has ecological functions such as regulating the atmosphere, conserving water source, water and soil conservation, etc., but does not have a production function. Therefore, understanding LUCs and their evolution characteristics can provide a scientific guide for the optimization of the PLES. Although the importance and necessity are acknowledged, there is still a lack of effective analysis of LUCs for UCC.

Compared with existing studies that focus on regional land use, and land-cover change, analyzing regional land-use conflicts can better reflect the interaction and relations between human and land resources [14]. As early as the 1970s, the study of LUCs began to attract wide attention over the world. It mainly studies the contradiction between human needs and economic development [15]. "Land management, land-use relations and conflicts" was one of the five main topics of the Urban Fringe Symposium organized by the English Countryside Association in 1977 [16]. With the ongoing concern for ecological civilization construction, the study of LUCs has gradually became the focus of scholars [17–19]. The number of studies on LUCs from different perspectives is gradually increasing [20,21]. The recent decade of research on LUCs has reached a new height. The research topics of LUCs are mainly as follows: (1) the LUCs patterns [22,23]; (2) the LUCs identification and intensity diagnosis [24,25]; and (3) the LUCs evolution and driving mechanism [1,26]. Based on the existing research, the LUCs identification methods mainly include game theory [27], participatory mapping [28], stress state response (PSR) model, multiobjective comprehensive assessment [2,29,30], and landscape ecological risk (LER) assessment [21,31]. These studies have provided many references for the identification of LUCs in this study, but there are few related studies on the central city area and from the PLES perspective.

Chongqing plays a vital role in the construction of ecological civilization in China. It plays a supporting position in the development of the western region in the new era, an exemplary role in the green development of the Yangtze River Economic Belt, and a key role in driving the development of the Belt and Road. Furthermore, Chongqing is also the youngest municipality with rapid urbanization in China. Due to the intensification of urbanization and rapid economic development, the dramatic changes in land-use structure have caused an imbalance of production, living, and ecological spaces [1]. The degree of LUCs in the UCC has further deteriorated. The man-land relations are becoming increasingly tense. Therefore, we selected the urban center of Chongqing (UCC) as a study area, because this area is the economic and cultural center of the whole of Chongqing. We analyzed the spatial evolution characteristics of LUCs and potential risk in the past 20 years in the UCC from the perspective of the PLES. Our study expanded the research perspective of LUCs identification, and provided a reference for those regions in the world whose development orientation is "ecological priority and green development", so that urban managers and policymakers may be better informed when developing pertinent land-use policies and strategies at different levels. It is helpful to relieve the level of LUCs in the rapid urbanization area, promote optimized land spatial patterns, the rational use of land resources, and the coordinated development of man-land relations in the world [11]. Our aim includes the following three objectives.

(1) Based on the LER assessment method, to construct the evaluation model of LUCs from the PLES perspective;

(2) To identify the LUCs zones and diagnosis the conflict intensity;

(3) To construct a land-use conflicts risk index to explore the potential LUCs.

2. Materials and Methods

2.1. The Study Area

Chongqing is the youngest municipality in China and is located inland in the southwest of China. Chongqing is an essential strategic fulcrum of China's western development. It is geographically located at the connection point of the "Belt and Road Initiative" and the Yangtze River Economic Belt. It has a unique and vital development navigation orientation in the pattern of national regional economic development, and opening to the outside world. The UCC is the political, economic, and cultural center of Chongqing, and this area is essential for the development of the city and surrounding areas. The UCC includes nine districts: Yuzhong District, Shapingba District, Beibei District, Yubei District, Jiangbei District, Nan'an District, Banan District, Dadukou District, and Jiulongpo District, with a total area of 5466 km^2 (Figure 1). The topography of the study area is complex and diverse, with hills and low mountains as the main ones and few flat dams. The principal rivers are mainly the Yangtze River and the Jialing River. According to the Seventh National Population census, Chongqing has a population of 32.0542 million, with a GDP of CNY 2500.279 billion. The UCC has a considerable population and a solid economic foundation. Its population and GDP accounted for 32.27% and 39.5% of the total urban population, respectively [32]. With the rapid growth of the economy and people, the LUCs of the PLES and the tension of man-land relations in the UCC are evident, and need to be paid more attention.

2.2. Data

The data of administrative districts of the study area were derived from the Remote Sensing Monitoring Institute of Chongqing Planning and Natural Resources Survey and Monitoring Institute. The land-use data used for three phases from 2000 to 2020 were obtained from the Data Center for Resources and Environment Science and Data Center, Chinese Academy of Sciences (http://www.resdc.cn/ (accessed on 9 July 2022)). The resolution of the land-use data is 30 m × 30 m. The types of land-use include six types of the first order: cultivated land, woodland, grassland, water area, residential land, and unused land, and 25 types of the second order. The social–economic data mainly come from the Chongqing Statistical Yearbook (2021).

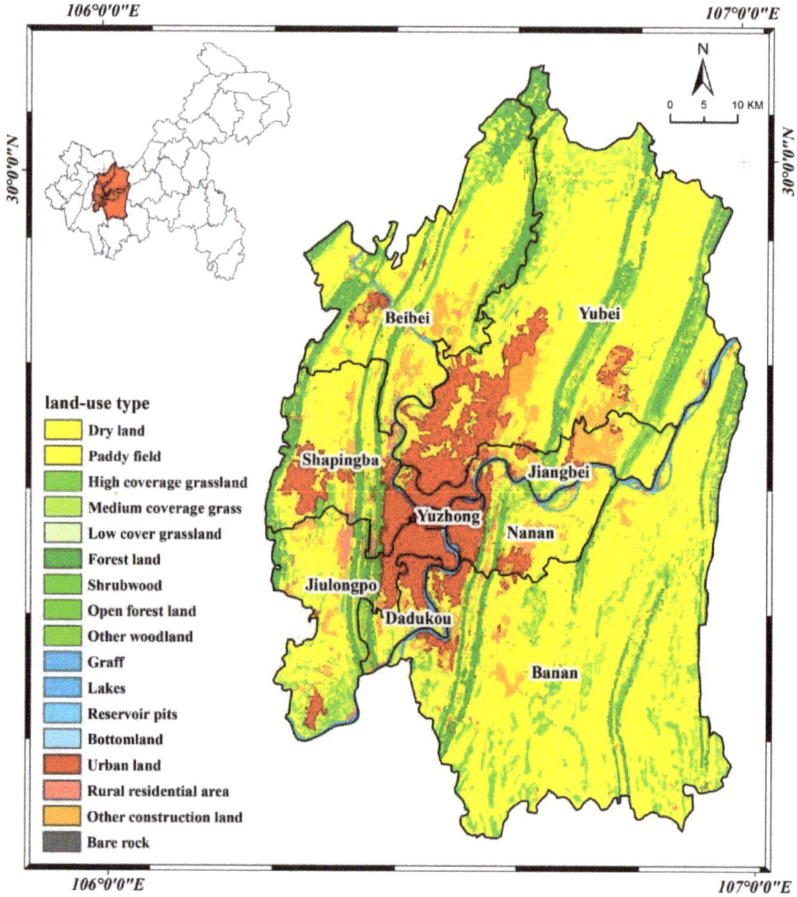

Figure 1. Location of the study area.

2.3. Methods

2.3.1. Construction of the PLES Classification System

The PLES has the characteristics of complex spatial functions, differences in spatial scales, and the heterogeneity of spatial land use [33]. Its classification is exceptionally complicated. Land resources are the lifeblood of promoting economic development, the source of all productive activities, and the interrelated and unified complex of the PLES. Land-use type has multiple functions, e.g., the construction land has two functions of production and living, and the cultivated land has two functions of ecological and production. Combined with the theory of the PLES and the previous research results [34–36], each land-use category was linked with the leading function of the PLES. Referring to relevant studies [37,38], the PLES classification system was constructed according to the study area's actual situation, as shown in Table 1.

2.3.2. Dynamic of Land–Use Change

The conversion between land-use types is mainly realized by the land-use transfer matrix, and the dynamic change process of land-use types is mainly expressed by single land-use dynamic and bidirectional land-use dynamic [38].

Table 1. The PLES classification system.

The PLES Types	Meaning	Land–Use Type	Classification Basis
Living–production space (LPS)	Meets the needs of human life and entertainment, and contains the highest economic value.	Urban, other construction lands, rural residential land.	Mainly refers to cities and towns and other places where human beings live, which are important places to live and rest.
Production–ecological space (PES)	It is mainly used for agricultural production functions and has ecological functions at the same time.	Cultivated land, orchard.	Other gardens and orchards are important garden land with ecological functions. Drylands and paddy fields are important agricultural lands, both of which have dual functions of ecological and production.
Ecological–production space (EPS)	Has ecological and production functions, and the main function is ecological.	Forest, shrub forest, sparse woodland, other woodlands, reservoir pond.	Shrub forest, other forest land, and forested land play an essential role in regulating climate and environment, their ecological functions are beyond doubt, and they can also provide wood. Reservoir pits and ponds have the function of maintaining water source, and also have production functions such as aquaculture.
Ecological space (ES)	Most of them have ecological functions such as regulating the atmosphere, conserving water source, water and soil conservation, etc., but do not have a production function.	High–coverage grassland, medium–coverage grassland, low–coverage grassland, river, lakes, bare land, bottomland.	Grasslands with low, medium, and high coverage have ecological values such as regulating climate, conserving water and soil, and conserving water resources. Rivers, lakes, beaches, and swamps have the functions of water conservation and climate regulation, and belong to the ecological land. Bare land has high vegetation coverage and ecological landscape effect, which belongs to ecological function.

The single land-use dynamic ($LCDI_i$) refers to the ratio of the area change of land-use types in the region to the study period, which mainly reflect the rate of change of a single land-use type in a certain period. The bidirectional land-use dynamic $(BLCDI_i)$ is a further supplement to $LCDI_i$ [38]. It can better describe the change process and the direction of a certain kind of land use, which mainly reflects the transfer intensity between land-use types. The formulas are below:

$$LCDI_i = \frac{(S_{ib} - S_{ia})}{S_{ia}} \times \frac{1}{T} \times 100\% \tag{1}$$

$$BLCDI_i = \frac{\left(\sum S_{ij} + \sum S_{ji}\right)}{S_i} \times \frac{1}{T} \times 100\% \tag{2}$$

where $LCDI_i$ is the single land-use dynamic of i–th PLES type; S_{ia} is the area of i–th PLES type at the beginning; S_{ib} is the area of i–th PLES type at the end; T is the study time interval; $BLCDI_i$ is the bidirectional land-use dynamic of i–th PLES type; $\sum S_{ij}$ is the sum of the area of i–th PLES type changing into other PLES types; $\sum S_{ji}$ is the sum of the area of other PLES types changing into i–th PLES type; S_i is the area of i–th PLES type at the beginning.

2.3.3. Construction of LUCs Assessment Model

Land-use systems are dynamic, fragile, and complex [39]. To avoid the fragmentation of regional spatial units, we consider the research scope, scale, spatial resolution, spatial patch status, and data types. The maximum horizontal distance of the study area between east and west is about 74 km, and the maximum vertical distance between north and south is about 112 km. To ensure that there are multiple spatial types in an assessment unit, this paper choose a 4 km × 4 km grid as the evaluation unit. The study area was divided into a total of 397 assessment units. We chose the method of the LER assessment to measure the intensity of LUCs because we consider that the LUCs and landscape ecological risk assessment are highly related (Figure 2).

The spatial complexity index (*CI*), spatial vulnerability index (*FI*), and spatial stability index (*SI*) were chosen to construct the LUCs assessment model to measure the conflict level of spatial units in a region from the ecological point of view. Referring to previous studies [37,40,41], the LUCs assessment model can be expressed as:

$$SCCI = CI + FI - SI \tag{3}$$

where *SCCI* is the LUCs level; *CI* is the landscape complexity index; *FI* is the landscape vulnerability index; SI is the landscape stability index.

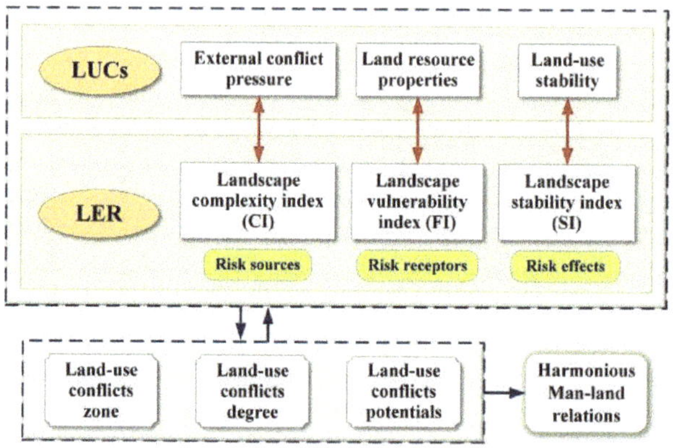

Figure 2. Conceptual framework of coupled LUCs and LER assessment.

(1) Landscape complexity index (*CI*)

CI is mainly due to the low efficiency of land use caused by rapid urbanization, which leads to the negative effects of system fragmentation and complexity [40]. *AWMPFD* index is called the area–weighted average patch fractal dimension in landscape ecology. The higher its value, the greater the degree of human disturbance, and vice versa. In this paper, the *AWMPFD* index is used to express *CI*. The formula is below:

$$AWMPFD = \sum_{i=1}^{m}\sum_{j=1}^{n}\left[\frac{2ln(0.25P_{ij})}{ln(a_{ij})}\left(\frac{a_{ij}}{A}\right)\right] \tag{4}$$

where P_{ij} is the perimeter of the patch; a_{ij} is the area of the patch; A is the total area of spatial units in the landscape; m is the total number of evaluation units in the study area; n is the number of three spatial types.

(2) Landscape vulnerability index (*FI*)

FI is mainly due to the vulnerability of a land-use system under the interference of external pressure, which will cause significant damage to the system. With time and space changes in land use, there are significant differences in maintaining ecosystem stability, protecting biodiversity, and improving system functions; that is to say, the various landscape elements have different responses to spatial conflicts, which are related to the stages in the natural succession process. In different locations, the ability of land-use types to resist external disturbance is other. *FI* is used to express it. The formula is below:

$$FI = \sum_{i=1}^{n} F_i \times \frac{a_i}{S}(n = 4) \tag{5}$$

where *FI* is the landscape vulnerability index; a_i is the area of various landscapes in the unit; S is the total area of units. Referring to the existing research [42], the order of landscape fragility of *FI* from strong to weak is LPS = 4; PES = 3; EPS = 2; ES = 1.

(3) Landscape stability index (*SI*)

SI refers to the phenomenon that landscape patches are fragmented under the interference of external pressure. The more fragmented the spatial form of the land-use unit, the stronger the dynamic, the worse the stability, and the stronger the conflict effect. The formula is below:

$$PD = \frac{n_i}{A}, \ SI = 1 - PD \tag{6}$$

where n_i represents the number of i–th PLES in the spatial unit; A is the space unit area; PD is the density of plaque. The larger the PD value, the higher the fragmentation degree, the worse the stability, and the worse the anti-interference ability of the unit. For the convenience of calculation, the numerical values in Formulas (3)–(6) are linearly standardized to (0, 1) by using Formula (7) to calculate the conflict in the later period. The standardized formula is below:

$$S = \frac{X - X_{\min}}{X_{\max} - X_{\min}} \quad (7)$$

where X is the value in (3), (4), (5), and (6), $X_{\min}$ is the minimum value, and $X_{\max}$ is the maximum value according to the existing research [43], based on the inverted "U" model. The LUCs of the study area were divided into four stages [9,23,44]: the stable control stage [0.0, 0.35), the basic control stage [0.35, 0.7), the basic out-of-control stage [0.7, 0.9), and the serious out-of-control stage [0.9, 1). In the stable control stage, the conflict has not yet formed or is in the potential stage, and will have no negative impact on the regional land use. In the basic control stage, conflict begins to form and gradually emerge, but is mostly constructive rather than destructive [45,46]. Appropriate measures should be taken to regulate and avoid or minimize the negative effects of conflict. In the basic out-of-control stage, the conflict broke out gradually, and the direction of land-use transformation gradually lost control. Effective measures must be taken to curb the conflict. Otherwise, the regional land use will gradually be unbalanced. In the serious out-of-control stage, the conflict completely breaks out, which requires the intervention of various administrative, economic, and legal measures. Otherwise, it may evolve from LUCs to a conflict of a social nature [47].

2.3.4. Spatial Relationship of LUCs

The land-use patches are often significantly affected by the land-use patterns around them, which is manifested in two aspects, including spatial agglomeration and spatial adjacency. On one side, the cold spots and hot spots are statistically significant spatial clusters of high values and low values, respectively [48], reflecting the spatial agglomeration relationship and the active degree of the LUCs zones. On another, if there is a more significant conflict difference between adjacent units, the interference of land-use patterns in its adjacent units will be stronger, and the possibility of causing LUCs will be higher [23]. Thus, the methods of cold- and hot-spot analysis and neighborhood analysis were used to reflect the spatial agglomeration relationship and potential risk of LUCs in the UCC. Taking the "3 × 3" rectangular component as the range, the standard deviation of the central element can be obtained through the neighborhood analysis function, to judge the influence degree of the surrounding units on the main unit [23,47]. The potential land-use conflicts risk index (PLUCRI) was constructed to analyze the potential LUCs. The specific formula is below:

$$L_i = \frac{\sum \left| G_{nbr_{ij}} - G_{scr_i} \right|}{N_{nbr_i}} \quad (8)$$

where L_i is the PLUCRI of the i-th evaluation unit, and the larger the value is, the greater the possibility of conflict in the evaluation unit would be. G_{scr_i} denotes the conflict intensity of the i-th evaluation unit, and it is represented by eight LUCs zones (for quantitative calculation, I–VIII is represented by 1, 2, ... 8, respectively). $G_{nbr_{ij}}$ is the conflict intensity of the j-th neighborhood unit's i-th evaluation unit. N_{nbr_i} is the number of neighborhood units of the i-th evaluation unit [23].

3. Results

3.1. Spatiotemporal Evolution Characteristic of the PLES

According to the classification system of the PLES, the land-use types in the UCC were divided into four spatial types. Spatiotemporal evolution characteristic of the PLES in the study area is shown in Figure 3. The spatial distribution of PES was the most extensive. ES was mainly distributed in the areas where the Yangtze River and Jialing River flow through. EPS was strip-shaped from north to south. The concentration of LPS was the highest in Yuzhong District and its vicinity.

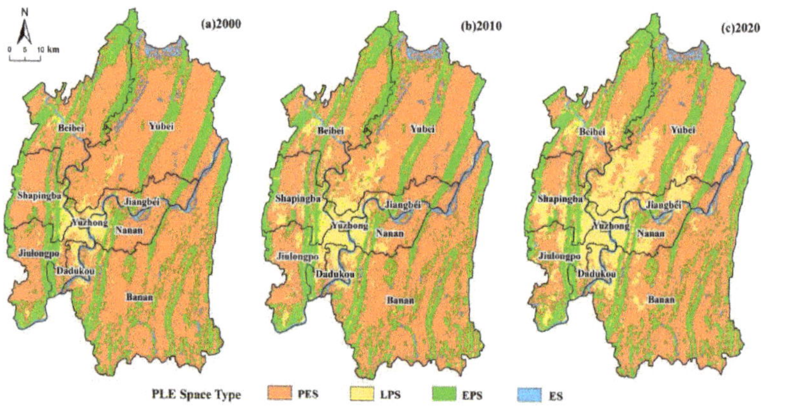

Figure 3. Spatial-temporal evolution of the PLES from 2000 to 2020.

As Table 2 shows, from 2000 to 2020, the LPS expanded significantly by 673.68 km^2 (+254.22%), while the PES was compressed by LPS and reduced by 671.35 km^2 (−17.12%). Due to the ecological protection policy, the proportion of ES increased sustainably in the past 20 years. It shows that rapid urbanization mainly eroded the PES and EPS in the UCC.

Table 2. The area ratio of the PLES in the UCC from 2000 to 2020.

Year	PES		EPS		LPS		ES	
	Area (km^2)	Ratio	Area (km^2)	Ratio	Area (km^2)	Ratio	Area (km^2)	Ratio
2000	3921.63	71.77%	1091.72	19.98%	265.00	4.85%	186.07	3.41%
2010	3653.07	66.85%	1089.26	19.93%	533.16	9.76%	189.10	3.46%
2020	3250.28	59.48%	1084.91	19.85%	938.68	17.18%	190.71	3.49%

To analyze the transformation of the PLES, the land transfer matrix was used. From 2000 to 2010, the LPS in the UCC increased by 268.15 km^2, while the PES decreased by 268.66 km^2. The percentage difference between the transfer-in and transfer–out ratio of PES was −6.82%, the LPS was 49.76%, the ES was −1.86%, and the EPS was −2.47%. The $LCDI_i$ of the LPS was the largest. It showed that the rate of change in the LPS was the greatest in this decade. In the process of rapid urbanization, it mainly aimed to meet the regional expansion of LPS. The $BLCDI_i$ of the LPS was 10.33%, and was the largest during this period. It showed that the transformation between LPS and other spaces was the strongest. The conversion of PES to LPS was the primary source of the increase in the LPS area. It showed that urban expansion was mainly based on the interconversion of LPS, ES, and PES, and the transfer area of LPS was more significant than the other two (Table 3, Figure 4).

In the transfer of the PLES in the UCC from 2010 to 2020, the area of the LPS and the ES still increased, while the area of the PES and the EPS still decreased. The percentage difference between the transfer-in and transfer-out ratio of PES was −10.84%, the LPS was 42.08%, the ES was 0.81%, and the EPS was −0.38%. The LCDI of the LPS was the largest, which showed that the growth of construction land was the primary trend in the process of urbanization in this period. The $BLCDI_i$ of the LPS was 8.12% and the largest during this period. It showed that the conversion between LPS and other spaces was the strongest and most frequent, although the $LCDI_i$ of the EPS is −0.04%, the $LCDI_i$ of the ES is 0.09%, the

$BLCDI_i$ of the EPS is 1.12%, and the $BLCDI_i$ of the ES is 0.99%. It shows that although the rate of change of the ES is faster than the rate of change of the EPS over these ten years; however, between the EPS and other spaces is stronger than the transfer intensity between the ES and other spaces (Table 4, Figure 4).

Table 3. Spatial transfer matrix of the PLES in the study area from 2000 to 2010 (unit: km^2).

2000–2010	PES	LPS	ES	EPS	Transfer-Out	Variation
PES	0.00	253.70	8.82	24.92	7.33%	−268.66
LPS	2.02	0.00	0.37	0.41	1.06%	268.15
ES	1.04	0.34	0.00	5.56	3.73%	3.02
EPS	15.72	16.91	0.77	0.00	3.06%	−2.51
Transfer-in	0.51%	50.82%	1.87%	2.84%	—	—
$LCDI_i$	−0.68%	10.12%	0.16%	−0.02%	—	—
$BLCDI_i$	0.78%	10.33%	0.91%	0.59%	—	—

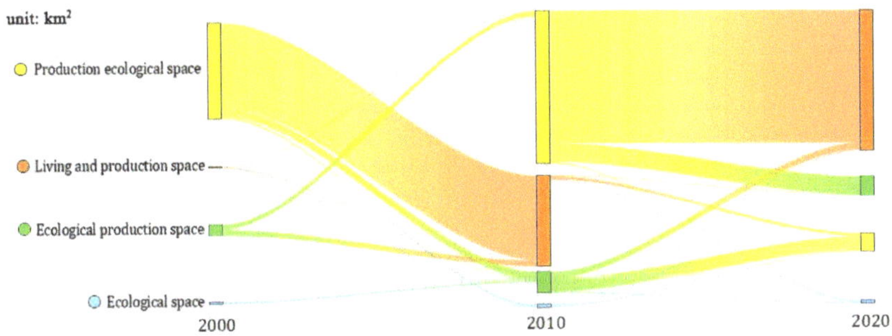

Figure 4. Sangji map of the PLES transfer from 2000 to 2020.

Table 4. Spatial transfer matrix of the PLES in the study area from 2010 to 2020 (unit: km^2).

2010–2020	PES	LPS	ES	EPS	Transfer-Out	Variation
PES	0.00	396.37	5.33	55.25	12.51%	−402.78
LPS	9.76	0.00	1.80	2.22	2.58%	405.51
ES	4.84	2.48	0.00	1.23	4.52%	1.62
EPS	39.57	20.44	3.04	0.00	5.79%	−4.35
Transfer-in	1.67%	44.67%	5.33%	5.41%	—	—
$LCDI_i$	−1.10%	7.61%	0.09%	−0.04%	—	—
$BLCDI_i$	1.40%	8.12%	0.99%	1.12%	—	—

3.2. Spatiotemporal Distribution Characteristics of the LUCs

According to Formulas (3)–(7), the LUCs in the UCC were calculated and visually expressed (Figure 5). We derived spatial patterns of the UCC's land-use conflicts from 2000 to 2020. The distribution pattern was "high in the south and low in the north" in 2000. The LUCs was higher on the southern side of the UCC, especially in the center area of the UCC and the northeastern of Banan District. The relatively high LUCs in the south of regions were dominated by the basic out-of-control zone and the serious out-of-control zone. From 2000 to 2010, the basic out-of-control zone had spread around. It is worth noting that by 2010, the level of the LUCs in the Yuzhong District changed from the basic out-of-control to the basic control. From 2010 to 2020, the basic out-of-control zone was widely distributed in the UCC gradually. The serious out-of-control zone was distributed on the northern side of the UCC, especially in the Yubei District and Beibei District. Therefore, the spatial distribution pattern was "high in the north and low in the south" in 2020. The relatively high LUCs in the northern regions were also dominated by the serious out-of-control zone, and the basic out-of-control zone. The main gathering areas of the basic out-of-control zone and the serious out-of-control zone gradually expanded from Dadukou District, Jiulongpo District, and Yuzhong District to Shapingba District, Beibei District, Yubei District, Nan'an District, and Banan District. From 2000 to 2020, Chongqing's economy developed rapidly, the urban traffic network changed from simple to complex, the medical facilities changed from few to many, and the construction land changed from scattered to concentrated. The rapid urbanization process caused serious conflict between the PLES. The area of controllable decreased, while the area of out-of-control increased in the UCC from 2000 to 2020.

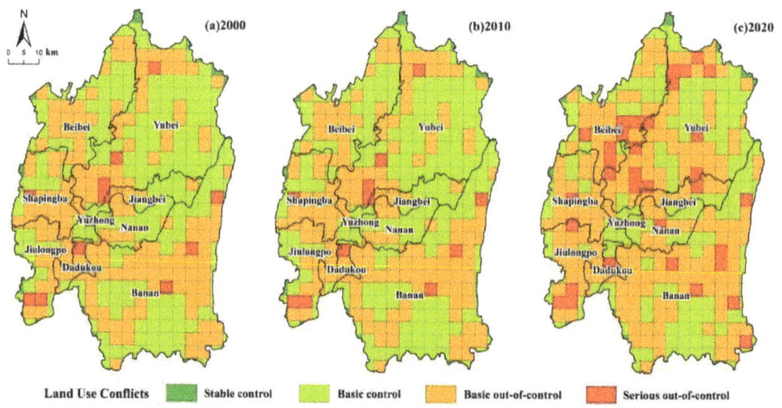

Figure 5. Spatial-temporal distribution characteristics of the LUCs from 2000 to 2020.

According to the result presented in Table 5, the average conflict level in Chongqing urban center from 2000 to 2020 increased from 0.62 to 0.69. In the process of rapid urbanization, the characteristics of "two decreases and two increases" appeared. The characteristic of "two decreases" means the area of the stable control zone and area of the basic control zone declined. The characteristic of "two increases" means the area of the basic out-of-control zone and the area of the serious out-of-control zone increased. From 2000 to 2020, the area of stable control zone was maintained at 6–7%. The area of basic control zone was reduced by 18.48%, while the area of out-of-control zone increased by 19.24%. The area of basic and serious out-of-control zone occupied almost 56% of the study area. It posed a serious threat to the ecological environment of Chongqing. Compared with the conflict change over the three periods, it showed that the out-of-control level was mainly clustered in the LPS and PES area. It is due to the conflicting living–ecological, production–ecological functions. The area far away from the LPS and PES was mainly the controllable level, and this type of area was dominated by ecological function land.

Table 5. Land-use conflicts level in the study area from 2000 to 2020.

Conflict Level	Conflict Value	Number of Spatial Units			Percentage of Space Units (%)		
		2000	2010	2020	2000	2010	2020
Stable control	0.0–0.35	27	24	24	6.80	6.05	6.03
Basic control	0.35–0.7	226	203	152	56.93	51.13	38.44
Basic out-of-control	0.7–0.9	135	159	187	34.01	40.05	46.98
Serious out-of-control	0.9–1.0	9	11	34	2.27	2.77	8.54
The average conflict value		0.62	0.64	0.69	—	—	—

The spatial distribution characteristics of the PLES in the LUCs zones in the UCC (Figure 6) and the corresponding area ratio changes (Table 6) were analyzed by spatial superposition analysis.

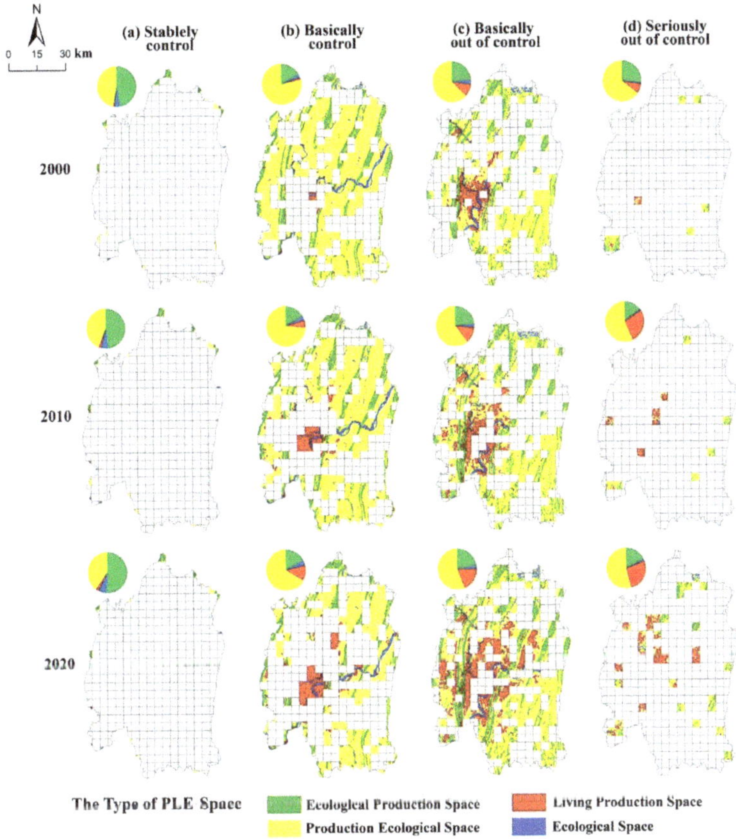

Figure 6. The evolution characteristics of the PLES in the LUCs zone from 2000 to 2020.

Table 6. The area ratio statistics of LUCs zones from 2000 to 2020.

Conflict Zone	Area Ratio (%)			PLES	Area Ratio (%)			Variation
	2000	2010	2020		2000	2010	2020	
Stable controllable Zone	1.23	0.97	0.89	EPS	47.55	48.22	50.99	3.44
				ES	4.67	5.97	6.48	1.81
				LPS	0.75	1.87	2.62	1.87
				PES	47.03	43.93	39.90	−7.13
Basic controllable Zone	57.54	50.17	35.85	EPS	16.79	16.64	16.90	0.11
				ES	3.22	3.76	4.52	1.3
				LPS	1.34	6.08	12.09	10.75
				PES	78.65	73.51	66.49	−12.16
Basic out-of-control zone	38.61	45.65	53.31	EPS	23.45	23.36	21.85	−1.6
				ES	3.78	3.23	3.04	−0.74
				LPS	9.99	12.69	19.02	9.03
				PES	62.78	60.71	56.09	−6.69
Serious out-of-control zone	2.62	3.22	9.95	EPS	26.12	14.49	17.24	−8.88
				ES	1.34	1.25	1.87	0.53
				LPS	7.88	27.54	26.78	8.90
				PES	64.66	56.72	54.11	−10.55

In the stable controllable zone, EPS occupied the absolute number, accounting for 50.99% in 2020. The land-use type of EPS was mainly forest, which has strong ecological functions. From 2000 to 2020, due to the decrease of PES and the increase of LPS, the range of the basic controllable areas was also shrinking.

From 2000 to 2020, in the basic out-of-control zone and the serious out-of-control zone, LPS increased by 27.93% in total, while the other three types decreased. Among them, PES decreased by 17.24%, EPS decreased by 10.48%, and ES decreased by 0.21%. The out-of-control zone was mainly distributed in the core and peripheral areas of urban economic development, which were the critical area of Chongqing's future production and living development.

In the past 20 years, Chongqing has been in rapid development of the economy and society. With the level of industrialization and urbanization further improved, the demand for the land for production and living increased. In the continuous improvement of transportation facilities and growing construction land, the ecological land and agriculture land were eroded heavily by urban land. The ES, EPS, and PES were gradually fragmented. The proportion of the out-of-control level in the study area was becoming heavier and heavier. The mutual occupation between the production, living, and ecological land made the LUCs more intense. It resulted in more prominent contradictions between humans and land. The man-land relations were becoming increasingly tense in the UCC.

3.3. Spatial Agglomeration Relationship of LUCs Zones

The spatial agglomeration relationships of LUCs zones were analyzed by using the methods of cold- and hot-spot analysis [49–51]. The location of high value and low value of LUCs clustering was obtained by cold- and hot-spot analysis. The results showed that the hot and cold spots of LUCs in the UCC showed an agglomerate spatial distribution pattern. The distribution pattern of hot spots, between 2000 and 2010, was mainly concentrated in Yuzhong and Dadukou, eastern Jiulongpo, southeastern Shapingba, southern Yubei, western Jiangbei, and eastern Banan. By 2020, it especially gathered at the intersection between Yubei and Beibei and the east part of Banan. Complex social and economic activities and convenient transportation facilities diversified the land-use function in these areas. This may be the main reason for the high-value aggregation of LUCs. The confrontation between various land-use functions was particularly fierce, and the conflicts became more and more frequent from 2000 to 2020. The cold spots represent stable and controllable. From the distribution pattern, they were mainly distributed in the boundary edge areas of the urban areas of Chongqing in the past 20 years, such as the south of Banan and the northeast of Yubei. These areas have higher elevations, and the land-use type is dominated by forests. Typical hilly and mountain cover characteristics and the remote geographical location may be the main reasons for the low-value aggregation of LUCs (Figure 7).

3.4. Potential Risk of LUCs

The potential land-use conflicts risk index (PLUCRI) was accessed by using the method of neighborhood analysis [23]. The results showed that the entire area faced the risk of the LUCs, but not seriously. In Figure 8, the areas of the low and general potential conflict area decreased in the last 20 years, while the areas of high and extreme potential conflict increased and concentrated in the north of the study area, especially in the north of Yubei District and Beibei District, and the junction of Shapingba District and Beibei District. The zones of low potential conflict, general potential conflict, high potential conflict, and extreme potential conflict account for 43.43%, 32.27%, 17.37%, and 6.93% in 2000, and 41.5%, 33.2%, 16.22%, and 9.08% in 2020. The cold- and hot-spot analysis of the PLUCRI showed that the hot spots of PLUCRI were also concentrated in the north of Yubei District and the junction of Shapingba District and Beibei District in Figure 9. These areas are not only land-use out-of-control areas, but also areas with the higher potential risk of LUCs, indicating that strict spatial boundary control measures need to be implemented in these areas as soon as possible. Therefore, the future regional land use needs to strictly implement the border control of spatial planning to reduce the influence of neighborhood units so that the intensification of conflict will not arise [47].

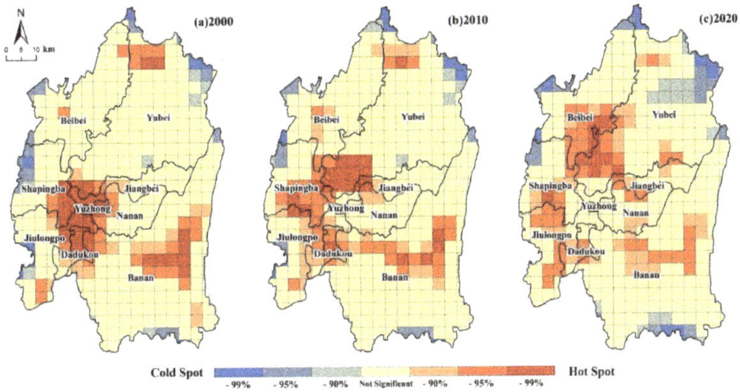

Figure 7. The cold- and hot-spot analysis of LUCs in the study area.

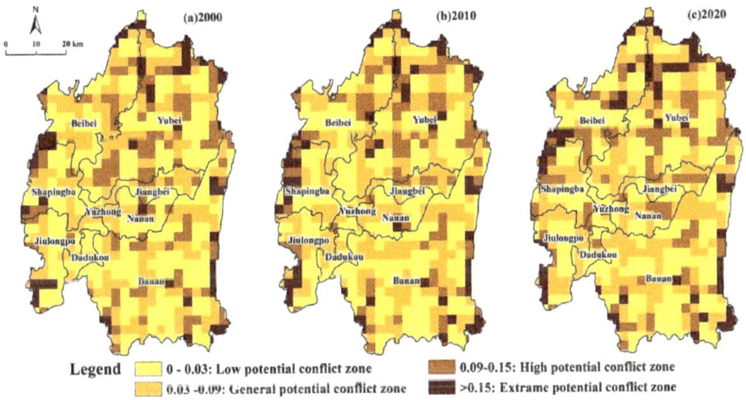

Figure 8. The potential risk of LUCs in the study area.

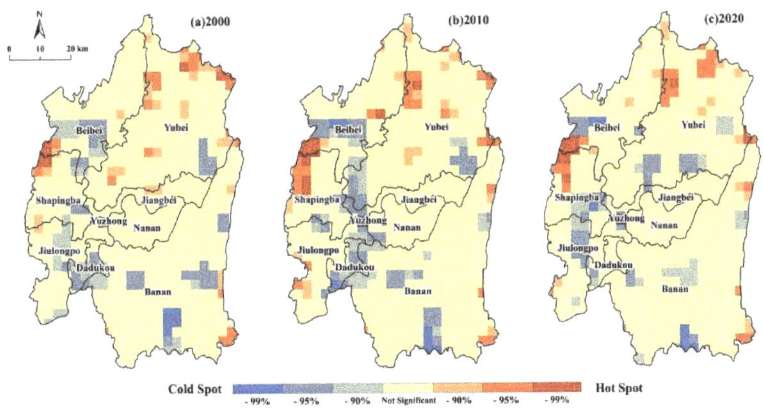

Figure 9. The cold- and hot-spot analysis of PLUCRI in the study area.

4. Discussion

4.1. Contributions and Limitations

Compared with previous studies, our paper has the following advantages. First, the theoretical basis is sufficient. In our study, we found that the multifunctional land use, land resource scarcity, and diversity of human needs are the fundamental causes of LUCs in the process of rapid urbanization. Rapid urbanization and unreasonable land use will also lead to the complexity and fragmentation of the spatial landscape and damage the stability of the system. Therefore, we believe that LUCs are highly correlated with LER. Based on this, we constructed a conceptual framework for LUCs assessment. Second, the LUCs evaluation framework we constructed can identify the intensity of LUCs at grid scale and regional scale. We find that the conflict zones are gathered in concentrated areas of social and economic activities, which is consistent with the previous findings [29]. This provides more accurate location information for land-use decisions and land management. Third, the identification of potential hotbeds of LUCs is easily overlooked in previous studies. In our study, we try to use domain analysis method to explore the potential LUCs areas. We find that the rural-urban border areas are particularly vulnerable to the potential risk of LUCs, because of high population movements and lack of government supervision. This is an interesting finding and is different from previous studies [47].

In this paper, the complexity, fragility, and stability of spatial patterns were taken as the basis for calculating conflict. However, it should be pointed out that methods and ideas of this study need to be improved:

(1) The method of LUCs we used was based on the LER assessment, so that the factor chosen in the calculation of conflict relied too much on the present situation of land-use function. However, the manifestation, formation mechanism, and influencing factors of spatial conflict are highly complex, involving resources, environment, society, and economy.

(2) The LUCs in our study are caused by the disorderly spatial pattern of PLES, which only reflects the incoordination of the land-use structure, but as per the discussions in the literature [52–57], the types of LUCs are various, such as regional conflicts, socioenvironmental conflicts, structure and function conflicts of an urban system, land system conflicts, and cultural conflicts. We will explore these types in the future.

(3) Our research carried the foundation analysis of the evolution and the potential risk of LUCs in the UCC from the view of the PLES. We did not simulate and predict the future development trend of land-use conflicts in the study area, so it is difficult to analyze and solve the conflict problem comprehensively. Therefore, this will be our future direction for further research.

4.2. Policy Recommendations of LUCs Optimization

The LUCs reflect the competition of stakeholders for scarce land resources and the land-use contradiction produced for the realization of their respective interests [58]. The ultimate goal of land-use conflict research is for contradiction reconciliation and relieving man-land relations. These are intended to develop reasonable reconciliation programs to promote sustainable development in the study area [59]. The key is to identify the distribution, manifestation, and degrees of conflicts, and take them as a basis to set up targeted governance strategies. In our paper, we hope to have provided targeted management strategies for different LUCs zones by combining different governance strategies for LUCs at international and national levels, to promote the harmonious development of regional man-land relations. In particular:

(1) In the stable controllable zone

The stable controllable zone in the UCC is mainly concentrated on the edge of Chongqing, where the economy is relatively backward, and the land-use type is dominated by forest. The degree and potential risk of LUCs are weak. These areas should be designated as ecological nature reserves, and production and construction activities should be strictly prohibited. The government should increase ecological subsidies and encourage inefficient farmland to restore forest in these areas. Thus, the self-evolution and regulatory function of regional ecosystem can be enhanced. In addition, the rural residential land in these areas should be gradually moved out according to the policy of increasing and decreasing balance, so as to reduce the interference of human activities to the environment.

(2) In the basic controllable zone

The area of basic controllable zone was reduced heavily by 18.48% in the last 20 years due to the decrease in PES and the increase in LPS. This zone is located in the transition area between human production–life and ecological protection, and does not pose a threat to the regional sustainable land use, but has the potential risk and possibility of LUCs. Therefore, in these regions, the government should adopt a slightly stronger management strategy and more diversified management means. On the premise of not destroying the regional natural ecosystem and farmland ecosystem, land should be properly utilized under the support and drive of policies to maximize the function of land. For example, the government can increase agricultural input, promote the planting and reprocessing of ecological agricultural products, and improve the development level of the primary industry. In addition, the government can also plan and build nature protection resorts to stimulate the living and production functions of land, and develop the tertiary industry while protecting the regional ecological environment.

(3) Basic out-of-control zone

The area of the basic out-of-control zone increased by nearly 13% from 2000 to 2020, occupying almost 47% of the study area in 2020. It posed a serious threat to the ecological environment of Chongqing. This zone was mainly distributed in the core and the peripheral regions of urban economic development, which were human production and living intensive activity areas. The LUCs between production, living, and ecological were very intense and basic out-of-control, and were the focus of governance areas. The government must take into account the fact that Chongqing is an ecologically fragile area. It is urgent to give priority to its ecological protection function and formulate a strong ecological protection policy. Therefore, the basic farmland protection line, ecological protection line, and urban expansion boundary line should be delimited from the macro level. The city scale should be reasonably controlled, and the construction land extraction mechanism should be formulated. There is also a need to strengthen the construction of ecological civilization in this area and enhance the awareness of ecological culture of local people to avoid further deterioration of the ecological environment and to prevent the reform area from further evolving into a serious out of control area.

(4) Serious out-of-control zone

The proportion of this region in 2020 is 8.54%, a 6.27% increase from 2000. Compared with the basic out-of-control zone, the LUCs in this zone reached a stage that is challenging to recover from human intervention. This is a very dangerous area that is constantly threatening the ecological stability and balance of the whole region. In these regions, the production, living, and ecological functions of land have been extensively developed and utilized, resulting in particularly intense land-use conflicts. In the process of governance, the definition of the status of PLES function is followed, in which the production function is the basis, the living function is the purpose, and the ecological function is the guarantee [46]. The strictest control policies and the most powerful protective measures must be adopted to reshape the regional ecological protection barrier and curb its impact on the surrounding land. Occupation and destruction of the ecological environment and illegal occupation of basic farmland are strictly prohibited. No new construction land is allowed within the ecological protection red line. On the premise of ensuring ecological security, the government will rationally arrange ecological communities and green enterprises according to the territorial space planning.

5. Conclusions

This paper selected the UCC as a new study area from the view of the PLES, based on the leading function of land use and the situation of the study area. The PLESs were divided into four types: EPS, PES, LPS, and ES. Based on the LER assessment method, we established the LUCs model to analyze the spatial relationship and potential risk of LUCs in the past 20 years. The LUCs in the study area were divided into four degrees: stable controllable, basic controllable, basic out-of-control, and serious out-of-control. The conclusions are as follows:

(1) The land-use types in the UCC were divided into four spatial types. From 2000 to 2020, the PES was in the dominant position in the study area, followed by EPS, LPS, and ES. The rate and transfer intensity of LPS was highest among the spaces. It showed that during the process of rapid urbanization, the LPS mainly eroded the PES and EPS.

(2) The average conflict level of the UCC increased over the past 20 years. It poses a serious threat to the ecological environment of Chongqing. In the process of rapid urbanization, the area of basic and serious out-of-control zones increased and occupied almost 56%, while the area of stable and basic controllable decreased. The mutual occupation between the production, living, and ecological land made the LUCs and the man-land relations become increasingly tense.

(3) LUCs typically occur in specific areas. The population agglomeration and regional economic development positioning will cause conflicts to become out of control. The out-of-control conflict zones are gathered in concentrated areas of social and economic activities, due to the complex social and economic activities and convenient transportation facilities. The controllable conflict zones were gathered in high-altitude forest areas. These areas were typically hilly and mountains with remote geographical location, and the economy was relatively backward.

(4) The UCC faced the potential risk of the LUCs, but not seriously. The zones of low potential conflict, general potential conflict, high potential conflict, and extreme potential conflict account for 43.43%, 32.27%, 17.37%, and 6.93% in 2000, and 41.5%, 33.2%, 16.22%, and 9.08% in 2020. The areas of high and extreme potential conflicts were increased. The hot spots for LUCs' potential distribution were concentrated in the western and northern areas of the study area, especially in the north of Yubei District and the junction of Shapingba District and Beibei District.

Author Contributions: The co-authors together contributed to the completion of this article. Specifically, their individual contributions are as follows: conceptualization, T.L. and P.D.; software, P.D., Y.S. and F.Y.; methodology, T.L., P.D. and Y.S.; validation, T.L. and F.Y.; formal analysis, T.L. and P.D.; resources, T.L.; data curation, P.D. and Y.S.; writing—original draft preparation, T.L., P.D. and Y.S.; writing—review and editing, T.L., F.Y., Y.L. and Y.W.; visualization, T.L., F.Y. and P.D.; supervision, T.L., F.Y., Y.L. and C.W. All authors have read and agreed to the published version of the manuscript.

Funding: This research was funded by the Key Program of the National Social Science Foundation of China (Grant No. 20&ZD095), the Science and Technology Research project of Chongqing Education Commission (Grant No. KJQN202102103), the Humanities and Social Sciences Research Project of Chongqing Education Commission (Grant No. 21SKGH308), the Chongqing Social Science Planning Social Organization Project (Grant No. 2021SZ26), and the National Social Science Youth Foundation of China (Grant No. 18CJL031).

Institutional Review Board Statement: Not applicable.

Informed Consent Statement: Not applicable.

Data Availability Statement: The data presented in this study are available on request from the author.

Conflicts of Interest: The authors declare no conflict of interest.

References

1. Song, J.; Meng, J.; Zhu, L.; Cheng, H. Spatial-temporal pattern of land use conflict in China and its multilevel driving mechanisms. *Sci. Total Environ.* **2021**, *801*, 149697.
2. Zou, L.; Liu, Y.; Wang, J.; Yang, Y.; Wang, Y. Land use conflict identification and sustainable development scenario simulation on China's southern cost. *J. Clean. Prod.* **2019**, *238*, 117899. [CrossRef]
3. Zhen, Y.; Jian, C.; Ting, W.; Tao, S. Promoting industrialization development by urbanization-reflection and analysis of urban industrialization road based on Shunde. *Adv. Geogr. Sci.* **2013**, *32*, 1814–1824.

4. Hui, E.C.M.; Bao, H. The logic behind conflicts in land acquisitions in contemporary China: A framework based upon game theory. *Land Use Policy* **2013**, *30*, 373–380. [CrossRef]
5. Hui, T.; Jia, L.; Kui, L.; He, Z. Study on Spatial Division of Tourism Urbanization Area Based on the Concept of PLE space—A Case Study of Mayangxi Eco-tourism Area. *Humanist. R.* **2016**, *31*, 153–160.
6. Song, J.; Meng, J.; Zhu, L. Spatial and temporal analyses of potential land use conflict under the constraints of water resources in the middle reaches of the Heihe River. *Land Use Policy* **2020**, *97*, 104773.
7. Zong, S.; Hu, Y.; Zhang, Y.; Wang, W. Identification of land use conflicts in China's coastal zones: From the perspective of ecological security. *Ocean Coast. Manag.* **2021**, *213*, 105841.
8. Li, S.; Zhu, C.; Lin, Y.; Dong, B.; Chen, B.; Si, B.; Li, Y.; Deng, X.; Gan, M.; Zhang, J.; et al. Conflicts between agricultural and ecological functions and their driving mechanisms in agroforestry ecotone areas from the perspective of land use functions. *J. Clean. Prod.* **2021**, *317*, 128453.
9. Zou, L.; Liu, Y.; Wang, Y. Research progress and prospect of land-use conflicts in China. *Prog. Geogr.* **2020**, *39*, 298–309. [CrossRef]
10. Zhu, C.; Xiang, F.; Zhi, H.; Bin, M.; Yi, L. Study on spatial conflict risk assessment and zoning optimization of land use in Nanchang city. *World Geogr. Res.* **2021**, *30*, 533–545.
11. Dong, G.; Ge, Y.; Jia, H.; Sun, C.; Pan, S. Land Use Multi-Suitability, Land Resource Scarcity and Diversity of Human Needs: A New Framework for Land Use Conflict Identification. *Land* **2021**, *10*, 1003.
12. Vonderdunk, A.; Gretregamey, A.; Dalang, T.; Hersperger, A.M. Defining a typology of peri-urban land-use conflicts: A case study from Switzerland. *Landsc. Urban Plan.* **2011**, *101*, 149–156. [CrossRef]
13. Liu, D.; Ma, X.; Gong, J.; Li, H. Functional identification and spatio-temporal pattern analysis of production-living-ecological space in watershed scale: A case study of Bailongjiang Watershed in Gansu. *Chin. J. Ecol.* **2018**, *37*, 1490–1497.
14. Mannarini, T.; Roccato, M.; Russo, S. The false consensus effect: A trigger of radicalization in locally unwanted land uses conflicts? *J. Environ. Psychol.* **2015**, *42*, 76–81.
15. Zhao, X.; Tang, F.; Zhang, P.T.; Hu, B.Y.; Xu, L. Dynamic simulation and characteristics analysis of county production-life-ecological space conflict based on CLUE-S model. *Acta Ecol. Sin.* **2019**, *39*, 5897–5908.
16. Wang, J.; Yu, D.; Lu, Y.; Zhang, T.; Zheng, Y. Recognition and analysis of land use conflicts at county level based on "Production-Living-Ecological" suitability. *J. Nat. Resour.* **2021**, *36*, 1238–1251.
17. Bao, W.; Yang, Y.; Zou, L. How to reconcile land use conflicts in mega urban agglomeration? A scene-based study in the Beijing-Tianjin-Hebei region, China. *J. Environ. Manag.* **2021**, *296*, 113168.
18. Reyes-Garcia, V.; Orta-Martinez, M.; Gueze, M.; Luz, A.C.; Paneque-Galvez, J.; Macia, M.J.; Pino, J.; Taps, B.S.T. Does participatory mapping increase conflicts? A random evaluation in the Bolivian Amazon. *Appl. Geogr.* **2012**, *34*, 650–658.
19. Lin, G.; Jiang, D.; Fu, J. Spatial conflict of production-living-ecological space and sustainable-development scenario simulation in Yangtze River Delta agglomerations. *Sustainability* **2020**, *12*, 2175.
20. Ma, X.; Tang, J. A review of the research on land use conflict in coastal zones based on CiteSpace atlas analysis. *Urban Plan. Des.* **2017**, *11*, 42–50.
21. Zhou, D.; Lin, Z.; Lim, S.H. Spatial characters and risk factor identification for land use spatial conflicts in a rapid urbanization region in China. *Environ. Monit. Assess.* **2019**, *191*, 677. [PubMed]
22. Calvert, K.; Mabee, W. More solar farms or more bioenergy crops? Mapping and assessing potential land-use conflicts among renewable energy technologies in eastern Ontario, Canada. *Appl. Geogr.* **2015**, *56*, 209–221.
23. Brown, G.; Raymond, C.M. Methods for identifying land use conflict potential using participatory mapping. *Landsc. Urban Plan.* **2014**, *122*, 196–208.
24. De Groot, R. Function-analysis and evaluation as a tool to assess land use conflicts in planning for sustainable, multi-functional landscapes. *Landsc. Urban Plan.* **2006**, *75*, 175–186.
25. Torre, A.; Melot, R.; Magsi, H.; Bossuet, L.; Cadoret, A.; Caron, A.; Darly, S.; Jeanneaux, P.; Kirat, T.; Pham, H.V.; et al. Identifying and measuring land-use and proximity conflicts: Methods and identification. *SpringerPlus* **2014**, *3*, 85.
26. Adam, Y.O.; Pretzsch, J.; Darr, D. Land use conflicts in central Sudan: Perception and local coping mechanisms. *Land Use Policy* **2015**, *42*, 1–6. [CrossRef]
27. Reuveny, R.; Maxwell, J.W.; Davis, J. On conflict over natural resources. *Ecol. Econ.* **2011**, *70*, 698–712.
28. Karimi, A.; Brown, G. Assessing multiple approaches for modelling land-use conflict potential from participatory mapping data. *Land Use Policy* **2017**, *67*, 253–267.
29. Zou, L.; Liu, Y.; Wang, J.; Yang, Y. An analysis of land use conflict potents based on ecological-production-living function in the southern coastal area of China. *Ecol. Indic.* **2021**, *122*, 107297.
30. Lin, G.; Fu, J.; Jiang, D. Production–Living–Ecological Conflict Identification Using a Multiscale Integration Model Based on Spatial Suitability Analysis and Sustainable Development Evaluation: A Case Study of Ningbo, China. *Land* **2021**, *10*, 383.
31. Gao, Y.; Wang, J.; Zhang, M.; Li, S. Measurement and prediction of land use conflict in an opencast mining area. *Resour. Policy* **2021**, *71*, 101999.
32. Li, Y.; Liu, C.; Li, Y. Identification of Urban Functional Areas and Their Mixing Degree Using Point of Interest Analyses. *Land* **2022**, *11*, 996.

33. Yan, H.; Cheng, T.; Guo, Z.; Sheng, H.; Yong, Q.; Liang, S.; Hao, Z. Spatial Conflict Measurement in Rapid Urbanization Areas Based on Geographical Perspective—A Case Study of Changsha-Zhuzhou-Xiangtan Urban Agglomeration. *Nat. J. Resour.* **2014**, *29*, 1660–1674.
34. Li, X.W.; Fang, C.L.; Huang, J.C.; Mao, H.Y. The urban land use transformations and associated effects on eco-environment in northwest China arid region: A case study in Hexi Region, Gansu Province. *Quat. Sci.* **2003**, *23*, 280–290.
35. Cui, J.; Zang, S.Y. Regional disparities of land use changes and their eco-environmental effects in Harbin-Daqing-Qiqihar Industrial Corridor. *Geogr. Res.* **2013**, *32*, 848–856.
36. Yang, Q.K.; Duan, X.J.; Wang, L.; Jin, J.Y. Land Use Transformation Based on Ecological-production-living Spaces and Associated Eco-environment Effects:A Case Study in the Yangtze River Delta. *Sci. Geogr. Sin.* **2018**, *38*, 97–106.
37. Li, L.; Wen, D.; Juan, C.; Juan, C.; Wan, H.; Fang, J.; Qiu, H. Analysis of spatial conflict of "production-life-ecological" in the process of rapid urbanization in Pingtan island. *Sci. Resour.* **2017**, *39*, 1823–1833.
38. Hong, Z.; Erqi, X.; Hui, Z. Classification and spatial pattern of ""production-life-ecological" land use" in China. *J. Resour. Sci.* **2015**, *37*, 1332–1338.
39. Zhu, C.; Xiang, F.; Xiao, W. Analysis of landscape pattern changes around expressways in Wuhan based on LUCC. *Soil Bull.* **2019**, *50*, 35–44.
40. Jia, P.; Guo, Z.; Cheng, T.; Yan, H. Spatial Conflict Measurement in Rapid Urbanization Areas Based on Ecological Security—A Case Study of Changsha-Zhuzhou-Xiangtan Urban Agglomeration. *Nat. J. Resour.* **2012**, *27*, 1507–1519.
41. Xiao, C.; Xiao, L.; Peng, W.; Fu, K. Study on Ecological Risk of Spatial Conflict of Ecological Migrants in Arid Area—A Case Study of Hongsibu District, Ningxia. *Hum. Geogr.* **2018**, *33*, 106–113.
42. Lovell, S.T.; DeSantis, S.; Nathan, C.A.; Olson, M.B.; Me'ndez, V.E.; Kominami, H.C.; Erickson, D.L.; Morris, K.S.; William, B. Morris Integrating agroecological and landscape multifunctionality in Vermont: An evolving framework to evaluate the design of agroecosystems. *Agric. Syst.* **2010**, *103*, 327–341.
43. Qiu, G.; Wang, Y.; Guo, S.; Niu, Q.; Qin, L.; Zhu, D.; Gong, Y. Assessment and Spatial-Temporal Evolution Analysis of Land Use Conflict with Urban Spatial Zone: Case of the Su-Xi-Chang Region. *Sustainability* **2022**, *14*, 2286. [CrossRef]
44. Jing, W.; Yu, K.; Wu, L.; Luo, P. Potential Land Use Conflict Identification Based on Improved Multi-Objective Suitability Evaluation. *Remote Sens.* **2021**, *13*, 2416.
45. Adams, W.M.; Brockington, D.; Dyson, J.; Vira, B. Managing tragedies: Understanding conflict over common pool resources. *Science* **2003**, *302*, 1915–1916. [PubMed]
46. Zhou, D.; Xu, J.; Lin, Z. Conflict or coordination? Assessing land use multi-functionalization using production-living-ecology analysis. *Sci. Total Environ.* **2017**, *577*, 136–147. [PubMed]
47. Cheng, Z.; Zhang, Y.; Wang, L.; Wei, L.; Wu, X. An Analysis of Land-Use Conflict Potential Based on the Perspective of Production–Living–Ecological Function. *Sustainability* **2022**, *14*, 5936.
48. Li, G.; Fang, C.; Wang, S. Exploring spatiotemporal changes in ecosystem-service values and hotspots in China. *Sci. Total Environ.* **2016**, *545–546*, 609–620.
49. Getis, A.; Ord, K. The Analysis of Spatial Association by Use of Distance Statistics. *Geogr. Anal.* **1992**, *24*, 189–206.
50. Moran, P.A.P. Notes on continuous stochastic phenomena. *Biometrika* **1950**, *37*, 17–23. [CrossRef]
51. Ord, J.K.; Getis, A. Local Spatial Autocorrelation Statistics: Distributed Issues and an Application. *Geogr. Anal.* **1995**, *27*, 286–306. [CrossRef]
52. Ho, P. The 'credibility thesis' and its application to property rights: (in) secure land tenure, conflict and social welfare in China. *Land Use Policy* **2014**, *40*, 13–27.
53. Yu, A.T.W.; Wu, Y.Z.; Shen, J.H.; Zhang, X.L.; Shen, L.Y.; Shan, L.P. The key causes of urban–rural conflict in China. *Habitat Int.* **2015**, *49*, 65–73.
54. Ide, T.; Frohlich, C. Socio-environmental cooperation and conflict? A discursive understanding and its application to the case of Israel and Palestine. *Earth Syst. Dyn.* **2015**, *6*, 659–671.
55. Zhou, X.; Lu, X.; Lian, H.; Chen, Y.; Wu, Y. Construction of a spatial planning system at city-level: Case study of "integration of multi-planning" in Yulin City, China. *Habitat Int.* **2017**, *65*, 32–48.
56. Scherwitz, E. The EU's approach towards Asia: Regional cooperation and regional conflicts in the past and today. *Asia Eur. J.* **2016**, *14*, 367–382.
57. Shen, H.J.; Li, X.; Luo, J.M.; Chau, K.Y. One country, two strata: Implications of social and cultural conflicts of Chinese outbound tourism to Hong Kong. *J. Destin. Mark. Manag.* **2017**, *6*, 166–169. [CrossRef]
58. Yu, B.; Lv, C.H. The progress and prospect of land use conflicts. *Prog. Geogr.* **2006**, *25*, 106–115.
59. Wang, J.Y.; Zou, L.L.; Li, M.G. Identification and governance of potential land use conflicts in tourist resorts based on "Production–Living–Ecological" suitability. *Trans. CSAE* **2019**, *35*, 279–288.

land

Article

Study on the Trade-Off Synergy Relationship of "Production-Living-Ecological" Functions in Chinese Counties: A Case Study of Chongqing Municipality

Ling Cheng [1,2], Haiyang Cui [1,*], Tian Liang [3], Dan Huang [3], Yuanxia Su [3], Zhiyong Zhang [4,5,*] and Chuanhao Wen [6]

1 School of Economics, Guizhou University, Guiyang 550025, China; chengling477447@mail.gufe.edu.cn
2 College of Business Administration, Guizhou University of Finance and Economics, Guiyang 550025, China
3 College of Public Management, Chongqing Finance and Economics College, Chongqing 401320, China; xhxhxlth@163.com (D.H.)
4 Upper Yangtze River Economic Research Center, Chongqing Technology and Business University, Chongqing 400067, China
5 Guizhou Academy of Social Sciences, Guiyang 550002, China
6 School of Economics, Yunnan University, Kunming 550002, China
* Correspondence: hosanna2004@163.com (H.C.); 2020651920@email.ctbu.edu.cn (Z.Z.)

Abstract: The trade-off and coordinated development of the "production-living-ecological" functions (PLEFs) of an area is an important basis for the optimization of the spatial pattern of the land, and helps to promote the harmonious symbiotic relationship between human beings and nature. This paper combines ecological niche theory, a coupling coordination model, and a trade-off synergy model to construct an evaluation index system for Chinese county PLEFs. Quantitative methods were used to measure spatiotemporal evolution characteristics, trade-off synergy of PLEFs in 38 counties in Chongqing, China, and the coupling coordination degree between PLEFs. The results showed that the ecological niche width of the "production-ecological" function revealed an overall growth trend. However, there was a mismatch in regional development of the "production-ecological" function, showing dislocation characteristics of "high in the west and low in the east" and "high in the east and low in the west." The niche width of the life function is similar to the comprehensive niche width of PLEFs, showing the characteristics of fluctuation and partition change. PLEFs and both the aforementioned functions showed distribution characteristics of "high in the west and low in the east," with the whole moving towards the stage of coordinated coupling, of which the "production-living" function has the highest coupling level. The functional coupling coordination degree of "production-living-ecological" is generally manifested as "high in the west and low in the east," and changing from the primary stage of imbalance to well-coordinated development. The "production-ecological" and "living-ecological" functions are in low-level imbalance in the primary and moderate coordination stages. Additionally, the evolution trend of the "production-ecological" and the "living-ecological" functions are similar, showing alternating and fluctuating development characteristics. Overall, in the past 20 years, Chongqing's "living-production" function has changed from a trade-off model to a collaborative development relationship, and the "living-ecological" function is generally based on a collaborative development relationship. The "ecological function" is manifested as a trade-off constraint relationship. Moreover, the coordinated development level of "living-production," "living-ecological," and "production-ecological" functions in the central urban area has been greatly improved, while counties have gradually shown different degrees of trade-offs.

Keywords: PLEFs; spatiotemporal evolution; ecological niche theory; coupling coordination; Chongqing City

Citation: Cheng, L.; Cui, H.; Liang, T.; Huang, D.; Su, Y.; Zhang, Z.; Wen, C. Study on the Trade-Off Synergy Relationship of "Production-Living-Ecological" Functions in Chinese Counties: A Case Study of Chongqing Municipality. *Land* **2023**, *12*, 1010. https://doi.org/10.3390/land12051010

Academic Editors: Li Ma, Yingnan Zhang, Muye Gan and Zhengying Shan

Received: 19 March 2023
Revised: 19 April 2023
Accepted: 26 April 2023
Published: 4 May 2023

1. Introduction

Land space is the environmental place for national survival and the primary carrier of all economic and social activities. According to the use of land space, it can be divided into three types: production-living-ecological space (PLES), which represent production functions, living functions, and ecological functions, respectively. Additionally, PLEFs are a coupling coordination outcome of the economy–society–nature system. With the acceleration of global economic and social development, industrialization, and urbanization, changes in land use patterns have led to a decline in global ecosystem services [1] coupled with increasing global climate change, population growth, resource scarcity, environmental pollution, ecological degradation, and land-use imbalance [2,3], with rising competition and conflict of the PLES [4,5]. The functional problems and contradictions of "production-living-ecological" are becoming increasingly prominent [6]. There is a serious impact on regional and even global sustainable development. Additionally, this may lead to the degradation of ecological well-being provided by the earth to humans [7,8]. A reasonable territorial and spatial development pattern is essential in estimating long-term sustainable development, harmony between human beings and nature, and coordination of economic and social activities in regions [9–11]. Therefore, strengthening the research on the effective trade-off and synergy between the PLEFs in a county and accurately grasping the synergistic relationship between PLEFs is crucial for carrying out territorial spatial planning. Promoting the coordinated and sustainable development of PLEFs is an important goal to optimize the country's spatial patterns. The multilateral cooperation mechanism of PLEFs is a significant path to a better future of harmonious coexistence between human beings and nature.

As research on this progresses, people realize that PLEFs are in a nonequilibrium state and have a synergistic or trade-off relationship [12]. Synergy refers to the two functions cooperating and gaining together under the control of key factors so that the PLEFs develop in an orderly way and have a spatial fusion effect [13]. The trade-off refers to the situation where the two functions are a trade-off, which easily causes the overall dysfunction of the "production-living-ecological" and have a spatial conflict or competition [14]. Due to the complexity of ecosystems and the diversity of human use and interference with ecosystems, there are complex and diverse dynamic interactions of different PLEFs [15]. This is usually manifested as mutually reinforcing synergies and trade-offs [16,17]. As the leading spatial carrier of regional development, land complements the high-quality development of the region and is scientifically reasonable [18]. The territorial spatial planning system is necessary for sustainable regional development [19]. Due to spatial functions' complex and overlapping nature, it is challenging to define spatial types [20], so more attention is paid to the discussion of functions [21]. The multifunctionality of land use can be divided into production function land, living function land, and ecological function land [22]. Scholars' research on PLEFs mainly focuses on the theoretical connotation [23,24]. Classification system construction [25,26], coordination [27,28], functional measurement [29], pattern evolution [30], and other aspects are studied. Some scholars have constructed "production-living" from the aspects of material space for land use production, social space for life security, and natural space for ecological supply—an "ecological" three-dimensional index system exploring the relationship between land use change and ecological and environmental benefits [31]. Many believe that any single type of land use is a superposition of multiple functions, and the choice between different land use types should be a game and conflict between different functions and goals [32,33].

Current research on the interaction of PLEFs mainly focuses on the quantity and spatial structure of PLEFs [34,35]. The spatiotemporal evolution law of PLEFs is mostly based on a qualitative perspective of the functional interaction relationship of PLEFs. A few scholars quantitatively express it with the help of the coupled coordination model. However, mathematical statistical methods ignore the spatial pattern process, and pay more attention to the quantitative association characteristics and insufficient expression of the spatial interaction relationship. Considering the difference and complexity of PLEFs in the same space, the interrelationship between PLEFs is subject to the stage of socioeconomic development and the law of geographical differentiation. The trade-offs and synergistic relationships are also significantly different [36]. Therefore, based on the perspective of synergy theory, this paper discusses the functional coupling and coordination relationship between PLEFs and analyzes its pattern evolution characteristics. There is a basis for promoting the integrated and coordinated development of regional PLEFs and optimizing the spatial layout of the national territory. The research on the coupling and coordination of PLEFs is to diagnose the utilization status and efficiency between production functions, living functions, and ecological functions, evaluate the implementation effect of spatial planning in a particular area, and improve the deficiencies of spatial planning to promote the harmonious integration between economy, society and ecology and ensure the green and healthy development of the region [37,38]. Land use is explored by applying coupled coordination models [39–43]. The study area involves countries and urban agglomerations at the macro level [44], provinces and municipalities at the medium level [45,46], and counties and townships at the micro level. The typical research areas of the coupling and coordination analysis of the PLES mainly include the Yellow River Basin [47] and the Yangtze River Economic Belt [48]. From the research status of scholars at home and abroad, most scholars focus on the static pattern evaluation of PLEFs, and need to reveal more about the dynamic evolution rule. It is challenging to portray the continuity and dynamics of PLEFs and the state of a particular moment and predict the development trend [49,50].

Based on the above discussion and review, this paper selects 38 districts (counties) of Chongqing as the research targets and collects data for 2000, 2010, and 2020. Based on the socioeconomic data, ecological environment data, and land use data, the coupling and coordination degree between PLEFs and the two functions was measured to analyze the spatiotemporal pattern characteristics of PLEFs and the interaction relationship between PLEFs in Chongqing, and combined with ecological niche theory and spatial autocorrelation analysis, explore its pattern evolution characteristics, reveal its temporal and spatial evolution law, and discuss the "production-living-ecological" from the spatial and quantitative levels The functional synergy/balance relationship provides a new perspective for clarifying the functional interaction relationship of "production-living-ecological," to provide practical reference and reference for the optimization of land space layout.

2. Materials and Methods

2.1. Study Area

Chongqing (28°10′–32°13′ N, 105°11′–110°11′ E) is located in southwest China, the fourth city under central government jurisdiction, a comprehensive transportation hub in southwest China, and an economic center in the upper reaches of the Yangtze River. Chongqing's terrain is dominated by mountains and hills, with a total area of 82,400 square kilometers. There are 38 districts and counties under Chongqing's jurisdiction. These 38 counties can be geographically divided into "one district and two groups," namely, the main urban area of Chongqing (UC), the urban agglomeration of Wulingshan District in southeast Chongqing (SE), and the urban agglomeration of the Three Gorges Reservoir Area in northeast Chongqing (NE) (Figure 1). Chongqing is an important area for the implementation of the strategy of large-scale development of the western region, a necessary strategic node for the construction of the national "Belt and Road" initiative, and an important modern manufacturing base with unique location advantages and resource advantages, rapid industrial development and obvious industrial process [51]. However, due to problems such as single production mode, disorderly spatial expansion, extensive land use, industrial homogenization, and unreasonable resource allocation, the dual structure of urban and rural areas is prominent, the ecological environment is severely damaged, and the development of the region is unbalanced and uncoordinated, so the level of sustainable economic, social and ecological development needs to be improved [52,53]. In recent years, Chongqing has adopted rural measures of large cities and large rural areas, which has enabled the towns and villages under the jurisdiction of Chongqing to achieve sound development. As of 2020, in Chongqing municipal towns, the chemical rate reached 69.46%. However, land degradation is severe due to its location in the mountainous area of southwest China, high ecological fragility, unreasonable development, and excessive use of pesticides. In 2020, the area of soil erosion in the city reached 25,142.46 km^2, which also caused a contradiction between production status, lifestyle, and ecological benefits in urban and rural development. Coordinating the interrelationship between the spatial PLEFs in urban and rural development is crucial, so Chongqing needs to use the PLES to clarify the spatial functions of the national territory. The spatiotemporal characteristics of development are replanned, the urban and rural ecological pattern is coordinated, and the relationship between the PLES in the county is reconstructed to enhance PLEFs effective synergy.

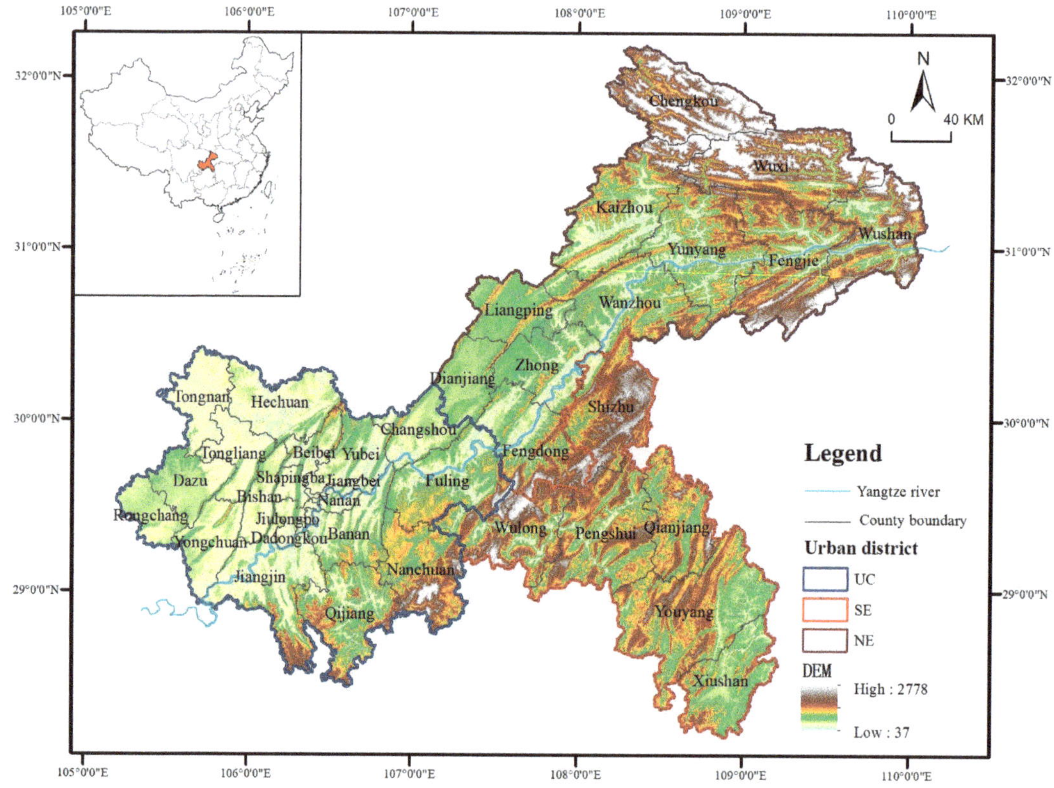

Figure 1. Location map of the study area.

2.2. Research Methods

2.2.1. Construction of PLEFs Evaluation Index System

The county land space is a complex community involving multiple factors such as society, economy, and geographical environment. Based on the connotation and classification criteria of the PLEF function [54,55], it is crucial to consider the interaction of PLEFs. Furthermore, the study explores the combined relationship between dominant functions and multiple functions [56], drawing on synergy theory and niche situation theory [57–59], combined with Chongqing's unique geographical location, land use characteristics, and socioeconomic development status, based on the principles of comprehensiveness, regional, scientific, and innovative, from the aspects of material space for land use production, social space for living security, and natural space for ecological supply. Aiming at intensive and efficient production space, moderate livable living space, and beautiful ecological space, reference is made to the research experience of relevant scholars [60–64]. This article combined with the science, representativeness, comprehensiveness, and data availability of indicators covering agricultural production, economic development, living security, social services, ecological pressure, and ecological bearing, the land space functions of the county were identified as PLEFs, and we selected 17 PLEFs indicators. After standardizing by the extreme value method, the preliminary weight of each index is calculated by the entropy method. Then, the index weight is corrected by the analytic hierarchy method (AHP). The final weighted sum yields the combined weights of each indicator (Table 1).

Table 1. Evaluation index system of PLEFs in counties in China.

Target Layer	Guidelines Layer	Metrics Layer	Indicator Interpretation and Calculation Method	Attribute	Entropy Method Weights	AHP Weight	Synthesis Weight
produce function	agriculture produce	Grain yield	Grain production/arable land area, t/km^2	+	0.0186	0.0606	0.0396
		Land reclamation rate	Arable land area/total area of regional land, %	+	0.0194	0.0606	0.0400
		Food availability per capita	Total food production/population, t/10,000 people	+	0.0256	0.0909	0.0583
	economy develop	Average gross industrial production value	Gross Industrial Production/Total Land Area of the Region, 100 million yuan/km^2	+	0.2026	0.2121	0.2074
		Financial contribution rate	Local general budget revenue/total regional land area, billion yuan/km^2	+	0.3641	0.2727	0.3184
		Economic density	Gross regional product/total land area of the region, 100 million yuan/km^2	+	0.3697	0.3030	0.3364
living function	living guarantee	Proportion of housing area	Total area of land used in settlements/area, %.	+	0.3873	0.2652	0.3262
		Density of land used for transportation	Highway mileage/total area of land in the area, km/km^2	+	0.0620	0.1189	0.0905
		population density	Total population/land area, people/km^2	–	0.0066	0.0324	0.0195
	society serve	There are hospitals per 10,000 people Number of beds	Number of hospital beds/total population, sheets/10,000	+	0.1455	0.1621	0.1538
		Total retail sales of consumer goods per capita	Total retail sales of consumer goods/total population 100 million yuan/10,000	+	0.1763	0.1945	0.1854
		Number of books in public libraries per capita	Number of books in public libraries/total population Book/person	+	0.2224	0.2269	0.2246
ecological function	ecological pressure	Agricultural fertilizer input intensity	Agricultural fertilizer application rate/cultivated land area, kg/km^2	–	0.0525	0.1154	0.0840
		Land degradation index	Degraded land area/total regional land area, %	–	0.0514	0.1154	0.0834
	ecological conservation	Water per capita	Total water resources/total population, m^3/person	+	0.5852	0.3847	0.4850
		Forest cover	Forest land area/total area of regional land, %	+	0.1694	0.2308	0.2001
		Habitat abundance index	Status of biodiversity in the region	+	0.1414	0.1538	0.1476

Note: "+" is a positive indicator, and "−" is a negative indicator.

(1) Production function refers to the ability of the county land space to provide products for human beings and increase social wealth. Therefore, this paper characterizes the production function from the two criteria of agricultural production and economic development and selects grain yield, land reclamation rate, per capita grain ownership, average industrial production value, financial contribution rate, economic density, and other indicators to characterize the production function, all of which are positive indicators, and the larger the value, the stronger the production function of the county land space.
(2) Living function refers to the ability of the county land space to provide a living environment for human beings, which can reflect the income, consumption structure,

and welfare of residents. Therefore, this paper characterizes the living function from the two standard layers of life security and social services and selects indicators such as the proportion of housing area, the density of transportation land, the population density, the number of hospitals per 10,000 people, the number of beds, the total retail sales of social consumer goods per capita, and the number of books in the per capita public library collection. Population density is a negative indicator, and the larger the value, the weaker the living function of the county. The rest are positive indicators, and the larger the value, the stronger the county living function.

(3) Ecological function refers to the ability of the county land space to provide ecological products and services for residents, as well as to respond to external interference, realize self-repair, maintain ecosystem stability, conserve water sources, and maintain ecological security. Therefore, this paper characterizes the strength of county ecological functions from the aspects of ecological pressure and ecological supply. The two indicators of agricultural fertilizer input intensity and land degradation index were selected to characterize the ecological stress function of the county, both of which were negative indicators. The larger the value, the weaker the ecological function of the county. The three indicators of per capita water resources, forest coverage, and habitat abundance index were selected to characterize the capacity of county ecological supply, all of which were positive indicators. The larger the value, the stronger the rural ecological function.

2.2.2. Functional Niche Width Evaluation Model of PLEFs

Niche situation theory is one of the critical theories of ecology, in which the width of the ecological niche indicates the degree of resource utilization by a species. It has been widely used in urban geography, urban economy, and other fields, mainly including urban competition research [65–67] and sustainable use of arable land [68], and less used in the field of "sunshine" space research. Referring to the methods of [69], a functional [70] "production-living-ecological" model was constructed. In this way, the competitiveness of PLEFs in the research area was explored, and the larger the niche width, the higher the dominant position and the more assertive the competitiveness.

$$N_i = \frac{(S_i + A_i P_i)}{\sum_{j=1}^{n}(S_j + A_j P_j)} \times W \tag{1}$$

In the formula, $i, j = 1, 2, \cdots n$; N_i indicates the ecological niche of the evaluation unit indicator i; S_i and P_i represent the state and potential of the research unit i; S_j and P_j represent the state and potential of the research unit j; A_i and A_j are dimensional conversion coefficients; $S_j + A_j P_j$ is an absolute niche; W is a weight. The indicators of PLEFs measure the state in 2000, 2010, and 2020. The potential is measured by each indicator's average annual growth rate from 2000 to 2020, with a study period interval of 10 years and a dimensional conversion coefficient of 0.1.

$$M_i = \sum_{j=1}^{n} N_{ij} \times W_j \tag{2}$$

In the formula, M_i indicates the comprehensive ecological niche of the PLEFs of the region i, weighted by the three dimensions of production, living, and ecological function; j represents the dimension, N_{ij} indicates the ecological niche of the region i in the dimension j, W_j represents the weight of the dimension j. Considering the importance of the synergistic development of each function, the weight of these three dimensions is set to $1/3$.

2.2.3. PLEFs Functional Coupling Coordination model

With the PLEFs coupling coordination model, the coupling coordination degree of PLEFs was measured to reflect PLEFs in the study area and spatially the level of collaborative development.

$$C = 3\left\{ \frac{P_i \times R_i \times E_i}{(P_i + R_i + E_i)^3} \right\}^{1/3} \tag{3}$$

$$T = \alpha P_i + \beta R_i + \gamma E_i \tag{4}$$

$$D = (CT)^{1/2} \tag{5}$$

where: C indicates the degree of coupling, $C \in [-1, 1]$. The higher the value of C, the stronger the correlation between functions; the lower the value of C, the weaker the correlation between functions. The coupling degree is divided into four stages [71] (Table 2). P_i, R_i, and E_i represent the total scores of PLEFs, respectively. T represents the coordination coefficient, and α, β, and γ represent the undetermined coefficients of PLEFs set to 1/3, respectively. D refers to coupled co-scheduling, reflecting the degree of coordination between various functions. The coupling coordination degree is divided into 7 levels (Table 3). The higher the value of D, the better the coupling and coordination between various functions, and the higher the overall efficiency; The lower the value of D, the poorer the coordination between various functions, and the more obvious the conflict.

Table 2. PLEFs and the coupling degree between the production-living.

Coupling Phase	Coupling Type	Degree of Coupling
Coordinated coupling period	Highly coupled	0.8–1.0
Run-in period	II Moderately coupled	0.5–0.8
Antagonistic period	III Primary coupling	0.3–0.5
Low coupling period	IV low-level coupling	0.0–0.3

Table 3. PLEFs and the coupling coordination type division between the two functions.

Coordination Phase	Coupling Coordination Type	Degree of Coordination
Coordinated period	Highly coordinated	0.8–1.0
	II Well coordinated	0.6–0.8
	III moderate coordination	0.5–0.6
	IV Essential coordination	0.4–0.5
Periods of misalignment	V Primary dysregulation	0.3–0.4
	VI moderate outrage	0.2–0.3
	VII Severe dysregulation	0.0–0.2

Upon further analysis of the coupling coordination degree between the two pairs of PLEFs, the coupling coordination model of PLEFs is refined into C_1, C_2, and C_3:

$$C_1 = 2\left\{ \frac{P_i \times R_i}{(P_i + R_i)^2} \right\}^{1/2} C_2 = 2\left\{ \frac{P_i \times E_i}{(P_i + E_i)^2} \right\}^{1/2} C_3 = 2\left\{ \frac{R_i \times E_i}{(R_i + E_i)^2} \right\}^{1/2} \tag{6}$$

$$T_1 = \alpha P_i + \beta R_i,\ T_2 = \alpha P_i + \gamma E_i,\ T_3 = \beta R_i + \gamma E_i \tag{7}$$

$$D = (CT)^{1/2} \tag{8}$$

Referring to existing research [72]. the "production-living" coupling coordination model T_1: $\alpha = \beta = 0.5$; In the "production-ecological" coupling coordination model T_2: $\alpha = 0.55$, $\gamma = 0.45$; In the "living-ecological" coupling coordination model T_3: $\beta = 0.55$, $\gamma = 0.45$.

2.2.4. The PLEFs Weigh the Degree of Synergy

Estimating the synergy/trade-off relationship between production, living, and ecology further clarifies the spatial heterogeneity and correlation between PLEFs in Chongqing and the synergistic/trade-off development relationship between them from the quantitative and spatial perspectives. Ecosystem services trade-off degree (*ESTD*) is based on linear data fitting and is a method for reflecting the interrelationships between ecosystem services. The calculation method is:

$$ESCI_i = \frac{(ES_{ia} - ES_{ib})}{ES_{ib}} \tag{9}$$

$$ESTD_{ij} = \frac{\left(\frac{ESCI_i}{ESCI_j} - \frac{ESCI_j}{ESCI_i}\right)}{2} \tag{10}$$

In the formula, ES_{ia} and ES_{ib} represent the values of the i type ES at moments a and b, respectively; $ESCI_i$ is the index of change for the i type ES; $ESCI_j$ is the index of change for the j type ES; $ESTD_{ij}$ represents the trade-off synergy between ES types i and j, $ESTD$ negative value indicates that the ES between categories i and j is a trade-off relationship, $ESTD$ positive value indicates that the ES between categories i and j is a synergistic relationship; The magnitude of the absolute value of $ESTD$ reflects the level of trade-off and synergy.

2.3. *Data Sources and Preprocessing*

2.3.1. Data Sources

Socioeconomic data are mainly from the Chongqing Statistical Yearbook (Website: http://tjj.cq.gov.cn/, accessed on 6 June 2022); The data on total water resources and forest coverage are from the Chongqing Bulletin of Water Resources (Website: http://slj.cq.gov.cn/, http://tjj.cq.gov.cn/, accessed on 6 June 2022) and the Chongqing Forest Resources Bulletin Website: http://lyj.cq.gov.cn/, http://tjj.cq.gov.cn/, accessed on 6 June 2022); land degradation index, habitat abundance index [8,69]) and other statistics are provided by the Geospatial Data Cloud of the Chinese Academy of Sciences (Website: http://www.gscloud.cn/, accessed on 6 June 2022). For land areas of 30 m × 30 m, data extraction calculations are performed. Due to the difference in statistical caliber, the data of Wansheng District and Shuangqiao District in 2000 and 2010 were classified as Qijiang District and Dazu District, respectively, based on the existing administrative district planning. For individual missing data, SPSS software was used.

2.3.2. Data Preprocessing

PLEFs index system involves 17 indicators, such as grain yield, land reclamation rate, and average industrial output value. Based on different efficacy, the extreme value method was used to eliminate the influence of dimensionality to normalize the positive and negative indicators and their average annual growth so that the processed value range was [−1,1]. The closer the value is to 1, the higher its power.

$$X'_{ij} = \frac{X_{ij} - Min_j}{Max_j - Min_j} \quad \text{(Positive indicator)} \tag{11}$$

$$X'_{ij} = \frac{Max_j - X_{ij}}{Max_j - Min_j} \quad \text{(Negative indicator)} \tag{12}$$

Formula: X'_{ij} indicates i the normalized value of j the regional indicator; X_{ij} indicates the actual value of i the regional j indicator; and Max_j indicates Min_j the j maximum and minimum values of the first indicator.

2.4. Technical Route

This paper adheres to the problem-oriented and goal-oriented and adopts the technical route of "problem-raising—previous preparation—problem analysis—problem-solving" (Figure 2). The problem-raising stage mainly explains the research background, status, objectives, method, and value. The previous preparation stage mainly includes literature review, data collection and preprocessing, research framework, and constructing the index system of PLEFs. The problem analysis stage mainly analyzed the spatiotemporal evolution characteristics, evenness and coordination degree, and trade-off synergy of PLEFs in the study area. The contributions and shortcomings of the research are analyzed to optimize the spatial pattern of the county land effectively. The problem-solving stage mainly includes research contributions and research conclusions. The solution measures that effectively weigh the functional relationship of PLEFs in the coordinated county are proposed. The contributions and shortcomings of the research are analyzed to optimize the spatial pattern of the county land effectively.

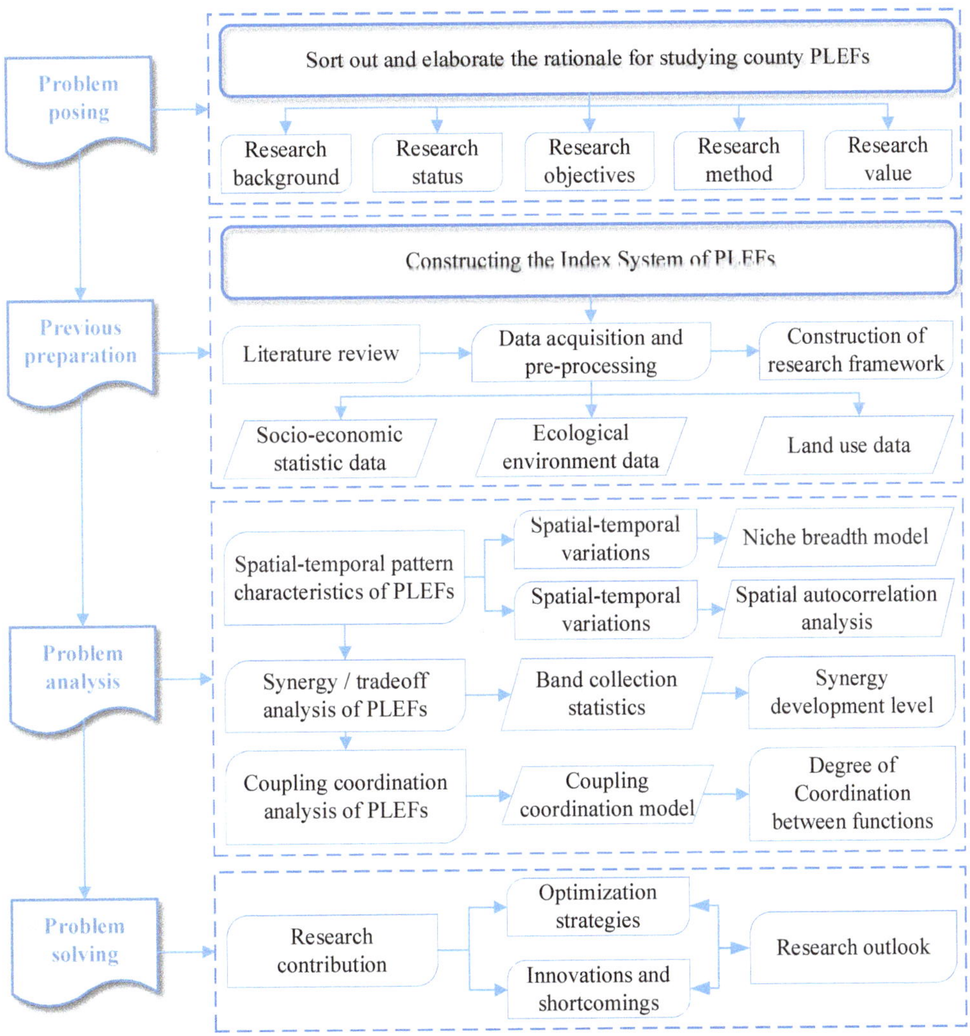

Figure 2. Research technology roadmap.

3. Results and Analysis

3.1. Characteristics of the Spatiotemporal Bureau of PLEFs in Chongqing City

3.1.1. Ecological Niche Evolution

The ecological niche width of PLEFs in 38 counties in Chongqing in 2000, 2010 and 2020 was measured by data visualization mapping analysis (Figure 3). To better present the mapping effect, the calculated values of a production function, living function, and total niche width of Yuzhong District are much higher than those of other counties, so the data of Yuzhong District are not added to the chart.

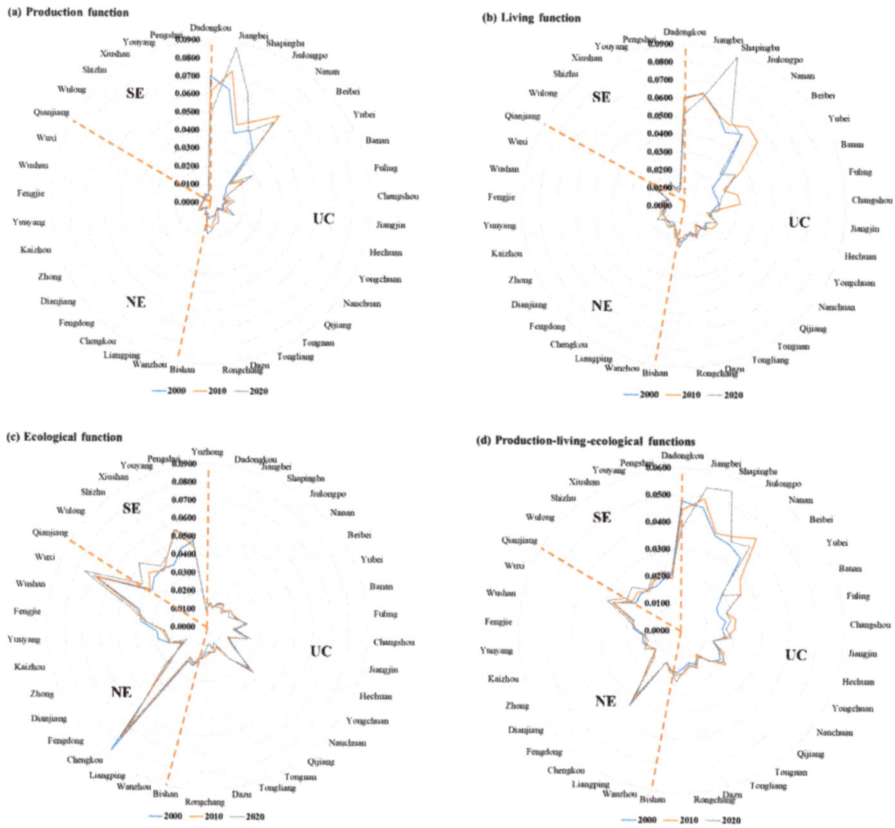

Figure 3. Functional niche width of "production-living-ecological" in Chongqing from 2000 to 2020.

1. Production function niche width

From 2000 to 2020, the ecological niche width of production functions in Chongqing generally showed a growth trend, and the high-value areas were mainly located in the main urban areas, of which Yuzhong District (0.5437–0.4104) consistently ranked first, as the only area in Chongqing that did not have agricultural production and achieved complete urbanization, with the financial industry, commerce and trade as the leading industries, the financial contribution rate, The economic density advantage is obvious. Dadukou District (0.0703–0.0480) production function niche width downward trend is significant, as the former bearing place of Chongqing iron and steel industry, due to the overall relocation of heavy steel, the industrial output value and fiscal revenue have a lot of impact, is trying to explore a new path for the transformation and development of the old industrial base in the urban area, and the rest of the areas show stable growth with slight fluctuations.

The low-value area is mainly located in the northeast, among which the well-developed and representative ones are Wanzhou (0.0060–0.0073), Matjiang (0.0062–0.0080), Liangping (0.0053–0.0073), as well as the southeast of Yu, of which Xiushan (0.0048–0.0059) is relatively prominent, but also limited by topography, traffic factors, although most of the production function ecological niche width continues to expand, in agricultural production is relatively good, and industrial development. In the future, the urban agglomeration of the Three Gorges Reservoir Area in the NE and the urban agglomeration in the Wuling Mountain Area in the SE and the main urban area will be further promoted, forming an industrial pattern of differentiated development and coordinated development.

2. Ecological niche width of living functions

From 2000 to 2010, most areas generally showed an expansion trend, while the main urban areas showed a contraction or flat trend from 2010 to 2020, and the southeast and NE regions continued to expand as a whole. Specifically, areas with high ecological niche width of living functions were concentrated in the main urban areas, including Yuzhong District (0.2351–0.1722), Jiangbei District (0.0636–0.0.0611), and Shapingba District (0.0553–0.0878). and Banan District (0.0185–0.0257) showed improved living functions significantly. The living security and social service levels in housing, transportation, medical care, culture, and other aspects of high-value areas are generally better, and the quality of living index is high, so it is livable. The low-value areas were mainly located in the southeast and northeast of Chongqing, among which the living function of Zhongxian (0.0120–0.0164) and Xiushan (0.0124–0.0154) was significantly improved. The variation in the ecological niche width of living functions showed two states. The main urban areas with a high urbanization rate and more reasonable industrial structures tended to shrink or have stable development. The living space was relatively saturated and may be squeezed by production and ecological space. Compared with the backward urban agglomerations in the SE and NE, the urbanization rate needs to be improved. Various living guarantees such as medical care, housing, and transportation need to be improved and living space will be further expanded.

3. Ecological function niche width

From 2000 to 2020, the ecological niche width of ecological functions in Chongqing showed a steady development trend. High-value areas are primarily concentrated in the SE and NE, represented by Chengkou County (0.0864–0.0705), Wuxi County (0.0681–0.0755), Youyang County (0.0455–0.0567), etc., with high forest coverage and abundant water resources. The ecological carrying capacity is relatively strong, and the ecological advantages are significant, consistent with its geographical advantages and development positioning along the mountains and rivers. They cultivate urban corridors along the river, promote the ecological priority green development of urban agglomerations in the Three Gorges Reservoir Area in NE, coordinate the development of characteristic resources such as ethnic customs, history and humanities, and ecological health care, and promote Chongqing integrated development of cultural tourism in urban groups in Wuling Mountainous Area in southeast China. Low-value areas are primarily located in the main urban area, limited by factors such as scarcity of land resources and industrial development positionings, such as Yuzhong District (0.0096–0.0091), Shapingba District (0.0150–0.0122), and Dazu District (0.0140–0.0093). Taking Dazu District as an example, it is a robust industrial area dominated by hardware, automobile, intelligence, and other industries. Natural resources are relatively scarce. Ecological carrying capacity is poor and ecological space is squeezed by production space: Nanchuan District (0.0332–0.0367), Fuling District (0.0223–0.0235). Nanchuan District is also known as the "urban back garden," with rich ecological resources and good ecological functions.

4. The comprehensive niche width of PLEFs

From 2000 to 2020, the total niche width of PLEFs in Chongqing showed the characteristics of fluctuating changes. The high-value areas were mainly concentrated in the main urban area, with Yuzhong District (0.2628–0.1972), Shapingba District (0.0370–0.0547), Jiangbei District (0.0460–0.0535) and other areas as the representative areas, mainly due to industrial agglomeration, high level of economic development, with production and living functions to drive the overall efficiency of the region, or create production and lifetime value with higher ecological functions and improve the overall coordination level of PLEFs, such as Chengkou County (0.0333–0.0293) and Wuxi County (0.0269–0.0293) and other parts of the NE. The comprehensive ecological niche range of PLEFs in most areas is similar. A trend of fluctuation and alternating development within a specific range shows that PLEFs constantly compete and squeeze each other. It is necessary to optimize the industrial structure further and promote "production-living-ecological" spatial synergy, orderly integration, and balanced development.

3.1.2. Temporal and Spatial Correlation Analysis

Through Formula (6), the global spatial autocorrelation values of Chongqing's PLEFs can be calculated to test whether the study area has an overall spatial correlation. Table 4 shows that the test statistic Z score of the Global Moran's I Index of PLEFs from 2000 to 2020 was significantly more significant than the test threshold of 1.65, and the significance test of 10% was passed. The results show that the spatial distribution of production, living, and ecology in Chongqing from 2000 to 2020 shows significant spatial autocorrelation, showing a significant agglomeration distribution trend. Among them, the Z score of ecological function fluctuated slightly.

Table 4. Global Moran's I Index of functional niche width of "production-living-ecological" in Chongqing, 2000–2020.

Year	Global Moran's I			Z-Score		
	Production Features	Living Function	Ecological Function	Production Features	Living Function	Ecological Function
2000	0.1056	0.2845	0.5872	4.1926	5.2513	5.9643
2010	0.1717	0.4645	0.6007	5.0186	5.8778	5.8525
2020	0.2119	0.4768	0.6405	5.1811	6.0668	6.0518

From the time change trend perspective, the global Moran's I Index of PLEFs from 2000 to 2020 was more significant than 0 and gradually increased. It shows a spatial similarity in the adjacent counties, and the autocorrelation and spatial agglomeration distribution phenomena are also increasing, showing a positive spatial correlation. Among them, the global Moran's I of ecological function increased from 0.5872 to 0.6405, which is more spatial correlation than production and living functions; The living function is second, and the production function is lower.

Due to the differences in the spatial autocorrelation level between different spatial units and adjacent areas in the study area, the Local Moran's I Index was further used to analyze the spatial correlation between the spatial distribution of PLEFs in Chongqing and the neighboring areas with the help of the significance LISA map.

- Production function

From 2000 to 2020, Chongqing's spatial distribution of production functions changed little. In 2000, the production function area had 5 hot and 12 cold spots. From 2010 to 2020, the number of cold spots increased to 14; the hot spot area remained unchanged. The proportion of the two increased, which showed that the agglomeration characteristics of production function space were strengthened. The low-value agglomeration areas of production functions are mainly located in the northeast and southeast of Chongqing. The high-value agglomeration areas are located in the main urban area.

- Living function

From 2000 to 2020, Chongqing's spatial distribution of living functions fluctuated. In 2000, there were 5 hot spots and 12 cold spots in the living function area. From 2010 to 2020, cold spots increased to 13 and then decreased to 11. The living function in SE showed the agglomeration characteristics of strengthening and weakening, and the hot spots remained unchanged. The agglomeration characteristics of living function space show a weakening trend. The high-value agglomeration area is located in the core area of the central city, and the proportion of the number of the two decreases.

- Ecological function

From 2000 to 2020, Chongqing's spatial distribution of ecological function changed significantly (Figure 4). In 2000, there were, respectively, ecological functions of 5 hot spot and 15 cold spot areas, and from 2010 to 2020, the number of hot spots increased from 6 to 7, and the number of cold spots decreased from 14 to 13 and added 1 sub-cold spot area—Kaizhou District. In general, the characteristics of low-value accumulation in the main urban area were weakened, and the ecological function improved significantly. The clusters of high-value agglomeration areas were distributed in the NE and SE. The agglomeration characteristics in the NE were weakened, the agglomeration characteristics in SE were strengthened, and the proportion of quantity strengthened the spatial agglomeration characteristics of ecological function.

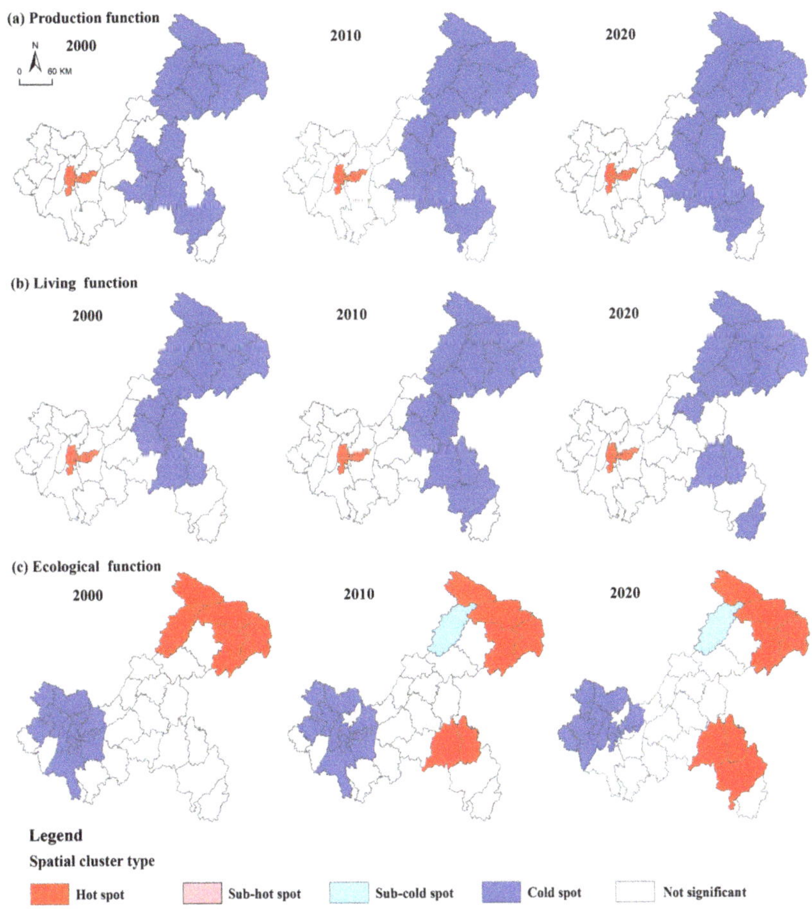

Figure 4. Spatial autocorrelation clustering diagram of PLEFs area in Chongqing from 2000 to 2020.

3.2. Coupling and Coordination Analysis of PLEFs in Chongqing City

The coordinated development level of PLEFs in various districts and counties in Chongqing is worthy of our in-depth understanding. This article calculates the PLEFs in 38 counties of Chongqing in 2000, 2010, and 2020, as well as the coupling degree and coordination scheduling between the two functions (Figures 5 and 6). Referring to the existing research results of Chen et al. (2006) [73] and Wang and Tang (2018) [74], the frequency statistical analysis method is used to group statistics on the coupling and coordination degree of PLEFs and the two functions to draw the evolution curve of the coupling and coordination degree of PLEFs in Chongqing from 2000 to 2020 (Figures 7 and 8).

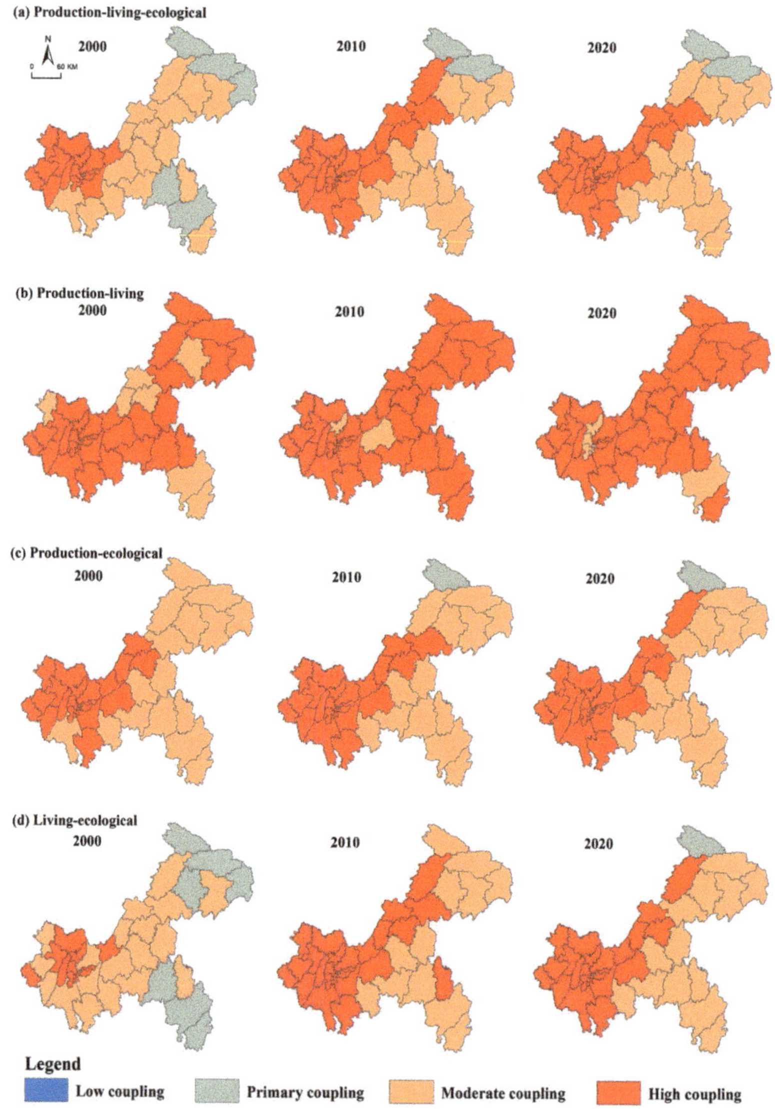

Figure 5. Spatial distribution of PLEFs function and coupling degree of two functions in Chongqing from 2000 to 2020.

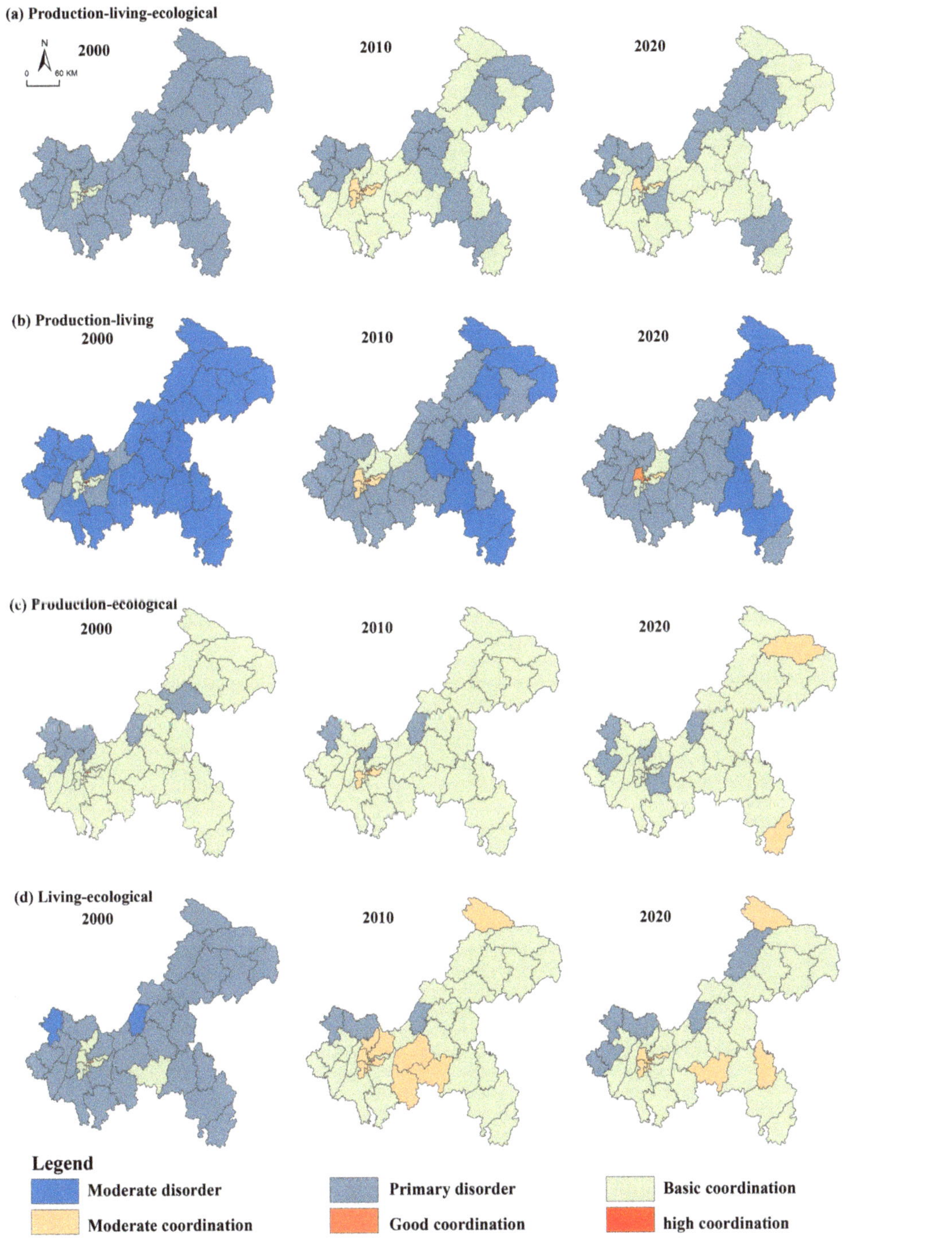

Figure 6. Spatial distribution of PLEFs and the coupling coordination degree of two functions in Chongqing from 2000 to 2020.

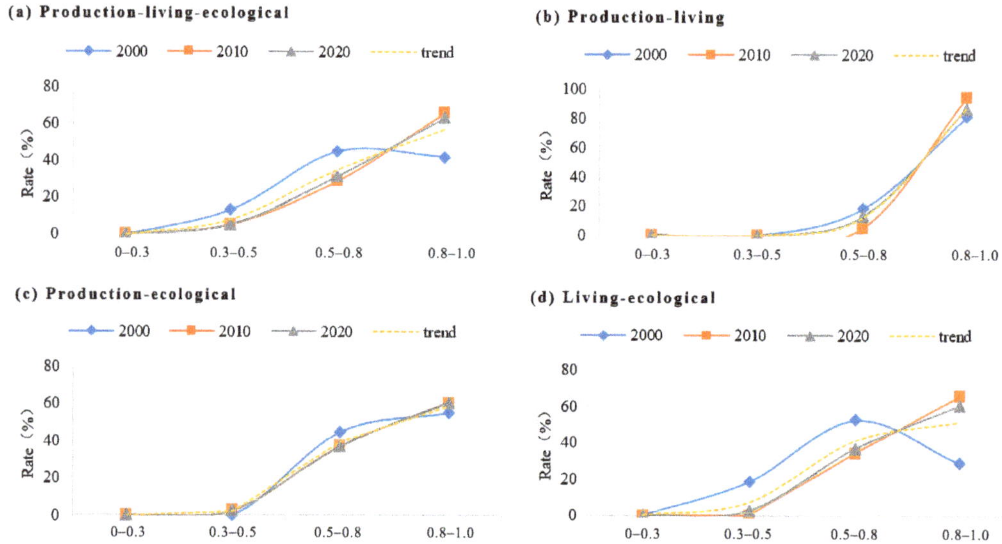

Figure 7. Evolution curve of PLEFs function and coupling degree of two functions in Chongqing from 2000 to 2020.

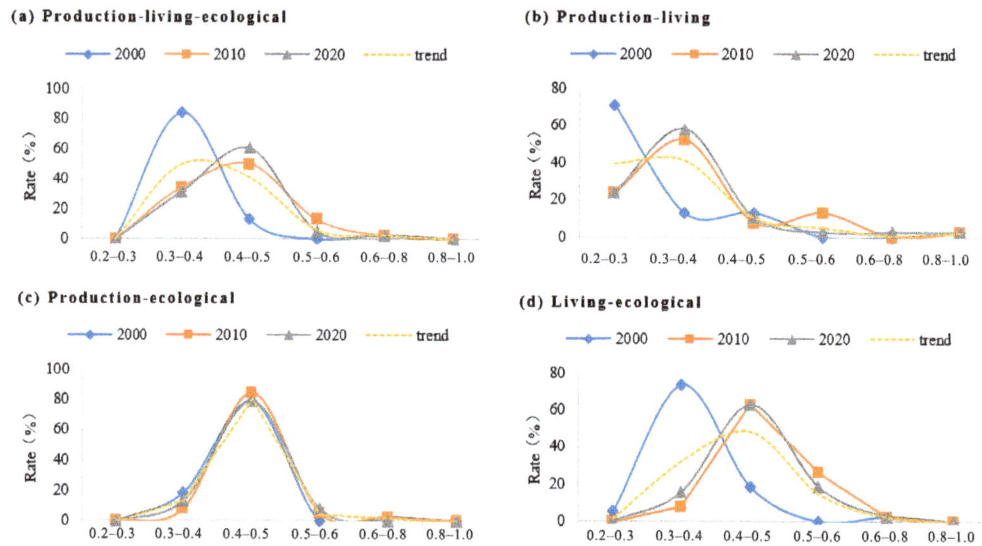

Figure 8. Evolution curve of PLEFs function and coupling coordination degree of two functions in Chongqing from 2000 to 2020.

3.2.1. Coupling and Coordination Analysis of PLEFs

From the spatial point of view, the functional coupling degree of "production-living-ecological" functions in Chongqing presents a distribution trend of "high in the west and low in the east," and the regional differentiation is significant. The areas with high coupling degrees are mainly concentrated in urban areas. The areas with low coupling degrees are concentrated in Chengkou County, Wuxi County, and Wushan County in the NE. From the time perspective, the overall functional coupling of "production-living-ecological" in Chongqing has improved. The highly coupled areas with the main urban area as the core have continuously spread to the NE. From 2000 to 2020, Chongqing's average coupling degree of "production-living-ecological" increased from 0.73 to 0.82. It gradually entered the coordinated coupling from the running-in period, showing a solid benign promotion effect. However, some counties have fluctuated, such as Kaizhou County, which has undergone the "first increase and then decrease" of "running-in-coordinated coupling-running-in."

Specifically, in 2000, the overall degree of medium and high coupling was the mainstay, the highly coupled areas were mainly concentrated in the main urban areas, and the degree of interaction between functions was substantial. Among them, the primary coupling area, the medium coupling area and the highly coupled area accounted for 13.16%, 44.74%, and 42.11%, respectively. At this stage, Chongqing focused on economic development and urban construction, which squeezed ecological functions into production and living functions. In most regions, the coupling degree of PLEFs is in the intermediate and advanced coupling stages. In 2010, the proportion of regions in the primary and medium coupling stages decreased to 5.26% and 28.92%, respectively. The proportion of highly coupled areas increased significantly to 65.79%. Dianjiang County, Liangping County, and Kaizhou County were the main increased areas. At this stage, Chongqing gradually realized the guaranteed role of ecological functions in production and living functions, further improved the intensity of ecological functions, and promoted the degree of coupling. In 2020, compared with 2010, although the coupling degree in local areas decreased, the highly coupled areas decreased to 63.16%. The moderately coupled areas increased to 31.58%, mainly dominated by the coordinated coupling period. Chongqing advocated sustainable development and combined economic development with ecological construction at this stage. It promoted the peaceful development of the coupling degree of PLEFs in Chongqing through ecological compensation policies such as "returning farmland to the forest."

The coupling and coordination degree of PLEFs shows the spatial distribution trend of "high in the west and low in the east" as a whole. From the primary offset to the well-coordinated evolution trend, the average level of coupling coordination increased from 0.45 to 0.48. High-value areas are mainly developed with the main urban area as the core and outer. Specifically, in 2000, the coupling coordination degree of Chongqing was between [0.31, 0.72]; the highest value was in Yuzhong District and the lowest was in Yunyang County. Yuzhong District mainly uses production and living functions to drive the overall efficiency of the area. Including three types: primary disorder, essential coordination, and good coordination, it accounts for 84.21%, 13.16%, and 2.63%, respectively. However, the coordination stage accounts for a small proportion, most of which are in the primary disorder stage. In 2010, the coupling coordination degree of Chongqing was between [0.35, 0.67], and the highest value was in Yuzhong District, but the degree of coordination decreased. It is mainly related to the fact that the ecological function is lower than that of production and ecological function, and the lowest value is in Tongnan District. It included four types: primary disorder, essential coordination, moderate coordination, and good coordination, accounting for 34.21%, 50.00%, 13.16%, and 2.63%, respectively. Compared with 2000, the proportion of primary dysregulation stages has decreased significantly, and moderate coordination phases have begun to appear. However, half of the regions are still in the primary coordination stage. In 2020, the coupling co-scheduling in Chongqing was between [0.35, 0.67], with the highest value being Yuzhong District and the lowest value being Dazu District. It included four types: primary disorder, essential coordination, moderate coordination, and good coordination, accounting for 31.58%, 60.53%, 5.26%, and 2.63%, respectively. Compared with 2010, the proportion of essential and moderate coordination has increased slightly, and the overall coordination is still in the primary coordination stage. Combined with the evolution curve of coupling coordination (Figures 3–6), the coupling coordination degree of PLEFs in Chongqing presents an inverted "U" shaped evolution trend. It shows that the gap in the degree of coupling and coordination of PLEFs will be further widened as a whole. When it reaches a certain extent, it will change into a shrinking trend, gradually coordinate the development, and finally, achieve spatial synergy.

3.2.2. Coupling and Coordinated Analysis of Two Functions

- "Production-living" function

From 2000 to 2020, the average coupling degree of the "production-living" function in Chongqing increased from 0.94 to 0.96, which has always shown a substantial mutual promotion effect. However, it showed a fluctuating development trend, and the proportion of moderately coupled areas changed from 18.42–5.26–13.16%, which was highly coupled. The regional proportion changed from 81.58% to 94.74% to 86.84%. It relates to relocating industries in Jiulongpo, Shapingba, and other areas or transforming the industrial center of gravity in Youyang and other places to tertiary industries such as tourism, reducing the production function, causing a decrease in the coupling degree. This decreases in coupling. "production-living" is the essential demand function of residents' living and economic development. The two promote each other and continue to develop in the direction of coordinated coupling.

The functional coupling coordination degree of "production-living" shows the development characteristics of "high in the west and low in the east" and from moderate imbalance to high coordination, showing a solid phenomenon of spatiotemporal differentiation. The average level of coupling coordination increased from 0.31 to 0.37, and the high-value areas mainly developed from the main urban area as the core to the outer layer. Specifically, in 2000, the coupling coordination degree was between [0.20, 0.96]; the highest value was Yuzhong District, and the lowest value was Wuxi County, among which there were mainly moderate imbalances, primary imbalance and essential coordination sum. There were four types of highly coordinated coordination, accounting for 71.05%, 13.16%, 13.16%, and 2.63%, respectively. The geographical differentiation is apparent, showing the development of fault-type and zonal stages of "essential coordination-high coordination," most of which are in the moderate imbalance stage. In 2010, the coupling coordination degree was between [0.25, 0.86]; the highest value was in Yuzhong District, but the coordination degree decreased. The lowest value was in Wuxi District, which mainly had moderate imbalance, primary imbalances, and essential coordination. There were five types, moderate coordination and high coordination, accounting for 23.68%, 52.63%, 7.89%, 13.16%, and 2.63%, respectively. Compared with 2000, the proportion of moderate dysregulation stage decreased significantly and began There is a moderate coordination phase, with most areas in the primary disorder stage; In 2020, the coupling coordination degree was between [0.26, 0.86]; the highest value was in Yuzhong District, and the lowest value was in Youyang County, among which there were mainly moderate imbalance, primary imbalance, essential coordination, moderate coordination, There are six types of good coordination and high coordination, accounting for 23.68%, 57.89%, 10.53%, 2.63%, 2.63%, and 2.63%, respectively. Compared with 2010, geographical differentiation was further intensified, and the scope of primary imbalance expanded. For example, the longevity zone regressed from the prior coordination stage to the primary disorder stage, and the proportion of good coordination increased. The evolution curve of coupling coordination also shows that the level of "production-living" function coordination has a downward trend. In general, the "production-living" function shows the difference between high coupling and low coordination; with the continuous improvement of urbanization and economic development levels, living conditions have been improved, and living standards have been further improved. The two complement each other and restrict each other. It is consistent with the conclusions obtained from the above production characteristics and ecological niche widths. The development speed between the two is not matched, and the living function needs to be stronger than the production function.

- "Production-ecological" function

From 2000 to 2020, the average coupling degree of the "production-ecological" function in Chongqing increased from 0.81 to 0.83, which was in the period of coordinated coupling. High-value areas are mainly concentrated in the main urban area and the northeast Yu region, such as Dianjiang County, Liangping County, and Zhong County. The overall number included three types: primary coupling, moderate coupling, and coordinated coupling, and the regional proportions varied from 0% to 2.63%, respectively −2.63%, 44.74–36.84–36.84%, 55.26–60.53–60.53%. 2010–2020, through environmental remediation, pay attention to ecological function restoration, the transformation of production methods, strict management of production pollutants, and encourage the development of green production methods so that the production the coupling degree of "ecological" function has been significantly improved.

The coupling coordination degree of the "production-ecological" function shows the development characteristics of "high in the east and low in the west" and from primary imbalance to medium coordination. The average level of coupling coordination increased from 0.43 to 0.44, and the high-value areas were mainly distributed in SE and NE. Specifically, in 2000, the coupling coordination degree was between [0.38, 0.64]; the highest value was Yuzhong District, and the lowest value was Beibei District, among which there were mainly primary imbalance, essential coordination, and moderate. The three types, accounting for 18.42%, 78.95% and 2.63% respectively, are mainly in the primary coordination stage. In 2010, the coupling coordination degree was between [0.38, 0.61]; the highest value was in Yuzhong District, but the coordination degree decreased, and the lowest value was in Tongnan District, among which there were mainly primary imbalances, essential and moderate coordination and good coordination of four types, accounting for 7.89%, 84.21%, 5.26%, and 2.63%, respectively. Spatial synergy has been further enhanced compared to 2000. In 2020, the coupling coordination degree was between [0.35, 0.59]; the highest value was in Yuzhong District, and the lowest value was in Dazu Region, among which there were mainly primary imbalance, essential coordination and medium. There were three types of degree coordination, accounting for 13.16%, 78.95%, and 7.89%, respectively. The phenomenon of geographical differentiation intensified, and the synergy between SE and NE was enhanced. Some areas of the main urban area show low-value synergy and scattered distribution, and the overall "production-ecological" function shows the characteristics of alternating changes and small fluctuations.

- "Living-ecological" function

From 2000 to 2020, the average functional coupling degree of the "living-ecological" in Chongqing increased from 0.69 to 0.84, gradually entering the period of coordinated coupling from the running-in period. Although there are fluctuations in local areas, such as Qianjiang, which has experienced the "running-in-coordinated coupling-running-in" process, it generally shows prominent growth characteristics. Among them are three types: primary coupling, moderate coupling, and coordinated coupling, and the regional proportion changes are 18.42–0–2.63%, 52.63–34.21–36.84%, 28.95–65.79–60.53%. With the development of the economy, people pay more and more attention to the impact of the ecological environment on the quality of living and then adopt measures such as comprehensive ecological improvement and the construction of a green and livable environment to promote harmonious coexistence between man and nature. However, in recent years, due to the construction of a livable environment, excessive emphasis on construction and neglect of management has weakened ecological functions and low-level development. Therefore, the coupling degree of the "living-ecological" function has a downward trend.

The functional coupling coordination degree of the "living-ecological" shows the overall development characteristics of "high in the east and low in the west" and from moderate imbalance to well-coordinated development. The average level of coupling coordination increased from 0.38 to 0.46. Compared with "production-living" and "production-Ecological" functions, the level of coupling coordination increased significantly, and the level of coordinated development of different regions in different years differed. In 2000, the coupling coordination degree was between [0.28, 0.66]. The highest value was in Yuzhong District, and the lowest was in Tongnan District, among which there were mainly moderate imbalances, primary imbalance, essential coordination, and sound. There were four types of coordination, accounting for 5.26%, 73.68%, 18.42%, and 2.63%, respectively. Mainly in the primary co-offset phase; In 2010, the coupling coordination degree was between [0.36, 0.61]; the highest value was Yuzhong District, but the coordination degree decreased, and the lowest value was Tongnan District, among which there were mainly primary imbalances, essential coordination and moderate coordination and good coordination of four types, accounting for 7.89%, 63.16%, 26.32%, and 2.63%, respectively. Compared with 2000, more than half of the regions have entered the primary coordination stage, and the spatial synergy relationship has been further enhanced; In 2020, the coupling coordination degree was between [0.34, 0.63]; the highest value was Yuzhong District, and the lowest value was Dazu Region, among which there were mainly primary imbalance, essential coordination and moderate There were four types of coordination and sound coordination, accounting for 15.79%, 63.16%, 18.42% and 2.63%, respectively, the phenomenon of geographical differentiation intensified, the synergy between the SE and NE increased. Some areas of the main urban area regressed to the primary disorder stage and showed the characteristics of agglomeration distribution; the evolution trend of the "living-ecological" and "production-ecological" functions are similar, showing the characteristics of alternating and fluctuating development.

3.3. *Synergy Trade-Off Analysis of PLEFs in Chongqing City and County*

To deeply explore the trade-off synergy between production, living, and ecological functions in Chongqing, 38 counties were used as the essential units to study the PLEFs of counties in 2000, 2010, and 2020. The matrix distribution of PS, LS, and ES values can better express the spatial interaction relationship between the two functions adjacent to the region and the spatiotemporal evolution characteristics (Figure 9).

3.3.1. The Degree of Trade-Off Synergy of PLEFs of the County

As shown in Figure 9, from 2000 to 2010, 20 counties in Chongqing weighed off the synergy between "living and production," accounting for 52.63%. Among them, Jiangbei District has the strongest trade-off (−17.71). 18 counties showed synergistic relationships, accounting for 47.36%. The Dadukou District has the highest degree of synergy (6.96). For the "living-ecological" functional trade-off synergy, 18 counties showed a trade-off relationship, accounting for 47.36%. Among them, Jiangbei District still has the most robust trade-off relationship (−17.71); 20 counties showed synergistic relationships, accounting for 52.63%. Among them, Yubei District has the highest synergy (54.11). For the "production-ecological" functional trade-off synergy, 24 counties showed the trade-off relationship, accounting for 63.15%. Among them, Yuzhong District had the strongest trade-off (−14.01); 14 counties showed synergistic relationships, accounting for 36.84%. Among them, Yubei District still has the highest synergy (48.91).

Between 2010 and 2020, 18 counties in Chongqing were regarded as the "living-production" functional trade-off synergy. The proportion was 47.36%. Among them, Qijiang District has the strongest trade-off (−11.18). 20 counties showed synergy, accounting for 52.63%. Wulong District has the highest degree of synergy (86.49). For the "living-ecological" functional trade-off synergy, 18 counties showed a trade-off relationship, accounting for 47.36%. Among them, Zhongxian had the strongest trade-off (−10.30); 20 counties showed synergistic relationships, accounting for 52.63%. Jiangjin District had the highest synergy (23.42). For the "production-ecological" functional trade-off synergy, 23 counties showed a trade-off relationship, accounting for 60.52%. Wanzhou District had the strongest trade-off (−29.42); 15 counties showed synergy, accounting for 39.47%. Among them, Wulong District has the highest synergy (78.05).

3.3.2. The Trade-Off Synergy Relationship of PLEFs in the County from 2000 to 2020

In the first ten years, among the 38 counties in Chongqing, the trade-off between "living-production" and "living-ecological" functions in Jiangbei District was the strongest, and the coordination level of "living-ecological" and "production-ecological" functions in Yubei District was the highest. The "living-production" function in Dadukou District has the highest degree of synergy, and the trade-off relationship between the "production-ecological" function in Yuzhong District is the strongest. In the following decade, the counties with the strongest trade-off between the "living-production" function, "living-ecological" function, and "production-ecological" function were Qijiang District, Zhongxian County, and Wanzhou District. Wulong District has the highest degree of "living-production" function and "production-ecological" function synergy, and Jiangjin District has the highest degree of "living-ecological" function. On the whole, in the past 20 years, the "living-production" function in Chongqing has changed from a trade-off to a collaborative development relationship, the "living-ecological" function is generally based on the collaborative development relationship, and the trade-off constraint relationship dominates the "production-ecological" function. Moreover, the coordinated development level of "living-production," "living-ecological," and "production-ecological" functions in the central urban area has been dramatically improved. In contrast, the counties in SE and NE have gradually shown different degrees of trade-offs.

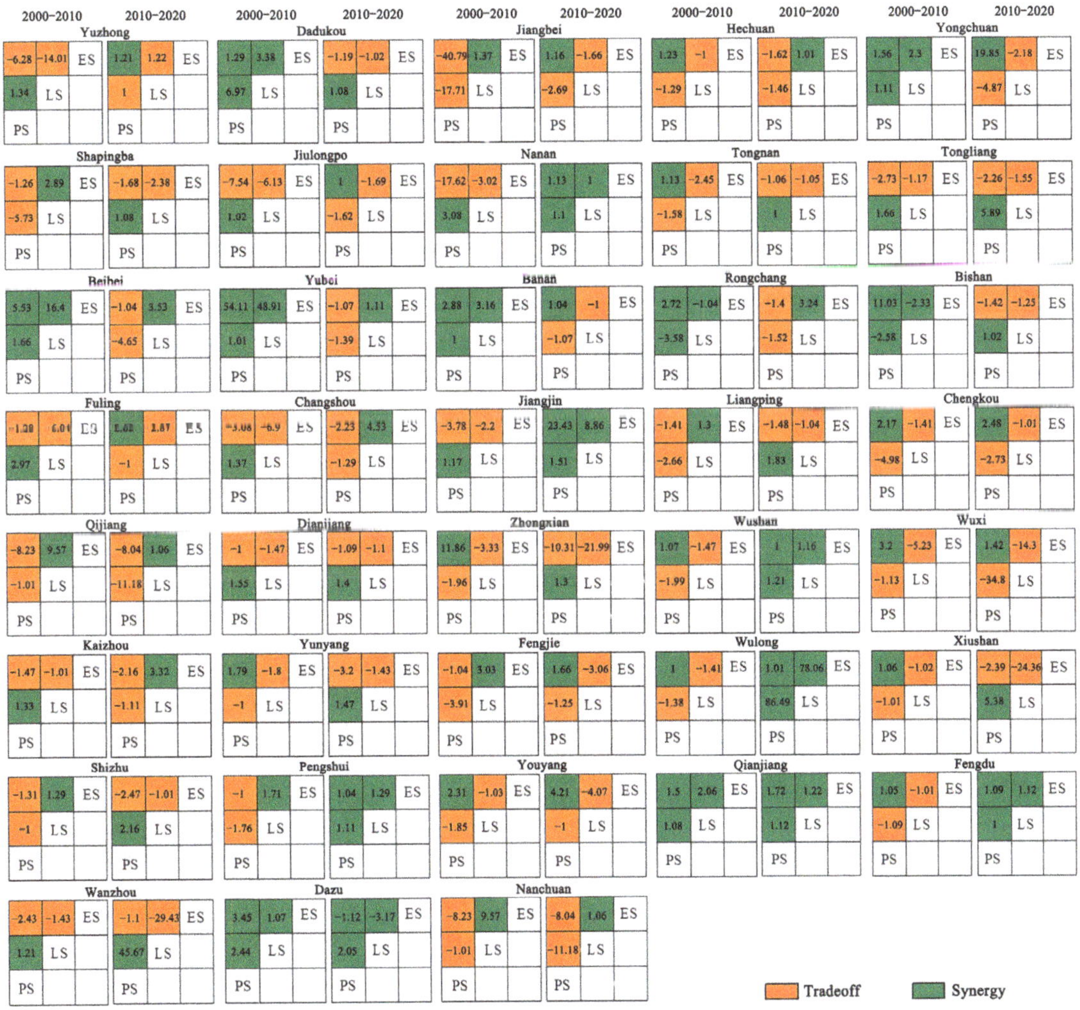

Figure 9. Evolution of PLEFs trade-off synergy in various counties in Chongqing.

4. Discussion

4.1. Contributions and Deficiencies

Based on the connotation and characteristics of PLEFs, this paper constructs the index system of PLEFs from various aspects, such as agricultural production, economic development, livelihood security, social services, ecological pressure, and ecological carrying capacity. Our research enriches the theoretical framework for optimal county land development and spatial patterns in China. Based on the connotation and characteristics of PLEFs, this paper constructs the index system of PLEFs from various aspects, such as agricultural production, economic development, livelihood security, social services, ecological pressure, and ecological carrying capacity. Previous studies have only emphasized the relationship between PLEFs, focusing on the impact of space function on land space utilization. However, the degree of synergy and balance between PLEFs needs to be discussed in depth [75]. Therefore, this paper focuses more on the discussion of the impact between functions, provides a new method for coordinating the relationship between human activity intensity in counties from the perspective of PLEFs [45,76,77], and provides a scientific basis for optimizing and adjusting the relationship between PLEFs according to local conditions.

(1) By introducing the ecological niche situation theory [78], the interrelationship between PLEFs was explored based on the microscale of the county [79]. Previous studies have explored PELFs from a single perspective, such as rural, regional, or urban ecological functions [80,81]. However, the territorial space of a given region involves both urban and rural areas. Therefore, this paper reveals the temporal and spatial evolution law from the entire county and combines GIS spatial analysis and correlation analysis to help explore the functional synergy/balance relationship of "production-living-ecological" from the spatial and quantitative levels, clarify the interaction of PLEFs, and provide a new perspective.

(2) As with many research results, the correlation between PLEFs in typical regions is measured. The dynamic changes and spatial differentiation patterns of the relationship between PLEFs in the study area are systematically studied. However, in this paper, the moderating effect model is further used to identify the driving role of several basic urbanization factors on the evolution of the relationship between PLEFs [82]. Adopting the composition of the interaction technology between different functions in PLEFs can help identify environmental factors that impact the spatial pattern of land development.

(3) In this paper, an evaluation model suitable for the functional niche width of "production-living-ecological" in the county was constructed; when measuring the width of a single ecological niche, each subsystem in the PLEFs is regarded as an independent unit, which effectively avoids interference and reveals the trade-off synergy between PLEFs in the county. The paper highlights the trade-offs and synergies between the PLEFs and supports assessing the spatial pattern of land development. Once again, the trade-offs and synergies between PLEFs are demonstrated, and the results are consistent with the current study [83]. This new technical framework can help provide theoretical guidance for regional sustainable development.

However, it should be pointed out that the research methods and research ideas of this paper still need to be improved:

(1) Due to the inconsistency of the statistical caliber of each county in the study area, coupled with the dilemma of data acquisition and limited technology, the data sources of some indicators have been replaced. Although explained in the text, there may also be the problem that the evaluation of PLEFs is not accurate enough, the research scale needs to be further refined, and measures and precise measures are implemented according to local conditions to make a systematic and overall evaluation of the coordinated development level of PLEFs.

(2) Dynamic changes with time accompany the functional relationship of "production-living-ecological" in the county. The selected indicators in this paper mainly focus on the current and past periods and the future-oriented indicators. The complex relationship between them also needs to be further explored, so the continuity and

dynamic transformation characteristics of the trade-off synergy of long-time series analysis must be further studied.

(3) Many biological processes and physical and chemical mechanisms in the PLES in the county. In the future, it is necessary to strengthen the exploration and analysis of biological, physical, and chemical processes and pay more attention to the interpretation of biological, physical, and chemical processes to clarify the driving mechanism of PLEFs better to improve the resilience and stability of the ecosystem and promote sustainable development under the effective synergy of PLEFs.

4.2. Policies and Recommendations

Based on the above analysis and demonstration, combined with the spatiotemporal pattern and evolution characteristics of PLEFs in Chongqing and the degree of coordination between functions, corresponding optimization strategies are proposed according to the economic level, human settlement environment, ecological pattern, and other conditions.

(1) For the main urban metropolitan areas with "production living" functional advantages, each district should take economic development as the premise, ecological protection as the foundation, and improving the quality of life as the goal, strengthen economic ties, coordinate the development, avoid low-level, repetitive construction of homogeneous and disorderly competition, and radiate and drive the economic development of the NE and SE. For the NE and SE, which has the functional advantages of "production ecology" and "living ecology," the NE and SE regions should utilize the advantages of the natural environment and landscape resources, encourage the development of a "low-carbon economy," complement and compete with the advantages of the main urban areas, and form a sizeable industrial pattern of differentiated and coordinated development, and explore the integrated development path of PLEFs.

(2) Overall, the development trend of PLEFs coupling coordination in Chongqing presents a benign trend. The "production-living" function is mainly low-level imbalance, so it is more feasible to improve the coordination of PLEFs. When formulating policy measures, we should focus on improving the people's living standards and strengthening infrastructure construction to improve the green living index. The main urban area should improve the livability of living space, promote the progress of new energy technology, upgrade and transform industrial industries, and promote supply-side structural reform. Use benign and green industrial development to meet people's growing needs for a better living. In terms of "production-ecological" and "living-ecological" functions, it is necessary to strengthen further the concept of "green water and green mountains are golden mountains and silver mountains," change the traditional industrial development model, adjust and optimize the economic structure, vigorously develop the ecological economy and circular economy, and continuously promote the normalization and institutionalization of ecological governance, to promote Chongqing's road of "ecological priority and green development."

5. Conclusions

Exploring the trade-off synergy between the three leading functions of production, living, and ecology in the county is the critical link to realizing the optimization and integration of PLES, rationalizing the order of spatial development, coordinating urban-rural relations, improving regional competitiveness, and promoting rural revitalization. This article combines the research methods of ecological niche situation theory, PLEF spatial theory, coupling coordination model, and trade-off synergy to quantitatively measure PLEFs and the coupling coordination degree between the three eight counties in Chongqing, and analyze the evolution characteristics of their spatiotemporal patterns. The following conclusions were reached:

(1) Significant diversity exists in Chongqing's spatiotemporal differentiation characteristics of PLEFs. During the study period, the overall growth trend of production niche width, the high-value area was mainly located in the UC, and the downward trend in

the Dadukou district was noticeable. The width of the living niche showed fluctuating and zonal growth, and most areas generally showed an expansion trend from 2000 to 2010, while the main urban areas showed a contraction or flat trend from 2010 to 2020.

(2) PLEFs and the coupling degree between the two functions in Chongqing City and county show the distribution characteristics of "high in the west and low in the east," and the whole is moving towards the stage of coordinated coupling. Among them, the "production-living" function has the highest coupling level and the most robust interaction between the two, while the "production-ecological" function and the "living-ecological" function are the second.

(3) The spatial and temporal differences in the coupling and coordination degree of PLEFs in Chongqing City and County are significant. Overall, it is manifested as "high in the west and low in the east" and from the primary imbalance to a well-coordinated development feature. It presents an inverted "U" shaped evolutionary trend. It means that with further development, the gap in the degree of coupling and coordination of PLEFs in each county will be further expanded and narrowed.

(4) On the whole, in the past 20 years, the "living-production" function in Chongqing has changed from a trade-off to a collaborative development relationship, the "living-ecological" function is generally based on the collaborative development relationship, and the trade-off constraint relationship dominates the "production-ecological" function; Moreover, the coordinated development level of "living-production," "living-ecological," and "production-ecological" functions in the central urban area has been dramatically improved, while the county has gradually shown different degrees of trade-offs.

Author Contributions: Conceptualization, L.C., Z.Z. and T.L.; methodology, T.L., D.H. and Y.S.; software, T.L., D.H. and Y.S.; validation, L.C., H.C. and Z.Z.; formal analysis, L.C. and T.L.; investigation, Z.Z.; resources, Z.Z. and D.H.; data curation, L.C., H.C., Y.S. and C.W.; writing—original draft preparation, Z.Z. and H.C.; writing—review and editing, L.C., Z.Z., T.L. and C.W.; visualization, D.H., Y.S., H.C. and C.W.; supervision, Z.Z., Y.S. and T.L.; project administration, Z.Z. and C.W.; funding acquisition, Z.Z. and H.C. All authors have read and agreed to the published version of the manuscript.

Funding: This research was funded by the Youth Program of the National Social Science Foundation of China (grant 21CJY044), the Key Program of the National Social Science Foundation of China (grant 20&ZD095), a key program of Guizhou Philosophy and Social Sciences Planning (grant 21GZD60), a general program of Guizhou Philosophy and Social Sciences Planning (grant 20GZYB10), the Humanities and Social Sciences Research Project of Chongqing Education Commission (grant 23SKGH403), and Humanities and Social Sciences Research Project of Chongqing Municipal Education Commission (grant 22SKJD111).

Institutional Review Board Statement: Not applicable.

Informed Consent Statement: Not applicable.

Data Availability Statement: The data presented in this study are available on request from the corresponding author.

Conflicts of Interest: The authors declare no conflict of interest.

References

1. Costanza, R.; de Groot, R.; Sutton, P.; vander Ploeg, S.; Anderson, S.J.; Kubiszewski, I.; Farber, S.; Turner, R.K. Changes in the global value of ecosystem services. *Glob. Environ. Chang.* **2014**, *26*, 152–158. [CrossRef]
2. Pauleit, S.; Duhme, F. Assessing the environmental performance of land cover types for urban planning. *Landsc. Urban Plan.* **2000**, *52*, 1–20. [CrossRef]
3. Smith, J.M.; Price, G.R. The Logic of Animal Conflict. *Nature* **1973**, *246*, 15–18. [CrossRef]
4. Xiao, P.; Xu, J.; Zhao, C. Conflict Identification and Zoning Optimization of “Production-Living-Ecological” Space. *Int. J. Environ. Res. Public Health* **2022**, *19*, 7990. [CrossRef] [PubMed]
5. Myers, N. Population, Environment, and Conflict. *Environ. Conserv.* **1987**, *14*, 15–22. [CrossRef]

6. Stehfest, E.; van Zeist, W.-J.; Valin, H.; Havlik, P.; Popp, A.; Kyle, P.; Tabeau, A.; Mason-D'Croz, D.; Hasegawa, T.; Bodirsky, B.L.; et al. Key determinants of global land-use projections. *Nat. Commun.* **2019**, *10*, 2166. [CrossRef]
7. Sutton, P.C.; Anderson, S.J.; Costanza, R.; Kubiszewski, I. The ecological economics of land degradation: Impacts on ecosystem service values. *Ecol. Econ.* **2016**, *129*, 182–192. [CrossRef]
8. Wang, H.; Long, H.; Li, X.; Yu, F. Evaluation of changes in ecological security in China's Qinghai Lake Basin from 2000 to 2013 and the relationship to land use and climate change. *Environ. Earth Sci.* **2014**, *72*, 341–354. [CrossRef]
9. Masoudi, M.; Tan, P.Y.; Fadaei, M. The effects of land use on spatial pattern of urban green spaces and their cooling ability. *Urban Clim.* **2021**, *35*, 100743. [CrossRef]
10. Xie, X.; Li, X.; Fan, H.; He, W. Spatial analysis of production-living-ecological functions and zoning method under symbiosis theory of Henan, China. *Environ. Sci. Pollut. Res.* **2021**, *28*, 69093–69110. [CrossRef]
11. Duan, Y.; Xu, Y.; Huang, A.; Lu, L.; Ji, Z. Progress and prospects of "production-living-ecological" functions evaluation. *J. China Agric. Univ.* **2021**, *26*, 113–124. [CrossRef]
12. Liu, Y. On the logical structure, check and balance mechanism and development principle of "three living spaces". *Hubei Soc. Sci.* **2016**, *3*, 5–9. [CrossRef]
13. Haken, H. *Advanced Synergetics*; Springer: Berlin/Heidelberg, Germany, 1983.
14. Feng, Q.; Zhao, W.; Fu, B.; Ding, J.; Wang, S. Ecosystem service trade-offs and their influencing factors: A case study in the Loess Plateau of China. *Sci. Total Environ.* **2017**, *607–608*, 1250–1263. [CrossRef] [PubMed]
15. Nelson, E.; Mendoza, G.; Regetz, J.; Polasky, S.; Tallis, H.; Cameron, D.; Chan, K.M.A.; Daily, G.C.; Goldstein, J.; Kareiva, P.M.; et al. Modeling multiple ecosystem services, biodiversity conservation, commodity production, and trade-offs at landscape scales. *Front. Ecol. Environ.* **2009**, *7*, 4–11. [CrossRef]
16. Liu, J.; Jiang, Z. *The Synergy Theory of Economic Growth*; Springer: Singapore, 2018.
17. Liu, Y.; Liu, X.; Zhao, C.; Wang, H.; Zang, F. The trade-offs and synergies of the ecological-production-living functions of grassland in the Qilian mountains by ecological priority. *J. Environ. Manag.* **2023**, *327*, 116883. [CrossRef]
18. Chen, C.; He, X.; Liu, Z.; Sun, W.; Dong, H.; Chu, Y. Analysis of regional economic development based on land use and land cover change information derived from Landsat imagery. *Sci. Rep.* **2020**, *10*, 12721. [CrossRef]
19. Chang, X.; Zhang, F.; Cong, K.; Liu, X. Scenario simulation of land use and land cover change in mining area. *Sci. Rep.* **2021**, *11*, 12910. [CrossRef]
20. Huang, j.; Lin, H.; Qi, X. A literature review on optimization of spatial development pattern based on ecological-production-living space. *Prog. Geogr.* **2017**, *36*, 378–391. [CrossRef]
21. Bliege Bird, R.; Nimmo, D. Restore the lost ecological functions of people. *Nat. Ecol. Evol.* **2018**, *2*, 1050–1052. [CrossRef]
22. Havel, M.B. Delineation of property rights as institutional foundations for urban land markets in transition. *Land Use Policy* **2014**, *38*, 615–626. [CrossRef]
23. Lu, L.; Zhou, S.; Zhou, B.; Dai, L.; Ting, C.; Bao, G.; Hua, Z.; Zhu, L. Land Use Transformation and Its Eco-environmental Response in Process of the Regional Development: A Case Study of Jiangsu Province. *Sci. Geogr. Sin.* **2013**, *33*, 1442–1449. [CrossRef]
24. Yang, Q.; Duan, X.; Wang, L.; Jin, Z. Land Use Transformation Based on Ecological-production-living Spaces and Associated Eco-environment Effects: A Case Study in the Yangtze River Delta. *Sci. Geogr. Sin.* **2018**, *38*, 97–106. [CrossRef]
25. Geng, S.; Zhu, W.; Shi, P. A Functional Land Use Classification for Ecological, Production and Living Spaces in the Taihang Mountains. *J. Resour. Ecol.* **2019**, *10*, 246. [CrossRef]
26. Moore, A.L.; McCarthy, M.A. Optimizing ecological survey effort over space and time. *Methods Ecol. Evol.* **2016**, *7*, 891–899. [CrossRef]
27. Li, J.; Sun, W.; Li, M.; Linlin, M. Coupling coordination degree of production, living and ecological spaces and it's influencing factors in the Yellow River Basin. *J. Clean. Prod.* **2021**, *298*, 126803. [CrossRef]
28. Yang, Y.; Bao, W.; Liu, Y. Coupling coordination analysis of rural production-living-ecological space in the Beijing-Tianjin-Hebei region. *Ecol. Indic.* **2020**, *117*, 106512. [CrossRef]
29. Li, F.; Wang, R.; Hu, D.; Ye, Y.; Yang, W.; Liu, H. Measurement methods and applications for beneficial and detrimental effects of ecological services. *Ecol. Indic.* **2014**, *47*, 102–111. [CrossRef]
30. Reynolds, C.S. Ecological pattern and ecosystem theory. *Ecol. Model.* **2002**, *158*, 181–200. [CrossRef]
31. Liu, C.; Xu, Y.; Lu, X.; Han, J. Trade-offs and driving forces of land use functions in ecologically fragile areas of northern Hebei Province: Spatiotemporal analysis. *Land Use Policy* **2021**, *104*, 105387. [CrossRef]
32. Zong, S.; Hu, Y.; Zhang, Y.; Wang, W. Identification of land use conflicts in China's coastal zones: From the perspective of ecological security. *Ocean Coast. Manag.* **2021**, *213*, 105841. [CrossRef]
33. Scheidel, A.; Temper, L.; Demaria, F.; Martínez-Alier, J. Ecological distribution conflicts as forces for sustainability: An overview and conceptual framework. *Sustain. Sci.* **2018**, *13*, 585–598. [CrossRef] [PubMed]
34. Perminova, T.; Sirina, N.; Laratte, B.; Baranovskaya, N.; Rikhvanov, L. Methods for land use impact assessment: A review. *Environ. Impact Assess. Rev.* **2016**, *60*, 64–74. [CrossRef]
35. LaGro, J.A. Land-use classification. In *Encyclopedia of Soils in the Environment*; Hillel, D., Ed.; Elsevier: Oxford, UK, 2005; pp. 321–328.
36. Liu, N.; Liu, C.; Xia, Y.; Da, B. Examining the coordination between urbanization and eco-environment using coupling and spatial analyses: A case study in China. *Ecol. Indic.* **2018**, *93*, 1163–1175. [CrossRef]

37. Ding, L.; Zhao, W.; Huang, Y.; Cheng, S.; Liu, C. Research on the Coupling Coordination Relationship between Urbanization and the Air Environment: A Case Study of the Area of Wuhan. *Atmosphere* **2015**, *6*, 1539–1558. [CrossRef]
38. Zhang, Y.; Shen, L.; Shuai, C.; Bian, J.; Zhu, M.; Tan, Y.; Ye, G. How is the environmental efficiency in the process of dramatic economic development in the Chinese cities? *Ecol. Indic.* **2019**, *98*, 349–362. [CrossRef]
39. Lu, H.; Zhou, L.; Chen, Y.; An, Y.; Hou, C. Degree of coupling and coordination of eco-economic system and the influencing factors: A case study in Yanchi County, Ningxia Hui Autonomous Region, China. *J. Arid Land* **2016**, *9*, 446–457. [CrossRef]
40. Zhou, D.; Xu, J.; Lin, Z. Conflict or coordination? Assessing land use multi-functionalization using production-living-ecology analysis. *Sci. Total Environ.* **2017**, *577*, 136–147. [CrossRef]
41. Song, Q.; Zhou, N.; Liu, T.; Siehr, S.A.; Qi, Y. Investigation of a "coupling model" of coordination between low-carbon development and urbanization in China. *Energy Policy* **2018**, *121*, 346–354. [CrossRef]
42. Lai, Z.; Ge, D.; Xia, H.; Yue, Y.; Wang, Z. Coupling coordination between environment, economy and tourism: A case study of China. *PLoS ONE* **2020**, *15*, e0228426. [CrossRef]
43. Hou, C.; Chen, H.; Long, R. Coupling and coordination of China's economy, ecological environment and health from a green production perspective. *Int. J. Environ. Sci. Technol.* **2022**, *19*, 4087–4106. [CrossRef]
44. Zeng, P.; Wei, X.; Duan, Z. Coupling and coordination analysis in urban agglomerations of China: Urbanization and ecological security perspectives. *J. Clean. Prod.* **2022**, *365*, 132730. [CrossRef]
45. Zhang, X.; Xu, Z. Functional Coupling Degree and Human Activity Intensity of Production–Living–Ecological Space in Underdeveloped Regions in China: Case Study of Guizhou Province. *Land* **2021**, *10*, 56. [CrossRef]
46. Wang, Y.; Geng, Q.; Si, X.; Kan, L. Coupling and coordination analysis of urbanization, economy and environment of Shandong Province, China. *Envrion. Dev. Sustain.* **2021**, *23*, 10397–10415. [CrossRef]
47. Cai, J.; Li, X.; Liu, L.; Chen, Y.; Wang, X.; Lu, S. Coupling and coordinated development of new urbanization and agroecological environment in China. *Sci. Total Environ.* **2021**, *776*, 145837. [CrossRef] [PubMed]
48. Tian, Y.; Huang, P.; Zhao, X. Spatial analysis, coupling coordination, and efficiency evaluation of green innovation: A case study of the Yangtze River Economic Belt. *PLoS ONE* **2020**, *15*, e0243459. [CrossRef] [PubMed]
49. Zhang, Y.; Wu, T.; Song, C.; Hein, L.; Shi, F.; Han, M.; Ouyang, Z. Influences of climate change and land use change on the interactions of ecosystem services in China's Xijiang River Basin. *Ecosyst. Serv.* **2022**, *58*, 101489. [CrossRef]
50. Cao, Y.; Kong, L.; Zhang, L.; Ouyang, Z. The balance between economic development and ecosystem service value in the process of land urbanization: A case study of China's land urbanization from 2000 to 2015. *Land Use Policy* **2021**, *108*, 105536. [CrossRef]
51. Wen, C.; Zhang, Z.; Zhao, B. Preliminary Study on the innovative Mode of "Governance in Five Domains" for the Great Ecological Protection in the Upper Reaches of Yangtze River. *Study Pract.* **2022**, *7*, 54–64. [CrossRef]
52. Wen, C.; Zhang, Z.; Cao, X. The Connotation, Strategy and Path of Ecological Protection in the Upper Reaches of the Yangtze River. *Reg. Econ. Rev.* **2021**, *1*, 123–130. [CrossRef]
53. Wang, R.; Cheng, J.; Zhu, Y.; Lu, P. Evaluation on the coupling coordination of resources and environment carrying capacity in Chinese mining economic zones. *Resour. Policy* **2017**, *53*, 20–25. [CrossRef]
54. Dussault, A.C. Two Notions of Ecological Function. *Philos. Sci.* **2022**, *89*, 171–179. [CrossRef]
55. Giller, P.S. Niche theory. In *Community Structure and the Niche*; Giller, P.S., Ed.; Springer: Dordrecht, The Netherlands, 1984; pp. 8–21.
56. Jax, K. Function and "functioning" in ecology: What does it mean? *Oikos* **2005**, *111*, 641–648. [CrossRef]
57. Laland, K.; Matthews, B.; Feldman, M.W. An introduction to niche construction theory. *Evol. Ecol.* **2016**, *30*, 191–202. [CrossRef] [PubMed]
58. Letten, A.D.; Ke, P.J.; Fukami, T. Linking modern coexistence theory and contemporary niche theory. *Ecol. Monogr.* **2017**, *87*, 161–177. [CrossRef]
59. Laland, K.N.; Odling-Smee, F.J.; Feldman, M.W. Evolutionary consequences of niche construction and their implications for ecology. *Proc. Natl. Acad. Sci. USA* **1999**, *96*, 10242–10247. [CrossRef] [PubMed]
60. Li, W.; Cai, Z.; Jin, L. Spatiotemporal characteristics and influencing factors of the coupling coordinated development of production-living-ecology system in China. *Ecol. Indic.* **2022**, *145*, 109738. [CrossRef]
61. Shi, J.-G.; Liu, M.; Duan, K. Spatiotemporal evolution of coupling coordination in population-land relationship and its influencing factors: A case study of 26 main cities in Yangtze River Delta, China. *Phys. Chem. Earth Parts A/B/C* **2022**, *128*, 103247. [CrossRef]
62. Yang, C.; Zeng, W.; Yang, X. Coupling coordination evaluation and sustainable development pattern of geo-ecological environment and urbanization in Chongqing municipality, China. *Sustain. Cities Soc.* **2020**, *61*, 102271. [CrossRef]
63. Jiang, Y.; Chen, M.; Zhang, J.; Sun, Z.; Sun, Z. The improved coupling coordination analysis on the relationship between climate, eco-environment, and socio-economy. *Environ. Ecol. Stat.* **2022**, *29*, 77–100. [CrossRef]
64. Hou, Y.; Zhang, Z.; Wang, Y.; Sun, H.; Xu, C. Function Evaluation and Coordination Analysis of Production–Living–Ecological Space Based on the Perspective of Type–Intensity–Connection: A Case Study of Suzhou, China. *Land* **2022**, *11*, 1954. [CrossRef]
65. Peng, Y.; Yan, L. Study on competition relationship among tourist cities in Zhejiang province based on the niche theories. *Acta Ecol. Sin.* **2013**, *35*, 2195–2205. [CrossRef]
66. Han, B.; Wang, R.; Tao, Y.; Gao, H. Urban population agglomeration in view of complex ecological niche: A case study on Chinese prefecture cities. *Ecol. Indic.* **2014**, *47*, 128–136. [CrossRef]

67. Qin, T.; Qi, W.; Li, Y.; Qu, Y. Suitability evaluation of rural residential land based on niche theory in mountainous area. *Acta Ecol. Sin.* **2011**, *32*, 5175–5183. [CrossRef]
68. Cai, H.; Chen, Y.; Zhang, X. Suitability evaluation and zoning method research on development and utilization of selenium-rich soil resources based on niche theory. *Acta Ecol. Sin.* **2020**, *40*, 9208–9219. [CrossRef]
69. Li, X.; Fang, B.; Yin, R.; Rong, H. Spatial-temporal change and collaboration/trade-off relationship of "production-living-ecological" functions in county area of Jiangsu province. *J. Nat. Resour.* **2019**, *34*, 2363. [CrossRef]
70. Nie, X.; Shi, P.; LÜ, R.; Zhang, X.; Liang, B.; Wei, W. Ecological niche-based competition and cooperation relationships among county-level cities in Hexi Corridor. *Acta Ecol. Sin.* **2016**, *38*, 841–851. [CrossRef]
71. Wu, Q.; Cao, Y.; Fang, X.; Wang, J.; Li, G. A systematic coupling analysis framework and multi-stage interaction mechanism between urban land use efficiency and ecological carrying capacity. *Sci. Total Environ.* **2022**, *853*, 158444. [CrossRef]
72. Zhang, Y.; Yang, Q.; Min, J. An analysis of coupling between the bearing capacity of the ecological environment and the quality of new urbanization in Chongqing. *Acta Geogr. Sin.* **2016**, *71*, 817–828. [CrossRef]
73. Chen, S.; Lin, J.; Yang, L.; Zhang, H. Study on urban competition strategy based on the theory of niche. *Hum. Geogr.* **2006**, *21*, 72–76. [CrossRef]
74. Wang, C.; Tang, N. Spatio-temporal characteristics and evolution of rural production-living-ecological space function coupling coordination in Chongqing Municipality. *Geogr. Res.* **2018**, *37*, 1100–1114. [CrossRef]
75. Zhao, J.; Zhao, Y. Synergy/trade-offs and differential optimization of production, living, and ecological functions in the Yangtze River economic Belt, China. *Ecol. Indic.* **2023**, *147*, 109925. [CrossRef]
76. Chen, H.; Yang, Q.; Su, K.; Zhang, H.; Lu, D.; Xiang, H.; Zhou, L. Identification and Optimization of Production-Living-Ecological Space in an Ecological Foundation Area in the Upper Reaches of the Yangtze River: A Case Study of Jiangjin District of Chongqing, China. *Land* **2021**, *10*, 863. [CrossRef]
77. Wei, L.; Zhang, Y.; Wang, L.; Mi, X.; Wu, X.; Cheng, Z. Spatiotemporal Evolution Patterns of "Production-Living-Ecological" Spaces and the Coordination Level and Optimization of the Functions in Jilin Province. *Sustainability* **2021**, *13*, 3192. [CrossRef]
78. Odling-Smee, J.; Erwin, D.H.; Palkovacs, E.P.; Feldman, M.W.; Laland, K.N. Niche construction theory: A practical guide for ecologists. *Q Rev. Biol.* **2013**, *88*, 4–28. [CrossRef]
79. Doncaster, C.P. Ecological Equivalence: A Realistic Assumption for Niche Theory as a Testable Alternative to Neutral Theory. *PLoS One* **2009**, *4*, e7460. [CrossRef]
80. Wang, J.-L.; Liu, B.; Zhou, T. The category identification and transformation mechanism of rural regional function based on SOFM model: A case study of Central Plains Urban Agglomeration, China. *Ecol. Indic.* **2023**, *147*, 109926. [CrossRef]
81. Breuste, J.H.; Anwar, M.M.; Nawaz, R.; Rani, M. *Urban Ecosystems: Functions, Value and Management*; Springer International Publishing: Berlin/Heidelberg, Germany, 2017.
82. Schwarz, N.; Hoffmann, F.; Knapp, S.; Strauch, M. Synergies or Trade-Offs? Optimizing a Virtual Urban Region to Foster Plant Species Richness, Climate Regulation, and Compactness Under Varying Landscape Composition. *Front. Environ. Sci.* **2020**, *8*, 16. [CrossRef]
83. Deng, X.; Xiong, K.; Yu, Y.; Zhang, S.; Kong, L.; Zhang, Y. A Review of Ecosystem Service Trade-Offs/Synergies: Enlightenment for the Optimization of Forest Ecosystem Functions in Karst Desertification Control. *Forests* **2023**, *14*, 88. [CrossRef]

land

MDPI

Article

Unsustainable Urban Development Based on Temporary Workers: A Study on the Changes of Immigration in Macau between 1992 and 2019

Bo Zhou [1,2] and Lei Jiang [2,3,*]

1 Public Administration School, Guangzhou University, Guangzhou 510006, China
2 Center for Human Geography and Urban Development, Guangzhou University, Guangzhou 510006, China
3 School of Geography and Remote Sensing, Guangzhou University, Guangzhou 510006, China
* Correspondence: jianglei@gzhu.edu.cn

Abstract: Macau's urban development model has many unique characteristics, including expansion of the city through sea reclamation, increasing population mainly through immigration, and economic development driven by the gaming industry. Based on data from the Macau Statistics and Census Service, this study uses the Error Correction representation of the Autoregressive Distributed Lag model (ARDL-ECM) to analyze the impact of urban development on the trends of immigration and labor migration in Macau between 1992 and 2019. Results show that both land area and wage level have positive effects on the number of migrant workers and negative effects on the number of immigrants, indicating that Macau is over-dependent on short-term migrant workers. Macau's land and human resources are tilted towards the gaming industry, resulting in a decreasing living environment and resident carrying capacity as the city develops. Therefore, this paper suggests that Macau should reduce the cost of city expansion and improve economic diversity through strengthening cooperation with neighboring mainland cities, hence sparing resources to absorb non-local talent and ensuring sustainable urban development.

Keywords: migrant workers; immigrants; land shortage; labor shortage

Citation: Zhou, B.; Jiang, L. Unsustainable Urban Development Based on Temporary Workers: A Study on the Changes of Immigration in Macau between 1992 and 2019. *Land* **2022**, *11*, 1985. https://doi.org/10.3390/land11111985

Academic Editor: Li Ma, Yingnan Zhang, Muye Gan, Zhengying Shan

Received: 16 October 2022
Accepted: 4 November 2022
Published: 5 November 2022

1. Introduction

The measurement of city development is multidimensional, and area, population and economic volume are often regarded as its most important indicators. Macau has shown its uniqueness in all these aspects. Around two-thirds of Macau's land was reclaimed from the sea and over half of its current population consists of immigrants or migrant workers [1]. Although its immigration policies have been gradually tightened in the recent decades, Macau's population density still fluctuates up to around 20,000 people per square kilometer, making it one of the most crowded regions in the world. On its 32.9 km^2 land, Macau has created an economic miracle and established the "Las Vegas of Asia", with its gross gaming revenue surpassing Las Vegas in 2006 and per capita GDP reaching the second place in the world in 2019 [2,3].

Due to the extreme scarcity of land and the decline of traditional competitive industries, Macau has chosen to focus on developing its gaming–tourism industry in recent decades [4]. The Macau government auctions off six gaming business licenses to large companies, developing a gaming–tourism industry in which over half of the employees are non-local workers and consumers are tourists. Such an economic model shares similarities with convention and exhibition economy: low cost, high flexibility and high output multipliers [5]. However, Macau's booming gaming industry has significantly more negative social outcomes and has suppressed the development of other industries [4,6]. With gaming–tourism industry as its only pillar, Macau's economy became strongly dependent on the external environment and lacks economic resilience [7]. The grand casino hotels occupying a huge area have also suppressed housing construction in Macau [8], which makes it difficult for Macau to accommodate new immigrants.

Despite the great contribution of migrant workers to Macau's urban development, Macau's migration policies have been rather controversial in view of its dependence on temporary workers and strict regulation on status changing from non-immigrant to immigrant. Since the enforcement of Law 3/2005 and adjustment on temporary residence system for investors, administrators and technicians, Macau has only approved 2084 applications for residence of qualified technicians [9]. Therefore, the chance for 0.2 million migrant workers in Macau to become settlers through talent programs is extremely trivial. Studies find that Macau's policies on recruitment of non-local talent have no details and their implementation lacks transparency [10,11]. Moreover, the development of the gaming–tourism industry relies heavily on migrant workers with low salary and low social security. Consequently, Macau's dependence on the gaming industry makes it difficult to change current migration policies and hence reverse the problem of talent shortage [12].

Based on the statistical yearbook and census reports from the Macau Statistics and Census Service, this paper examines the influential factors of two types of migration in Macau and discusses how Macau's current policies on migrant workers and immigrants might hinder its sustainable development. The current study not only helps understand Macau's immigration-driven urban development process, but also provides reference for future studies on the development trend of other cities. Immigration-driven population growth is becoming more and more common around the world. For instance, the population growth of San Francisco, New York, and Tokyo bay areas in the last decade has mainly depended on immigration [13]. Macau has historically been built up and developed by immigrants [14], hence findings on Macau's trend of immigration and its consequences can be applicable to many other research sites as well.

In the following section, this paper reviews the relationship between immigration and the population, land and economic development process of Macau. Then we explain how the Error Correction Model (ECM) is applied to analyze the main influential factors on Macau's number of immigrants and migrant workers between 1992 and 2019. A section on the findings from data analyses follows. Finally, we discuss and conclude how the case of Macau can deepen our understanding of immigration-driven urban development, the impact of immigration on small economies, and the man–land relationship in land-scarce cities.

2. Background

2.1. Immigrants and Migrant Workers

Migration is one of the most fundamental sources of demographic change. In the 19th century, Ravenstein already proposed some general laws on population migration [15]. As the scale, frequency and distance of migration continue to rise, the theoretical explanations for the occurrence of migration also continue to evolve. Push–pull theory points out that there are all sorts of positive and negative factors (related to the economy, environment, population structure, etc.) in the sending and receiving regions which influence the migration decision of potential migrants, but potential migrants also need to overcome the obstacle factors in-between [16]. Neo-classical migration theory uses economic equilibrium to explain population mobility [17,18]. It suggests that on the one hand, the relative labor shortage among regions is the main driving force of migration, while on the other hand, migration will lead to more balanced labor demand between the sending and receiving regions. According to new economics of labor migration, migrations in developing countries need to be understood from the perspective of families rather than individuals, and that individual migration is more often a way to minimize family risk than to maximize personal benefits [19,20]. An aspiration–(cap)ability framework encourages researchers to pay attention to both individual's willingness and capability to migrate, and suggests that migration actually occurs when these two aspects match with each other [21–23]. The combination of this framework on micro level migration on the one hand, and regional development at the macro level on the other, can further form an interaction theory of migration and development that helps explain the scale of regional immigration and emigration [24].

Transnational migrant workers, whose settlement opportunities are severely restricted, have proliferated in recent decades as national controls have tightened. After World War II, due to the lack of labor, many Western European countries adopted the Guest Worker system to import laborers in batches from other countries through national agreements, resulting in matching problems and immigration problems [25]. In modern times, especially in the industrializing countries, the import mode of foreign low-skilled labor has been changed to the short-term contract labor system, which is characterized by the regulation of import methods by the state and the importing of foreign workers organized by enterprises [26]. This kind of cross-country labor migration with a clear path is often called the "point-to-point" globalization in immigration research. Studies on transnational corporations in Africa point out that "point-to-point" globalization allows companies to establish relatively independent production departments away from local societies, thus maintaining "social thinness" and smoothness of transnational capital [27]. Many East Asian countries are important exporters of migrant workers. For instance, Indonesia exported 700,000 regulated workers in 2007 [28], China sent overseas a total of 800,000 contract workers in 2011 [29], and the cross-border workers from Vietnam to China alone exceeded 200,000 in 2018 [30]. Research based on migrant workers in East Asia shows that with the stricter supervision of the state and the embedding of labor intermediary services between the sending and receiving countries, conditions of individualized labor transplanting has been clearly improved [29,31].

In recent decades, as the attention of migration scholars has been dispersed by the diversity of population migration, studies are shifting from inducing big-picture migration theory to conceptualizing the lives, identities and experiences of migrants from an insider's perspective [32]. This change of focus caused by the complexity of research objects has hindered the progress of migration-related theory development [33]. We believe that Macau's large population of immigrants and migrant workers makes it a strong candidate site to encourage and promote the development of migration theory. Macau used to absorb a large number of immigrants and multi-ethnic immigrants who integrated well into its diversified business-port culture [34]. However, as we will introduce in the following two sections, Macau has now become increasingly dependent on migrant workers due to the limitation of land and its economic development strategy, which in turn leads to a reduced immigration quota and a lower chance for non-local talents to settle.

2.2. Immigrant-Driven Population Growth

Historically speaking, Macau is a city built and fueled up by immigrants. The population record of the Ming Dynasty shows that there were only 400 residents in Macau in the year 1555 [14]. In 1563, Macau's population grew to over 5000, among which there were 900 Portuguese merchants. By the end of the Ming Dynasty, Macau had become a center harbor in east Asia, with mature trade routes starting from there to Japan, Mexico and Portugal. In 1640, the population of Macau reached a peak of 40,000, but than merchants left due to the sea ban in the early and mid-Qing Dynasty, and Macau's population shrank sharply to under 4000. In the late Qing Dynasty, China's foreign trade resumed and Macau became one of the starting points of the Chinese coolie trade. In addition, the Portuguese government occupied the Chinese territory of Macau, and brought the original residents under its jurisdiction. Under its comprehensive influence, the total population of Macau reached 75,000 in 1910. In the 1940s, Macau, being a relatively safe place, had attracted a large number of refugees from Chinese mainland, and its population temporarily surged to 0.25 million, but most refugees moved back to their hometown after the Second Sino-Japanese War. Due to China's Reform and Opening-up policy, the number of immigrants from the mainland to Macau has increased rapidly since the 1980s, boosting Macau's total population to 0.68 million within four decades [1].

In the 1970s, the garment export industry became the pillar industry of Macau [35,36]. Local laborers only met 50 to 70 percent of export orders, so factories in Macau hired large numbers of skilled textile workers from the adjacent Pearl River Delta region. During that period, the Macau government had no legislative regulations on labor migration. Through the two greatest amnesties in 1982 and 1990, around 60 thousand undocumented immigrants became local residents of Macau [37]. Research shows that these relatively young immigrants from the mainland alleviated Macau's low fertility problems to some extent, and postponed aging problems [38,39].

In 1988, Dispatch 12/GM/88 permitted Macau's enterprises to hire unskilled migrant workers legally, but for these temporary workers, the chance of becoming local residents was insignificant. Through enforcement of Order 2/90/M, Order 55/95/M, Law 4/2003 and Law 16/2021, the Macau government gradually distinguished between different purposes of entry and tightened up the granting of the right of abode [40]. In Macau's current official documents, migrant workers are recorded as non-resident workers, while immigrants are recorded as new arrivals from mainland China with a one-way permit (i.e., immigrants through the family reunion program) or individuals granted the right of abode (i.e., immigrants through other programs). Both types of immigrants are granted permanent residency after residing in Macau for over seven years, while non-resident workers do not have such rights. Highly skilled migrant workers can apply for immigration through the employer sponsored category, yet this quota is extremely limited in Macau [9]. At the end of 2019, migrant workers and immigrants who had arrived after the mid-1980s accounted for 26% and 27% of the Macau population, respectively; thus, around half of the city's population consists of recent arrivals [1].

2.3. *Booming Gaming Industry and Shortage of Land and Labor*

Macau used to have a much higher degree of economic diversification in the 1970s than it does today [15]. During its golden age, the gross product value of Macau's export processing industries increased by over 20% per year, and the most successful industries included textiles, clothing and toy manufacturing. However, neighboring regions with abundant human resources, such as Guangdong province and Southeast Asian countries, joined the competition of processing trade in the 1980s, forcing Macau to shift its development focus from manufacturing to infrastructure construction [37]. Driven by a large influx of investment and speculators, the land and housing prices in Macau skyrocketed in the early-1990s and the construction industry was extremely prosperous [41]. However, since the number of newly-constructed properties far exceeded the actual market demand, the housing prices dropped sharply from the second half of 1993. Although Macau's Order 14/95/M boosted the construction industry a little through attracting investment immigrants, this industry was later struck by the 1997 Asian Financial Crisis and 2008 Global Financial Crisis and has continued to shrink ever since [42].

With neighboring countries and regions banning gambling, its legal gambling service has become a unique advantage for Macau. The gross product value of the gambling–tourism industry has surpassed the export processing industry and become the largest industry in Macau since the 1980s, reaching 70 percent of the total GDP in the 2010s [4]. The gaming–tourism industry is relatively insensitive to the global economic environment, such that its gross product value increased by 9.6% during the 2008 Global Financial Crisis [22]. However, this industry is highly dependent on the flow of gamblers, and after the Chinese mainland carried out an anti-corruption campaign and tightened the issue of transit visas to Macau, Macau's gross gaming revenue dropped by one third in 2015 [2,4].

Studies in the UK note that demands for migrant workers often rise because of certain types of domestic labor shortage caused by welfare policies and other system effects [43]. Similarly, system effects including high welfare for local workers and unbalanced industry development have resulted in the long lasting labor shortage in Macau [35]. During one decade of rapid development, the employment population of the gaming industry increased from 22,900 in 2004 to 83,300 in 2013 [44]. While some high payment and low skill requirement positions directly involved in gambling (such as the casino dealers) are preserved for local residents, tens of thousands of migrant workers are hired as administrative or service staff in the grand casino hotels [45].

The mass labor migration in Macau has not been without controversy. Studies show that the wage below market rate in Macau and the insufficient cultivation of local labor potential have led to structural labor insufficiency [11,46]. Approval of the entry of migrant workers met the demands and interests of employers, but suppressed the wage level of local employees. Some studies further claim that using labor migration as the solution to short-term labor shortage has aggravated Macau's long-term problems including slow industrial restructuring, lagging technological upgrading and disconnection between local talent training and economic demand [11]. Therefore, local scholars have accused the Macau government of adopting a "nonintervention" attitude towards migrant workers, sacrificing the rights and interests of both local and immigrant labor for the benefit of its short-term economic interests [47].

Macau's gaming industry also has enormous demands on land and has been a strong driver of sea reclamation [37]. Macau is a city built on small islands located at the mouth of the Pearl River [4]. With numerous coastal shores, the water depth around the port is only one to three meters, so the cost of sea reclamation is relatively low [48]. Still, the cost of land expansion for Macau is much higher than expansion costs for non-island cities. Through sea reclamation projects between 1863 and 2020, the land area of Macau grew from 10.37 to 32.9 km^2. In the last two decades, Macau began to acquire land from neighboring mainland cities through cooperative development, such as the cross-border industrial zone established in Gongbei, Zhuhai, and the new University of Macau campus established in Hengqin, Zhuhai [49]. While cross-border cooperation is gradually becoming an important way for Macau to expand and develop, the gaming industry is still constrained within Macau SAR due to legal restrictions.

Through examining the influence of booming sectors in small economies (i.e., the Dutch disease), scholars have concluded that a booming sector would make other sectors uncompetitive and unsustainable through increasing the price of non-tradable goods [50]. Such a finding is also applicable to Mediterranean and Atlanta islands, in that the dominance of tourism on these islands have made these economies excessively dependent on foreign countries and lose the motivation to improve education levels, innovation and technology [51]. Consistent with these prior studies, studies on Macau show that the gaming industry absorbs the scarce land and human resources, promoting deindustrialization through increasing the prices of various resources and increasing the city's dependency on tourists [4]. Because of its economic pillar role in Macau, the gaming industry takes priority over other industries in the allocation of land resources [44]. Even residents' demands on housing and living environment have been suppressed to make way for expansion of the gaming industry. Following the current development trend, the gaming–tourism industry will continuously drain Macau's newly increased land and labor resources, making it difficult for Macau to achieve economic transformation and sustainable development or to attract non-local elites.

2.4. Research Hypotheses

The sea reclamation projects call for a large number of construction workers. With the casino hotels and souvenir shops being built on the new land, Macau's demand for other types of laborers also keeps increasing [44]. Meanwhile, the abundance of low-wage migrant workers with no social security obligations has kept Macau from training local man power to keep up with economic development [45]. Hence, this study hypothesizes that in the long run, Macau's dependence on migrant workers would increase along with the expansion of Macau's urban scale:

Hypothesis 1. *The number of migrant workers increases if the land area of Macau increases.*

The sea reclamation projects have been fueled by the gaming–tourism industry's hunger for land. However, sea reclamation results in destruction of the landscape and affects the region's degree of habitability. In addition, with a disproportion of Macau's new land area occupied by casino hotels, the growth of housing and amenities failed to keep pace with the city's economic development. Accordingly, this study assumes that when the land scale of Macau expands, the city becomes less attractive to potential immigrants:

Hypothesis 2. *The number of immigrants decreases if the land area of Macau increases.*

Macau has long provided higher wages than surrounding regions, which is a key pull factor for migrant workers. The rise in wage level in the past decade has largely resulted from development of the gaming industry, which has a great demand for manpower [4]. Since the wage level in general reflects the demand for labor, this study puts forward the following research hypothesis:

Hypothesis 3. *The number of migrant workers increases if the median wage in Macau increases.*

Different from migrant workers who pursue higher income, the driving forces of immigrants' arrival in Macau are more diverse. For instance, before the 2005 revision of policies on investment immigration, a large proportion of immigrants were investors being bullish on Macau's real estate development. After 2005, the majority of new immigrants are holders of one-way permits for the purpose of family reunion. Since migrant workers and immigrants are in fact competing for quotas in Macau due to its limited natural and social resources, the rising demand of the city for labor usually results in restrictions on immigrants. Therefore, this study hypothesizes that as an indicator of labor demand, wage level would have a negative effect on the number of immigrants:

Hypothesis 4. *The number of immigrants decreases if the median earning of Macau increases.*

As is shown in Figure 1, we hypothesize that along with the urban development of Macau, which is featured by increasing land through sea reclamation and rising wage level, the number of migrant workers would keep increasing, while the number of immigrants would keep decreasing. While we do not directly examine the relationship between the numbers of migrant workers and immigrants, we would assume that Macau's reliance on migrant workers is a main driving force of reducing quotas for immigrants. Employers in Macau can now avoid high employee welfare for locals and immigrants and fulfill their demand for laborer through hiring more obedient migrant workers [11]. However, as we will discuss more specifically in later sections, this strategy is not a solution to Macau's problem of insufficient manpower, but only postpones while deteriorating its consequences.

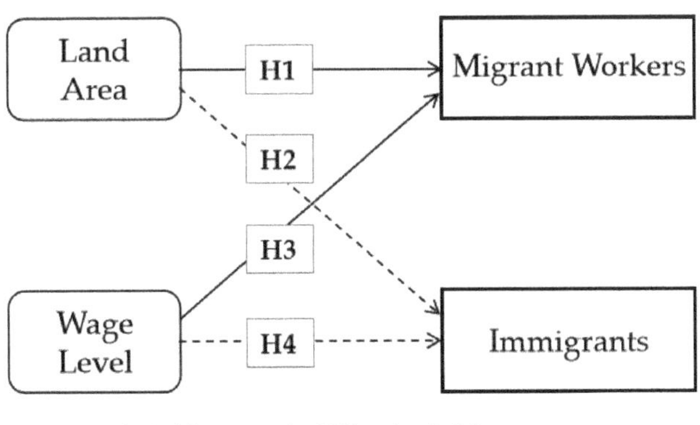

Figure 1. Analytical Framework of Migration in Macau.

3. Data and Methodology

3.1. Data

This study examines the trends of immigration in Macau based on data from the Statistical Yearbook compiled by the Macau Statistics and Census Service [1]. In addition, this study compares population structure and educational attainment between Macau's local residents (accounting for 83.4% of the total population) and non-local residents (including migrant workers who account for 13.9% of the population and non-local students accounting for 2.8% of the population) based on Macau census report data.

Due to the lack of statistical data on Macau before the 1990s and the dramatic change in 2020 caused by Covid-19, this paper analyzes the impact of city development on the number of migrant workers and immigrants between 1992 and 2019. There are two dependent variables in this paper (Table 1). The first dependent variable is the number of migrant workers, i.e., the number of non-resident workers at the end of each statistical year in Macau. The second variable is the number of immigrants, i.e., the cumulative number of individuals granted the right of abode either through family reunion, investment or employer sponsored programs since 1984 (There are no official records on numbers of foreign residents and new immigrants from Chinese mainland before 1984 nor any official statistics on year-end numbers of new immigrants; we therefore constructed a variable to measure the accumulated number of new immigrants in Macau since 1984. In fact, Macau's population consists basically of immigrants, so this dependent variable only measures the cumulative number of new immigrants after Macau established formal immigration policies. The other independent variable, i.e. year-end number of migrant workers, is also to some extent accumulative, in that it measures the number of migrant workers who arrived in the recent years and stayed until the end of a certain year). Independent variables include the total urban land area of Macau (unit: km^2) and the median annual wage (unit: 1000 MOP or Macau Pataca). Because there are a total of 28 analysis time points in the autoregression models, the number of independent variables that can be included is rather limited (In an ARDL model, the number of variables plus their lag terms should not exceed half of the degrees of freedom, so only 1 to 3 independent variables can be included in our regression model. Moreover, statistics on unit housing price, wage of migrant workers, etc., are only available since the 2000s, so inclusion of related variables would further reduce time points in regression analysis. Therefore, even though we examined potential indicators such as housing price, housing space, population density and wage gap, we finally chose land area and wage level as best reflecting Macau's urban development and yielding a higher R^2. We do, however, plan to examine further indicators in future studies after collecting individual data).

Table 1. Descriptive Statistics (n = 28).

	Variable Name	Mean	S.D.	Min.	Max.
Dependent Variables	Number of Migrant Workers	79,812	62,354	21,088	196,538
	Number of Immigrants	94,516	51,603	24,974	179,055
Independent Variables	Land Area (Unit: km^2)	26.92	4.24	18.70	32.90
	Median Annual Wage (Unit: 1000 MOP)	99.32	51.66	41.998	204.00

The number of migrant workers in Macau at the end of each year has increased from 21,088 in 1990 to 196,538 in 2019 (Figure 2). The number of migrant workers is strongly influenced by the local economic environment. After the collapse of Macau's real estate industry in 1993, Macau's demand for construction workers remained extremely low for nearly a decade, resulting in a smaller flow of labor migration. The 2008 Global Financial Crisis also led to a clear decline in labor migration to Macau. Before the COVID-19 outbreak, Macau had witnessed a decade of prosperity fueled by the gaming industry, while about half of its labor force were migrant workers. Among non-local employees, 60% were from the mainland, 10% from the Philippines, 10% from Vietnam, 5% from Hong Kong and 5% from Indonesia.

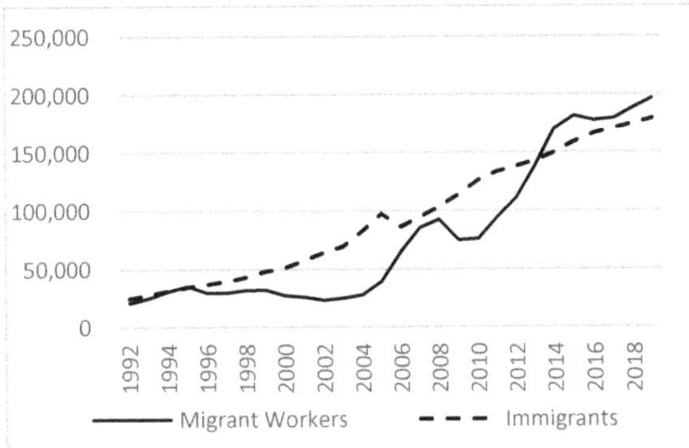

Figure 2. The Number of Non-Immigrant Workers and immigrants in Macau, 1992–2019.

Between 1992 and 2019, the number of immigrants in Macau increased quite steadily from around 24,974 to 179,055 (Figure 2). The only sharp change is related to the enforcement of Law 3/2005, which significantly raised the threshold for investment immigration. Since then, the number of new foreign residents granted right of abode by the Macau government each year dropped from several thousand to around one thousand [52]. Immigration from the mainland to Macau with the purpose of family reunification is under the strict regulation of the mainland government, and the cumulative number of immigrants with a one-way permit has increased by an average of 3500 per year.

The median annual wage in Macau increased from MOP 42,000 to MOP 204,000 between 1992 and 2019, with an annual growth rate of 5.8% (Figure 3). The salary in Macau fell sharply after the Asian Financial Crisis, and the absolute value of median wage did not recover to the 1997 level until 2004. Wages have since risen rapidly as the gaming industry has boomed, except for the stagnant years between 2014 and 2016 when the gaming industry was trying to recover from the economic shock caused by the mainland's anti-corruption campaign [3]. Meanwhile, Macau's land area grew relatively steadily from 18.7 km^2 in 1992 to 32.9 km^2 in 2019 through sea reclamation, with an average annual growth rate of 2 percent.

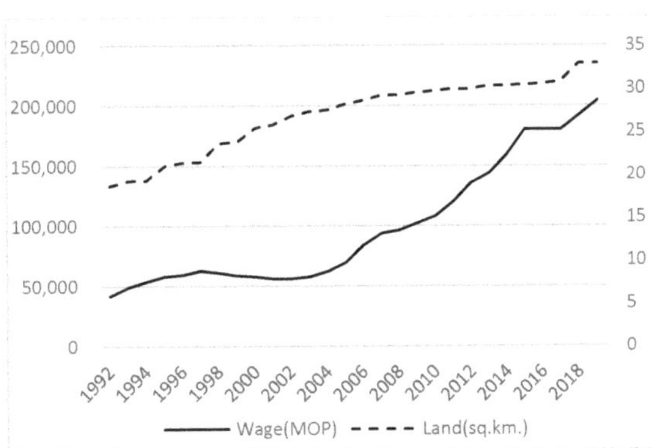

Figure 3. The Median of Annual Wage and Land Area in Macau, 1992–2019.

3.2. Analytical Methods

This paper uses the Error Correction representation of the Autoregressive Distributed Lag model (ARDL-ECM) to analyze the impact of Macau's land area and wage level on the number of migrant workers and immigrants. The EC model is a special application of the ARDL model that separates the long-term and short-term impacts through the inclusion of an error correction term [53]. The reason for using an autoregressive model is that the number of migrants of either type in Macau is not only affected by the current effect of land area and wage level, but is also affected by the number of migrants in the previous years and the lagged effect of land area and wage level.

ECM suffers from several limitations. Specifically, the biggest one is that it requires the variables to be cointegrated. Otherwise, the disequilibrium error term will not be a stationary variable, because the errors in the long-run relationship become larger and larger. Moreover, in terms of model specifications, ECM can be explained as a reparameterization of the general ARDL or dynamic linear regression (DLR) models because it focuses on the importance of general to specific modeling [54]. Although the ARDL-ECM imposes no restrictions on the DLR, there are no statistical criteria to distinguish between the two model specifications because they are observationally equivalent. Hence for these coefficients, the interpretation of the ECM is heavily dependent on explicit theory [55].

On the other hand, ECM has much more advantages than limitations in empirical studies. The biggest one is that it does not suffer from the problem of serial autocorrelation since it was formulated in terms of first differences, which eliminates trends from the variables. In this sense, it can overcome the problem of spurious regressions. In addition, the estimated coefficients have explicit interpretation [56]. Specifically, for our study it is able to analyze the short-term and long-term dynamics of wage and urban area changes on migrants, producing better forecasts for migration policy analysis. Hence, since it can measure the correction from disequilibrium of the previous period with an explicit implication, it is a well-recognized convenient model in many applications. To conclude, the advantages of the ECM outweigh the disadvantages, making it a suitable model for our study.

This study applies the ARDL-ECM on the Macau data to specify the long-term influence of land area and wage level on the number of migrant workers and immigrants:

$$Migrant = \alpha_1 Area + \alpha_2 Wage + \alpha_0 \,, \tag{1}$$

$$\Delta Migrant_t = \sum_{i=1}^{p-1} \theta_{1,\,i} \Delta Migrant_{t-i} + \sum_{i=0}^{q-1} \theta_{2,\,i} \Delta Area_{t-i} + \sum_{i=0}^{r-1} \theta_{3,\,i} \Delta Wage_{t-i} + \delta * ecm_{t-1} + \varepsilon_t \,, \tag{2}$$

$$ecm_{t-1} = Migrant_{t-1} - \alpha_1 Area_{t-1} - \alpha_2 Wage_{t-1} - \alpha_0 \,. \tag{3}$$

Equation (1) demonstrates the long-run equilibrium relationship between dependent and independent variables, where Area measures Macau's land area, and Wage measures median annual wage. Migrant in Equation (1) stands for the dependent variable, that is, the number of migrant workers in the first EC model and the number of immigrants in the second EC model. Equation (2) demonstrates the short-run relationship, where each of the symbols with Δ represents the first difference of the corresponding variable. A symbol with ti subscript represents the *i*th lag term. The *p*, *q* and *r* represent the optimal lag orders of the dependent variable and the two independent variables in each EC model. In Equation (2), each θ indicates a short-term regression coefficient, εt indicates the residual term, and δ indicates the coefficient of the error correction term "ecmt-1". Equation (3) shows how the error correction term is calculated. As is shown in Figures 1 and 2, indicators of population, economy and city scale in Macau all share a rising trend during the last few decades. Therefore, this study also includes the variable year in the models as a way to detrend and to avoid spurious regression, forming two EC models with unrestricted trends and unrestricted constants [57].

All variables included in a ARDL-ECM are required to be either I(0) or I(1) variables [53]. A stationary time series, i.e., a time series without trend or seasonality, is said to be integrated of order zero and denoted as I(0). A non-stationary time series with its first difference being stationary is said to be integrated of order one and denoted as I(1). The results of the Augmented Dickey–Fuller Test (ADF) in Table 2 show that the two dependent variables and two independent variables in this study are all I(1) variable (Normally the land area of a city over time would be a stationary series; however, the land area of Macau increased rather fast in the recent decades due to large-scale sea reclamation projects, to such an extent that it does have a trend and is not integrated of order zero). Each group of variables needs to pass a cointegration test to ensure cointegration relationship, i.e., a long-term equilibrium relationship, before being included in an ARDL-EC model [57]. The results of the bounds test for cointegration are also shown in Table 2. The variables in both EC models have F values and t values significant at 0.05, indicating that long-run relationships of both groups of variables have been confirmed. Optimal lag lengths of variables in these two EC models are selected based on the Akaike Information Criterion (AIC).

Table 2. Results of the ADF Test and Bounds Test.

ADF Test	**Migrant Workers**	**Immigrants**	**Land Area**	**Median Wage**
ADF Value	1.125	2.564	0.229	2.183
ADF Value of First Difference	−2.616 *	−4.826 ***	−6.181 ***	−2.667 *
Integration Order	I(1)	I(1)	I(1)	I(1)
Bounds Test	ECM for Migrant Workers		ECM for Immigrants	
F	12.253***		6.677*	
t	−5.680***		−4.172*	
Optimal Lag Lengths (AIC)	[3,0,3]		[2,2,2]	

Note: Significance: *< 0.05; **< 0.01; ***< 0.001.

4. Findings

4.1. Influential Factors on the Number of Migrant Workers

Table 3 demonstrates the results of regression analysis on the number of migrant workers in Macau using ECM. Results at the Long Run section clearly show that both land area and median wage have long-term positive effects on the number of migrant workers, which are significant at 0.01 and 0.001, respectively. In the long term, increasing 1 km^2 of Macau's land area would lead to an increase in the number of migrant workers by 8050, and each MOP 1000 increase in the median annual wage would result in the number of migrant workers increasing by 1642. These results offer support to our research hypotheses 1 and 3, that the sea reclamation project and the rising wage level are significant driving forces of the growth of labor migration in Macau.

The adjusted R^2 is 0.854, indicating that over 80% of the variance of the number of migrant workers in Macau can be explained by this model. Figure 4 shows the model fitting results, where the curve of the annual variation of labor migration fitted by the model (dashed line) is very close to the curve of the actual annual variation (solid line), further indicating that the model fits well.

4.2. Influential Factors on the Number of Immigrants

Table 4 shows the results of regression analysis of the number of immigrants in Macau. Both land area and median wage have long-term negative effects on the number of immigrants, which are significant at 0.001 and 0.01, respectively. In the long run, increasing 1 km^2 of Macau's land area leads to a decrease in immigrants by 11,855, and each MOP

1000 increase in the median annual wage reduces the number of immigrants by 527. These results provide support to the research hypotheses 2 and 4. They also imply that Macau has allocated so much land to the gaming–tourism industry that the growth of its living space failed to keep pace with the city's development: hence, the expansion of the city leads to a higher dependency on migrant workers with small demand for land resources, and to a reduced number of immigrants for whom long-term occupancy of land and housing is essential.

Table 3. ECM Predicting the Number of Migrant Workers in Macau (n = 25, Adj. R^2 = 0.854).

	Coef.	S.E.	t	$p > t$
EC Term	−1.400 ***	0.247	−5.68	0
Long Run				
Land Area (km^2)	8050 **	2353	3.42	0.004
Median Wage (1000 MOP)	1642 ***	122	13.46	0
Short Run				
Non-Immigrant Worker				
First Lag of First-Difference	0.391 *	0.183	2.13	0.05
Second Lag of First-Difference	0.346	0.225	1.54	0.145
Median Wage (1000 MOP)				
First-Difference	−561	348	−1.61	0.128
First Lag of First-Difference	−742 *	334	−2.22	0.042
Second Lag of First-Difference	1160 **	332	3.49	0.003
Year	−9466**	2928	−3.23	0.006
Constant	18,600,000 **	5,751,746	3.23	0.006

Note: Significance: *< 0.05; **< 0.01; ***< 0.001.

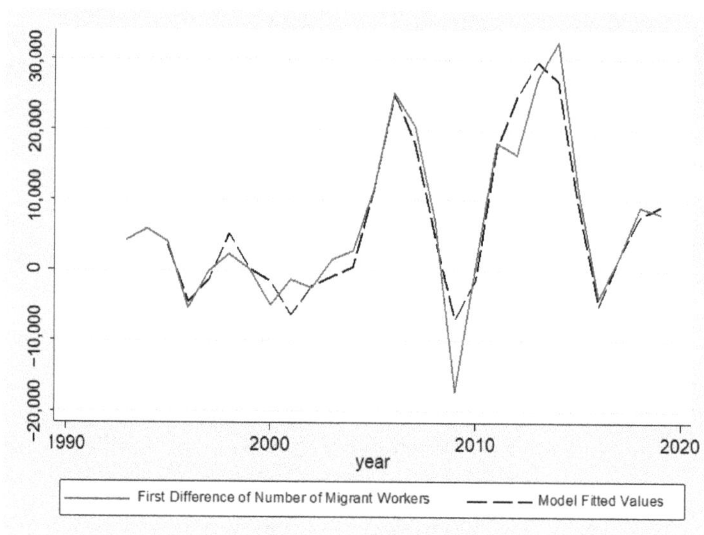

Figure 4. The Fitted Curve of The First-Difference of The Number of Migrant Workers.

The adjusted R^2 of this model is 0.465, indicating that nearly half of the variation in the number of immigrants in Macau has been explained. As is shown in Figure 5, the annual variation in the model-fitted number of immigrants (dashed line) is relatively closer to the actual annual variation (solid line) after the year 2000. The poor fit before 1999 may be caused by the great changes in the immigration situation of Macau before its return to China.

Table 4. ECM Predicting the Number of Immigrants in Macau ($n = 25$, Adj. $R^2 = 0.465$).

	Coef.	S.E.	t	$p > t$
EC Term	−1.555 ***	0.373	−4.17	0.001
Long Run				
Land Area (km^2)	−11,855 ***	2578	−4.60	0
Median Wage (1000 MOP)	−527 **	169	−3.11	0.001
Short Run				
Non-Immigrant Worker				
First Lag of First-Difference	0.452$^+$	0.246	1.84	0.086
Land Area (km^2)				
First Lag of First-Difference	12,655 *	4927	2.57	0.021
Second Lag of First-Difference	5361$^+$	2988	1.76	0.093
Median Wage (1000 MOP)				
First-Difference	712 *	289	2.46	0.026
First Lag of First-Difference	525 *	245	2.14	0.049
Year	23,491 **	6394	3.67	0.002
Constant	−46,400,000 ***	12,600,000	−3.68	0.002

Note: Significance: $^+$< 0.1; *< 0.05; **< 0.01; ***< 0.001.

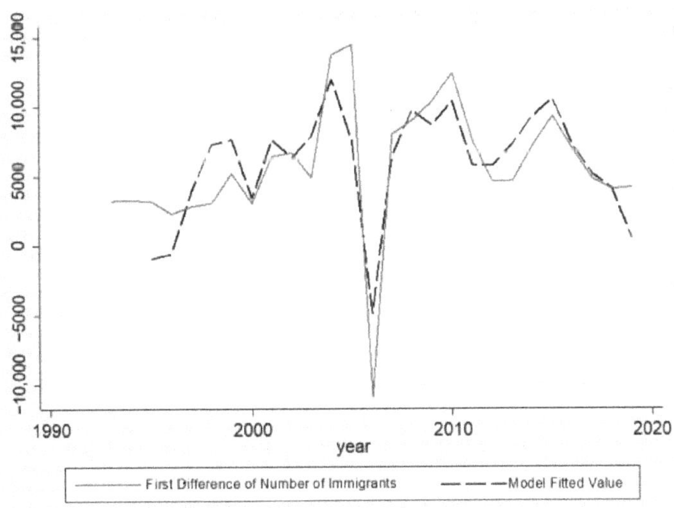

Figure 5. The Fitted Curve of the First-Difference of the Number of Immigrants.

4.3. Structural Difference between Locals and Non-Locals

In the 2021 Macau census, the non-local population numbers a total of 113,438, and includes 94,637 migrant workers and 18,771 non-local students who face policy restrictions and are not allowed to become permanent residents. Immigrants, on the contrary, are counted as part of the local population in the census. As is shown in Table 5, the labor force population (people aged 15–64) accounted for 99.5% of the non-local population, while it only accounted for 68.1% of the local population.

Table 5. Age Structure of Local and Non-Local Populations in Macau.

AGE GROUP	Local, N	Local, Per.	Non-Local, N	Non-Local, Per.	Total, n	Total, Per.
0–14	98,981	17.41%	30	0.03%	99,011	14.52%
15–19	23,643	4.16%	4532	4.00%	28,175	4.13%
20–24	27,878	4.90%	13,912	12.26%	41,790	6.13%
25–29	38,688	6.80%	15,587	13.74%	54,275	7.96%
30–34	51,094	8.98%	19,791	17.45%	70,885	10.39%
35–39	46,797	8.23%	17,674	15.58%	64,471	9.45%
40–44	34,400	6.05%	14,501	12.78%	48,901	7.17%
45–49	37,278	6.56%	12,010	10.59%	49,288	7.23%
50–54	37,476	6.59%	8376	7.38%	45,852	6.72%
55–59	45,309	7.97%	4850	4.28%	50,159	7.35%
60–64	44,829	7.88%	1652	1.46%	46,481	6.81%
≧65	82,289	14.47%	523	0.46%	82,812	12.14%
Total	568,662	100%	113,438	100%	682,100	100%

Since those people belonging to the non-local population are not allowed to settle down, they only work or study for a few years and are unlikely to reproduce in Macau, and statistics excluding the non-local population can reflect the actual population situation. The proportion of the non-local population in 2021 is extremely large in Macau, equivalent to 24% of the local population. The non-native population had been equivalent to 14% of the local population in the 2011 census. When non-locals are excluded from the statistics, the labor share in Macau falls from 73.3% to 69.1%, and the proportion of the elderly (aged 65 and up) rises from 12.1% to 14.5%. The proportion of women of prime reproductive age in Macau (20–34 years old) also falls from 27.7% to 24.4% when non-natives are excluded from the statistics (Figure 6).

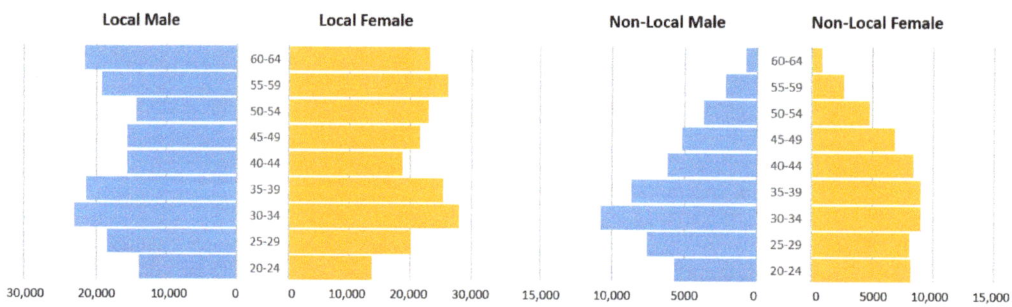

Figure 6. Age Structure of the 20–64 Year-Old Local and Non-Local Populations in Macau.

Table 6 demonstrates the educational attainment of Macau's local and non-local populations aged 25 and above. The proportion of people with higher education in the local population is 31.9%, a bit higher than that for the non-local population (27.7%). Benefiting from funding from the rich government and local universities hiring large numbers of non-local professors, the higher education coverage rate among local residents has more than doubled in Macau since 1999 [6]. However, 64.4% of the non-local population has a high school education or above, which is 8.3 percent points higher than high school coverage rate of the local population. Given the intensifying competition for talent in the Pearl River Delta [58], and the impact of post-pandemic deglobalization [59,60], it remains to be seen whether Macau will be able to attract another wave of migrant workers with similar educational attainment as a succession of the current wave. Assuming that Macau fails in the competition for non-local laborers, it might gradually lose 16.5% of its workforce with higher education and 25.6% of its workforce with high school diplomas, resulting in an enormous number of job vacancies that will not be able to be filled with the local workforce.

Table 6. Education Attainment of 25+ Local and Non-Local Populations in Macau.

Education	Local, N	Local, %	Non-Local, N	Non-Local, %	Total, N	Total, %
Lower than Elementary School	38,815	9.28%	2997	3.16%	41,812	8.15%
Elementary School	73,246	17.52%	10,030	10.56%	83,276	16.23%
Middle School	71,364	17.07%	20,736	21.84%	92,100	17.95%
High School	101,236	24.21%	34,902	36.75%	136,138	26.53%
Higher Education	133,499	31.93%	26,299	27.69%	159,798	31.14%
Total	418,160	100%	94,964	100%	513,124	100%

Prior studies show that localization of labor migration used to be an important approach for Macau to improve its population structure [37–39]. Results of regression analysis and comparison of non-local to local populations in this study also imply that converting a larger proportion of migrant workers to immigrants would still be a potential way for Macau to ensure sustainable population and manpower development.

Our four research hypotheses are supported by results of regression analysis, which implies that the current development strategies of Macau would only lead to higher dependency on migrant workers and a smaller quota for new immigrants. A vicious circle might lie within Macau's current urban development process: (A) due to rapid economic development and insufficient local labor accumulation, Macau must import some non-local manpower; (B) since the economic and social cost of hiring migrant workers is much lower than hiring immigrants, the government and employers are more in favor of non-resident migrant workers; (C) relying on migrant workers suppresses Macau's accumulation of local human resources, resulting in more serious shortages of local laborers in the future; (D) Macau has to import more non-local employers in the next round, because of the increasing vacancies in labor market due to economic growth and aging local population. Furthermore, once migrant labor supply declines, Macau's economic development would be interrupted and its population problems would deteriorate.

5. Discussion

Through applying the ARDL-ECM on population studies, this paper presents a more accurate long-term relationship between Macau's immigration, land and economy. Our study shows that the number of migrant workers increases as the land area and median wage increase. This is not a surprise, since Macau has long regarded temporary migrant workers as an important solution to its scarcity of land and social resources. Migrant workers are only allowed to work in Macau for a few years, so they can flexibly fill vacancies in the labor market without causing long-term pressure on the city's welfare and social resources [38]. Moreover, a large number of migrant workers from the Chinese mainland have become cross-border commuters, who work in Macau but reside in neighboring cities such as Zhuhai or Zhongshan, sparing the scarce housing resources for residents in Macau [61].

Our study also shows that land area and wage level are negatively correlated with the number of immigrants in Macau. This finding implies that Macau has been suppressing the chance for migrant workers to become residents during its urban development process. Although a large proportion of migrant workers are young and highly educated, they are not able to settle in Macau due to the restrictions of the immigration policy [10]. Unlike earlier immigrants who improved the population structure of Macau and alleviated aging, low fertility and other demographic problems [11], migrant workers' contributions to

Macau are relatively short-term. In fact, the findings of this study imply that under Macau's current migration policy, an increase in the number of migrant workers would result in a negative impact on sustainable population development.

Consistent with prior studies [6], this study reveals a critical problem of Macau's current development strategy, in that after allocating most of the new land reclaimed from the sea to the gaming–tourism industry, the city fails to reserve sufficient land resources for increased economic diversification or improvement in the living environment. The booming gaming–tourism industry not only affects Macau's attractiveness to non-local talent, but its high-salary and low-skill-requirement positions reserved for local residents also drives local students to concentrate on majors related to gambling, resulting in extremely high homogeneity of local manpower [45]. Macau's GDP plunged by more than half and gross income of the gaming industry dropped by 79.2% in 2020 due to the Covid-19 pandemic, clearly showing the vulnerability of an economic system solely based on gaming–tourism [62]. Meanwhile, the neighboring mainland cities in the Pearl River Delta have significantly increased the welfare for the non-local high-skilled population in recent years in order to compete for human resources [55,63]. Macau has become relatively less attractive to the non-local manpower that it relies on, and its strategy of rotating high-end workers has become harder to sustain. Therefore, Macau urgently needs to improve its immigration policies and increase opportunities for skilled migrant workers to gain residency, so as to attract non-local talent essential for its economy.

These findings not only help deepen the understanding of the driving factors within Macau immigration, but also allow for theoretical extension in the following aspects. First, Macau has long been depending on a migrant population, making it a suitable site for analyzing the long-term impact of migration on urban development. There is a rather consistent correlation between the decline in fertility and economic growth. At present, the total fertility rates (TFR) of most developed countries have dropped to below the replacement level (i.e., 2.1 children per woman) [64]. Cross-country comparisons of fertility policies show that policies supporting gender equality in the workforce are more effective than traditional family supports for stay-at-home mothers [65,66]. The low fertility issue in Confucian countries is even more severe, such that China's TFR was only 1.4 in the early-2000s, Korea's TFR in 2005 was 1.08, Japan's TFR in 2013 was 1.43, and Singapore's TFR in 2010 was 1.15 [63]. Fertility support programes in these countries have only slowed down the decline of TFR. In fact, except for a few Northern European countries that have achieved fertility growth through ultra-high social welfare [67], countries with low fertility level often reverse the declining TFR through influx of immigrants with higher fertility [68]. The TFR of Macau dropped to below 2.1 in 1970 and decreased by 22% in the following two decades [38]. This relatively slow decline was mainly due to the absorption of immigrants. In 1991 the fertility rate of mainland immigrants in Macau was twice that of locals. However, with the implementation of Dispatch 12/GM/88 and Order 2/90/M, Macau imposed restrictions on settlement of migrant workers, leading to an increase in the average age of new immigrants and a decrease in their fertility rate. From 1991 to 1996, Macau's TFR plummeted by 22% [38]. Macau can be regarded as a preview of future cities, where, in common with Macau, immigration is the main driving force of population growth [69], and offers important insights into the long-term demographic consequences of certain migrant policies.

Second, prior studies on tourism economies primarily focus on places that developed on the basis of comparative advantages in natural resources [51]. Macau, by contrast, makes more use of man-made landscapes, offering tourism, entertainment and consumption activities inside grand casino hotels. Among them, the Venetian Resort Hotel is the largest, occupying an area of 0.975 km^2, that is, 3% of Macau's total land area [44]. The advantage of Macau's economic model lies in its relatively low dependence on natural landscape, while its disadvantage is the very neglect of the natural environment. Macau's land reclamation has aggravated multiple environmental problems, such as inland sea pollution, coastal ecosystem degradation, flood resistance degradation, soil salinity increase, natural landscape destruction, and heat island effect [8]. Therefore, in addition to the unsustainable population growth and the unstable economic development [70,71], Macau's economic pattern is also environmentally unfriendly and thus conflicts with many United Nations Sustainable Development Goals. This study presents more facets of the development problems of tourism economies and urges the Macau government to restore the balance between short- and long-term development.

Third, the extremely high population density of Macau makes it a good research site to rethink the man–land relationship. Previous studies have shown that institutional resistance to migration mainly comes from local people's rejection of outsiders [72]. For example, after Chinese food buffets challenged the local restaurant industry in Germany, the German government judged that Chinese food buffets did not belong to the ethnic restaurant category and turned down work visa applications of all their employees [73]. While the local residents' view on immigrants also has a policy impact in Macau [11], due to the extreme shortage of land resources, economic development and improvement of housing conditions to some extent become mutually exclusive options, which in turn leads to competition between migrant workers and immigrants in the resource allocation process. The relationship between migration and land is highlighted under such conditions of tension.

6. Conclusions

Based on the annual data from the Macau Statistics and Census Service, this study analyzes the influencing factors in immigration trends in Macau between 1992 and 2019. Through analysis using ARDL-ECM, our study shows that two important indicators of Macau's urban development, its land area and wage level, have both positive long-term influence on the number of migrant workers, and negative long-term impact on the number of immigrants.

Our study suggests that one possible way out of Macau's current dilemma is to reduce the cost of urban expansion and improve economic diversity through cross-border cooperation with mainland cities. Macau has mainly expanded its city scale through sea reclamation, the high cost of reclamation programs driving it to become more dependent on the tax revenue from the gaming–tourism industry and hence to allocate yet more land and manpower to this pillar industry. Consequently, Macau cannot fundamentally solve its land shortage through sea reclamation, which only postpones the problem by relying on migrant workers who make low demands on land and social resources. Facing increasing short-term pressures, Macau is unable to address its long-term problems through recruiting skilled immigrants or nurturing local talent. Through transferring a larger proportion of its non-gaming sectors to cooperation zones, Macau might be able to spare land and other resources to absorb a larger amount of non-local talent as well, which would in turn increase the sustainability of its urban development.

Our study is not without limitations. Although Covid-19 and subsequent sudden changes have had a significant impact on the long-term patterns of immigration in Macau, it is not possible for us to include the pandemic period into the analysis of this paper due to data constraints. Both the Statistical Yearbook data and Census data provided by the Macau Statistics and Census Service have rather long intervals between survey time points, which put a limit on their ability to capture sudden changes. Thus, further study of the influential factors on Macau's immigration trends during and after the pandemic is essential. Moreover, censuses often sacrifice survey comprehensiveness for their coverage. Surveys and interviews on immigrants and migrant workers in Macau in the future should allow for deepened exploration of the changing mechanism. We also notice that regional factors have yet to be taken into account in the causal analysis of migration flows in Macau. In the future, we expect there to be studies combining data from Macau, Hong Kong and mainland cities in the Pearl River Delta which will present a more comprehensive picture of migration in this bay area.

Author Contributions: Conceptualization, methodology, and formal analysis, writing—original draft preparation, B.Z.; writing—review and editing, L.J. All authors have read and agreed to the published version of the manuscript.

Funding: This research was funded by the National Social Science Foundation of China, grant number 20CSH088.

Data Availability Statement: The data analyzed in this study are publicly available at the website of the Macau Statistics and Census Service: https://www.dsec.gov.mo/en-US/ (accessed on 5 November 2022).

Conflicts of Interest: The authors declare no conflict of interest.

References

1. Macau Statistics and Census Service (DSEC). *Macau Yearbook of Statistics 2019*; Welfare Printing: Macau, China, 2020; pp. 58–60.
2. Gu, X.H.; Li, G.Q.; Chang, X.; Guo, H.Z. Casino tourism, economic inequality, and housing bubbles. *Tour. Manag.* **2017**, *62*, 253–263. [CrossRef]
3. Sheng, M.; Gu, C. Economic growth and development in Macau (1999–2016): The role of the booming gaming industry. *Cities* **2018**, *1*, 72–80. [CrossRef]
4. Liu, M.T.; Dong, S.; Chang, S.; Tan, F. Macau gambling industry's quick V-shape rebound from 2014 to 2019. *Asia Pac. J. Mark. Logist.* **2021**, *2*, 449–473. [CrossRef]
5. Kim, S.S.; Park, J.Y.; Lee, J. Predicted Economic Impact Analysis of a Mega-Convention Using Multiplier Effects. *J. Conv. Event Tour.* **2010**, *1*, 42–61. [CrossRef]
6. Chen, P.; Huang, J. Macao's Resilient City Construction. *Glob. Urban Stud.* **2021**, *4*, 148–158.
7. Li, S. The transformation of island city politics: The case of Macau. *Isl. Stud. J.* **2016**, *2*, 521–536.
8. Yang, J.; Chen, Z.; Cha, F. Evaluation on the Effects of Encircling the Sea to Make Land to Macau's Tourist Environment—Based on HRP Methodology. *J. Huzhou Teach. Coll.* **2013**, *6*, 75–81.
9. Macau Trade and Investment Promotion Institute (IPIM). IPIM Processes Applications for and Renewals of Temporary Residence Permits for Specially Qualified Technicians in Accordance with the Law. Available online: https://www.gov.mo/zh-hans/news/327039 (accessed on 10 February 2021).
10. Yu, E.; Tam, N. The puzzle of Macao's talent development. *Asian Educ. Dev. Stud.* **2016**, *4*, 423–437. [CrossRef]
11. Zhao, L. An analysis of Macao's policy of importing foreign labor. *Macau Stud.* **1998**, *8*, 97–150.
12. Ming, L. Labor Shortage and the Development of Gaming Industry in Macau. *UNLV Theses Diss. Prof. Pap. Capstones* **2011**, *1144*, 1–38.
13. Wang, J. Global Bay Area Factors Flows Index and a Comparative Analysis on Flows of Six Bay Area Cities. *Shenzhen Soc. Sci.* **2020**, *6*, 5–20.
14. Huang, Q. Changes in Macao's population over the past 400 years. *South China Popul.* **1987**, *2*, 44–46.
15. Ravenstein, E.G. The laws of migration. *J. R. Stat. Soc.* **1885**, *48*, 167–227. [CrossRef]
16. Lee, E.S. A theory of migration. *Demography* **1966**, *1*, 47–57. [CrossRef]
17. Todaro, M.P. A model of labor migration and urban unemployment in less developed countries. *Am. Econ. Rev.* **1969**, *1*, 138–148.
18. Harris, J.R.; Todaro, M.P. Migration, unemployment and development: A two-sector analysis. *Am. Econ. Rev.* **1970**, *1*, 126–142.
19. Stark, O.; Levhari, D. On Migration and Risk in LDCs. *Econ. Dev. Cult. Chang.* **1982**, *31*, 191–196. [CrossRef]
20. Rosenzweig, M.R.; Stark, O. Consumption Smoothing, Migration and Marriage: Evidence from Rural India. *J. Political Econ.* **1989**, *4*, 905–926. [CrossRef]
21. Carling, J. Migration in the age of involuntary immobility: Theoretical reflections and Cape Verdean experiences. *J. Ethn. Migr. Stud.* **2002**, *1*, 5–42. [CrossRef]
22. de Haas, H. The internal dynamics of migration processes: A theoretical inquiry. *J. Ethn. Migr. Stud.* **2010**, *10*, 1587–1617. [CrossRef]
23. Schewel, K. Understanding immobility: Moving beyond the mobility bias in migration studies. *Int. Migr. Rev.* **2020**, *2*, 328–355. [CrossRef]
24. de Haas, H. Migration and development: A theoretical perspective. *Int. Migr. Rev.* **2010**, *1*, 227–264. [CrossRef] [PubMed]
25. Basok, T. He Came, He Saw, He Stayed: Guest Worker Programmes and the Issue of Non-Return. *Int. Migr.* **2000**, *2*, 215–238.
26. Seol, D.; Skrentny, J. Why Is There So Little Migrant Settlement in East Asia? *Int. Migr. Rev.* **2009**, *3*, 578–620. [CrossRef]
27. Ferguson, J. *Global Shadows: Africa in the Neoliberal World Order*; Duke University Press: Durham, NC, USA, 2006; p. 14.
28. Hugo, G. International Labour Migration and Migration Policies in Southeast Asia. *Asian J. Soc. Sci.* **2012**, *40*, 392–418. [CrossRef]
29. Xiang, B. Labor Transplant 'Point-to-Point' Transnational Labor Migration in East Asia. *South Atl. Q.* **2012**, *4*, 721–739. [CrossRef]
30. Zhou, B.; Zhong, Y. Instability in the Cross-Border Labor Market: A Study on the High Job Turnover of Migrant Workers from Rural Vietnam to Rural China. *Sustainability* **2022**, *12*, 7447. [CrossRef]
31. Xiang, B. Return and the Reordering of Transnational Mobility in Asia. In *Return: Nationalizing Transnational Mobility in Asia*; Xiang, B., Yeoh, B., Toyota, M., Eds.; Duke University Press: Durham, NC, USA, 2013; pp. 1–20.

32. de Haas, H. A theory of migration: The aspirations-capabilities framework. *Comp. Migr. Stud.* **2021**, *9*, 8–42. [CrossRef]
33. Massey, D.S.; Arango, J.; Hugo, G.; Kouaouci, A.; Pellegrino, A.; Taylor, J.E. Theories of international migration: A review and appraisal. *Popul. Dev. Rev.* **1993**, *3*, 431–466. [CrossRef]
34. Hao, Z. Social Problems in Macau. *China Perspect.* **2005**, *62*, 1–18. [CrossRef]
35. Huang, Q.; Zheng, W. *Four Hundred Years of Macao's Economy*; Macau Foundation: Macau, China, 1994; pp. 163–167.
36. Lin, F. *Walking the Needle Line: Oral History of Garment Workers in Macau*; Guangxi Normal University Press: Guilin, China, 2018; p. 100.
37. Edmonds, R.L.; Kyle, W.J. Land use in Macau: Changes between 1972 and 1994. *Land Use Policy* **1988**, *4*, 271–292. [CrossRef]
38. Gao, Z. A brief analysis of the characteristics of population birth in Macau in the past 30 years. *Popul. Econ.* **1999**, *6*, 11–16.
39. Macau Statistics and Census Service (DSEC). *Trends and Challenges of Population Aging*; DSEC: Macau, China, 2013; pp. 7–17.
40. Macau Printing Bureau. Law 16/2021 Legal system for entry and exit control, stay and residence permit in the Macau SAR. Available online: https://bo.io.gov.mo/bo/i/2021/33/lei16_cn.asp (accessed on 5 November 2022).
41. Chen, D. *Hongkong-Macau Economic Yearbook 2000*; Hongkong-Macau Economic Yearbook Press: Beijing, China, 2001; pp. 153–157.
42. Cai, C. *Hongkong-Macau Economic Yearbook 2010*; Hongkong-Macau Economic Yearbook Press: Beijing, China, 2011; pp. 197–212.
43. Ruhs, M.; Bridget, A. *Who Needs Migrant Workers? Labour Shortages, Immigration, and Public Policy*; Oxford University Press: Oxford, UK, 2010; pp. 15–56.
44. Yuan, C.; Jie, Y. Research on the Adaptability between Gaming Development and the Carrying Capacity of Macao. *Hong Kong Macao J.* **2014**, *4*, 55–61.
45. Liu, Z.; Sun, X. Economic Structure and Education Structure: Research on Education Structure Unbalance in Macau. *J. Asia Pac. Econ.* **2012**, *5*, 144–148.
46. Zhang, S. Macao labor problem and its solution. *J. Asia Pac. Econ.* **1988**, *6*, 49–53.
47. Cai, X. Migrant workers in Macau: Globalization and labor. In *The Selection of Studies in Humanities and Social Sciences of Macau, Society*; Cheng, T., Ed.; Social Sciences Academic Press: Beijing, China, 2004; pp. 197–221.
48. Hu, Z.; Li, Y.; Ruan, X. Research on Macao's Man-Land Relationship. *Acta Geogr. Sin.* **1999**, *6*, 481–486.
49. Guo, Y. A Discussion on Zhuhai-Macau joint development of Hengqin. *J. Guangdong Inst. Public Adm.* **2010**, *4*, 84–88.
50. Corden, W.M.; Neary, J.P. Booming sector and de-industrialisation in a small open economy. *Econ. J.* **1982**, *92*, 825–848. [CrossRef]
51. Capo, J.; Font, A.R.; Nadal, J.R. Dutch disease in tourism economies: Evidence from the Balearics and the Canary Islands. *J. Sustain. Tour.* **2007**, *6*, 615–627. [CrossRef]
52. Yan, J. Research on the Immigration Policy for Macao. Master's Thesis, Huazhong University of Science and Technology, Wuhan, China, 2009; pp. 15–29.
53. Pesaran, M.H.; Shin, Y.; Smith, R.J. Bounds testing approaches to the analysis of level relationships. *J. Appl. Econom.* **2001**, *1*, 289–326. [CrossRef]
54. Mizon, G.E. Model selection procedures. In *Studies in Modern Economic Analysis*; Artis, M., Nobay, R., Eds.; Basil Blackwell: Oxford, UK, 1977; pp. 97–120.
55. Alogoskoufis, G.; Smith, R. On error correction models: Specification, interpretation, estimation. *J. Econ. Surv.* **1991**, *1*, 97–128. [CrossRef]
56. Pinshi, C.P. Rethinking error correction model in macroeconometric analysis: A relevant review. *J. Appl. Econ. Sci.* **2020**, *15*, 267–274.
57. Hassler, U.; Wolters, J. Autoregressive Distributed Lag Models and Cointegration. *Allg. Stat. Arch.* **2006**, *90*, 59–74.
58. China Daily. Zhuhai Talent Program—Unsurpassed in the Pearl River Delta. Available online: https://govt.chinadaily.com.cn/a/201904/24/WS5cc01132498e079e6801ede3.html (accessed on 26 April 2018).
59. Leblang, D.; Peters, M.E. Immigration and Globalization (and Deglobalization). *Annu. Rev. Political Sci.* **2022**, *25*, 377–399. [CrossRef]
60. Dionne, K.; Turkmen, F. The politics of pandemic othering: Putting COVID-19 in global and historical context. *Int. Organ.* **2020**, *1*, 213–230. [CrossRef]
61. Ju, B.; Sandel, T. Perceived understandings of home and self identity among Mainland Chinese dual migrants in Macao. *J. Intercult. Stud.* **2020**, *1*, 99–115. [CrossRef]
62. Macau Statistics and Census Service (DSEC). *Analysis Report of Statistical Indicator System for Moderate Economic Diversification of Macao 2020*; DSEC: Macau, China, 2021; pp. 8–9.
63. Wang, X.; Wang, X.; Liu, F. Social Welfare, Hukou Promise for Talents and Regional Innovation Performance: Re-examination of the "Talent Scramble Battle". *Econ. Sci.* **2022**, *3*, 65–78.
64. Wesolowski, K.; Ferrarini, T. Family policies and fertility: Examining the link between family policy institutions and fertility rates in 33 countries 1995–2011. *Int. J. Sociol. Soc. Policy* **2018**, *38*, 1057–1070. [CrossRef]
65. Billingsley, S.; Ferrarini, T. Family policy and fertility intentions in 21 European countries. *J. Marriage Fam.* **2014**, *2*, 428–445. [CrossRef]
66. Tang, Z.; Deng, H. Experience and implications of family support policies in Japan, Korea and Singapore. *Soc. Sci. Abroad* **2018**, *2*, 36–42.
67. Grill, A. The myth of unpaid family leave: Can the united states implement paid leave policy based on the swedish model. *Comp. Labor Law J.* **1996**, *2*, 373–397.

68. Volant, S.; Pison, G.; Héran, F. French fertility is the highest in Europe. Because of its immigrants? *Popul. Soc.* **2019**, *7*, 1–4.
69. Chen, Y.; Guo, F.; Wang, J.; Cai, W.; Wang, C.; Wang, K. Provincial and gridded population projection for China under shared socioeconomic pathways from 2010 to 2100. *Sci. Data* **2020**, *7*, 83–95. [CrossRef] [PubMed]
70. Macau Statistics and Census Service (DSEC). *Macau Yearbook of Statistics 2021*; San Hip Wo Printing: Macau, China, 2022; pp. 60–62.
71. Alves, J.C.; Lok, T.C.; Luo, Y.B.; Hao, W. Crisis Management for Small Business during the COVID-19 Outbreak:Survival, Resilience and Renewal Strategies of Firms in Macau. *Front. Bus. Res. China* **2020**, *14*, 26. [CrossRef]
72. Seol, D.H. Global Dimensions in Mapping the Foreign Labor Policies of Korea: A Comparative and Functional Analysis. *Dev. Soc.* **2005**, *1*, 75–124.
73. Zhu, Q. Discussion on the scale Development of Chinese Restaurants in Germany and the "competition" of ethnic economy. In *Mobility and Governance: Global Talent, Migration and Immigration Law*; Wang, H., Liu, G., Eds.; World Affairs Press: Beijing, China, 2019; pp. 235–244.

land

Article

Rural Transformation Development and Its Influencing Factors in China's Poverty-Stricken Areas: A Case Study of Yanshan-Taihang Mountains

Yuanzhi Guo [1,2,*] and Wenyue Zhong [1,2,3]

1 Institute of Geographic Sciences and Natural Resources Research, Chinese Academy of Sciences, Beijing 100101, China; zhongwenyue@webmail.hzau.edu.cn
2 Key Laboratory of Regional Sustainable Development Modeling, Chinese Academy of Sciences, Beijing 100101, China
3 University of Chinese Academy of Sciences, Beijing 100049, China
* Correspondence: guoyz.16b@igsnrr.ac.cn

Abstract: Rural China has undergone a rapid transformation in the past few decades, especially the poverty-stricken areas, making a historic leap from inadequate subsistence to full well-off status. Based on rural regional system theory, this study analyzes the connotation of rural transformation development (RTD), explores the spatio-temporal patterns of RTD in the Yanshan-Taihang Mountainous, and diagnoses its influencing factors using a geographically and temporally weighted regression model. The results show that RTD is a dynamic process of qualitative changes in rural regional systems based on the accumulation of quantitative changes of elements, and the key to its measurement lies in analyzing the coupling coordination degree between quantitative changes of elements. From 2000 to 2020, the rapid development of urban population share, non-agricultural industry share, construction land share and NDVI in the Yanshan-Taihang Mountains contributed to a leap in RTD status, and the proportion of counties in a coupling coordination state increased from 24.24% to 96.97%. Spatially, the RTD level in the Taihang Mountains was significantly superior to that in the Yanshan Mountains. Average years of schooling, road density, per capita GDP and urban–rural dual structure were the main influencing factors of RTD, of which the first three were positive factors and the last one presented a negative correlation. To promote RTD to a higher level, it is an urgent matter to boost the high-quality development of county economy and rural education, improve public transportation infrastructure and innovate the policy system.

Keywords: rural regional system; rural transformation development; poverty-stricken areas; human–land relationship; rural revitalization; Yanshan-Taihang Mountains

Citation: Guo, Y.; Zhong, W. Rural Transformation Development and Its Influencing Factors in China's Poverty-Stricken Areas: A Case Study of Yanshan-Taihang Mountains. *Land* **2023**, *12*, 1080. https://doi.org/10.3390/land12051080

Academic Editor: Le Yu

Received: 14 April 2023
Revised: 12 May 2023
Accepted: 15 May 2023
Published: 17 May 2023

1. Introduction

The countryside is a geographical complex with natural and human characteristics and multiple functions such as production, living, ecology and culture outside the built-up urban areas [1,2]. Since the 1950s, the continuous advancement of globalization as well as the rapid development of industrialization and urbanization have driven rural transformation from production-based to consumption-based, and then to multifunctional and globalized-based [3]. During this process, the characteristics and functions of rural areas have changed dramatically. Meanwhile, international research on rural transformation development (RTD) has shown a clear divergence. One is to continue the previous framework of political economy, that is, exploring rural areas from a political–economic perspective and policy framework and analyzing the influences of policy adjustments on RTD [4]. The other is to redefine the spatial composition and values of rural areas, advocating themes such as pastoralism, post-ruralism, multifunctionality and gentrification [5,6], and to shift the understanding of rural areas from a physical to a socially constructed theory. Essentially,

RTD refers to the changes in rural socio-economic forms and rural spatial reconstruction under the combined effects of internal and external forces, and the core lies in the transformation of industrial–agricultural and urban–rural relations [7,8]. Due to the mutagenicity of RTD, it has a profound impact on national food security, farmers' income increases and social equality (especially poverty alleviation).

China is a developing country with a long history of farming and a large rural population [9,10]. During the various historical periods of China's revolution, construction and reform, promoting RTD has been an important mainline of national economic and social development [11]. The introduction of a series of policies to support agricultural and rural development has enabled rural China to achieve world-renowned achievements, realizing a historical leap from national liberation to getting rid of poverty and then to a well-off society in an all-round way [12]. Therefore, transformation is one of the most prominent features of China's rural development in the past few decades, and it is also the core issue of China's rural development in the current and future period [7,11].

The spatial carrier of rural development is the rural regional system [13,14]. Therefore, RTD is a complex process involving elemental changes, structural adjustment and functional transformation of rural regional system, in which elemental changes cause the structural adjustment of the system, and structural adjustment determines the transformation of the overall function and state of the system [2,7,13]. Meanwhile, the spatial heterogeneity of physical geography and regional imbalance of socioeconomics determine the complexity and diversity of RTD patterns and types [15]. Due to the important roles of population, land and industry in rural development, most studies on China's RTD focus on demographic transition [16,17], industrial transition [18–21] and land use transition [22–24], aiming to reveal the general characteristics of the evolution of rural regional system through the analysis of elemental changes. Given the complexity of RTD, comprehensive indexes such as rurality [25,26], rural development level [8] and rural multifunctional index [14,27] measure the status of RTD from multiple dimensions. Additionally, analytical frameworks such as vulnerability and resilience have also been employed to analyze the level and status of RTD across time and space [28,29].

Since entering the 21st century, China has actively promoted the implementation of major strategies such as the new socialist countryside construction and targeted poverty alleviation, especially the latter, which has completely solved the long-standing problem of absolute poverty in rural Central and Western China [12,30]. For the poverty-stricken areas, getting rid of poverty is the biggest transformation of their development. In line with the current shift of China's rural development focus from poverty alleviation to rural revitalization [31], an in-depth analysis of the regional differences and influencing factors of RTD in typical poverty-stricken areas is of great significance in guiding China's poverty-stricken areas to consolidate and expand antipoverty achievements and comprehensively promote rural revitalization. On the other hand, the global antipoverty process has suffered a setback for the first time in over 20 years due to multiple challenges such as the COVID-19 pandemic, climate change and geopolitical conflicts [32,33]. If the current trends continue, nearly 7% of the world's population, or nearly 600 million people, will still be struggling in extreme poverty by 2030 [34,35]. China is the first developing country in the world to achieve UN Sustainable Development Goal (SDG) 1 [31]. The study of antipoverty practices in China's typical poverty-stricken areas will help summarize antipoverty experiences and contribute Chinese wisdom to global poverty alleviation, thereby supporting the timely achievement of the 2030 SDGs.

Following the introduction, Section 2 constructs a cognitive framework of RTD, selects targeted indicators to feature RTD, and develops a model to diagnose the factors influencing the regional divergence of RTD. Section 3 briefly introduces the study area and the data used in this study. Section 4 presents the spatial and temporal patterns of RTD in the Yanshan-Taihang Mountains from 2000 to 2020 and identifies the influencing factors of RTD using the geographically and temporally weighted regression (GTWR) model. Section 5 re-examines RTD and its influencing factors and explores key measures for rural revitalization in the Yanshan-Taihang Mountains and the implications of this study for global poverty alleviation and development. Conclusions are drawn in Section 6.

2. Theory and Methodology

2.1. Theoretical Framework

Transformation refers to the process of fundamental changes in the way things operate and the form of their structures, reflecting the change of things from one qualitative state to another. From the perspective of dialectical materialism, this transformation is also known as qualitative change, and is one of the basic states of the movement of things [36]. Another state of the movement of things corresponding to it is quantitative change, which reveals the increase/decrease in the number of things, the adjustment of places and the changes in the arrangement of various components within things. In essence, quantitative change is a continuous, universal, inconspicuous, gradual and objective change [37,38]. When the accumulated quantitative change does not exceed a specific threshold, it does not change the nature of things; while when it exceeds this threshold, it triggers a change in the nature of things. Generally, the development of things begins with quantitative change, which is the premise of qualitative change, and qualitative change is the result of quantitative change. Meanwhile, qualitative change opens the way for new quantitative change, making things start new quantitative change on the basis of qualitative change. In this way, things continue to develop from a low to a high level, promoting the emergence of new things and the demise of old things.

The rural regional system is a spatial system with a certain structure, function, and inter-regional connection, which is composed of human, economic, resource and environmental elements [15]. Among them, human elements mainly refer to social matters closely related to human development; economic elements mainly refer to various economic activities in rural areas; resource elements mainly refer to water, land, atmosphere, and other natural resources that can be developed and used by human beings; and environmental elements mainly refer to the earth's surface environment composed of the atmosphere, hydrosphere, lithosphere, biosphere, etc. The changes in the elements within the system constitute the initial dynamics of RTD [7]. They are linked through material circulation, energy conversion and information transmission, and constantly interact with external systems, allowing rural regional system to evolve from simple to complex. In this process, the organizational structure and operational mode of rural regional system are also constantly improved, promoting the evolution of system function and state. Poverty is a state presented by element shortages and structural imbalances in rural regional system, and the purpose of poverty alleviation is to improve the function and state of rural regional system through targeted measures to compensate for element shortages and optimize structural organization [39]. When compensating for shortcomings in the period of quantitative change, rural areas are still in poverty; only when the optimization and adjustment of the system makes rural socio-economic development reach a certain level can poverty-stricken areas get rid of poverty. With the improvement in sustainable development capability, poverty-stricken areas can achieve the goal of eliminating absolute poverty and gradually transform to the stage of affluence (Figure 1).

RTD reveals the dynamic process of qualitative changes based on accumulated quantitative changes in the rural regional system. Therefore, to scientifically understand RTD, the core is to analyze the changes in system function and state based on the changes in system element, where the changes in elements reveal the quantitative changes of the system, the evolution of the function and state reflects the qualitative changes of the system, and the key to bridging the change of element and the evolution of function and state is the optimization and adjustment of system structure.

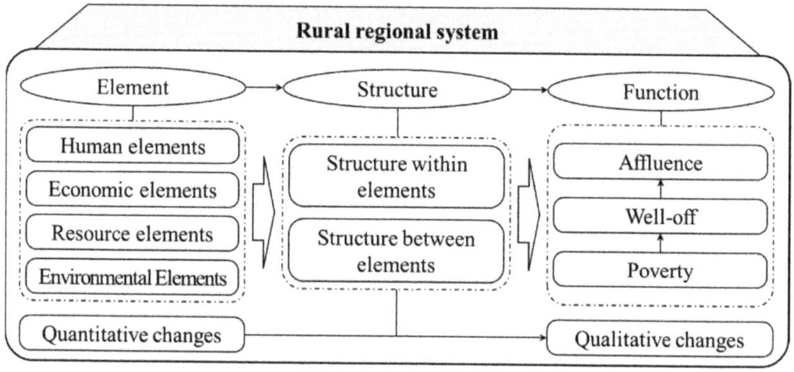

Figure 1. Theoretical framework of RTD.

2.2. Research Methods

2.2.1. Measurement of RTD

According to the theoretical understanding of RTD, the development of human, economic, resource and environmental elements in rural regional system is the quantitative change process of RTD. Based on the availability of data and drawing on existing studies [7,8,40–42], the urban population share, non-agricultural industry ratio, construction land share and normalized difference vegetation index (NDVI) are employed to represent the changes of human, economic, resource and environmental elements in rural regional system, respectively, thereby revealing the quantitative changes of rural regional system (Table 1). In terms of the connotation of each indicator, they are all positive, i.e., the higher their values, the more quantitative changes RTD accumulates, and the easier it is for the function of the rural regional system to leap forward and achieve qualitative changes.

Table 1. Indexes characterizing the quantitative changes of rural regional system.

Dimension	Indicator	Description
Human elements	Urban population share	Urban resident population/total resident population
Economic elements	Non-agricultural industry share	Added-value of secondary and tertiary industries/GDP
Resource elements	Construction land share	(Total of urban and rural residential areas, transportation, industrial and mining and other land)/total area of the region
Environmental elements	NDVI	An index reflecting the status of land vegetation cover

Systems theory suggests that the function and state of a system depend not only on the number of elements, but also on the structural organization formed by the interaction between elements [43]. Therefore, this study uses the coupling coordination degree between elements to characterize the function and state of rural regional system and reveal the qualitative changes of rural regional system, i.e., the RTD level at a certain point in time. In general, the coupling coordination degree is calculated as follows:

$$D_{it} = \sqrt{T_{it} \times C_{it}} = \sqrt{(aHum_{it} + bEcon_{it} + cRes_{it} + dEnv_{it}) \times \left(4 \times \frac{\sqrt[4]{Hum_{it} \times Econ_{it} \times Res_{it} \times Env_{it}}}{Hum_{it} + Econ_{it} + Res_{it} + Env_{it}}\right)} \quad (1)$$

where D_{it} is the coupling coordination degree of human (*Hum*), economic (*Econ*), resource (*Res*) and environmental (*Env*) elements of the rural regional system *i* in period *t*; C_{it} and T_{it} are the coupling degree and comprehensive coordination index of the four elements in the rural regional system, respectively; and *a*, *b*, *c* and *d* are the weights of human, economic, resource and environmental elements in the rural regional system, respectively. The weight values obtained with the combined Delphi method [44] and entropy weight method [45] are 0.3259, 0.2155, 0.2447 and 0.2139, respectively. Additionally, the RTD state characterized by the coupling coordination degree of elements can be divided into eight levels: severe disorder (0.00~0.20), moderate disorder (0.21~0.30), mild disorder (0.31~0.40), imminent disorder (0.41~0.50), barely coupling coordination (0.51~0.60), primary coupling coordination (0.61~0.70), intermediate coupling coordination (0.71~0.80) and advanced coupling coordination (0.81~1.00).

2.2.2. Pre-Section of Influencing Factors and Model Construction

The dynamics of rural development has been at the core of research on issues concerning agriculture, rural areas and farmers. Since the 1950s, classical rural development theories such as exogenous rural development theory, endogenous rural development theory and comprehensive rural development theory have been developed [46,47]. Exogenous rural development theory places rural areas in the macro context of regional socioeconomic development and argues that regional industrialization and urbanization can affect rural development through the spillover of their polarization effects, thus realizing "top-down" rural development [48]. Endogenous rural development theory focuses on the natural resources and human elements of rural areas and believes that the potential of rural resources can be fully explored through appropriate methods to achieve "bottom-up" rural development [49]. Comprehensive rural development theory overcomes the deficiencies of the previous two theories by focusing on the linkage and balance of internal and external factors in rural areas and pointing out that rural development is the result of a combination of local endogenous forces and external driving forces [50,51].

Essentially, RTD is the result of the interaction between the internal and external subsystems of rural regional system. The key to the former lies in the strength of rural self-development capabilities, while the latter mainly depends on regional industrialization and urbanization [52,53]. Based on data availability and a review of existing studies [8,52,54,55], four variables, namely average altitude, per capita farmland area, average years of schooling, and road density, were selected to feature the level of rural self-development capacity, and four variables, namely, per capita GDP, level of agricultural mechanization, per capita income of rural households, and urban–rural dual structure, were employed to characterize the development of regional industrialization and urbanization (Table 2).

Table 2. Pre-selection of variables influencing RTD level.

Variables	Description
Average altitude	The average value of DEM
Per capita farmland area	Total farmland area/Rural registered population
Average years of schooling	Average years of schooling for the resident population aged 6 and above
Road density	Total length of road/Total area of the region
Per capita GDP	GDP/Total resident population
Level of agricultural mechanization	Total power of agricultural machinery/Total sown area
Per capita income of rural households	Per capita disposable income of rural households
Urban–rural dual structure	Ratio of per capita disposable income of urban and rural households

Among the commonly used regression models, the ordinary least squares (OLS) model ignores spatial effects, and the geographically weighted regression (GWR) model ignores temporal effects although spatial effects are considered. Therefore, the estimation results of these models are biased. Based on the GWR model, Huang et al. [56] proposed the GTWR model, which takes both temporal and spatial effects into consideration, making the model estimation results more accurate and providing an analytical basis for dealing with spatio-temporal non-stationarity. Here, the GTWR model is employed to explore the factors influencing RTD in the Yanshan-Taihang Mountains from 2000 to 2020, and its expression is as follows:

$$y_i = \beta_0(\mu_i, v_i, t_i) + \sum_{k=1}^{p} \beta_k(\mu_i, v_i, t_i)x_{ik} + \varepsilon_i \quad (2)$$

where y_i is the RTD level of county i; x_{ik} is the value of the kth independent variable of county i; u_i and v_i are the longitude and latitude of the center of gravity, respectively; t_i is the time sequence; β_0 (u_i, v_i, t_i) is the regression constant, which is also the intercept of the GTWR model; β_k (u_i, v_i, t_i) is the kth regression coefficient; p is the total number of explanatory variables; and ε_i is the residual term of the model.

3. Study Area and Data Sources

3.1. Study Area

The Yanshan-Taihang Mountains are located in the hinterland of the Yanshan and Taihang Mountains, with a total area of 93,000 km^2 (Figure 2). In terms of administrative division, it covers 33 counties in Hebei, Shanxi and Inner Mongolia, including 25 old revolutionary counties and 5 ethnic autonomous counties. Due to the constraints such as fragile ecology, poor infrastructure and frequent disasters, poverty has long been a prominent problem and challenge for rural development in the Yanshan-Taihang Mountains. If calculated at the constant price of 2300 yuan in 2010, the rural poverty-stricken population in the Yanshan-Taihang Mountains reached 2.19 million in 2013, with a poverty incidence of 31.58%. Generally, rural poverty in this region presents the main characteristics of large amount, wide distribution, deep degree and difficulty in poverty alleviation [57]. Over the past decades, the Yanshan-Taihang Mountains have actively promoted poverty alleviation and development based on rapid economic development and continuous policy investment. In 2020, the elimination of absolute poverty in the rural Yanshan-Taihang Mountains indicated that rural development in this region has entered a new stage of consolidating and expanding antipoverty achievements and comprehensively promoting rural revitalization.

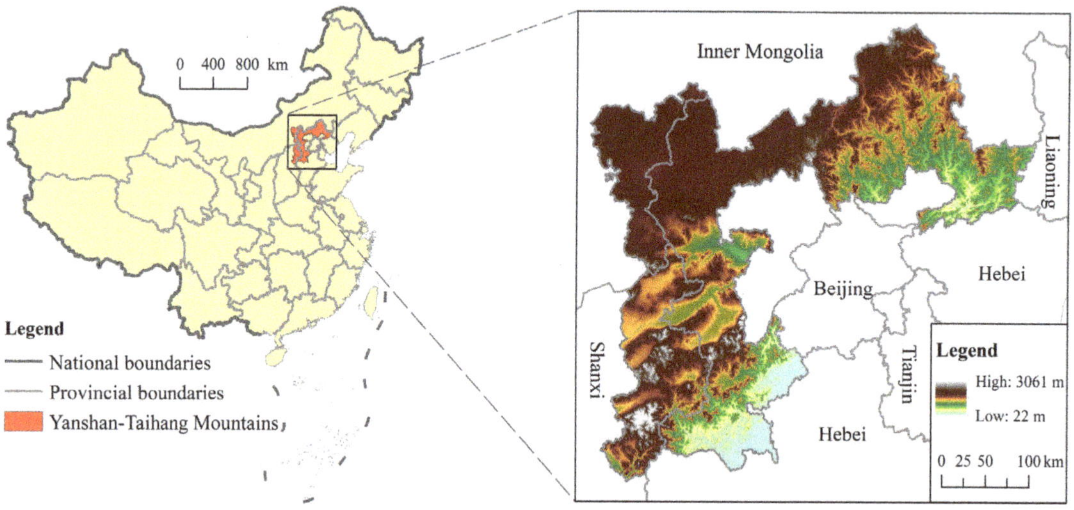

Figure 2. Location and DEM of Yanshan-Taihang Mountains.

3.2. Data Sources and Processing

This study takes a county as the basic spatial analysis unit. The administrative division data come from the National Geomatics Center of China (http://www.ngcc.cn/ngcc/) (accessed on 15 June 2022), and the DEM, NDVI and land use data are downloaded from the Resource and Environment Science and Data Center (https://www.resdc.cn/) (accessed on 15 June 2022). Socioeconomic data were collected from the *China County Statistical Yearbook*, *China Statistical Yearbook for Regional Economy*, *Hebei Economic Yearbook*, *Shanxi Statistical*

Yearbook, *Inner Mongolia Statistical Yearbook*, and the statistical bulletins on national economic and social development of each county. Population data are obtained from tabulation on the 2000 and 2010 population census of the People's Republic of China by county, and the main data bulletins of the seventh census of each county. In terms of the missing data, they are replaced by the data of adjacent years or the average of prefecture-level municipalities. Table 3 presents statistics of the indicators measuring RTD level and the pre-selected influencing factors. To eliminate the influence of dimensional differences on RTD level, the max–min normalization method is employed to standardize each indicator. Additionally, all data were treated as a whole and then further analyzed by year to achieve comparable results across years.

Table 3. Statistics of the indicators measuring RTD and the pre-selected influencing factors.

Type	Indicator/Variable	Unit	Minimum	Maximum	Mean	Standard Deviation
Measurement indicators	Urban population share	%	3.86%	62.23%	29.91%	15.39%
	Non-agricultural industry share	%	39.75%	94.09%	72.87%	10.94%
	Construction land share	%	0.17%	15.76%	3.33%	2.89%
	NDVI	-	0.4321	0.8541	0.6687	0.0992
Pre-selected influencing factors	Average altitude	m	44	1500	1044	426
	Per capita farmland area	mu/person	0.65	11.44	3.73	2.53
	Average years of schooling	year	6.02	10.10	8.19	0.86
	Road density	km/km^2	0.0782	1.7014	0.5949	0.3217
	Per capita GDP	yuan/person	1369	83,402	16,632	15,094
	Level of agricultural mechanization	kWh/ha	0.60	17.37	4.85	3.07
	Per capita income of rural households	yuan/person	958	16,543	5536	4574
	Urban–rural dual structure	-	1.85	5.23	3.16	0.78

4. Empirical Results

4.1. Spatio-Temporal Patterns of RTD in Yanshan-Taihang Mountains

4.1.1. Spatio-Temporal Patterns of Quantitative Changes

Using the four indicators of urban population share, non-agricultural industry share, construction land share and NDVI, the evolution of rural human, economic, resource and environmental elements of the Yanshan-Taihang Mountains from 2000 and 2020 is revealed to feature the quantitative changes of RTD.

(1) Urban population share

In 2000, the urban population share in the Yanshan-Taihang Mountains was generally at a low level, with most counties having a ratio of less than 20% and relatively high-value counties being scattered (Figure 3a). In 2010, the urban population share increased rapidly, with relatively high-value counties mainly located in Zhangjiakou (Figure 3e). In 2020, only Guangling, Linqiu, Tangxian and Longhua had a ratio below 40%, and the relatively high-value counties concentrated on the Hebei side of Taihang Mountains (Figure 3i). In terms of changes, most counties had higher increases in 2010–2020 than in 2000–2010. Spatially, counties with small increases were mainly distributed in the Yanshan Mountains, and counties with large increases were mainly distributed in Hebei; the highest average increase was in Hebei, and the lowest was in Shanxi (Figure 3m).

(2) Non-agricultural industry share

In the three selected time sections, the non-agricultural industry share in the counties of the Taihang Mountains was generally higher than that in the Yanshan Mountains. Specifically, the county non-agricultural industry share in 2000 and 2010 showed a decreasing trend from east to west and from south to north (Figure 3e,f). In 2020, the decreasing trend in this ratio from south to north was still obvious, but the feature of decreasing from east to west was not obvious (Figure 3j). From 2000 to 2020, counties with negative growth in the non-agricultural industry share were mainly distributed in the Yanshan Mountains, especially Weichang, Pingquan and Chengde, with a decline of more than 12 percentage points; the increase in the non-agricultural industry share in the Taihang Mountains generally showed a gradual upward trend from south to north (Figure 3n).

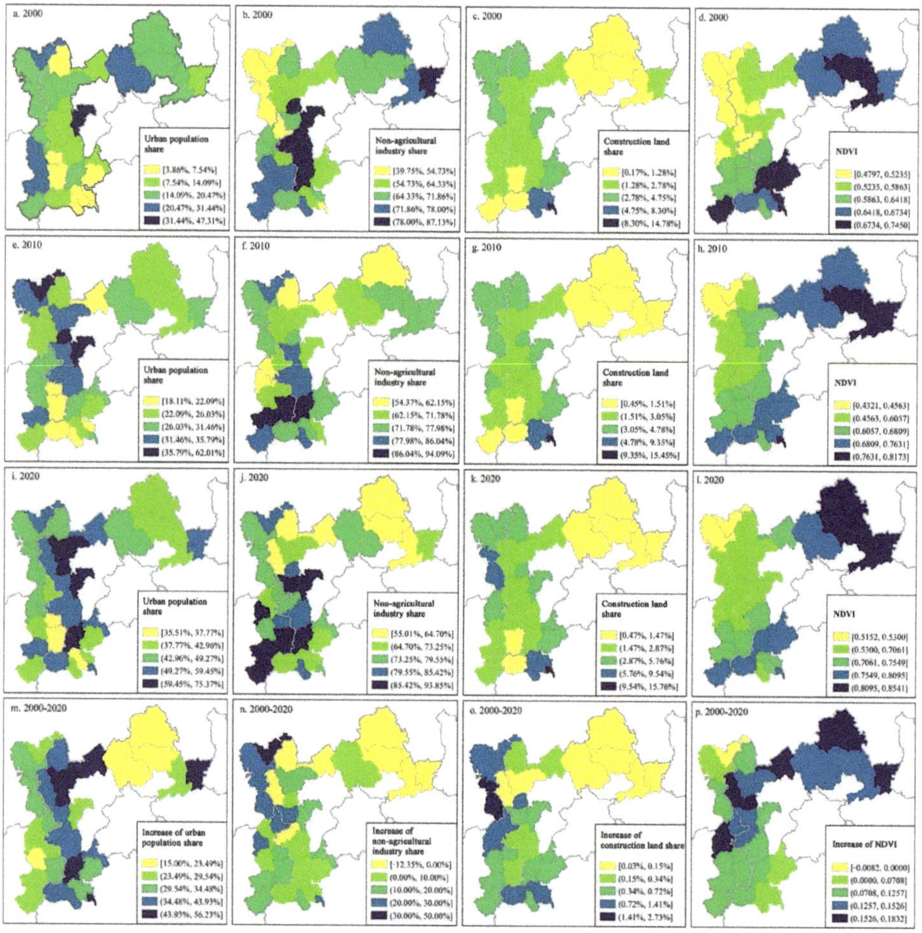

Figure 3. Spatio-temporal patterns of the quantitative changes of rural regional system in Yanshan-Taihang Mountains.

(3) Construction land share

Due to the location in mountainous areas and the positioning of national key ecological function zone, most counties in the Yanshan-Taihang Mountains ran at a low construction land share, and only a few counties located in the transition zone from the Taihang Mountains to the North China Plain had a relatively high construction land share. Specifically, the county construction land share showed an increasing trend from east to west and from south to north in 2000, 2010 and 2020, and the low-value counties were concentrated in the Yanshan Mountains and the deep Taihang Mountains (Figure 3c,g,k). The increase in the county construction land share in the Yanshan Mountains was generally small, while the Taihang Mountains presented a spatial characteristic of being smaller in the east than in the west, i.e., the county increase in Hebei was generally smaller than that in Shanxi and Inner Mongolia (Figure 3o).

(4) NDVI

As an important ecological security barrier in the Beijing–Tianjin–Hebei region, the NDVI of the study area has been at a relatively high level for a long time. Spatially, the county NDVI in 2000, 2010 and 2020 showed a decreasing from east to west and from north to south (Figure 3d,h,l). In terms of changes, the increase in NDVI in the Yanshan Mountains was significantly greater than that in the Taihang Mountains; except for the counties in Inner Mongolia, the increase in NDVI generally shows a divergence of increasing from south to north (Figure 3p).

4.1.2. Spatial and Temporal Patterns of RTD

Using Equation (1), this study measures the coupling coordination degree among the elements of rural regional system in each county, thus revealing RTD level in the Yanshan-Taihang Mountains from 2000 to 2020 (Figure 4). In 2000, RTD in most counties was in a disorderly state, accounting for 75.76% of the total. Therefore, rural development in the Yanshan-Taihang Mountains generally faced problems such as insufficient endogenous power, fragile ecological environment and prominent urban–rural dual structure, which determined the prevalence of rural poverty. In 2010, most counties achieved a leap in RTD status, and the ratio of counties in the state of barely and primary coupling coordination increased from 24.24% in 2000 to 84.85%. With the victory in the decisive fight against poverty, most counties in the Yanshan-Taihang Mountains reached a RTD state of coupling coordination in 2020, especially in Xuanhua and Wangdu, where RTD entered a state of advanced coupling coordination. Spatially, in 2000, the state of county RTD in the south of the study area was significantly better than that in the north; in 2010 and 2020, the county RTD showed a similar spatial pattern, i.e., RTD in the Taihang Mountains generally outperformed that in the Yanshan Mountains.

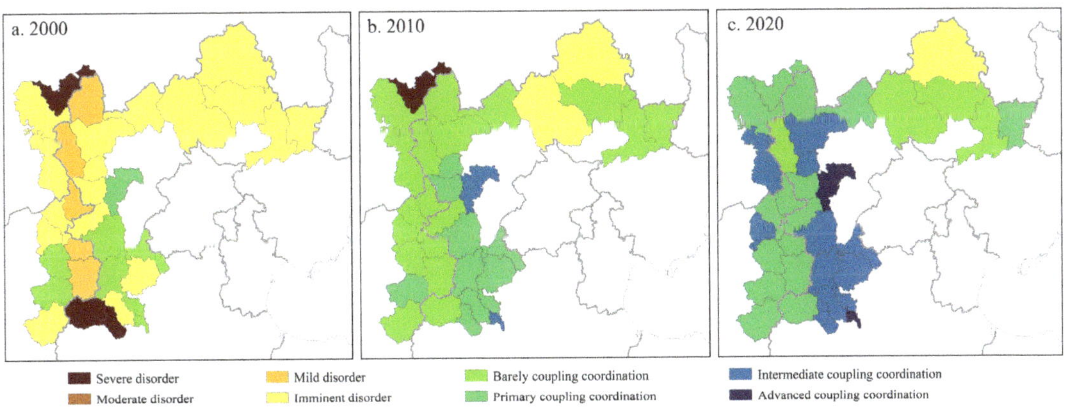

Figure 4. Spatio-temporal patterns of RTD level in Yanshan-Taihang Mountains.

In terms of the changes in RTD level from 2000 to 2020, all counties in the Yanshan-Taihang Mountains, except Weichang, achieved a leap in RTD level (Table 4). Specifically, the number of counties with an RTD level upgraded from imminent disorder to primary coupling coordination was the largest at seven, followed by six counties with an RTD level upgraded from imminent disorder to intermediate coupling coordination. The number of counties that upgraded from mild disorder to primary coupling coordination and from barely coupling coordination to intermediate coupling coordination was both four, and the number of counties with other state transitions did not exceed three.

Table 4. State changes of RTD in Yanshan-Taihang Mountains.

		2020					
		Imminent Disorder	**Barely Coupling Coordination**	**Primary Coupling Coordination**	**Intermediate Coupling Coordination**	**Advanced Coupling Coordination**	**Total**
2000	Primary coupling coordination					1	1
	Barely coupling coordination			2	4	1	7
	Imminent disorder	1	3	7	6		17
	Mild disorder		1	4			5
	Severe disorder			2	1		3
	Total	1	4	15	11	2	33

Further, Geoda was employed to analyze the spatial characteristics of county RTD level in the Yanshan-Taihang Mountains. In 2000, 2010 and 2020, the global Moran's *I* of county RTD level was 0.154, 0.269 and 0.489, respectively, indicating a positive spatial correlation of RTD, and this correlation showed an increasing evolutionary trend. The Moran scatterplot showed that the number of counties in the first quadrant (high-value counties surrounded by high-value counties) decreased from 18 in 2000 to 13 in 2010 and further to 11 in 2020; the number of counties in the third quadrant (low-value counties surrounded by low-value counties) increased from 6 in 2000 to 10 in 2010 and further to 14 in 2020; the number of counties in the second quadrant (low-value counties surrounded by high-value counties) was 4, 5 and 2 in 2000, 2010 and 2020, respectively; and the number of counties in the fourth quadrant (high-value counties surrounded by low-value counties) was 5 in both 2000 and 2010, increasing to 6 in 2020 (Figure 5). In general, nearly 70% of the counties were distributed in the first and third quadrants, indicating that most counties had similar RTD level to their neighboring counties during the study period.

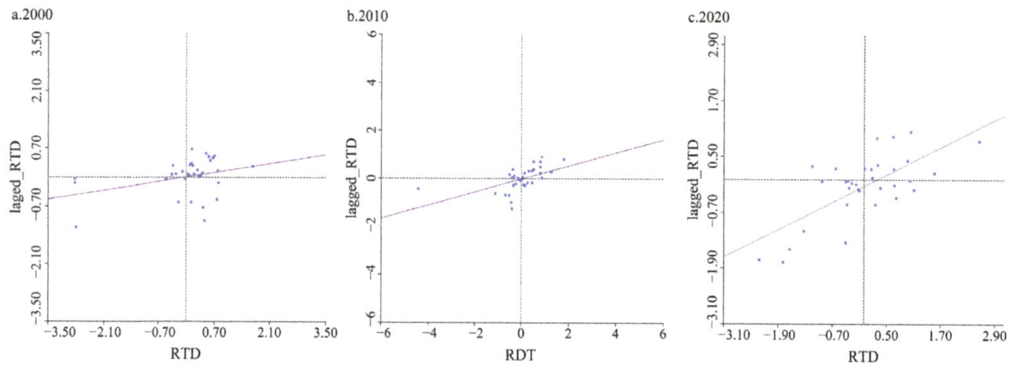

Figure 5. Moran scatterplot of county RTD level in Yanshan-Taihang Mountains.

4.2. Factors Influencing RTD in Yanshan-Taihang Mountains

Due to the large values of average altitude, per capita GDP and per capita income of rural households, the logarithmic operation of the three independent variables was performed to facilitate calculation. Meanwhile, the data were diagnosed for multicollinearity using SPSS 23.0. The results showed that after excluding the variable of per capita income of rural households, the variance inflation coefficients of the remaining independent variables were less than 10, and the tolerance was greater than 0.10, indicating that there was no significant collinearity among the remaining independent variables and can be used for model fitting analysis. Additionally, the key parameters of the GTWR and OLS model regression were given in Table 5. According to the Akaike information criterion (AIC) and

R^2, the GTWR model significantly outperforms the OLS model, i.e., the GTWR model can better reveal the influence of the pre-selected independent variables on RTD.

Table 5. Estimated parameters of GTWR and OLS models.

	Bandwidth	Residual Squares	Sigma	AIC	R^2	Adjusted R^2	Spatio-Temporal Distance Ratio
GTWR	0.2859	0.7762	0.0885	−136.706	0.6814	0.6569	0.5418
OLS				−140.1425	0.5035		

As can be seen from Figure 6, there were significant regional differences in the effects of the remaining seven independent variables on RTD of the Yanshan-Taihang Mountains.

(1) Average altitude (*altitude*). Most counties had a negative effect of average elevation on RTD, but only a few counties passed the significance test. Especially in 2020, all counties failed the 10% significance test. The main reason for this was that the development of productivity made the role of natural conditions in RTD insignificant.

(2) Per capita farmland area (*farm*). In 2000 and 2010, the effect of per capita farmland area on RTD was mainly negative, and the counties in the western Yanshan Mountains and northern Taihang Mountains both passed the significance test. In 2020, the impact of per capita farmland area was mainly positive, with few counties passing the significance test, concentrated in eastern Yanshan Mountains. This change reflects the diminishing role of natural factors in human activities as productivity develops.

(3) Average years of schooling (*school*). The effect of average years of schooling on RTD was positive, and the counties passing the significance test in 2000 and 2010 were mainly distributed in the central and southern Taihang Mountains, and most counties in the Yanshan-Taihang Mountains passed the significance test in 2020. The average years of schooling reflect the accumulation of human capital in rural areas, and its role in RTD is becoming increasingly prominent.

(4) Road density (*road*). The effect of road density on RTD was mainly positive, but only a few counties passed the significance test, mainly concentrated in the Yanshan Mountains. This was mainly because there were many cross-border roads distributed in the Taihang Mountains, which did not have a significant driving effect on RTD; while the roads in the Yanshan Mountains better played their tandem role to promote RTD.

(5) Per capita GDP (*pgdp*). The effect of per capita GDP on RTD was mainly positive, and most counties passed the significance test, indicating that the county economy was an important support for RTD. Specifically, the impact of per capita GDP on RTD in 2000 was significantly positive in most counties except for those in the eastern Yanshan Mountains and southern Taihang Mountains; in 2010 and 2020, the counties with significantly positive effects were mainly located in the Taihang Mountains.

(6) Level of agricultural mechanization (*mecha*). The effect of the level of agricultural mechanization on RTD was mainly negative, but the counties whose effects passed the significance test were concentrated in 2000, and all counties failed the significance test in 2010 and 2020. This was mainly due to the declining share of agriculture in rural economy caused by the restructuring of rural industrial during rapid industrialization and urbanization, which in turn led to an increasingly insignificant role of agriculture in RTD.

(7) Urban–rural dual structure (*dual*). The effect of urban–rural dual structure on RTD was negative, and most counties passed the significance test, indicating that the more pronounced the urban–rural dual structure, the lower the RTD level. Spatially, the counties that passed the significance test in 2000 were mainly distributed in the Taihang Mountains, while in 2010 and 2020, they were mainly distributed in the western Yanshan Mountains and northern Taihang Mountains.

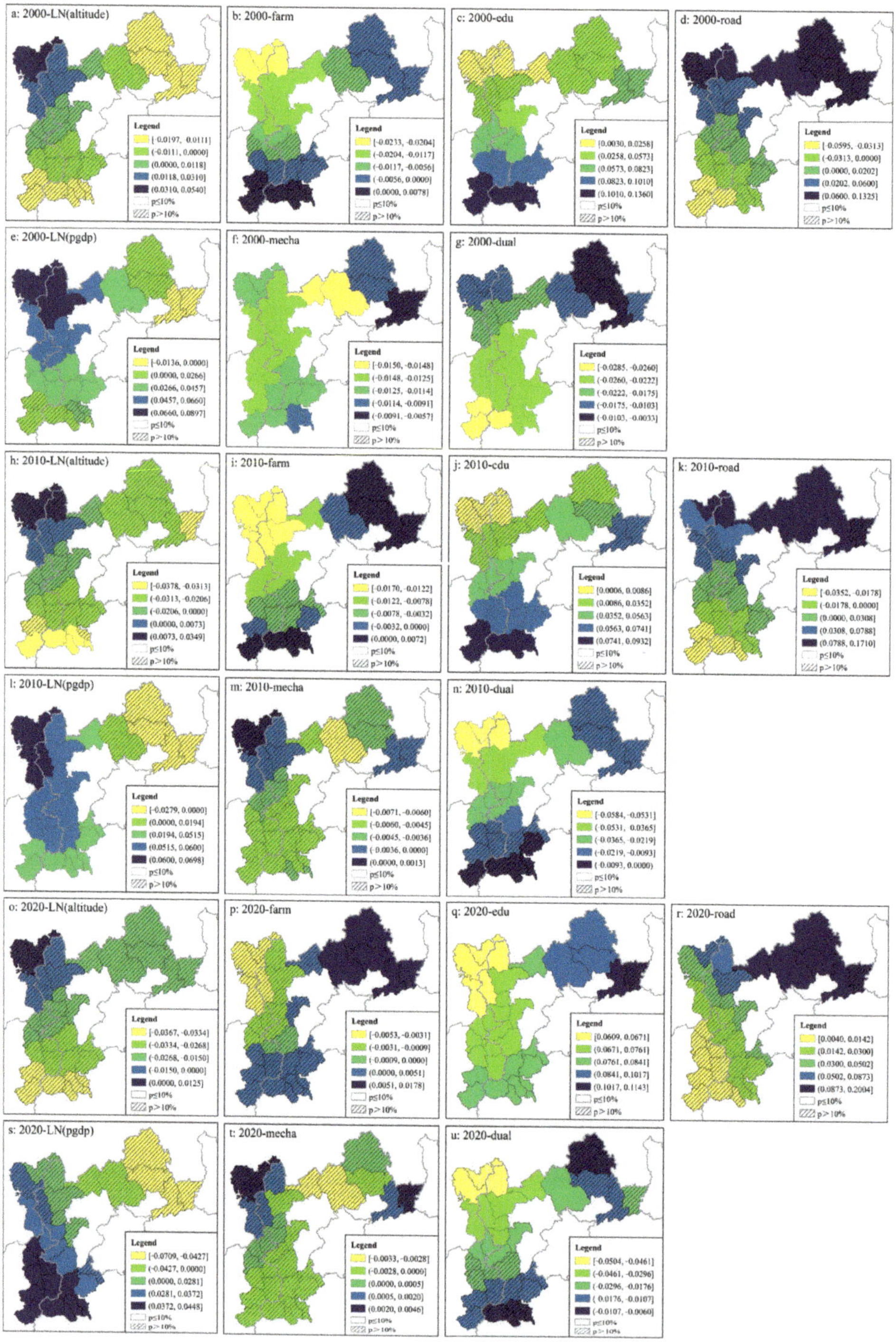

Figure 6. Spatial distribution of the regression coefficient of GTWR model.

5. Discussion

5.1. Reexamining RTD and Its Influencing Factors in Poverty-Stricken Areas

Rural development has its own laws and follows a life cycle of germination, growth, maturity, and decline/leap [58,59]. The first-nature geography, dominated by natural conditions [60], plays a decisive role in the early stages of rural development. Due to the low resource and environmental carrying capacity, conflicts between rural people and environment often arise when the intensity of human activities increases, exacerbating the unsustainability of rural development and rural decline. As rural development level increases, the role of first-nature geography continues to diminish, but its fundamental supporting role remains significant [61]. Correspondingly, the role of second-nature geography, referring to human conditions [60], is increasing, i.e., rural development is shifting from a resource orientation to a knowledge and technology orientation. During this process, many rural populations are transformed into urban populations, the rural industrial structure changes from predominantly agricultural to non-agricultural, and a large amount of agricultural land is transformed into construction land. As a result, rural areas have changed from small to large in both amount and scale.

Sachs [62] once stated that "geography is destiny. If a country is geographically closed, inaccessible, environmentally vulnerable to disease and extreme weather, and has poor soil, it will be poor". Although this statement has been challenged for overemphasizing the decisive role of natural conditions, it is true that poverty is spatially uneven. In China, most of the poor are distributed in rural areas, especially those with remote location, scarce resources, fragile ecology and poor infrastructure [57,63,64]. Meanwhile, large-scale poverty alleviation and development have led to an increasing concentration of poor people in areas with poor location conditions such as deep rocky mountains, alpine areas and ecologically fragile areas, and the effectiveness of poverty reduction achieved with the same amount of antipoverty resources has also shown a diminishing trend [57].

Escaping poverty is the greatest transformation of rural development in poverty-stricken areas, and it is also a prerequisite for their agricultural and rural modernization. Due to the multidimensional nature of poverty [65], comprehensive measures should be taken to fill the shortcomings and strengthen the weaknesses to enhance the sustainable development capacities of rural areas; on the other hand, regional industrialization and urbanization should be promoted to consolidate the support for rural poverty alleviation and development. Meanwhile, institutional reforms and policy innovations are needed to avoid excluding poor individuals and regions from socio-economic development. Over the past few decades, China's large-scale poverty alleviation has aimed to ensure that the poor are well fed and clothed, and that their needs for compulsory education, basic healthcare and safe housing are guaranteed. Moreover, the development levels of industrialization and urbanization in poverty-stricken areas are improved to increase support for the poor to escape from poverty and promote the transformation and development of poor villages to a higher level.

5.2. Paths for RTD in Yanshan-Taihang Mountains under the Background of Rural Revitalization

Currently, the Yanshan-Taihang Mountains have achieved the leap from inadequate subsistence to full well-off status. However, the shortcomings of rural development remain prominent, resulting in the persistence of relative poverty and the risk of returning to poverty. With the transformation of China's principal social contradictions, the rural revitalization strategy has become the general grasp of issues concerning agriculture, rural areas and farmers in the Yanshan-Taihang Mountains. In this context, targeted measures are needed to enhance the sustainable development capacities of rural areas and help modernize agriculture and rural areas and the high-quality rural development in the Yanshan-Taihang Mountains.

First, the development level of rural education needs to be improved. The government should strengthen investment in rural education, especially actively promote the development of "internet + education", accelerate the digital transformation and intelligent upgrading of rural education, and promote the sharing of high-quality educational resources and the effective matching of educational demand and supply.

Second, the county transportation infrastructure needs to be improved. The government should promote the construction of transportation infrastructure, build a transportation network with clear functions, reasonable layout, and moderate scale, support the extension of industrial chains and the upgrading of value chains, improve the capacity and level of rural roads, and support rural development in the Yanshan-Taihang Mountains.

Third, the level and quality of county economic development need to be improved. The main function assigned by the central government determines that economic development in this region must abandon the traditional development path and rely on its ecological advantages to promote ecological industrialization and industrial ecologization. Meanwhile, it is necessary to actively promote the integrated development of primary, secondary, and tertiary industries, and build a modern industrial system through chain extension, technology penetration and policy innovation.

Fourth, system reform is needed to promote urban–rural integrated development. Focusing on improving the property rights system and optimizing resource allocation, the government should deepen the reform of land and *hukou* systems and accelerate the construction of mechanism for urban–rural integrated development. Meanwhile, there is an urgent need to actively promote the development of new-type urbanization, with the county-town as the carrier, and give full play to the important role of the county town in connecting cities and serving rural areas.

5.3. Suggestions for Global Poverty Alleviation and Development

Poverty means the lack of livelihood capital, and its accompanying problems such as hunger, disease and social conflict seriously hinder people's pursuit of a better life. Therefore, poverty eradication is not only a common aspiration of all people, but also an important goal that governments strive to achieve [57,66]. From the Millennium Development Goals (MDGs) in 2000 to the Sustainable Development Goals (SDGs) in 2015, a series of programmatic documents guiding poverty alleviation and development have greatly contributed to the development of poverty-stricken areas and the reduction in the poverty-stricken population. Globally, the proportion of people living in extreme poverty declined from 36% in 1990 to 10% in 2015 [67]. However, global poverty reduction is extremely uneven due to the spatial heterogeneity of locations, with East Asia and the Pacific achieving the most notable success in poverty reduction, especially China, which contributes more than 70% of total global poverty reduction [68,69]. In South Asia and sub-Saharan Africa, factors such as conflict, a lack of natural resources, being deep inland and surrounded by hostile neighbors, and poor governance have contributed to the large number of poverty-stricken population and the slow process of poverty reduction in these regions [70]. Moreover, the COVID-19 pandemic has put global poverty reduction gains at risk, and the goal of no poverty by 2030 will not be achieved without rapid and substantial policies and measures [71]. In this context, China's antipoverty experience provides important policy implications for other countries to overcome poverty and achieve the 2030 SDGs.

First, industrialization should be actively promoted to improve the level of regional economic development. At the same time, industrialization can also support the development of urbanization, and gradually solve the long-standing slum problems in less developed regions and countries through coordinated development with urbanization. Second, investment in education should be strengthened to transform the quantitative advantage of the population in poverty-stricken areas into a human capital advantage. On the one hand, education improves total factor productivity through human capital accumulation; on the other hand, education promotes industrial transformation and upgrading by improving people's knowledge and skills and cultivating workers needed for industrial development. Further, the construction of transportation networks should be strengthened. It is necessary to promote the integration and coordination of various transportation, such as railways and highways, to realize smooth passenger transportation and efficient freight transportation, and improve comprehensive transportation efficiency. As a result, it is conducive to the import of external factors to make up for the shortcomings of the region and the export of superior production factors, improving the effectiveness of regional economic development. Finally, institutional innovation is needed to realize the sharing of development results. An important reason why the poor are poor is that they are excluded from socio-economic development. Therefore, it is necessary to optimize regional relations and urban–rural relations through political, economic and social system innovation, so that the poor can equally enjoy the fruits of economic development.

6. Conclusions

RTD is a dynamic process in which a rural regional system changes from one qualitative state to another on the basis of continuous quantitative changes in the elements, and the key to understanding it lies in diagnosing the structural and functional changes of the system on the basis of exploring element changes. Therefore, the quantitative analysis of RTD can be characterized by measuring the coupling coordination degree of the human, economic, resource and environmental elements of the rural regional system. From 2000 to 2020, the county urban population share, non-agricultural industry share, construction land share and NDVI in the Yanshan-Taihang Mountains increased significantly, contributing to the leap in RTD level, and the proportion of counties with RTD in a coupling coordination state increased from 24.24% to 96.97%. Spatially, there was an increasing positive spatial correlation between the county RTD levels in the Yanshan-Taihang Mountains, where the RTD status in the Taihang Mountains was significantly better than that in the Yanshan Mountains. The estimation results of the GTWR model showed that the average years of schooling, road density, per capita GDP and urban–rural dual structure were the main factors affecting RTD in the Yanshan-Taihang Mountains, and there were obvious differences in the regression coefficients of each factor. Specifically, the urban–rural dual structure had a negative impact on RDT, while the average years of education was positive, and road density and per-capita GDP were mainly positive.

The experience of developed countries shows that without the modernization of agriculture and rural areas, there is no modernization of the whole country [72]. Therefore, developing countries should strengthen factor inputs to promote agricultural and rural development. Focusing on areas that have been lifted out of poverty, the government should strengthen the sustainability of antipoverty policies and measures, consolidate antipoverty achievements, and improve the sustainable development capacity of rural areas. Meanwhile, since most of these areas are ecologically fragile areas, actively promoting RTD to a higher level is also of great significance for realizing regional ecological protection and high-quality development.

Author Contributions: Conceptualization, Y.G.; methodology, Y.G. and W.Z.; software, W.Z.; validation, Y.G.; formal analysis, Y.G.; investigation, Y.G. and W.Z.; resources, Y.G.; data curation, Y.G.; writing—original draft, Y.G. and W.Z.; writing—review and editing, Y.G.; visualization, Y.G.; supervision, Y.G.; project administration, Y.G.; funding acquisition, Y.G. All authors have read and agreed to the published version of the manuscript.

Funding: This research was supported by the National Natural Science Foundation of China (Grant No. 42001203, 41931293).

Data Availability Statement: The associated dataset of the study is available upon request to the corresponding author.

Acknowledgments: The authors are thankful to all the reviewers and editors.

Conflicts of Interest: The authors declare no conflict of interest.

References

1. Bai, R.; Shi, Y.; Pan, Y. Land-use classifying and identification of the production-living-ecological space of island villages: A case study of islands in the western sea area of Guangdong province. *Land* **2022**, *5*, 705. [CrossRef]
2. Long, H.; Tu, S.; Ge, D.; Li, T.; Liu, Y. The allocation and management of critical resources in rural China under restructuring: Problems and prospects. *J. Rural Stud.* **2016**, *47*, 392–412. [CrossRef]
3. Hu, S.; Yu, B.; Wang, M. Rural restructuring and transformation: Western experience and its enlightenment to China. *Geogr. Res.* **2019**, *38*, 2833–2845.
4. Cloke, P. State deregulation and New Zealand's agricultural sector. *Sociol. Rural.* **1989**, *1*, 34–48. [CrossRef]
5. Cloke, P. Country backwater to virtual village? Rural studies and 'the cultural turn'. *J. Rural Stud.* **1997**, *4*, 367–375. [CrossRef]
6. Gurr, G.M.; Wratten, S.D.; Luna, J.M. Multi-function agricultural biodiversity: Pest management and other benefits. *Basic Appl. Ecol.* **2003**, *2*, 107–116. [CrossRef]
7. Liu, Y. *Urban-Rural Transformation Geography*; Springer: Singapore, 2021.
8. Long, H.; Zou, J.; Pykett, J.; Li, Y. Analysis of rural transformation development in China since the turn of the new millennium. *Appl. Geogr.* **2011**, *3*, 1094–1105. [CrossRef]
9. Lu, F.; Wang, X.; Han, B.; Ouyang, Z.; Duan, X.; Zheng, H.; Miao, H. Soil carbon sequestrations by nitrogen fertilizer application, straw return and no-tillage in China's cropland. *Glob. Chang. Biol.* **2009**, *2*, 281–305. [CrossRef]
10. Sattari, S.Z.; Van Ittersum, M.K.; Giller, K.E.; Zhang, F.; Bouwman, A.F. Key role of China and its agriculture in global sustainable phosphorus management. *Environ. Res. Lett.* **2014**, *5*, 054003. [CrossRef]
11. Liu, Y.; Zang, Y.; Yang, Y. China's rural revitalization and development: Theory, technology and management. *J. Geogr. Sci.* **2020**, *12*, 1923–1942. [CrossRef]
12. Guo, Y.; Zhou, Y.; Liu, Y. Targeted poverty alleviation and its practices in rural China: A case study of Fuping county, Hebei Province. *J. Rural Stud.* **2022**, *93*, 430–440. [CrossRef]
13. Guo, Y.; Liu, Y. The process of rural development and paths for rural revitalization in China. *Acta Geogr. Sin.* **2021**, *6*, 1408–1421.
14. Long, H.; Ma, L.; Zhang, Y.; Qu, L. Multifunctional rural development in China: Pattern, process and mechanism. *Habitat Int.* **2022**, *121*, 102530. [CrossRef]
15. Liu, Y. Research on the urban-rural integration and rural revitalization in the new era in China. *Acta Geogr. Sin.* **2018**, *4*, 637–650.
16. Guo, Y.; Zhou, Y.; Han, Y. Population aging in rural China: Spatial-temporal pattern and countermeasures for rural revitalization. *Geogr. Res.* **2019**, *3*, 667–683.
17. Cai, F.; Giles, J.; O'Keefe, P.; Wang, D. *The Elderly and Old Age Support in Rural China*; World Bank Publications: Washington, DC, USA, 2012.
18. Ge, D.; Long, H.; Qiao, W.; Wang, Z.; Sun, D.; Yang, R. Effects of rural–urban migration on agricultural transformation: A case of Yucheng City, China. *J. Rural Stud.* **2020**, *76*, 85–95. [CrossRef]
19. Liu, Y.; Dai, L.; Long, H.; Woods, M.; Fois, F. Rural vitalization promoted by industrial transformation under globalization: The case of Tengtou village in China. *J. Rural Stud.* **2022**, *95*, 241–255. [CrossRef]
20. Meessen, B.; Bloom, G. Economic transition, institutional changes and the health system: Some lessons from rural China. *J. Econ. Policy Reform* **2007**, *3*, 209–231. [CrossRef]
21. Pei, X. The contribution of collective landownership to China's economic transition and rural industrialization: A resource allocation model. *Mod. China* **2002**, *3*, 279–314. [CrossRef]
22. Liu, Y.; Yang, R.; Long, H.; Gao, J.; Wang, J. Implications of land-use change in rural China: A case study of Yucheng, Shandong Province. *Land Use Policy* **2014**, *40*, 111–118. [CrossRef]
23. Long, H.; Qu, Y.; Tu, S.; Zhang, Y.; Jiang, Y. Development of land use transitions research in China. *J. Geogr. Sci.* **2020**, *30*, 1195–1214. [CrossRef]
24. Zhou, Y.; Li, X.; Liu, Y. Land use change and driving factors in rural China during the period 1995–2015. *Land Use Policy* **2020**, *99*, 105048. [CrossRef]
25. Chung, H. Rural transformation and the persistence of rurality in China. *Eurasian Geogr. Econ.* **2013**, *5–6*, 594–610. [CrossRef]
26. Zhang, R.; Li, H.; Yuan, Y. Analyzing space-time evolution of rural transition in a rapidly urbanizing region: A case study of Suzhou, China. *J. Geogr. Sci.* **2022**, *7*, 1343–1356. [CrossRef]
27. Jiang, T.; Shi, G.; Zhang, Y. Multifunctional rural transition along China's Yangtze River Transect: Spatial differentiation and economic relevance. *Environ. Sci. Pollut. Res.* **2021**, *43*, 60726–60740. [CrossRef] [PubMed]
28. Wilson, G.A.; Hu, Z.; Rahman, S. Community resilience in rural China: The case of Hu Village, Sichuan Province. *J. Rural Stud.* **2018**, *60*, 130–140. [CrossRef]
29. Yang, R.; Pan, Y. Rural vulnerability in China: Evaluation theory and spatial patterns. *J. Geogr. Sci.* **2021**, *10*, 1507–1528. [CrossRef]
30. Guo, Y.; Wang, J. Poverty alleviation through labor transfer in rural China: Evidence from Hualong County. *Habitat Int.* **2021**, *116*, 102402. [CrossRef]
31. Guo, Y.; Liu, Y. Poverty alleviation through land assetization and its implications for rural revitalization in China. *Land Use Policy* **2021**, *105*, 105418. [CrossRef]
32. World Bank. *Poverty and Shared Prosperity 2020: Reversals of Fortune*; The World Bank: Washington, DC, USA, 2020.

33. Moyer, J.D.; Verhagen, W.; Mapes, B.; Bohl, D.K.; Xiong, Y.; Yang, V.; McNeil, K.; Solórzano, J.; Irfan, M.; Carter, C.; et al. How many people is the COVID-19 pandemic pushing into poverty? A long-term forecast to 2050 with alternative scenarios. *PLoS ONE* **2022**, *7*, e0270846. [CrossRef]
34. World Bank. *Poverty and Shared Prosperity 2022: Correcting Course*; The World Bank: Washington, DC, USA, 2022.
35. United Nations. *The Sustainable Development Goals Report 2021*; United Nations: New York, NY, USA, 2021.
36. Kangal, K. *Friedrich Engels and the Dialectics of Nature*; Palgrave Macmillan: Cham, Switzerland, 2020.
37. Dafermos, M. The metaphysics of psychology and a dialectical perspective. *Theory Psychol.* **2021**, *3*, 355–374. [CrossRef]
38. Roth, W.M. Neoformation: A dialectical approach to developmental change. *Mind Cult. Act.* **2017**, *4*, 368–380. [CrossRef]
39. Guo, Y.; Zhou, Y.; Cao, Z. Geographical patterns and anti-poverty targeting post-2020 in China. *J. Geogr. Sci.* **2018**, *12*, 1810–1824.
40. Chigbu, U.E. Ruralisation: A tool for rural transformation. *Dev. Pract.* **2015**, *7*, 1067–1073. [CrossRef]
41. Liu, Y.; Long, H.; Chen, Y.; Wang, J.; Li, Y.; Li, Y.; Yang, Y.; Zhou, Y. Progress of research on urban-rural transformation and rural development in China in the past decade and future prospects. *J. Geogr. Sci.* **2016**, *8*, 1117–1132. [CrossRef]
42. Ohlan, R. Rural transformation in India in the decade of miraculous economic growth. *J. Land Rural Stud.* **2016**, *2*, 188–205. [CrossRef]
43. Bertalanffy, L.V. General System Theory: Foundations, Development, Applications. G. Braziller, 1968. Available online: https://repository.vnu.edu.vn/handle/VNU_123/90608 (accessed on 6 March 2022).
44. Okoli, C.; Pawlowski, S.D. The Delphi method as a research tool: An example, design considerations and applications. *Inf. Manag.* **2004**, *1*, 15–29. [CrossRef]
45. Zhu, Y.; Tian, D.; Yan, F. Effectiveness of entropy weight method in decision-making. *Math. Probl. Eng.* **2020**, *2020*, 3564835. [CrossRef]
46. Gkartzios, M.; Scott, M. Placing housing in rural development: Exogenous, endogenous and neo-endogenous approaches. *Sociol. Rural.* **2014**, *3*, 241–265. [CrossRef]
47. Terluin, I.J. Differences in economic development in rural regions of advanced countries: An overview and critical analysis of theories. *J. Rural Stud.* **2003**, *3*, 327–344. [CrossRef]
48. Aryeetey, E. Decentralization for rural development: Exogenous factors and semi-autonomous programme units in Ghana. *Community Dev. J.* **1990**, *3*, 206–214. [CrossRef]
49. Ray, C. Towards a meta-framework of endogenous development: Repertoires, paths, democracy and rights. *Sociol. Rural.* **1999**, *1*, 522–537. [CrossRef]
50. High, C.; Nemes, G. Social learning in LEADER: Exogenous, endogenous and hybrid evaluation in rural development. *Sociol. Rural.* **2007**, *2*, 103–119. [CrossRef]
51. Lowe, P.; Murdoch, J.; Ward, N. Networks in rural development: Beyond exogenous and endogenous models. *Beyond Mod.* **1995**, *5*, 87–105.
52. Liu, Y. *Geography of New Countryside Construction in China*; Science Press: Beijing, China, 2011.
53. Tu, S.; Long, H. Rural restructuring in China: Theory, approaches and research prospect. *J. Geogr. Sci.* **2017**, *10*, 1169–1184. [CrossRef]
54. Irwin, E.G.; Isserman, A.M.; Kilkenny, M.; Partridge, M.D. A century of research on rural development and regional issues. *Am. J. Agric. Econ.* **2010**, *2*, 522–553. [CrossRef]
55. Long, H.; Li, T., Zou, J. Analysis of dynamical mechanism of rural transformation development in typical regions of China. *Econ. Geogr.* **2011**, *12*, 2080–2085.
56. Huang, B.; Wu, B.; Barry, M. Geographically and temporally weighted regression for modeling spatio-temporal variation in house prices. *Int. J. Geogr. Inf. Sci.* **2010**, *3*, 383–401. [CrossRef]
57. Liu, Y.; Liu, J.; Zhou, Y. Spatio-temporal patterns of rural poverty in China and targeted poverty alleviation strategies. *J. Rural Stud.* **2017**, *52*, 66–75. [CrossRef]
58. Liu, Y.; Long, H.; Chen, Y.; Wang, J. *Research Report on Rural Development in China: Hollowed Village and Its Renovation Strategy*; Science Press: Beijing, China, 2011.
59. Moretti, M.; De Boni, A.; Roma, R. Economic and environmental sustainability of forestry measures in Apulia Region Rural Development Plan: An application of life cycle approach. *Land Use Policy* **2014**, *41*, 284–289. [CrossRef]
60. Krugman, P. First nature, second nature, and metropolitan location. *J. Reg. Sci.* **1993**, *2*, 129–144. [CrossRef]
61. Marsden, T.; Banks, J.; Bristow, G. The social management of rural nature: Understanding agrarian-based rural development. *Environ. Plan. A* **2002**, *5*, 809–825. [CrossRef]
62. Sachs, J.D. *The End of Poverty: Economic Possibilities for Our Time*; Penguin: New York, NY, USA, 2006.
63. Barbier, E.B. Poverty, development, and environment. *Environ. Dev. Econ.* **2010**, *6*, 635–660. [CrossRef]
64. Thiede, B.; Kim, H.; Valasik, M. The spatial concentration of America's rural poor population: A postrecession update. *Rural Sociol.* **2018**, *1*, 109–144. [CrossRef]
65. Alkire, S.; Foster, J. Counting and multidimensional poverty measurement. *J. Public Econ.* **2011**, *7–8*, 476–487. [CrossRef]
66. Bray, R.; de Laat, M.; Godinot, X.; Ugarte, A.; Walker, R. Realising poverty in all its dimensions: A six-country participatory study. *World Dev.* **2020**, *134*, 105025. [CrossRef]
67. Jolliffe, D.; Prydz, E.B. Societal poverty: A relative and relevant measure. *World Bank Econ. Rev.* **2021**, *1*, 180–206. [CrossRef]

68. Liu, M.; Feng, X.; Wang, S.; Qiu, H. China's poverty alleviation over the last 40 years: Successes and challenges. *Aust. J. Agric. Resour. Econ.* **2020**, *1*, 209–228. [CrossRef]
69. Omar, M.A.; Inaba, K. Does financial inclusion reduce poverty and income inequality in developing countries? A panel data analysis. *J. Econ. Struct.* **2020**, *1*, 37. [CrossRef]
70. Collier, P. *The Bottom Billion: Why the Poorest Countries Are Failing and What Can Be Done about It*; Oxford University Press: Oxford, MI, USA, 2008.
71. Shulla, K.; Voigt, B.F.; Cibian, S.; Scandone, G.; Martinez, E.; Nelkovski, F.; Salehi, P. Effects of COVID-19 on the sustainable development goals (SDGs). *Discov. Sustain.* **2021**, *2*, 15. [CrossRef]
72. Van der Ploeg, J.D.; Renting, H.; Brunori, G.; Knickel, K.; Mannion, J.; Marsden, T.; De Roest, K.; Sevilla-Guzmán, E.; Ventura, F. Rural development: From practices and policies towards theory. *Sociol. Rural.* **2000**, *4*, 391–408. [CrossRef]

MDPI
St. Alban-Anlage 66
4052 Basel
Switzerland
www.mdpi.com

Land Editorial Office
E-mail: land@mdpi.com
www.mdpi.com/journal/land

www.ingramcontent.com/pod-product-compliance
Lightning Source LLC
Chambersburg PA
CBHW061552220926
51850CB00005B/315
* 9 7 8 3 0 3 6 5 8 6 6 9 4 *